# RELIABILITY OF COMPUTER SYSTEMS AND NETWORKS

# RELIABILITY OF COMPUTER SYSTEMS AND NETWORKS

## Fault Tolerance, Analysis, and Design

**MARTIN L. SHOOMAN**
Polytechnic University
and
Martin L. Shooman & Associates

A Wiley-Interscience Publication
**JOHN WILEY & SONS, INC.**

This book is printed on acid-free paper. ∞

Copyright © 2002 by John Wiley & Sons, Inc., New York. All rights reserved.

Published simultaneously in Canada.

No part of this publication may be reproduced, stored in a retrieval system or transmitted in any form or by any means, electronic, mechanical, photocopying, recording, scanning or otherwise, except as permitted under Sections 107 or 108 of the 1976 United States Copyright Act, without either the prior written permission of the Publisher, or authorization through payment of the appropriate per-copy fee to the Copyright Clearance Center, 222 Rosewood Drive, Danvers, MA 01923, (978) 750-8400, fax (978) 750-4744. Requests to the Publisher for permission should be addressed to the Permissions Department, John Wiley & Sons, Inc., 605 Third Avenue, New York, NY 10158-0012, (212) 850-6011, fax (212) 850-6008, E-Mail: PERMREQ@WILEY.COM.

For ordering and customer service, call 1-800-CALL-WILEY.

*Library of Congress Cataloging-in-Publication Data is available.*

ISBN 0-471-29342-3

Printed in the United States of America

10 9 8 7 6 5 4 3 2

*To Danielle Leah and Aviva Zissel*

# CONTENTS

**Preface** xix

**1 Introduction** 1

    1.1 What is Fault-Tolerant Computing?, 1
    1.2 The Rise of Microelectronics and the Computer, 4
        1.2.1 A Technology Timeline, 4
        1.2.2 Moore's Law of Microprocessor Growth, 5
        1.2.3 Memory Growth, 7
        1.2.4 Digital Electronics in Unexpected Places, 9
    1.3 Reliability and Availability, 10
        1.3.1 Reliability Is Often an Afterthought, 10
        1.3.2 Concepts of Reliability, 11
        1.3.3 Elementary Fault-Tolerant Calculations, 12
        1.3.4 The Meaning of Availability, 14
        1.3.5 Need for High Reliability and Safety in Fault-Tolerant Systems, 15
    1.4 Organization of the Book, 18
        1.4.1 Introduction, 18
        1.4.2 Coding Techniques, 19
        1.4.3 Redundancy, Spares, and Repairs, 19
        1.4.4 *N*-Modular Redundancy, 20
        1.4.5 Software Reliability and Recovery Techniques, 20
        1.4.6 Networked Systems Reliability, 21
        1.4.7 Reliability Optimization, 22
        1.4.8 Appendices, 22

General References, 23
References, 25
Problems, 27

## 2 Coding Techniques    30

2.1 Introduction, 30
2.2 Basic Principles, 34
    2.2.1 Code Distance, 34
    2.2.2 Check-Bit Generation and Error Detection, 35
2.3 Parity-Bit Codes, 37
    2.3.1 Applications, 37
    2.3.2 Use of Exclusive OR Gates, 37
    2.3.3 Reduction in Undetected Errors, 39
    2.3.4 Effect of Coder–Decoder Failures, 43
2.4 Hamming Codes, 44
    2.4.1 Introduction, 44
    2.4.2 Error-Detection and -Correction Capabilities, 45
    2.4.3 The Hamming SECSED Code, 47
    2.4.4 The Hamming SECDED Code, 51
    2.4.5 Reduction in Undetected Errors, 52
    2.4.6 Effect of Coder–Decoder Failures, 53
    2.4.7 How Coder–Decoder Failures Effect SECSED Codes, 56
2.5 Error-Detection and Retransmission Codes, 59
    2.5.1 Introduction, 59
    2.5.2 Reliability of a SECSED Code, 59
    2.5.3 Reliability of a Retransmitted Code, 60
2.6 Burst Error-Correction Codes, 62
    2.6.1 Introduction, 62
    2.6.2 Error Detection, 63
    2.6.3 Error Correction, 66
2.7 Reed–Solomon Codes, 72
    2.7.1 Introduction, 72
    2.7.2 Block Structure, 72
    2.7.3 Interleaving, 73
    2.7.4 Improvement from the RS Code, 73
    2.7.5 Effect of RS Coder–Decoder Failures, 73
2.8 Other Codes, 75
References, 76
Problems, 78

## 3 Redundancy, Spares, and Repairs    83

3.1 Introduction, 85
3.2 Apportionment, 85

3.3 System Versus Component Redundancy, 86
3.4 Approximate Reliability Functions, 92
    3.4.1 Exponential Expansions, 92
    3.4.2 System Hazard Function, 94
    3.4.3 Mean Time to Failure, 95
3.5 Parallel Redundancy, 97
    3.5.1 Independent Failures, 97
    3.5.2 Dependent and Common Mode Effects, 99
3.6 An $r$-out-of-$n$ Structure, 101
3.7 Standby Systems, 104
    3.7.1 Introduction, 104
    3.7.2 Success Probabilities for a Standby System, 105
    3.7.3 Comparison of Parallel and Standby Systems, 108
3.8 Repairable Systems, 111
    3.8.1 Introduction, 111
    3.8.2 Reliability of a Two-Element System with Repair, 112
    3.8.3 MTTF for Various Systems with Repair, 114
    3.8.4 The Effect of Coverage on System Reliability, 115
    3.8.5 Availability Models, 117
3.9 RAID Systems Reliability, 119
    3.9.1 Introduction, 119
    3.9.2 RAID Level 0, 122
    3.9.3 RAID Level 1, 122
    3.9.4 RAID Level 2, 122
    3.9.5 RAID Levels 3, 4, and 5, 123
    3.9.6 RAID Level 6, 126
3.10 Typical Commercial Fault-Tolerant Systems: Tandem and Stratus, 126
    3.10.1 Tandem Systems, 126
    3.10.2 Stratus Systems, 131
    3.10.3 Clusters, 135
References, 137
Problems, 139

## 4 $N$-Modular Redundancy     145

4.1 Introduction, 145
4.2 The History of $N$-Modular Redundancy, 146
4.3 Triple Modular Redundancy, 147
    4.3.1 Introduction, 147
    4.3.2 System Reliability, 148
    4.3.3 System Error Rate, 148
    4.3.4 TMR Options, 150

4.4 $N$-Modular Redundancy, 153
    4.4.1 Introduction, 153
    4.4.2 System Voting, 154
    4.4.3 Subsystem Level Voting, 154
4.5 Imperfect Voters, 156
    4.5.1 Limitations on Voter Reliability, 156
    4.5.2 Use of Redundant Voters, 158
    4.5.3 Modeling Limitations, 160
4.6 Voter Logic, 161
    4.6.1 Voting, 161
    4.6.2 Voting and Error Detection, 163
4.7 $N$-Modular Redundancy with Repair, 165
    4.7.1 Introduction, 165
    4.7.2 Reliability Computations, 165
    4.7.3 TMR Reliability, 166
    4.7.4 $N$-Modular Reliability, 170
4.8 $N$-Modular Redundancy with Repair and Imperfect Voters, 176
    4.8.1 Introduction, 176
    4.8.2 Voter Reliability, 176
    4.8.3 Comparison of TMR, Parallel, and Standby Systems, 178
4.9 Availability of $N$-Modular Redundancy with Repair and Imperfect Voters, 179
    4.9.1 Introduction, 179
    4.9.2 Markov Availability Models, 180
    4.9.3 Decoupled Availability Models, 183
4.10 Microcode-Level Redundancy, 186
4.11 Advanced Voting Techniques, 186
    4.11.1 Voting with Lockout, 186
    4.11.2 Adjudicator Algorithms, 189
    4.11.3 Consensus Voting, 190
    4.11.4 Test and Switch Techniques, 191
    4.11.5 Pairwise Comparison, 191
    4.11.6 Adaptive Voting, 194
        References, 195
        Problems, 196

## 5 Software Reliability and Recovery Techniques    202

5.1 Introduction, 202
    5.1.1 Definition of Software Reliability, 203
    5.1.2 Probabilistic Nature of Software Reliability, 203
5.2 The Magnitude of the Problem, 205

5.3 Software Development Life Cycle, 207
    5.3.1 Beginning and End, 207
    5.3.2 Requirements, 209
    5.3.3 Specifications, 209
    5.3.4 Prototypes, 210
    5.3.5 Design, 211
    5.3.6 Coding, 214
    5.3.7 Testing, 215
    5.3.8 Diagrams Depicting the Development Process, 218
5.4 Reliability Theory, 218
    5.4.1 Introduction, 218
    5.4.2 Reliability as a Probability of Success, 219
    5.4.3 Failure-Rate (Hazard) Function, 222
    5.4.4 Mean Time To Failure, 224
    5.4.5 Constant-Failure Rate, 224
5.5 Software Error Models, 225
    5.5.1 Introduction, 225
    5.5.2 An Error-Removal Model, 227
    5.5.3 Error-Generation Models, 229
    5.5.4 Error-Removal Models, 229
5.6 Reliability Models, 237
    5.6.1 Introduction, 237
    5.6.2 Reliability Model for Constant Error-Removal Rate, 238
    5.6.3 Reliability Model for Linearly Decreasing Error-Removal Rate, 242
    5.6.4 Reliability Model for an Exponentially Decreasing Error-Removal Rate, 246
5.7 Estimating the Model Constants, 250
    5.7.1 Introduction, 250
    5.7.2 Handbook Estimation, 250
    5.7.3 Moment Estimates, 252
    5.7.4 Least-Squares Estimates, 256
    5.7.5 Maximum-Likelihood Estimates, 257
5.8 Other Software Reliability Models, 258
    5.8.1 Introduction, 258
    5.8.2 Recommended Software Reliability Models, 258
    5.8.3 Use of Development Test Data, 260
    5.8.4 Software Reliability Models for Other Development Stages, 260
    5.8.5 Macro Software Reliability Models, 262
5.9 Software Redundancy, 262
    5.9.1 Introduction, 262
    5.9.2 *N*-Version Programming, 263
    5.9.3 Space Shuttle Example, 266

5.10 Rollback and Recovery, 268
    5.10.1 Introduction, 268
    5.10.2 Rebooting, 270
    5.10.3 Recovery Techniques, 271
    5.10.4 Journaling Techniques, 272
    5.10.5 Retry Techniques, 273
    5.10.6 Checkpointing, 274
    5.10.7 Distributed Storage and Processing, 275
    References, 276
    Problems, 280

# 6 Networked Systems Reliability   283

6.1 Introduction, 283
6.2 Graph Models, 284
6.3 Definition of Network Reliability, 285
6.4 Two-Terminal Reliability, 288
    6.4.1 State-Space Enumeration, 288
    6.4.2 Cut-Set and Tie-Set Methods, 292
    6.4.3 Truncation Approximations, 294
    6.4.4 Subset Approximations, 296
    6.4.5 Graph Transformations, 297
6.5 Node Pair Resilience, 301
6.6 All-Terminal Reliability, 302
    6.6.1 Event-Space Enumeration, 302
    6.6.2 Cut-Set and Tie-Set Methods, 303
    6.6.3 Cut-Set and Tie-Set Approximations, 305
    6.6.4 Graph Transformations, 305
    6.6.5 $k$-Terminal Reliability, 308
    6.6.6 Computer Solutions, 308
6.7 Design Approaches, 309
    6.7.1 Introduction, 310
    6.7.2 Design of a Backbone Network Spanning-Tree Phase, 310
    6.7.3 Use of Prim's and Kruskal's Algorithms, 314
    6.7.4 Design of a Backbone Network: Enhancement Phase, 318
    6.7.5 Other Design Approaches, 319
    References, 321
    Problems, 324

# 7 Reliability Optimization   331

7.1 Introduction, 331
7.2 Optimum Versus Good Solutions, 332

7.3 A Mathematical Statement of the Optimization Problem, 334
7.4 Parallel and Standby Redundancy, 336
    7.4.1 Parallel Redundancy, 336
    7.4.2 Standby Redundancy, 336
7.5 Hierarchical Decomposition, 337
    7.5.1 Decomposition, 337
    7.5.2 Graph Model, 337
    7.5.3 Decomposition and Span of Control, 338
    7.5.4 Interface and Computation Structures, 340
    7.5.5 System and Subsystem Reliabilities, 340
7.6 Apportionment, 342
    7.6.1 Equal Weighting, 343
    7.6.2 Relative Difficulty, 344
    7.6.3 Relative Failure Rates, 345
    7.6.4 Albert's Method, 345
    7.6.5 Stratified Optimization, 349
    7.6.6 Availability Apportionment, 349
    7.6.7 Nonconstant-Failure Rates, 351
7.7 Optimization at the Subsystem Level via Enumeration, 351
    7.7.1 Introduction, 351
    7.7.2 Exhaustive Enumeration, 351
7.8 Bounded Enumeration Approach, 353
    7.8.1 Introduction, 353
    7.8.2 Lower Bounds, 354
    7.8.3 Upper Bounds, 358
    7.8.4 An Algorithm for Generating Augmentation Policies, 359
    7.8.5 Optimization with Multiple Constraints, 365
7.9 Apportionment as an Approximate Optimization Technique, 366
7.10 Standby System Optimization, 367
7.11 Optimization Using a Greedy Algorithm, 369
    7.11.1 Introduction, 369
    7.11.2 Greedy Algorithm, 369
    7.11.3 Unequal Weights and Multiple Constraints, 370
    7.11.4 When Is the Greedy Algorithm Optimum?, 371
    7.11.5 Greedy Algorithm Versus Apportionment Techniques, 371
7.12 Dynamic Programming, 371
    7.12.1 Introduction, 371
    7.12.2 Dynamic Programming Example, 372
    7.12.3 Minimum System Design, 372
    7.12.4 Use of Dynamic Programming to Compute the Augmentation Policy, 373

       7.12.5 Use of Bounded Approach to Check Dynamic
             Programming Solution, 378
7.13 Conclusion, 379
    References, 379
    Problems, 381

## Appendix A  Summary of Probability Theory    384

A1 Introduction, 384
A2 Probability Theory, 384
A3 Set Theory, 386
    A3.1 Definitions, 386
    A3.2 Axiomatic Probability, 386
    A3.3 Union and Intersection, 387
    A3.4 Probability of a Disjoint Union, 387
A4 Combinatorial Properties, 388
    A4.1 Complement, 388
    A4.2 Probability of a Union, 388
    A4.3 Conditional Probabilities and
         Independence, 390
A5 Discrete Random Variables, 391
    A5.1 Density Function, 391
    A5.2 Distribution Function, 392
    A5.3 Binomial Distribution, 392
    A5.4 Poisson Distribution, 395
A6 Continuous Random Variables, 395
    A6.1 Density and Distribution Functions, 395
    A6.2 Rectangular Distribution, 397
    A6.3 Exponential Distribution, 397
    A6.4 Rayleigh Distribution, 399
    A6.5 Weibull Distribution, 399
    A6.6 Normal Distribution, 400
A7 Moments, 401
    A7.1 Expected Value, 401
    A7.2 Moments, 402
A8 Markov Variables, 403
    A8.1 Properties, 403
    A8.2 Poisson Process, 404
    A8.3 Transition Matrix, 407
        References, 409
        Problems, 409

## Appendix B  Summary of Reliability Theory    411

B1 Introduction, 411
    B1.1 History, 411

B1.2 Summary of the Approach, 411
B1.3 Purpose of This Appendix, 412
B2 Combinatorial Reliability, 412
B2.1 Introduction, 412
B2.2 Series Configuration, 413
B2.3 Parallel Configuration, 415
B2.4 An $r$-out-of-$n$ Configuration, 416
B2.5 Fault-Tree Analysis, 418
B2.6 Failure Mode and Effect Analysis, 418
B2.7 Cut-Set and Tie-Set Methods, 419
B3 Failure-Rate Models, 421
B3.1 Introduction, 421
B3.2 Treatment of Failure Data, 421
B3.3 Failure Modes and Handbook Failure Data, 425
B3.4 Reliability in Terms of Hazard Rate and Failure Density, 429
B3.5 Hazard Models, 432
B3.6 Mean Time To Failure, 435
B4 System Reliability, 438
B4.1 Introduction, 438
B4.2 The Series Configuration, 438
B4.3 The Parallel Configuration, 440
B4.4 An $r$-out-of-$n$ Structure, 441
B5 Illustrative Example of Simplified Auto Drum Brakes, 442
B5.1 Introduction, 442
B5.2 The Brake System, 442
B5.3 Failure Modes, Effects, and Criticality Analysis, 443
B5.4 Structural Model, 443
B5.5 Probability Equations, 444
B5.6 Summary, 446
B6 Markov Reliability and Availability Models, 446
B6.1 Introduction, 446
B6.2 Markov Models, 446
B6.3 Markov Graphs, 449
B6.4 Example—A Two-Element Model, 450
B6.5 Model Complexity, 453
B7 Repairable Systems, 455
B7.1 Introduction, 455
B7.2 Availability Function, 456
B7.3 Reliability and Availability of Repairable Systems, 457
B7.4 Steady-State Availability, 458
B7.5 Computation of Steady-State Availability, 460

B8 Laplace Transform Solutions of Markov Models, 461
    B8.1 Laplace Transforms, 462
    B8.2 MTTF from Laplace Transforms, 468
    B8.3 Time-Series Approximations from Laplace Transforms, 469
    References, 471
    Problems, 472

## Appendix C  Review of Architecture Fundamentals     475

C1 Introduction to Computer Architecture, 475
    C1.1 Number Systems, 475
    C1.2 Arithmetic in Binary, 477
C2 Logic Gates, Symbols, and Integrated Circuits, 478
C3 Boolean Algebra and Switching Functions, 479
C4 Switching Function Simplification, 484
    C4.1 Introduction, 484
    C4.2 K Map Simplification, 485
C5 Combinatorial Circuits, 489
    C5.1 Circuit Realizations: SOP, 489
    C5.2 Circuit Realizations: POS, 489
    C5.3 NAND and NOR Realizations, 489
    C5.4 EXOR, 490
    C5.5 IC Chips, 491
C6 Common Circuits: Parity-Bit Generators and Decoders, 493
    C6.1 Introduction, 493
    C6.2 A Parity-Bit Generator, 494
    C6.3 A Decoder, 494
C7 Flip-Flops, 497
C8 Storage Registers, 500
    References, 501
    Problems, 502

## Appendix D  Programs for Reliability Modeling and Analysis     504

D1 Introduction, 504
D2 Various Types of Reliability and Availability Programs, 506
    D2.1 Part-Count Models, 506
    D2.2 Reliability Block Diagram Models, 507
    D2.3 Reliability Fault Tree Models, 507
    D2.4 Markov Models, 507
    D2.5 Mathematical Software Systems: Mathcad, Mathematica, and Maple, 508
    D2.6 Fault-Tolerant Computing Programs, 509
    D2.7 Risk Analysis Programs, 510
    D2.8 Software Reliability Programs, 510
D3 Testing Programs, 510

D4 Partial List of Reliability and Availability Programs, 512
D5 An Example of Computer Analysis, 514
   References, 515
   Problems, 517

**Name Index**     **519**
**Subject Index**     **523**

# PREFACE

**INTRODUCTION**

This book was written to serve the needs of practicing engineers and computer scientists, and for students from a variety of backgrounds—computer science and engineering, electrical engineering, mathematics, operations research, and other disciplines—taking college- or professional-level courses. The field of high-reliability, high-availability, fault-tolerant computing was developed for the critical needs of military and space applications. NASA deep-space missions are costly, for they require various redundancy and recovery schemes to avoid total failure. Advances in military aircraft design led to the development of electronic flight controls, and similar systems were later incorporated in the Airbus 330 and Boeing 777 passenger aircraft, where flight controls are triplicated to permit some elements to fail during aircraft operation. The reputation of the Tandem business computer is built on *NonStop computing*, a comprehensive redundancy scheme that improves reliability. Modern computer storage uses redundant array of independent disks (RAID) techniques to link 50–100 disks in a fast, reliable system. Various ideas arising from fault-tolerant computing are now used in nearly all commercial, military, and space computer systems; in the transportation, health, and entertainment industries; in institutions of education and government; in telephone systems; and in both fossil and nuclear power plants. Rapid developments in microelectronics have led to very complex designs; for example, a luxury automobile may have 30–40 microprocessors connected by a local area network! Such designs must be made using fault-tolerant techniques to provide significant software and hardware reliability, availability, and safety.

Computer networks are currently of great interest, and their successful operation requires a high degree of reliability and availability. This reliability is achieved by means of multiple connecting paths among locations within a network so that when one path fails, transmission is successfully rerouted. Thus the network topology provides a complex structure of redundant paths that, in turn, provide fault tolerance, and these principles also apply to power distribution, telephone and water systems, and other networks.

Fault-tolerant computing is a generic term describing redundant design techniques with duplicate components or repeated computations enabling uninterrupted (tolerant) operation in response to component failure (faults). Sometimes, system disasters are caused by neglecting the principles of redundancy and failure independence, which are obvious in retrospect. After the September 11th, 2001, attack on the World Trade Center, it was revealed that although one company had maintained its primary system database in one of the twin towers, it wisely had kept its backup copies at its Denver, Colorado office. Another company had also maintained its primary system database in one tower but, unfortunately, kept its backup copies in the other tower.

## COVERAGE

Much has been written on the subject of reliability and availability since its development in the early 1950s. Fault-tolerant computing began between 1965 and 1970, probably with the highly reliable and widely available AT&T electronic-switching systems. Starting with first principles, this book develops reliability and availability prediction and optimization methods and applies these techniques to a selection of fault-tolerant systems. Error-detecting and -correcting codes are developed, and an analysis is made of the probability that such codes might fail. The reliability and availability of parallel, standby, and voting systems are analyzed and compared, and such analyses are also applied to modern RAID memory systems and commercial Tandem and Stratus fault-tolerant computers. These principles are also used to analyze the primary avionics software system (PASS) and the backup flight control system (BFS) used on the Space Shuttle. Errors in software that control modern digital systems can cause system failures; thus a chapter is devoted to software reliability models. Also, the use of software redundancy in the BFS is analyzed.

Computer networks are fundamental to communications systems, and local area networks connect a wide range of digital systems. Therefore, the principles of reliability and availability analysis for computer networks are developed, culminating in an introduction to network design principles. The concluding chapter considers a large system with multiple possibilities for improving reliability by adding parallel or standby subsystems. Simple apportionment and optimization techniques are developed for designing the highest reliability system within a fixed cost budget.

Four appendices are included to serve the needs of a variety of practitioners and students: Appendices A and B, covering probability and reliability principles for readers needing a review of probabilistic analysis; Appendix C, covering architecture for readers lacking a computer engineering or computer science background; and Appendix D, covering reliability and availability modeling programs for large systems.

## USE AS A REFERENCE

Often, a practitioner is faced with an initial system design that does not meet reliability or availability specifications, and the techniques discussed in Chapters 3, 4, and 7 help a designer rapidly evaluate and compare the reliability and availability gains provided by various improvement techniques. A designer or system engineer lacking a background in reliability will find the book's development from first principles in the chapters, the appendices, and the exercises ideal for self-study or intensive courses and seminars on reliability and availability. Intuition and quick analysis of proposed designs generally direct the engineer to a successful system; however, the efficient optimization techniques discussed in Chapter 7 can quickly yield an optimum solution and a range of good suboptima.

An engineer faced with newly developed technologies needs to consult the research literature and other more specialized texts; the many references provided can aid such a search. Topics of great importance are the error-correcting codes discussed in Chapter 2, the software reliability models discussed in Chapter 5, and the network reliability discussed in Chapter 6. Related examples and analyses are distributed among several chapters, and the index helps the reader to trace the evolution of an example.

Generally, the reliability and availability of *large* systems are calculated using fault-tolerant computer programs. Most industrial environments have these programs, the features of which are discussed in Appendix D. The most effective approach is to preface a computer model with a simplified analytical model, check the results, study the sensitivity to parameter changes, and provide insight if improvements are necessary.

## USE AS A TEXTBOOK

Many books that discuss fault-tolerant computing have a broad coverage of topics, with individual chapters contributed by authors of diverse backgrounds using different notations and approaches. This book selects the most important fault-tolerant techniques and examples and develops the concepts from first principles by using a consistent notation-and-analytical approach, with probabilistic analysis as the unifying concept linking the chapters.

To use this book as a teaching text, one might: (a) cover the material sequentially—in the order of Chapter 1 to Chapter 7; (b) preface approach

(a) by reviewing probability; or (c) begin with Chapter 7 on optimization and cover Chapters 3 and 4 on parallel, standby, and voting reliability; then augment by selecting from the remaining chapters. The sequential approach of (a) covers all topics and increases the analytical level as the course progresses; it can be considered a bottom-up approach. For a college junior- or senior-undergraduate–level or introductory graduate–level course, an instructor might choose approach (b); for an experienced graduate–level course, an instructor might choose approach (c). The homework problems at the end of each chapter are useful for self-study or classroom assignments.

At Polytechnic University, fault-tolerant computing is taught as a one-term graduate course for computer science and computer engineering students at the master's degree level, although the course is offered as an elective to senior-undergraduate students with a strong aptitude in the subject. Some consider fault-tolerant computing as a computer-systems course; others, as a second course in architecture.

## ACKNOWLEDGMENTS

The author thanks Carol Walsh and Joann McDonald for their help in preparing the class notes that preceded this book; the anonymous reviewers for their useful suggestions; and Professor Joanne Bechta Dugan of the University of Virginia and Dr. Robert Swarz of Mitre Corporation (Bedford, Massachusetts) and Worcester Polytechnic for their extensive, very helpful comments. He is grateful also to Wiley editors Dr. Philip Meyler and Andrew Prince who provided valuable advice. Many thanks are due to Dr. Alan P. Wood of Compaq Corporation for providing detailed information on Tandem computer design, discussed in Chapter 3, and to Larry Sherman of Stratus Computers for detailed information on Stratus, also discussed in Chapter 3. Sincere thanks are due to Sylvia Shooman, the author's wife, for her support during the writing of this book; she helped at many stages to polish and improve the author's prose and diligently proofread with him.

MARTIN L. SHOOMAN

*Glen Cove, NY*
*November 2001*

# 1

# INTRODUCTION

The central theme of this book is the use of reliability and availability computations as a means of comparing fault-tolerant designs. This chapter defines fault-tolerant computer systems and illustrates the prime importance of such techniques in improving the reliability and availability of digital systems that are ubiquitous in the 21st century. The main impetus for complex, digital systems is the microelectronics revolution, which provides engineers and scientists with inexpensive and powerful microprocessors, memories, storage systems, and communication links. Many complex digital systems serve us in areas requiring high reliability, availability, and safety, such as control of air traffic, aircraft, nuclear reactors, and space systems. However, it is likely that planners of financial transaction systems, telephone and other communication systems, computer networks, the Internet, military systems, office and home computers, and even home appliances would argue that fault tolerance is necessary in their systems as well. The concluding section of this chapter explains how the chapters and appendices of this book interrelate.

## 1.1 WHAT IS FAULT-TOLERANT COMPUTING?

Literally, fault-tolerant computing means computing correctly despite the existence of errors in a system. Basically, any system containing redundant components or functions has some of the properties of fault tolerance. A desktop computer and a notebook computer loaded with the same software and with files stored on floppy disks or other media is an example of a redundant system. Since either computer can be used, the pair is tolerant of most hardware and some software failures.

The sophistication and power of modern digital systems gives rise to a host of possible sophisticated approaches to fault tolerance, some of which are as effective as they are complex. Some of these techniques have their origin in the analog system technology of the 1940s–1960s; however, digital technology generally allows the implementation of the techniques to be faster, better, and cheaper. Siewiorek [1992] cites four other reasons for an increasing need for fault tolerance: harsher environments, novice users, increasing repair costs, and larger systems. One might also point out that the ubiquitous computer system is at present so taken for granted that operators often have few clues on how to cope if the system should go down.

Many books cover the architecture of fault tolerance (the way a fault-tolerant system is organized). However, there is a need to cover the techniques required to analyze the reliability and availability of fault-tolerant systems. A proper comparison of fault-tolerant designs requires a trade-off among cost, weight, volume, reliability, and availability. The mathematical underpinnings of these analyses are probability theory, reliability theory, component failure rates, and component failure density functions.

The obvious technique for adding redundancy to a system is to provide a duplicate (backup) system that can assume processing if the operating (on-line) system fails. If the two systems operate continuously (sometimes called hot redundancy), then either system can fail first. However, if the backup system is powered down (sometimes called cold redundancy or standby redundancy), it cannot fail until the on-line system fails and it is powered up and takes over. A standby system is more reliable (i.e., it has a smaller probability of failure); however, it is more complex because it is harder to deal with synchronization and switching transients. Sometimes the standby element does have a small probability of failure even when it is not powered up. One can further enhance the reliability of a duplicate system by providing repair for the failed system. The average time to repair is much shorter than the average time to failure. Thus, the system will only go down in the rare case where the first system fails and the backup system, when placed in operation, experiences a short time to failure before an unusually long repair on the first system is completed.

Failure detection is often a difficult task; however, a simple scheme called a voting system is frequently used to simplify such detection. If three systems operate in parallel, the outputs can be compared by a voter, a digital comparator whose output agrees with the majority output. Such a system succeeds if all three systems or two or the three systems work properly. A voting system can be made even more reliable if repair is added for a failed system once a single failure occurs.

Modern computer systems often evolve into networks because of the flexible way computer and data storage resources can be shared among many users. Most networks either are built or evolve into topologies with multiple paths between nodes; the Internet is the largest and most complex model we all use.

If a network link fails and breaks a path, the message can be routed via one or more alternate paths maintaining a connection. Thus, the redundancy involves alternate paths in the network.

In both of the above cases, the redundancy penalty is the presence of extra systems with their concomitant cost, weight, and volume. When the transmission of signals is involved in a communications system, in a network, or between sections within a computer, another redundancy scheme is sometimes used. The technique is not to use duplicate equipment but increased transmission time to achieve redundancy. To guard against undetected, corrupting transmission noise, a signal can be transmitted two or three times. With two transmissions the bits can be compared, and a disagreement represents a detected error. If there are three transmissions, we can essentially vote with the majority, thus detecting and correcting an error. Such techniques are called error-detecting and error-correcting codes, but they decrease the transmission speed by a factor of two or three. More efficient schemes are available that add extra bits to each transmission for error detection or correction and also increase transmission reliability with a much smaller speed-reduction penalty.

The above schemes apply to digital hardware; however, many of the reliability problems in modern systems involve software errors. Modeling the number of software errors and the frequency with which they cause system failures requires approaches that differ from hardware reliability. Thus, software reliability theory must be developed to compute the probability that a software error might cause system failure. Software is made more reliable by testing to find and remove errors, thereby lowering the error probability. In some cases, one can develop two or more independent software programs that accomplish the same goal in different ways and can be used as redundant programs. The meaning of independent software, how it is achieved, and how partial software dependencies reduce the effects of redundancy are studied in Chapter 5, which discusses software.

Fault-tolerant design involves more than just reliable hardware and software. System design is also involved, as evidenced by the following personal examples. Before a departing flight I wished to change the date of my return, but the reservation computer was down. The agent knew that my new return flight was seldom crowded, so she wrote down the relevant information and promised to enter the change when the computer system was restored. I was advised to confirm the change with the airline upon arrival, which I did. Was such a procedure part of the system requirements? If not, it certainly should have been.

Compare the above example with a recent experience in trying to purchase tickets by phone for a concert in Philadelphia 16 days in advance. On my Monday call I was told that the computer was down that day and that nothing could be done. On my Tuesday and Wednesday calls I was told that the computer was still down for an upgrade, and so it took a week for me to receive a call back with an offer of tickets. How difficult would it have been to print out from memory files seating plans that showed seats left for the next week so that tickets could be sold from the seating plans? Many problems can be avoided at little cost if careful plans are made in advance. The planners must always think "what do we do if ...?" rather than "it will never happen."

This discussion has focused on system reliability: the probability that the system *never fails* in some time interval. For many systems, it is acceptable for them to go down for short periods if it happens infrequently. In such cases, the system availability is computed for those involving repair. A system is said to be highly available if there is a low probability that a system *will be down* at any instant of time. Although reliability is the more stringent measure, both reliability and availability play important roles in the evaluation of systems.

## 1.2 THE RISE OF MICROELECTRONICS AND THE COMPUTER

### 1.2.1 A Technology Timeline

The rapid rise in the complexity of tasks, hardware, and software is why fault tolerance is now so important in many areas of design. The rise in complexity has been fueled by the tremendous advances in electrical and computer technology over the last 100–125 years. The low cost, small size, and low power consumption of microelectronics and especially digital electronics allow practical systems of tremendous sophistication but with concomitant hardware and software complexity. Similarly, the progress in storage systems and computer networks has led to the rapid growth of networks and systems.

A timeline of the progress in electronics is shown in Shooman [1990, Table K-1]. The starting point is the 1874 discovery that the contact between a metal wire and the mineral galena was a rectifier. Progress continued with the vacuum diode and triode in 1904 and 1905. Electronics developed for almost a half-century based on the vacuum tube and included AM radio, transatlantic radiotelephony, FM radio, television, and radar. The field began to change rapidly after the discovery of the point contact and field effect transistor in 1947 and 1949 and, ten years later in 1959, the integrated circuit.

The rise of the computer occurred over a time span similar to that of microelectronics, but the more significant events occurred in the latter half of the 20th century. One can begin with the invention of the punched card tabulating machine in 1889. The first analog computer, the mechanical differential analyzer, was completed in 1931 at MIT, and analog computation was enhanced by the invention of the operational amplifier in 1938. The first digital computers were electromechanical; included are the Bell Labs' relay computer (1937–40), the Z1, Z2, and Z3 computers in Germany (1938–41), and the Mark I completed at Harvard with IBM support (1937–44). The ENIAC developed at the University of Pennsylvania between 1942 and 1945 with U.S. Army support is generally recognized as the first electronic computer; it used vacuum tubes. Major theoretical developments were the general mathematical model of computation by Alan Turing in 1936 and the stored program concept of computing published by John von Neuman in 1946. The next hardware innovations were in the storage field: the magnetic-core memory in 1950 and the disk drive in 1956. Electronic integrated circuit memory came later in 1975. Software improved greatly with the development of high-level languages: FORTRAN (1954–58), ALGOL (1955–56), COBOL (1959–60), PASCAL (1971), the C language (1973), and the Ada language (1975–80). For computer advances related to cryptography, see problem 1.25.

The earliest major computer systems were the U.S. Airforce SAGE air defense system (1955), the American Airlines SABER reservations system (1957–64), the first time-sharing systems at Dartmouth using the BASIC language (1966) and the MULTICS system at MIT written in the PL-I language (1965–70), and the first computer network, the ARPA net, that began in 1969. The concept of RAID fault-tolerant memory storage systems was first published in 1988. The major developments in operating system software were the UNIX operating system (1969–70), the CM operating system for the 8086 Microprocessor (1980), and the MS-DOS operating system (1981). The choice of MS-DOS to be the operating system for IBM's PC, and Bill Gates' fledgling company as the developer, led to the rapid development of Microsoft.

The first home computer design was the Mark-8 (Intel 8008 Microprocessor), published in *Radio-Electronics* magazine in 1974, followed by the Altair personal computer kit in 1975. Many of the giants of the personal computing field began their careers as teenagers by building Altair kits and programming them. The company then called Micro Soft was founded in 1975 when Gates wrote a BASIC interpreter for the Altair computer. Early commercial personal computers such as the Apple II, the Commodore PET, and the Radio Shack TRS-80, all marketed in 1977, were soon eclipsed by the IBM PC in 1981. Early widely distributed PC software began to appear in 1978 with the Wordstar word processing system, the VisiCalc spreadsheet program in 1979, early versions of the Windows operating system in 1985, and the first version of the Office business software in 1989. For more details on the historical development of microelectronics and computers in the 20th century, see the following sources: Ditlea [1984], Randall [1975], Sammet [1969], and Shooman [1983]. Also see www.intel.com and www.microsoft.com.

This historical development leads us to the conclusion that today one can build a very powerful computer for a few hundred dollars with a handful of memory chips, a microprocessor, a power supply, and the appropriate input, output, and storage devices. The accelerating pace of development is breathtaking, and of course all the computer memory will be filled with software that is also increasing in size and complexity. The rapid development of the microprocessor—in many ways the heart of modern computer progress—is outlined in the next section.

### 1.2.2 Moore's Law of Microprocessor Growth

The growth of microelectronics is generally identified with the growth of the microprocessor, which is frequently described as "Moore's Law" [Mann, 2000]. In 1965, *Electronics* magazine asked Gordon Moore, research director

**TABLE 1.1  Complexity of Microchips and Moore's Law**

| Year | Microchip Complexity: Transistors | Moore's Law Complexity: Transistors |
|---|---|---|
| 1959 | 1 | $2^0 = 1$ |
| 1964 | 32 | $2^5 = 32$ |
| 1965 | 64 | $2^6 = 64$ |
| 1975 | 64,000 | $2^{16} = 65,536$ |

of Fairchild Semiconductor, to predict the future of the microchip industry. From the chronology in Table 1.1, we see that the first microchip was invented in 1959. Thus the complexity was then one transistor. In 1964, complexity had grown to 32 transistors, and in 1965, a chip in the Fairchild R&D lab had 64 transistors. Moore projected that chip complexity was doubling every year, based on the data for 1959, 1964, and 1965. By 1975, the complexity had increased by a factor of 1,000; from Table 1.1, we see that Moore's Law was right on track. In 1975, Moore predicted that the complexity would continue to increase at a slightly slower rate by doubling every two years. (Some people say that Moore's Law complexity predicts a doubling every 18 months.)

In Table 1.2, the transistor complexity of Intel's CPUs is compared with

**TABLE 1.2  Transistor Complexity of Microprocessors and Moore's Law Assuming a Doubling Period of Two Years**

| Year | Microchip Complexity CPU | Transistors | Moore's Law Complexity: Transistors |
|---|---|---|---|
| 1971.50 | 4004 | 2,300 | $(2^0) \times 2,300 = 2,300$ |
| 1978.75 | 8086 | 31,000 | $(2^{7.25/2}) \times 2,300 = 28,377$ |
| 1982.75 | 80286 | 110,000 | $(2^{4/2}) \times 28,377 = 113,507$ |
| 1985.25 | 80386 | 280,000 | $(2^{2.5/2}) \times 113,507 = 269,967$ |
| 1989.75 | 80486 | 1,200,000 | $(2^{4.5/2}) \times 269,967 = 1,284,185$ |
| 1993.25 | Pentium (P5) | 3,100,000 | $(2^{3.5/2}) \times 1,284,185 = 4,319,466$ |
| 1995.25 | Pentium Pro (P6) | 5,500,000 | $(2^{2/2}) \times 4,319,466 = 8,638,933$ |
| 1997.50 | Pentium II (P6 + MMX) | 7,500,000 | $(2^{2.25/2}) \times 8,638,933 = 18,841,647$ |
| 1998.50 | Merced (P7) | 14,000,000 | $(2^{3.25/2}) \times 8,638,933 = 26,646,112$ |
| 1999.75 | Pentium III | 28,000,000 | $(2^{1.25/2}) \times 26,646,112 = 41,093,922$ |
| 2000.75 | Pentium 4 | 42,000,000 | $(2^{1/2}) \times 41,093,922 = 58,115,582$ |

Note: This table is based on Intel's data from its *Microprocessor Report:* http://www.physics.udel.edu/wwwusers.watson.scen103/intel.html.

Moore's Law, with a doubling every two years. Note that there are many closely spaced releases with different processor speeds; however, the table records the first release of the architecture, generally at the initial speed. The Pentium P5 is generally called Pentium I, and the Pentium II is a P6 with MMX technology. In 1993, with the introduction of the Pentium, the Intel microprocessor complexities fell slightly behind Moore's Law. Some say that Moore's Law no longer holds because transistor spacing cannot be reduced rapidly with present technologies [Mann, 2000; Markov, 1999]; however, Moore, now Chairman Emeritus of Intel Corporation, sees no fundamental barriers to increased growth until 2012 and also sees that the physical limitations on fabrication technology will not be reached until 2017 [Moore, 2000].

The data in Table 1.2 is plotted in Fig. 1.1 and shows a close fit to Moore's Law. The three data points between 1997 and 2000 seem to be below the curve; however, the Pentium 4 data point is back on the Moore's Law line. Moore's Law fits the data so well in the first 15 years (Table 1.1) that Moore has occupied a position of authority and respect at Fairchild and, later, Intel. Thus, there is some possibility that Moore's Law is a self-fulfilling prophecy: that is, the engineers at Intel plan their new projects to conform to Moore's Law. The problems presented at the end of this chapter explore how Moore's Law is faring in the 21st century.

An article by Professor Seth Lloyd of MIT in the September 2000 issue of *Nature* explores the fundamental limitations of Moore's Law for a laptop based on the following: Einstein's Special Theory of Relativity ($E = mc^2$), Heisenberg's Uncertainty Principle, maximum entropy, and the Schwarzschild Radius for a black hole. For a laptop with one kilogram of mass and one liter of volume, the maximum available power is 25 million megawatt hours (the energy produced by all the world's nuclear power plants in 72 hours); the ultimate speed is $5.4 \times 10^{50}$ hertz (about $10^{43}$ the speed of the Pentium 4); and the memory size would be $2.1 \times 10^{31}$ bits, which is $4 \times 10^{30}$ bytes ($1.6 \times 10^{22}$ times that for a 256 megabyte memory) [Johnson, 2000]. Clearly, fabrication techniques will limit the complexity increases before these fundamental limitations.

### 1.2.3 Memory Growth

Memory size has also increased rapidly since 1965, when the PDP-8 minicomputer came with 4 kilobytes of core memory and when an 8 kilobyte system was considered large. In 1981, the IBM personal computer was limited to 640,000 kilobytes of memory by the operating system's nearsighted specifications, even though many "workaround" solutions were common. By the early 1990s, 4 or 8 megabyte memories for PCs were the rule, and in 2000, the standard PC memory size has grown to 64–128 megabytes. Disk memory has also increased rapidly: from small 32–128 kilobyte disks for the PDP 8e

**8** INTRODUCTION

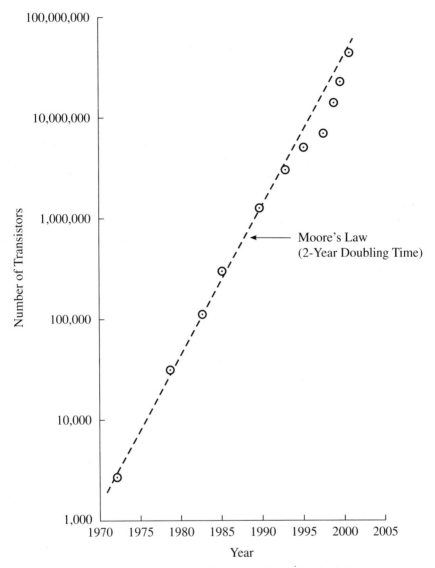

**Figure 1.1** Comparison of Moore's Law with Intel data.

computer in 1970 to a 10 megabyte disk for the IBM XT personal computer in 1982. From 1991 to 1997, disk storage capacity increased by about 60% per year, yielding an eighteenfold increase in capacity [Fisher, 1997; Markoff, 1999]. In 2001, the standard desk PC came with a 40 gigabyte hard drive. If Moore's Law predicts a doubling of microprocessor complexity every two years, disk storage capacity has increased by 2.56 times each two years, faster than Moore's Law.

## 1.2.4 Digital Electronics in Unexpected Places

The examples of the need for fault tolerance discussed previously focused on military, space, and other large projects. There is no less a need for fault tolerance in the home now that electronics and most electrical devices are digital, which has greatly increased their complexity. In the 1940s and 1950s, the most complex devices in the home were the superheterodyne radio receiver with 5 vacuum tubes, and early black-and-white television receivers with 35 vacuum tubes. Today, the microprocessor is ubiquitous, and, since a large percentage of modern households have a home computer, this is only the tip of the iceberg. In 1997, the sale of embedded microcomponents (simpler devices than those used in computers) totaled 4.6 billion, compared with about 100 million microprocessors used in computers. Thus computer microprocessors only represent 2% of the market [Hafner, 1999; Pollack, 1999].

The bewildering array of home products with microprocessors includes the following: clothes washers and dryers; toasters and microwave ovens; electronic organizers; digital televisions and digital audio recorders; home alarm systems and elderly medic alert systems; irrigation systems; pacemakers; video games; Web-surfing devices; copying machines; calculators; toothbrushes; musical greeting cards; pet identification tags; and toys. Of course this list does not even include the cellular phone, which may soon assume the functions of both a personal digital assistant and a portable Internet interface. It has been estimated that the typical American home in 1999 had 40–60 microprocessors—a number that could grow to 280 by 2004. In addition, a modern family sedan contains about 20 microprocessors, while a luxury car may have 40–60 microprocessors, which in some designs are connected via a local area network [Stepler, 1998; Hafner, 1999].

Not all these devices are that simple either. An electronic toothbrush has 3,000 lines of code. The Furby, a $30 electronic–robotic pet, has 2 main processors, 21,600 lines of code, an infrared transmitter and receiver for Furby-to-Furby communication, a sound sensor, a tilt sensor, and touch sensors on the front, back, and tongue. In short supply before Christmas 1998, Web site prices rose as high as $147.95 plus shipping! [*USA Today*, 1998]. In 2000, the sensation was Billy Bass, a fish mounted on a wall plaque that wiggled, talked, and sang when you walked by, triggering an infrared sensor.

Hackers have even taken an interest in Furby and Billy Bass. They have modified the hardware and software controlling the interface so that one Furby controls others. They have modified Billy Bass to speak the hackers' dialog and sing their songs.

Late in 2000, Sony introduced a second-generation dog-like robot called Aibo (Japanese for "pal"); with 20 motors, a 32-bit RISC processor, 32 megabytes of memory, and an artificial intelligence program. Aibo acts like a frisky puppy. It has color-camera eyes and stereo-microphone ears, touch sensors, a sound-synthesis voice, and gyroscopes for balance. Four different "personality" modules make this $1,500 robot more than a toy [Pogue, 2001].

What is the need for fault tolerance in such devices? If a Furby fails, you discard it, but it would be disappointing if that were the only sensible choice for a microwave oven or a washing machine. It seems that many such devices are designed without thought of recovery or fault-tolerance. Lawn irrigation timers, VCRs, microwave ovens, and digital phone answering machines are all upset by power outages, and only the best designs have effective battery backups. My digital answering machine was designed with an effective recovery mode. The battery backup works well, but it "locks up" and will not function about once a year. To recover, the battery and AC power are disconnected for about 5 minutes; when the power is restored, a 1.5-minute countdown begins, during which the device reinitializes. There are many stories in which failure of an ignition control computer stranded an auto in a remote location at night. Couldn't engineers develop a recovery mode to limp home, even if it did use a little more gas or emit fumes on the way home? Sufficient fault-tolerant technology exists; however, designers have to use it. Fortunately, the cellular phone allows one to call for help!

Although the preceding examples relate to electronic systems, there is no less a need for fault tolerance in mechanical, pneumatic, hydraulic, and other systems. In fact, almost all of us need a fault-tolerant emergency procedure to heat our homes in case of prolonged power outages.

## 1.3 RELIABILITY AND AVAILABILITY

### 1.3.1 Reliability Is Often an Afterthought

The attainment of high reliability and availability is very difficult to achieve in very complex systems. Thus, a system designer should formulate a number of different approaches to a problem and weigh the pluses and minuses of each design before recommending an approach. One should be careful to base conclusions on an analysis of facts, not on conjecture. Sometimes the best solution includes simplifying the design a bit by leaving out some marginal, complex features. It may be difficult to convince the authors of the requirements that sometimes "less is more," but this is sometimes the best approach. Design decisions often change as new technology is introduced. At one time any attempt to digitize the Library of Congress would have been judged infeasible because of the storage requirement. However, by using modern technology, this could be accomplished with two modern RAID disk storage systems such as the EMC Symmetrix systems, which store more than nine terabytes ($9 \times 10^{12}$ bytes) [EMC Products-At-A-Glance, www.emc.com]. The computation is outlined in the problems at the end of this chapter.

Reliability and availability of the system should always be two factors that are included, along with cost, performance, time of development, risk of failure, and other factors. Sometimes it will be necessary to discard a few design objectives to achieve a good design. The system engineer should always keep in mind that the design objectives generally contain a list of key features and a list of desirable features. The design must satisfy the key features, but if one or two of the desirable features must be eliminated to achieve a superior design, the trade-off is generally a good one.

### 1.3.2 Concepts of Reliability

Formal definitions of reliability and availability appear in Appendices A and B; however, the basic ideas are easy to convey without a mathematical development, which will occur later. Both of these measures apply to how good the system is and how frequently it goes down. An easy way to introduce reliability is in terms of test data. If 50 systems operate for 1,000 hours on test and two fail, then we would say the probability of failure, $P_f$, for this system in 1,000 hours of operation is 2/50 or $P_f(1,000) = 0.04$. Clearly the probability of success, $P_s$, which is known as the reliability, $R$, is given by $R(1,000) = P_s(1,000) = 1 - P_f(1,000) = 48/50 = 0.96$. Thus, *reliability is the probability of no failure within a given operating period*. One can also deal with a failure rate, $fr$, for the same system that, in the simplest case, would be $fr = 2$ failures/$(50 \times 1,000)$ operating hours—that is, $fr = 4 \times 10^{-5}$ or, as it is sometimes stated, $fr = z = 40$ failures per million operating hours, where $z$ is often called the hazard function. The units used in the telecommunications industry are fits (failures in time), which are failures per billion operating hours. More detailed mathematical development relates the reliability, the failure rate, and time. For the simplest case where the failure rate $z$ is a constant (one generally uses $\lambda$ to represent a constant failure rate), the reliability function can be shown to be $R(t) = e^{-\lambda t}$. If we substitute the preceding values, we obtain

$$R(1,000) = e^{-4 \times 10^{-5} \times 1,000} = 0.96$$

which agrees with the previous computation.

It is now easy to show that complexity causes serious reliability problems. The simplest system reliability model is to assume that in a system with $n$ components, all the components must work. If the component reliability is $R_c$, then the system reliability, $R_{sys}$, is given by

$$R_{sys}(t) = [R_c(t)]^n = [e^{-\lambda t}]^n = e^{-n\lambda t}$$

Consider the case of the first supercomputer, the CDC 6600 [Thornton, 1970]. This computer had 400,000 transistors, for which the estimated failure rate was then $4 \times 10^{-9}$ failures per hour. Thus, even though the failure rate of each transistor was very small, the computer reliability for 1,000 hours would be

$$R(1,000) = e^{-400,000 \times 4 \times 10^{-9} \times 1,000} = 0.20$$

**12**   INTRODUCTION

If we repeat the calculation for 100 hours, the reliability becomes 0.85. Remember that these calculations do not include the other components in the computer that can also fail. The conclusion is that the failure rate of devices with so many components must be very low to achieve reasonable reliabilities. Integrated circuits (ICs) improve reliability because each IC replaces hundreds of thousands or millions of transistors and also because the failure rate of an IC is low. See the problems at the end of this chapter for more examples.

### 1.3.3  Elementary Fault-Tolerant Calculations

The simplest approach to fault tolerance is classical redundancy, that is, to have an additional element to use if the operating one fails. As a simple example, let us consider a home computer in which constant usage requires it to be always available. A desktop will be the primary computer; a laptop will be the backup. The first step in the computation is to determine the failure rate of a personal computer, which will be computed from the author's own experience. Table 1.3 lists the various computers that the author has used in the home. There has been a total of 2 failures and 29 years of usage. Since each year contains 8,766 hours, we can easily convert this into a failure rate. The question becomes whether to estimate the number of hours of usage per year or simply to consider each year as a year of average use. We choose the latter for simplicity. Thus the failure rate becomes $2/29 = 0.069$ failures per year, and the reliability of a single PC for one year becomes $R(1) = e^{-0.069} = 0.933$. This means there is about a 6.7% probability of failure each year based on this data.

If we have two computers, both must fail for us to be without a computer. Assuming the failures of the two computers are *independent*, as is generally the case, then the system failure is the product of the failure probabilities for

**TABLE 1.3  Home Computers Owned by the Author**

| Computer | Date of Ownership | Failures | Operating Years |
|---|---|---|---|
| IBM XT Computer: Intel 8088 and 10 MB disk | 1983–90 | 0 failures | 7 years |
| Home upgrade of XT to Intel 386 Processor and 65 MB disk | 1990–95 | 0 failures | 5 years |
| IBM XT Components (repackaged in 1990) | Repackaged plus added new components used: 1990–92 | 1 failure | 2 years |
| Digital Equipment Laptop 386 and 80 MB disk | 1992–99 | 0 failures | 7 years |
| IBM Compatible 586 | 1995–2001 | 1 failure | 6 years |
| IBM Notebook 240 | 1999–2001 | 0 failures | 2 years |

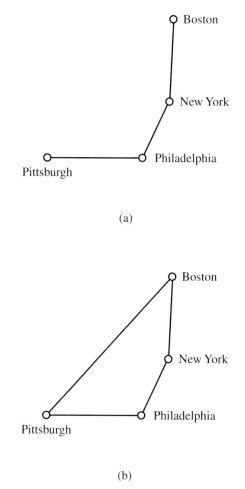

**Figure 1.2** Examples of simple computer networks: (a), a tree network connecting the four cities; (b), a Hamiltonian network connecting the four cities.

computer 1 (the primary) and computer 2 (the backup). Using the preceding failure data, the probability of one failure within a year should be 0.067; of two failures, $0.067 \times 0.067 = 0.00449$. Thus, the probability of having at least one computer for use is 0.9955 and the probability of having no computer at some time during the year is reduced from 6.7% to 0.45%—a decrease by a factor of 15. The probability of having no computer will really be much less since the failed computer will be rapidly repaired.

As another example of reliability computations, consider the primitive computer network as shown in Fig. 1.2(a). This is called a tree topology because all the nodes are connected and there are no loops. Assume that $p$ is the reliability for some time period for each link between the nodes. The probability that Boston and New York are connected is the probability that one link is good, that is, $p$. The same probability holds for New York–Philadelphia and for Philadelphia–Pittsburgh, but the Boston–Philadelphia connection requires two links to work, the probability of which is $p^2$. More commonly we speak of the all-terminal reliability, which is the probability that all cities are connected—$p^3$ in this example—because all three links must be working. Thus if $p = 0.9$, the all-terminal reliability is 0.729.

The reliability of a network is raised if we add more links so that loops are created. The Hamiltonian network shown in Fig. 1.2(b) has one more link than the tree and has a higher reliability. In the Hamiltonian network, all nodes are connected if all four links are working, which has a probability of $p^4$. All nodes are still connected if there is a single link failure, which has a probability of three successes and one failure given by $p^3(1-p)$. However, there are 4 ways for one link to fail, so the probability of one link failing is $4p^3(1-p)$. The reliability is the probability that there are zero failures plus the probability that there is one failure, which is given by $[p^4 + 4p^3(1-p)]$. Assuming that $p = 0.9$ as before, the reliability becomes 0.9477—a considerable improvement over the tree network. Some of the basic principles for designing and analyzing the reliability of computer networks are discussed in this book.

### 1.3.4 The Meaning of Availability

Reliability is the probability of no failures in an interval, whereas availability is the probability that an item is up at any point in time. Both reliability and availability are used extensively in this book as measures of performance and "yardsticks" for quantitatively comparing the effectiveness of various fault-tolerant methods. Availability is a good metric to measure the beneficial effects of repair on a system. Suppose that an air traffic control system fails on the average of once a year; we then would say that the mean time to failure (MTTF), was 8,766 hours (the number of hours in a year). If an airline's reservation system went down 5 times in a year, we would say that the MTTF was 1/5 of the air traffic control system, or 1,753 hours. One would say that, based on the MTTF, the air traffic control system was much better; however, suppose we consider repair and calculate typical availabilities. A simple formula for calculating the system availability (actually, the steady-state availability), based on the Uptime and Downtime of the system, is given as follows:

$$A = \frac{\text{Uptime}}{\text{Uptime} + \text{Downtime}}$$

If the air traffic control system goes down for about 1 hour whenever it fails, the availability would be calculated by substitution into the preceding formula yielding $A = (8,765)/(8,765 + 1) = 0.999886$. In the case of the airline reservation system, let us assume that the outages are short, averaging 1 minute each. Thus the cumulative downtime per year is five minutes = 0.083333 hours, and the availability would be $A = (8{,}765.916666)/(8{,}766) = 0.9999905$. Comparing the unavailabilities ($U = 1 - A$), we see $(1 - 0.999886)/(1 - 0.9999905) = 12$. Thus, we can say that based on availability the reservation system is 12 times better than the air traffic control system. Clearly one must use both reliability and availability to compare such systems.

A mathematical technique called Markov modeling will be used in this book to compute the availability for various systems. Rapid repair of failures in redundant systems greatly increases both the reliability and availability of such systems.

### 1.3.5 Need for High Reliability and Safety in Fault-Tolerant Systems

Fault-tolerant systems are generally required in applications involving a high level of safety, since a failure can injure or kill many people. A number of specifications, field failure data, and calculations are listed in Table 1.4 to give the reader some appreciation of the ranges of reliability and availability required and realized for various fault-tolerant systems.

A pattern emerges after some study of Table 1.4. The availability of several of the highly reliable fault-tolerant systems is similar. The availability requirement for the ESS telephone switching system (0.9999943), which is spoken of as "5 nines 43" in shorthand fashion, is seen to be equaled or bettered by actual performance of "5 nines 05" for (3B, 1A) and "5 nines 62" for (3A). Often one will compare system availability by quoting the downtime: for example, 5.7 hours per million for ESS requirements, 0.5 hours per million for (3B, 1A), and 3.8 hours per million for (3A). The Tandem goal was "5 nines 60" and the Stratus quote was "5 nines 05." Lastly, a standby system (if one could construct a fault-tolerant standby architecture) using 1985 technology would yield an availability of "5 nines 11." It is interesting to speculate whether this represents some level of performance one is able to achieve under certain limitations or whether the only proven numbers (the ESS switching systems) have become the goal others are quoting. The reader should remember that neither Tandem nor Stratus provides data on their field-demonstrated availability.

In the aircraft field there are some established system safety standards for the probability of catastrophe. These are extracted in Table 1.5, which also shows data on avionics-software-problem occurrence rates.

The two standards plus the software data quoted in Table 1.5 provide a rough but "overlapping" hierarchy of values. Some researchers have been pessimistic about the possibility of proving before use the reliability of hardware or software with reliabilities of $< 10^{-9}$. To demonstrate such a probability, we would need to test 10,000 systems for 10 years (about 100,000 hours) with 1 or 0 failures. Clearly this is not feasible, and one must rely on modeling and test data accumulated for major systems. However, from Shooman [1996], we can estimate that the U.S. air fleet of larger passenger aircraft flew about 12,000,000 flight hours in 1994 and today must fly about 20,000,000 hours. Thus if it were commercially feasible to install a new piece of equipment in every aircraft for

**TABLE 1.4  Comparison of Reliability and Availability for Various Fault-Tolerant Applications**

| Application | Reliability: $R$(hr), Unless Otherwise Stated | Availability (Steady State) | Comments or Source |
|---|---|---|---|
| 1964 NASA *Saturn Launch* computer | $R(250) = 0.99$ | — | [Pradhan, 1966, p. XIII] |
| *Apollo* (NASA) Moon Mission | $R$(mission) = 15/16 = 0.9375 (point estimate) | — | One failure (*Apollo* 13) in 16 missions |
| Space Shuttle (NASA) | $R$(mission) = 99/100 = 0.99 (point estimate) | — | One failure in 100 missions by end of 2000 |
| Bell Labs' ESS telephone switching system | — | *Requirement* of 2 hr of downtime in 40 yr or 3 min per year: 0.9999943  *Demonstrated* downtime per yr: ESS 3B (5 min) ESS 3A (2 min) ESS 1A (5 min) 0.9999905 (3B, 1A) 0.9999962 (3A) | [Pradhan, 1966, p. 438]; also Section 3.10.2 of this book  [Siewiorek, 1992, Fig. 8.30, p. 572] |
| Software-Implemented Fault Tolerance | Design *requirements*: $R(10) = 1 - 10^{-9}$ | — | [Siewiorek, 1992, pp. 710–735] |

| System | | Reliability | Reference |
|---|---|---|---|
| (SIFT): A research study conducted by SRI International with NASA support | | — | [Pradhan, 1966, pp. 460–463] |
| Fault-Tolerant Multiprocessor (FTMP): Experimental system, Draper Labs at MIT | Design *requirements*: $R(10) = 1 - 10^{-9}$ | — | [Siewiorek, 1992, pp. 184–187]; [Pradhan, 1966, pp. 460–463] |
| Tandem computer | — | 0.999996 | Based on Tandem goals; see Section 3.10.1 |
| Stratus computer | — | 0.9999905 | Based on Stratus Web site quote; see Section 3.10.2 |
| Vintage 1985 single CPU transaction-processing system | — | 0.997 | [Siewiorek, 1992, p. 586]; see also Section 3.10.1 |
| Vintage 1985 CPU 2 in parallel transaction-processing system | — | 0.999982 | See Section 4.9.2 |
| Vintage 1985 CPU 2 in standby transaction-processing system | — | 0.9999911 | See Section 4.9.2 |

**18** INTRODUCTION

**TABLE 1.5 Aircraft Safety Standards and Data**

| System Criticality | Likelihood | Probability of Failure/Flight Hr |
|---|---|---|
| Nonessential[a] | Probable | $> 10^{-5}$ |
| Essential[a] | Improbable | $10^{-5}$–$10^{-9}$ |
| Flight control[b] (e.g., bombers, transports, cargo, and tanker) | Extremely remote | $5 \times 10^{-7}$ |
| Critical[a] | Extremely improbable | $< 10^{-9}$ |
| Avionics software failure rates | — | Average failure rate of $1.5 \times 10^{-7}$ failures/hr for 6 major avionics systems |

[a]FAA, AC 25.1309-1A.
[b]MIL-F-9490.
*Source:* [Shooman, 1996].

one year and test it, but not have it connected to aircraft systems, one could generate 20,000,000 test hours. If no failures are observed, the statistical rule is to use 1/3 as the equivalent number of failures (see Section B3.5), and one could demonstrate a failure rate as low as $(1/3)/20{,}000{,}000 = 1.7 \times 10^{-8}$. It seems clear that the $10^{-9}$ probabilities given in Table 1.5 are the reasons why $10^{-9}$ was chosen for the goals of SIFT and FTMP in Table 1.4.

## 1.4 ORGANIZATION OF THIS BOOK

### 1.4.1 Introduction

This book was written for a diverse audience, including system designers in industry and students from a variety of backgrounds. Appendices A and B, which discuss probability and reliability principles, are included for those readers who need to deepen or refresh their knowledge of these topics. Similarly, because some readers may need some background in digital electronics, there is Appendix C that discusses digital electronics and architecture and provides a systems-level summary of these topics. The emphasis of this book is on analysis of systems and optimum design approaches. For large industrial problems, this emphasis will serve as a prelude to complement and check more comprehensive and harder-to-interpret computer analysis. Often the designer has to make a trade-off among several proposed designs. Many of the examples and some of the theory in this text address such trade-offs. The theme of the analysis and the trade-offs helps to unite the different subjects discussed in the various chapters. In many ways, each chapter is self-contained when it is accompanied by supporting appendix material; hence a practitioner can read sections of the book pertinent to his or her work, or an instructor can choose a selected group of chapters for a classroom presentation. This first chapter has described the complex nature of modern system design, which is one of the primary reasons that fault tolerance is needed in most systems.

### 1.4.2 Coding Techniques

A standard technique for guarding the veracity of a digital message/signal is to transmit the message more than once or to attach additional check bits to the message to detect and sometimes correct errors caused by "noise" that have corrupted some bits. Such techniques, called *error-detecting* and *error-correcting* codes, are introduced in Chapter 2. These codes are used to detect and correct errors in communications, memory storage, and signal transmission within computers and circuitry. When errors are sparse, the standard parity-bit and Hamming codes, developed from basic principles in Chapter 2, are very successful. The effectiveness of such codes is compared based on the probabilities that the codes fail to detect multiple errors. The probability that the coding and decoding chips may fail catastrophically is also included in the analysis. Some original work is introduced to show under which circumstances the chip failures are significant. In some cases, errors occur in groups of adjacent bits, and an introductory development of burst error codes, which are used in such cases, is presented. An introduction to more sophisticated Reed–Solomon codes concludes this chapter.

### 1.4.3 Redundancy, Spares, and Repairs

One way of improving system reliability is to reduce the failure rate of pivotal individual components. Sometimes this is not a feasible or cost-effective approach to meeting very high reliability requirements. Chapter 3 introduces another technique—redundancy—and it considers the fundamental techniques of system and component redundancy. The standard approach is to have two (or more) units operating *in parallel* so that if one fails the other(s) take over. Parallel *components* are generally more efficient than parallel *systems* in improving the resulting reliability; however, some sort of "coupling device" is needed to parallel the units. The reliability of the coupling device is modeled, and under certain circumstances failures of this device may significantly degrade system reliability. Various approximations are developed to allow easy comparison of different approaches and, in addition, the system mean time to failure (MTTF) is also used to simplify computations. The effects of common-cause failures, which can negate much of the beneficial effects of redundancy, are discussed.

The other major form of redundancy is *standby* redundancy, in which the redundant component is powered down until the on-line system fails. This is often superior to parallel reliability. In the standby case, the sensing system that detects failures and switches is more complex, and the reliability of this device is studied to assess the degradation in predicted reliability caused by the standby switch. The study of standby systems is based on Markov probability

**20** INTRODUCTION

models that are introduced in the appendices and deliberately developed in Chapter 3 because they will be used throughout the book.

Repair improves the reliability of both parallel and standby systems, and Markov probability models are used to study the relative benefits of repair for both approaches. Markov modeling generates a set of differential equations that require a solution to complete the analysis. The Laplace transform approach is introduced and used to simplify the solution of the Markov equations for both reliability and availability analysis.

Several computer architectures for fault tolerance are introduced and discussed. Modern memory storage systems use the various RAID architectures based on an array of redundant disks. Several of the common RAID techniques are analyzed. The class of fault-tolerant computer systems called nonstop systems is introduced. Also introduced and analyzed are two other systems: the Tandem system, which depends primarily on software fault tolerance, and the Stratus system, which uses hardware fault tolerance. A brief description of a similar system approach, a Sun computer system cluster, concludes the chapter.

### 1.4.4 *N*-Modular Redundancy

The problem of comparing the proper functioning of parallel systems was discussed earlier in this chapter. One of the benefits of a digital system is that all outputs are strings of 1s or 0s so that the comparison of outputs is simplified. Chapter 4 describes an approach that is often used to compare the outputs of three identical digital circuits processing the same input: triple modular redundancy (TMR). The most common circuit output is used as the system output (called majority voting). In the case of TMR, we assume that if outputs disagree, those two that are the same will together have a much higher probability of succeeding rather than failing. The voting device is simple, and the resulting system is highly reliable. As in the case of parallel or standby redundancy, the voting can be done at the system or subsystem level, and both approaches are modeled and compared.

Although the voter circuit is simple, it can fail; the effect of voter reliability, much like coupler reliability in a parallel system, must then be included. The possibility of using redundant voters is introduced. Repair can be used to improve the reliability of a voter system, and the analysis utilizes a Markov model similar to that of Chapter 3. Various simplified approximations are introduced that can be used to analyze the reliability and availability of repairable systems. Also introduced are more advanced voting and consensus techniques. The redundant system of Chapter 3 is compared with the voting techniques of Chapter 4.

### 1.4.5 Software Reliability and Recovery Techniques

Programming of the computer in early digital systems was largely done in complex machine language or low-level assembly language. Memory was limited, and the program had to be small and concise. Expert programmers often used tricks to fit the required functions into the small memory. Software errors—then as now—can cause the system to malfunction. The failure mode is different but no less disastrous than catastrophic hardware failures. Chapter 5 relates these program errors to resulting system failures.

This chapter begins by describing in some detail the way programs are now developed in modern higher-level languages such as FORTRAN, COBOL, ALGOL, C, C+ +, and Ada. Large memories allow more complex tasks, and many more programmers are involved. There are many potential sources of errors, such as the following: (a), complex, error-prone specifications; (b), logic errors in individual modules (self-contained sections of the program); and (c), communications among modules. Sometimes code is incorporated from previous projects without sufficient adaptation analysis and testing, causing subtle but disastrous results. A classical example of the hazards of reused code is the Ariane-5 rocket. The European Space Agency (ESA) reused guidance software from Ariane-4 in Ariane-5. On its maiden flight, June 4, 1996, Ariane-5 had to be destroyed 40 seconds into launch—a $500 million loss. Ariane-5 developed a larger horizontal velocity than Ariane-4, and a register overflowed. The software detected an exception, but instead of taking a recoverable action it shut off the processor as the specifications required. A more appropriate recovery action might have saved the flight. To cite the legendary Murphy's Law, "If things can go wrong, they will," and they did. Even better, we might devise a corollary that states "then plan for it" [Pfleeger, 1998, pp. 37–39].

Various mathematical models describing errors are introduced. The introductory model is based on a simple assumption: the failure rate (error discovery rate) is proportional to the number of errors remaining in the software after it is tested and released. Combining this software failure rate with reliability theory leads to a software reliability model. The constants in such models are evaluated from test data recorded during software development. Applying such models during the test phase allows one to predict the reliability of the software once it is released for operational use. If the predicted reliability appears unsatisfactory, the developer can improve testing to remove more errors, rewrite certain problem modulus, or take other action to avoid the release of an unreliable product.

Software redundancy can be utilized in some cases by using independently developed but functionally identical software. The extent to which common errors in independent software reduces the reliability gains is discussed; as a practical example, the redundant software in the NASA Space Shuttle is considered.

### 1.4.6 Networked Systems Reliability

Networks are all around us. They process our telephone calls, connect us to the Internet, and connect private industry and government computer and information systems. In general, such systems have a high reliability and availability

**22** INTRODUCTION

because there is more than one path that connects all of the terminals in the network. Thus a single link failure will seldom interrupt communications because a duplicate path will exist. Since network geometry (topology) is usually complex, there are many paths between terminals, and therefore computation of network reliability is often difficult. Computer programs are available for such computations, two of which are referenced in the chapter. This chapter systematically develops methods based on graph theory (cut-sets and tie-sets) for analysis of a network. Alternate methods for computation are also discussed, and the chapter concludes with the application of such methods to the design of a reliable backbone network.

### 1.4.7 Reliability Optimization

Initial design of a large, complex system focuses on several issues: (a), how to structure the project to perform the required functions; (b), how to meet the performance requirements; and (c), how to achieve the required reliability. Designers always focus on issues (a) and (b), but sometimes, at the peril of developing an unreliable system, they spend a minimum of effort on issue (c). Chapter 7 develops techniques for optimizing the reliability of a proposed design by parceling out the redundancy to various subsystems. Choice among optimized candidate designs should be followed by a trade-off among the feasible designs, weighing the various pros and cons that include reliability, weight, volume, and cost. In some ways, one can view this chapter as a generalization of Chapter 3 for larger, more complex system designs.

One simplified method of achieving optimum reliability is to meet the overall system reliability goal by fixing the level of redundancy for the various subsystems according to various apportionment rules. The other end of the optimization spectrum is to obtain an exact solution by means of exhaustively computing the reliability for all the possible system combinations. The Dynamic Programming method was developed as a way to eliminate many of the cases in an exhaustive computation scheme. Chapter 7 discusses the above methods as well as an effective approximate method—a greedy algorithm, where the optimization is divided into a series of steps and the best choice is made for each step.

The best method developed in this chapter is to establish a set of upper and lower bounds on the number of redundancies that can be assigned for each subsystem. It is shown that there is a modest number of possible cases, so an exhaustive search *within the allowed bounds* is rapid and computationally feasible. The bounded method displays the optimal configuration as well as many other close-to-optimum alternatives, and it provides the designer with a number of good solutions among which to choose.

### 1.4.8 Appendices

This book has been written for practitioners and students from a wide variety of disciplines. In cases where the reader does not have a background in either probability or digital circuitry, or needs a review of principles, these appendices provide a self-contained development of the background material of these subjects.

Appendix A develops probability from basic principles. It serves as a tutorial, review, or reference for the reader.

Appendix B summarizes reliability theory and develops the relationships among reliability theory, conventional probability density and distributions functions, and the failure rate (hazard) function. The popular MTTF metric, as well as sample calculations, are given. Availability theory and Markov models are developed.

Appendix C presents a concise introduction to digital circuit design and elementary computer architecture. This will serve the reader who needs a background to understand the architecture applications presented in the text.

Appendix D discusses reliability, availability, and risk-modeling programs. Most large systems will require such software to aid in analysis. This appendix categorizes these programs and provides information to aid the reader in contacting the suppliers to make an informed choice among the products offered.

## GENERAL REFERENCES

The references listed here are a selection of textbooks and proceedings that apply to several chapters in this book. Specific references for Chapter 1 appear in the following section.

Aktouf, C. et al. *Basic Concepts and Advances in Fault-Tolerant Design*. World Scientific Publishing, River Edge, NJ, 1998.

Anderson, T. *Resilient Computing Systems*, vol. 1. Wiley, New York, 1985.

Anderson, T., and P. A. Lee. *Fault Tolerance: Principles and Practice*. Prentice-Hall, New York, 1981.

Arazi, B. *A Commonsense Approach to the Theory of Error-Correcting Codes*. MIT Press, Cambridge, MA, 1988.

Avizienis, A. *The Evolution of Fault-Tolerant Computing*. Springer-Verlag, New York, 1987.

Avresky, D. R. (ed.). *Fault-Tolerant Parallel and Distributed Systems*. Kluwer Academic Publishers, Hingham, MA, 1998.

Bolch, G., S. Greiner, H. de Meer, and K. S. Trivedi. *Queueing Networks and Markov Chains: Modeling and Performance Evaluation with Computer Science Applications*. Wiley, New York, 1998.

Breuer, M. A., and A. D. Friedman. *Diagnosis and Reliable Design of Digital Systems*. Computer Science Press, Woodland Hills, CA, 1976.

Christian, F. (ed.). *Dependable Computing for Critical Applications*. Springer-Verlag, New York, 1995.

Special Issue on Fault-Tolerant Systems. *IEEE Computer Magazine*, New York (July 1990).

Special Issue on Dependability Modeling. *IEEE Computer Magazine*, New York (October 1990).

Dacin, M. et al. *Dependable Computing for Critical Applications*. IEEE Computer Society Press, New York, 1997.

Davies, D. W. *Distributed Systems—Architecture and Implementation*, Lecture Notes in Computer Science. Springer-Verlag, New York, 1981, ch. 8, 10, 13, 17, and 20.

Dougherty, E. M. Jr., and J. R. Fragola. *Human Reliability Analysis*. Wiley, New York, 1988.

Echte, K. Dependable Computing—EDCC-1. *Proceedings of the First European Dependable Computing Conference*, Berlin, Germany, 1994.

*Fault-Tolerant Computing Symposium, 25th Anniversary Compendium*. IEEE Computer Society Press, New York, 1996. (Author's note: Symposium proceedings are published yearly by the IEEE.)

Gibson, G. A. *Redundant Disk Arrays*. MIT Press, Cambridge, MA, 1992.

Hawicska, A. et al. Dependable Computing—EDCC-2. *Second European Dependable Computing Conference*, Taormina, Italy, 1996.

Johnson, B. W. *Design and Analysis of Fault Tolerant Digital Systems*. Addison-Wesley, Reading, MA, 1989.

Kanellakis, P. C., and A. A. Shvartsman. *Fault-Tolerant Parallel Computation*. Kluwer Academic Publishers, Hingham, MA, 1997.

Kaplan, G. The X-29: Is it Coming or Going? *IEEE Spectrum*, New York (June 1985): 54–60.

Lala, P. K. *Self-Checking and Fault-Tolerant Digital Design*. Academic Press, San Diego, CA, 2000.

Lee, P. A., and T. Anderson. *Fault Tolerance, Principles and Practice*, 2d ed. Springer-Verlag, New York, 1990.

Lyu, M. R. (ed.). *Handbook of Software Reliability Engineering*. McGraw-Hill, New York, 1996.

McCormick, N. *Reliability and Risk Analysis*. Academic Press, New York, 1981.

Ng, Y. W., and A. A. Avizienis. A Unified Reliability Model for Fault-Tolerant Computers. *IEEE Transactions on Computers* C-29, New York, no. 11 (November 1980): 1002–1011.

Osaki, S., and T. Nishio. *Reliability Evaluation of Some Fault-Tolerant Computer Architectures*, Lecture Notes in Computer Science. Springer-Verlag, New York, 1980.

Patterson, D., R. Katz, and G. Gibson. A Case for Redundant Arrays of Inexpensive Disks (RAID). *Proceedings of the 1988 ACM SIG on Management of Data (ACM SIGMOD)*, Chicago, IL, June 1988, pp. 109–116.

Pham, H. *Fault-Tolerant Software Systems, Techniques and Applications*. IEEE Computer Society Press, New York, 1992.

Pierce, W. H. *Fault-Tolerant Computer Design*. Academic Press, New York, 1965.

Pradhan, D. K. *Fault-Tolerant Computing Theory and Technique*, vols. I and II. Prentice-Hall, Englewood Cliffs, NJ, 1986.

Pradhan, D. K. *Fault-Tolerant Computing*, vol. I, 2d ed. Prentice-Hall, Englewood Cliffs, NJ, 1993.

Rao, T. R. N., and E. Fujiwara. *Error-Control Coding for Computer Systems*. Prentice-Hall, Englewood Cliffs, NJ, 1989.

Shooman, M. L. *Software Engineering, Design, Reliability, Management*. McGraw-Hill, New York, 1983.

Shooman, M. L. *Probabilistic Reliability: An Engineering Approach*, 2d ed. Krieger, Melbourne, FL, 1990.

Siewiorek, D. P., and R. S. Swarz. *The Theory and Practice of Reliable System Design*. The Digital Press, Bedford, MA, 1982.

Siewiorek, D. P., and R. S. Swarz. *Reliable Computer Systems Design and Evaluation*, 2d ed. The Digital Press, Bedford, MA, 1992.

Siewiorek, D. P., and R. S. Swarz. *Reliable Computer Systems Design and Evaluation*, 3d ed. A. K. Peters, www.akpeters.com, 1998.

Smith, B. T. *The Fault-Tolerant Multiprocessor Computer*. Noyes Data Corporation, 1986.

Trivedi, K. S. *Probability and Statistics with Reliability, Queuing and Computer Science Applications*, 2d ed. Wiley, New York, 2002.

*Workshop on Defect and Fault-Tolerance in VLSI Systems*. IEEE Computer Society Press, New York, 1995.

## REFERENCES

Anderson, T. *Resilient Computing Systems*. Wiley, New York, 1985.

Bell, C. G. *Computer Structures: Readings and Examples*. McGraw-Hill, New York, 1971.

Bell, T. (ed.). Special Report: Designing and Operating a Minimum-Risk System. *IEEE Spectrum*, New York (June 1989): pp. 22–52.

Braun, E., and S. McDonald. *Revolution in Miniature—The History and Impact of Semiconductor Electronics*. Cambridge University Press, London, 1978.

Burks, A. W., H. H. Goldstine, and J. von Neuman. *Preliminary Discussion of the Logical Design of an Electronic Computing Instrument*. Report to the U.S. Army Ordinance Department, 1946. Reprinted in Randell (p. 371–385) and Bell (1971, p. 92–119).

Clark, R. *The Man Who Broke Purple the Life of W. F. Friedman*. Little, Brown and Company, Boston, 1977.

Ditlea, S. (ed.). *Digital Deli*. Workman Publishing, New York, 1984.

Federal Aviation Administration Advisory Circular, AC 25.1309-1A.

Fisher, L. M. "IBM Plans to Announce Leap in Disk-Drive Capacity." *New York Times*, December 30, 1997, p. D2.

Fragola, J. R. Forecasting the Reliability and Safety of Future Space Transportation Systems. *Proceedings, Annual Reliability and Maintainability Symposium*, 2000. IEEE, New York, NY, pp. 292–298.

Friedman, M. B. RAID keeps going and going and.... *IEEE Spectrum*, New York (1996): pp. 73–79.

Giloth, P. K. No. 4 ESS—Reliability and Maintainability Experience. *Proceedings, Annual Reliability and Maintainability Symposium*, 1980. IEEE, New York, NY.

Hafner, K. "Honey, I Programmed the Blanket—The Omnipresent Chip has Invaded Everything from Dishwashers to Dogs." *New York Times*, May 27, 1999, p. G1.

Iaciofano, C. Computer Time Line, in *Digital Deli*, Ditlea (ed.). Workman Publishing, New York, 1984, pp. 20–34.

Johnson, G. "The Ultimate, Apocalyptic Laptop." *New York Times*, September 5, 2000, p. F1.

Lewis, P. H. "With 2 New Chips, the Gigahertz Decade Begins." *New York Times*, March 9, 2000, p. G1.

Mann, C. C. The End of Moore's Law? *Technology Review*, Cambridge, MA (May–June 2000): p. 42.

Markoff, J. "IBM Sets a New Record for Magnetic-Disk Storage." *New York Times*, May 12, 1999.

Markoff, J. "Chip Progress may soon Be Hitting Barrier." *New York Times* (on the Internet), October 9, 1999.

Markoff, J. "A Tale of the Tape from the Days when it Was Still Micro Soft." *New York Times*, September, 2000, p. C1.

Military Standard. General Specification for Flight Control Systems—Design, Install, and Test of Aircraft. MIL-F-9490, 1975.

Moore, G. E. Intel Developers Forum, 2000 [http://developer.intel.com/update/archive/issue2/feature.html].

Norwall, B. D. FAA Claims Progress on ATC Improvements. *Aviation Week and Space Technology* (September 25, 1995): p. 44.

Patterson, D., R. Katz, and G. Gibson. A Case for Redundant Arrays of Inexpensive Disks (RAID). *Proceedings of the 1988 ACM SIG on Management of Data (ACM SIGMOD)*, Chicago, IL, June 1988, pp. 109–116.

Pfleeger, S. L. *Software Engineering Theory and Practice*. Prentice Hall, Upper Saddle River, NJ, 1998.

Pogue, D. "Who Let the Robot Out?" *New York Times*, January 25, 2001, p. G1.

Pollack, A. "Chips are Hidden in Washing Machines, Microwaves and Even Reservoirs." *New York Times*, January 4, 1999, p. C17.

Randall, B. *The Origins of Digital Computers*. Springer-Verlag, New York, 1975.

Rogers, E. M., and J. K. Larsen. *Silicon Valley Fever—Growth of High-Technology Culture*. Basic Books, New York, 1984.

Sammet, J. E. *Programming Languages: History and Fundamentals*. Prentice-Hall, Englewood Cliffs, NJ, 1969.

Shooman, M. L. *Probabilistic Reliability: An Engineering Approach*, 2d ed. Krieger, Melbourne, FL, 1990.

Shooman, M. L. *Software Engineering, Design, Reliability, Management*. McGraw-Hill, New York, 1983.

Shooman, M. L. Avionics Software Problem Occurrence Rates. *Proceedings of Software Reliability Engineering Conference, ISSRE '96*, 1996. IEEE, New York, NY, pp. 55–64.

Siewiorek, D. P., and R. S. Swarz. *The Theory and Practice of Reliable System Design.* The Digital Press, Bedford, MA, 1982.

Siewiorek, D. P., and R. S. Swarz. *Reliable Computer Systems Design and Evaluation,* 2d ed. The Digital Press, Bedford, MA, 1992.

Siewiorek, D. P., and R. S. Swarz. *Reliable Computer Systems Design and Evaluation,* 3d ed. A. K. Peters, www.akpeters.com, 1998.

Stepler, R. "Fill it Up, with RAM—Cars Get More Megs Under the Hood." *New York Times*, August 27, 1998, p. G1.

Turing, A. M. On Computable Numbers, with an Application to the Entscheidungs problem. *Proc. London Mathematical Soc.*, 42, 2 (1936): pp. 230–265.

Turing, A. M. Corrections. *Proc. London Mathematical Soc.*, 43 (1937): pp. 544–546.

Wald, M. L. "Ambitious Update of Air Navigation Becomes a Fiasco." *New York Times*, January 29, 1996. p. 1.

Wirth, N. The Programming Language PASCAL. *Acta Informatica* 1 (1971): pp. 35–63.

Zuckerman, L., and M. L. Wald. "Crisis for Air Traffic Systems: More Passengers, More Delays." *New York Times*, September 5, 2000, front page.

*USA Today*, December 2, 1998, p. 1D.

www.emc.com (the EMC Products-At-A-Glance Web site).

www.intel.com (the Intel Web site).

www.microsoft.com (the Microsoft Web site).

## PROBLEMS

**1.1.** Show that the combined capacity of several (two or three) modern disk storage systems, such as the EMC Symmetrix System that stores more than nine terabytes ($9 \times 10^{12}$ bytes) [EMC Products-At-A-Glance, www.emc.com], could contain all the 26 million texts in the Library of Congress [Web search, Library of Congress].

   **(a)** Assume that the average book has 400 pages.

   **(b)** Estimate the number of lines per page by counting lines in three different books.

   **(c)** Repeat (b) for the number of words per line.

   **(d)** Repeat (b) for the number of characters per word.

   **(e)** Use the above computations to find the number of characters in the 26 million books.

   Assume that one character is stored in one byte and calculate the number of Symmetrix units needed.

**1.2.** Estimate the amount of storage needed to store all the papers in a standard four-drawer business filing cabinet.

**1.3.** Estimate the cost of digitizing the books in the Library of Congress. How would you do this?

**1.4.** Repeat problem 1.3 for the storage of problem 1.2.

**1.5.** Visit the Intel Web site and check the release dates and transistor complexities given in Table 1.2.

**1.6.** Repeat problem 1.5 for microprocessors from other manufacturers.

**1.7.** Extend Table 1.2 for newer processors from Intel and other manufacturers.

**1.8.** Search the Web for articles about the change of mainframes in the air traffic control system and identify the old and new computers, the past problems, and the expected improvements from the new computers. Hint: look at IEEE *Computer* and *Spectrum* magazines and the *New York Times*.

**1.9.** Do some research and try to determine if the storage density for optical copies (one page of text per square millimeter) is feasible with today's optical technology. Compare this storage density with that of a modern disk or CD-ROM.

**1.10.** Make a list of natural, human, and equipment failures that could bring down a library system stored on computer disks. Explain how you could incorporate design features that would minimize such problems.

**1.11.** Complex solutions are not always needed. There are many good programs for storing cooking recipes. Many cooks use a few index cards or a cookbook with paper slips to mark their favorite recipes. Discuss the pros and cons of each approach. Under what circumstances would you favor each approach?

**1.12.** An improved version of Basic, called GW Basic, followed the original Micro Soft Basic. "GW" did not stand for our first president or the university that bears his name. Try to find out what GW stands for and the origin of the software.

**1.13.** Estimate the number of failures per year for a family automobile and compute the failure rate (failures per mile). Assuming 10,000 miles driven per year, compute the number of failures per year. Convert this into failures per hour assuming that one drives 10,000 miles per year at an average speed of 40 miles per hour.

**1.14.** Assume that an auto repair takes 8 hours, including drop-off, storage, and pickup of the car. Using the failure rate computed in problem 1.13 and this information, compute the availability of an automobile.

**1.15.** Make a list of safety critical systems that would benefit from fault tolerance. Suggest design features that would help fault tolerance.

**1.16.** Search the Web for examples of the systems in problem 1.15 and list the details you can find. Comment.

**1.17.** Repeat problems 1.15 and 1.16 for systems in the home.

**1.18.** Repeat problems 1.15 and 1.16 for transportation, communication, power, heating and cooling, and entertainment systems in everyday use.

**1.19.** To learn of a 180 terabyte storage project, search the EMC Web site for the movie producer Steven Spielberg, or see the *New York Times*: Jan. 13, 2001, p. B11. Comment.

**1.20.** To learn of some of the practical problems in trying to improve an existing fault-tolerant system, consider the U.S. air traffic control system. Search the Web for information on the current delays, the effects of deregulation, and former President Ronald Reagan's dismissal of striking air traffic controllers; also see Zuckerman [2000]. A large upgrade to the system failed and incremental upgrades are being planned instead. Search the Web and see [Wald, 1996] for a discussion of why the upgrade failed.

(a) Write a report analyzing what you learned.

(b) What is the present status of the system and any upgrades?

**1.21.** Devise a scheme for emergency home heating in case of a prolonged power outage for a gas-fired, hot-water heating system. Consider the following: (a), fireplace; (b), gas stove; (c), emergency generator; and (d), other. How would you make your home heating system fault tolerant?

**1.22.** How would problem 1.21 change for the following:

(a) An oil-fired, hot-water heating system?

(b) A gas-fired, hot-air heating system?

(c) A gas-fired, hot-water heating system?

**1.23.** Present two designs for a fault-tolerant voting scheme.

**1.24.** Investigate the *speed* of microprocessors and how rapidly it has increased over the years. You may wish to use the microprocessors in Table 1.2 or others as data points. A point on the curve is the 1.7 gigahertz Pentium 4 microprocessor [*New York Times*, April 23, 2001, p. C1]. Plot the data in a format similar to Fig. 1.1. Does a law hold for speed?

**1.25.** Some of the advances in mechanical and electronic computers occurred during World War II in conjunction with message encoding and decoding and cryptanalysis (code breaking). Some of the details were, and still are, classified as secret. Find out as much as you can about these machines and compare them with those reported on in Section 1.2.1. Hint: Look in Randall [1975, pp. 327, 328] and Clark [1977, pp. 134, 135, 140, 151, 195, 196]. Also, search the Web for key words: Sigaba, Enigma, T. H. Flowers, William F. Friedman, Alan Turing, and any patents by Friedman.

# 2

# CODING TECHNIQUES

## 2.1 INTRODUCTION

Many errors in a computer system are committed at the bit or byte level when information is either transmitted along communication lines from one computer to another or else within a computer from the memory to the microprocessor or from microprocessor to input/output device. Such transfers are generally made over high-speed internal buses or sometimes over networks. The simplest technique to protect against such errors is the use of error-detecting and error-correcting codes. These codes are discussed in this chapter in this context. In Section 3.9, we see that error-correcting codes are also used in some versions of RAID memory storage devices.

The reader should be familiar with the material in Appendix A and Sections B1–B4 before studying the material of this chapter. It is suggested that this material be reviewed briefly or studied along with this chapter, depending on the reader's background.

The word *code* has many meanings. Messages are commonly coded and decoded to provide secret communication [Clark, 1977; Kahn, 1967], a practice that technically is known as cryptography. The municipal rules governing the construction of buildings are called building codes. Computer scientists refer to individual programs and collections of programs as software, but many physicists and engineers refer to them as computer codes. When information in one system (numbers, alphabet, etc.) is represented by another system, we call that other system a code for the first. Examples are the use of binary numbers to represent numbers or the use of the ASCII code to represent the letters, numerals, punctuation, and various control keys on a computer keyboard (see

# INTRODUCTION    31

Table C.1 in Appendix C for more information). The types of codes that we discuss in this chapter are *error-detecting* and *-correcting codes*. The principle that underlies error-detecting and -correcting codes is the addition of specially computed redundant bits to a transmitted message along with added checks on the bits of the received message. These procedures allow the detection and sometimes the correction of a modest number of errors that occur during transmission.

The computation associated with generating the redundant bits is called *coding*; that associated with detection or correction is called *decoding*. The use of the words *message*, *transmitted*, and *received* in the preceding paragraph reveals the origins of error codes. They were developed along with the mathematical theory of information largely from the work of C. Shannon [1948], who mentioned the codes developed by Hamming [1950] in his original article. (For a summary of the theory of information and the work of the early pioneers in coding theory, see J. R. Pierce [1980, pp. 159–163].) The preceding use of the term *transmitted bits* implies that coding theory is to be applied to digital signal transmission (or a digital model of analog signal transmission), in which the signals are generally pulse trains representing various sequences of 0s and 1s. Thus these theories seem to apply to the field of communications; however, they also describe information transmission in a computer system. Clearly they apply to the signals that link computers connected by modems and telephone lines or local area networks (LANs) composed of transceivers, as well as coaxial wire and fiber-optic cables or wide area networks (WANs) linking computers in distant cities. A standard model of computer architecture views the central processing unit (CPU), the address and memory buses, the input/output (I/O) devices, and the memory devices (integrated circuit memory chips, disks, and tapes) as digital signal (computer word) transmission, storage, manipulation, generation, and display devices. From this perspective, it is easy to see how error-detecting and -correcting codes are used in the design of modems, memory stems, disk controllers (optical, hard, or floppy), keyboards, and printers.

The difference between error detection and error correction is based on the use of redundant information. It can be illustrated by the following electronic mail message:

> Meet me in Manhattan at the information desk at Senn Station on July 43. I will arrive at 12 noon on the train from Philadelphia.

Clearly we can *detect an error* in the date, for extra information about the calendar tells us that there is no date of July 43. Most likely the digit should be a 1 or a 2, but we can't tell; thus the error can't be corrected without further information. However, just a bit of extra knowledge about New York City railroad stations tells us that trains from Philadelphia arrive at Penn (Pennsylvania) Station in New York City, not the Grand Central Terminal or the PATH Terminal. Thus, Senn is not only *detected* as an error, but is also *corrected* to Penn. Note that in all cases, error detection and correction required additional (redundant) information. We discuss both error-detecting and error-correcting codes in the sections that follow. We could of course send return mail to request a retransmission of the e-mail message (again, redundant information is obtained) to resolve the obvious transmission or typing errors.

In the preceding paragraph we discussed retransmission as a means of correcting errors in an e-mail message. The errors were detected by a redundant source and our knowledge of calendars and New York City railroad stations. In general, with pulse trains we have no knowledge of "the right answer." Thus if we use the simple brute force redundancy technique of transmitting each pulse sequence twice, we can compare them to detect errors. (For the moment, we are ignoring the rare situation in which both messages are identically corrupted and have the same wrong sequence.) We can, of course, transmit three times, compare to detect errors, and select the pair of identical messages to provide error correction, but we are again ignoring the possibility of identical errors during two transmissions. These brute force methods are inefficient, as they require many redundant bits. In this chapter, we show that in some cases the addition of a single redundant bit will greatly improve error-detection capabilities. Also, the efficient technique for obtaining error correction by adding more than one redundant bit are discussed. The method based on triple or $N$ copies of a message are covered in Chapter 4. The coding schemes discussed so far rely on short "noise pulses," which generally corrupt only one transmitted bit. This is generally a good assumption for computer memory and address buses and transmission lines; however, disk memories often have sequences of errors that extend over several bits, or *burst errors*, and different coding schemes are required.

The measure of performance we use in the case of an error-detecting code is the *probability of an undetected error*, $P_{ue}$, which we of course wish to minimize. In the case of an error-correcting code, we use the *probability of transmitted error*, $P_e$, as a measure of performance, or the *reliability*, $R$, (*probability of success*), which is $(1 - P_e)$. Of course, many of the more sophisticated coding techniques are now feasible because advanced integrated circuits (logic and memory) have made the costs of implementation (dollars, volume, weight, and power) modest.

The type of code used in the design of digital devices or systems largely depends on the types of errors that occur, the amount of redundancy that is cost-effective, and the ease of building coding and decoding circuitry. The source of errors in computer systems can be traced to a number of causes, including the following:

1. Component failure
2. Damage to equipment
3. "Cross-talk" on wires
4. Lightning disturbances

5. Power disturbances
6. Radiation effects
7. Electromagnetic fields
8. Various kinds of electrical noise

Note that we can roughly classify sources 1, 2, and 3 as causes that are internal to the equipment; sources 4, 6, and 7 as generally external causes; and sources 5 and 6 as either internal or external. Classifying the source of the disturbance is only useful in minimizing its strength, decreasing its frequency of occurrence, or changing its other characteristics to make it less disturbing to the equipment. The focus of this text is what to do to protect against these effects and how the effects can compromise performance and operation, assuming that they have occurred. The reader may comment that many of these error sources are rather rare; however, our desire for ultrareliable, long-life systems makes it important to consider even rare phenomena.

The various types of interference that one can experience in practice can be illustrated by the following two examples taken from the aircraft field. Modern aircraft are crammed full of digital and analog electronic equipment that are generally referred to as avionics. Several recent instances of military crashes and civilian troubles have been noted in modern electronically controlled aircraft. These are believed to be caused by various forms of electromagnetic interference, such as passenger devices (e.g., cellular telephones); "cross-talk" between various onboard systems; external signals (e.g., Voice of America Transmitters and Military Radar); lightning; and equipment malfunction [Shooman, 1993]. The systems affected include the following: autopilot, engine controls, communication, navigation, and various instrumentation. Also, a previous study by Cockpit (the pilot association of Germany) [Taylor, 1988, pp. 285–287] concluded that the number of soft fails (probably from alpha particles and cosmic rays affecting memory chips) increased in modern aircraft. See Table 2.1 for additional information.

**TABLE 2.1 Increase of Soft Fails with Airplane Generation**

| Airplane Type | Altitude (1,000s feet) | | | | Total Reports | No. of Aircraft | Soft Fails per $a/c$ |
| --- | --- | --- | --- | --- | --- | --- | --- |
| | Ground-5 | 5–20 | 20–30 | 30+ | | | |
| B707 | 2 | 0 | 0 | 2 | 4 | 14 | 0.29 |
| B727/737 | 11 | 7 | 2 | 4 | 24 | 39/28 | 0.36 |
| B747 | 11 | 0 | 1 | 6 | 18 | 10 | 1.80 |
| DC10 | 21 | 5 | 0 | 29 | 55 | 13 | 4.23 |
| A300 | 96 | 12 | 6 | 17 | 131 | 10 | 13.10 |

*Source:* [Taylor, 1988].

**34** CODING TECHNIQUES

It is not clear how the number of flight hours varied among the different airplane types, what the computer memory sizes were for each of the aircraft, and the severity level of the fails. It would be interesting to compare this data to that observed in the operation of the most advanced versions of B747 and A320 aircraft, as well as other more recent designs.

There has been much work done on coding theory since 1950 [Rao, 1989]. This chapter presents a modest sampling of theory as it applies to fault-tolerant systems.

## 2.2 BASIC PRINCIPLES

Coding theory can be developed in terms of the mathematical structure of *groups, subgroups, rings, fields, vector spaces, subspaces, polynomial algebra*, and *Galois fields* [Rao, 1989, Chapter 2]. Another simple yet effective development of the theory based on algebra and logic is used in this text [Arazi, 1988].

### 2.2.1 Code Distance

We will deal with strings of binary digits (0 or 1), which are of specified length and called the following synonymous terms: binary *block*, binary *vector*, binary *word*, or just *code word*. Suppose that we are dealing with a 3-bit message ($b_1$, $b_2$, $b_3$) represented by the bits $x_1$, $x_2$, $x_3$. We can speak of the eight combinations of these bits—see Table 2.2(a)—as the code words. In this case they are assigned according to the sequence of binary numbers. The *distance* of a code is the *minimum* number of bits by which any one code word differs from another. For example, the first and second code words in Table 2.2(a) differ only in the right-most digit and have a distance of 1, whereas the first and the last code words differ in all 3 digits and have a distance of 3. The total number of comparisons needed to check all of the word pairs for the minimum code distance is the number of combinations of 8 items taken 2 at a time $\binom{8}{2}$, which is equal to $8!/2!6! = 28$.

A simpler way of visualizing the distance is to use the "cube method" of displaying switching functions. A cube is drawn in three-dimensional space ($x$, $y$, $z$), and a main diagonal goes from $x = y = z = 0$ to $x = y = z = 1$. The distance is the number of cube edges between any two code words that represent the vertices of the cube. Thus, the distance between 000 and 001 is a single cube edge, but the distance between 000 and 111 is 3 since 3 edges must be traversed to get between the two vertices. (In honor of one of the pioneers of coding theory, the code distance is generally called the *Hamming distance*.) Suppose that noise changes a single bit of a code word from 0 to 1 or 1 to 0. The first code word in Table 2.2(a) would be changed to the second, third, or fifth, depending on which bit was corrupted. Thus there is no way to detect a single-bit error (or a multibit error), since any change in a code word transforms it

**TABLE 2.2 Examples of 3- and 4-Bit Code Words**

| (a) 3-Bit Code Words | | | (b) 4-Bit Code Words: 3 Original Bits plus Added Even-Parity (Legal Code Words) | | | | (c) Illegal Code Words for the Even-Parity Code of (b) | | | |
|---|---|---|---|---|---|---|---|---|---|---|
| $x_1$ $b_1$ | $x_2$ $b_2$ | $x_3$ $b_3$ | $x_1$ $p_1$ | $x_2$ $b_1$ | $x_3$ $b_2$ | $x_4$ $b_3$ | $x_1$ $p_1$ | $x_2$ $b_1$ | $x_3$ $b_2$ | $x_4$ $b_3$ |
| 0 | 0 | 0 | 0 | 0 | 0 | 0 | 1 | 0 | 0 | 0 |
| 0 | 0 | 1 | 1 | 0 | 0 | 1 | 0 | 0 | 0 | 1 |
| 0 | 1 | 0 | 1 | 0 | 1 | 0 | 0 | 0 | 1 | 0 |
| 0 | 1 | 1 | 0 | 0 | 1 | 1 | 1 | 0 | 1 | 1 |
| 1 | 0 | 0 | 1 | 1 | 0 | 0 | 0 | 1 | 0 | 0 |
| 1 | 0 | 1 | 0 | 1 | 0 | 1 | 1 | 1 | 0 | 1 |
| 1 | 1 | 0 | 0 | 1 | 1 | 0 | 1 | 1 | 1 | 0 |
| 1 | 1 | 1 | 1 | 1 | 1 | 1 | 0 | 1 | 1 | 1 |

into another legal code word. One can create error-detecting ability in a code by adding *check bits*, also called *parity bits*, to a code.

The simplest coding scheme is to add one redundant bit. In Table 2.2(b), a single check bit (parity bit $p_1$) is added to the 3-bit code words $b_1$, $b_2$, and $b_3$ of Table 2.2(a), creating the eight new code words shown. The scheme used to assign values to the parity bit is the coding rule; in this case, $p_1$ is chosen so that the number of one bits in each word is an even number. Such a code is called an *even-parity* code, and the words in Table 2.1(b) become legal code words and those in Table 2.1(c) become illegal code words. Clearly we could have made the number of one bits in each word an odd number, resulting in an *odd-parity* code, and so the words in Table 2.1(c) would become the legal ones and those in 2.1(b) become illegal.

### 2.2.2 Check-Bit Generation and Error Detection

The code generation rule (even parity) used to generate the parity bit in Table 2.2(b) will now be used to design a parity-bit generator circuit. We begin with a Karnaugh map for the switching function $p_1$ ($b_1$, $b_2$, and $b_3$) where the parity bit is a function of the three code bits as given in Fig. 2.1(a). The resulting Karnaugh map is given in this figure. The top left cell in the map corresponds to $p_1 = 0$ when $b_1$, $b_2$, and $b_3 = 000$, whereas the top right cell represents $p_1 = 1$ when $b_1$, $b_2$, and $b_3 = 001$. These two cells represent the first two rows of Table 2.2(b); the other cells in the map represent the other six rows in the table. Since none of the ones in the Karnaugh map touch, no simplification is possible, and there are four minterms in the circuit, each generated by the four gates shown in the circuit. The OR gate "collects" these minterms, generating a parity check bit $p_1$ whenever a sequence of pulses $b_1$, $b_2$, and $b_3$ occurs.

**36**   CODING TECHNIQUES

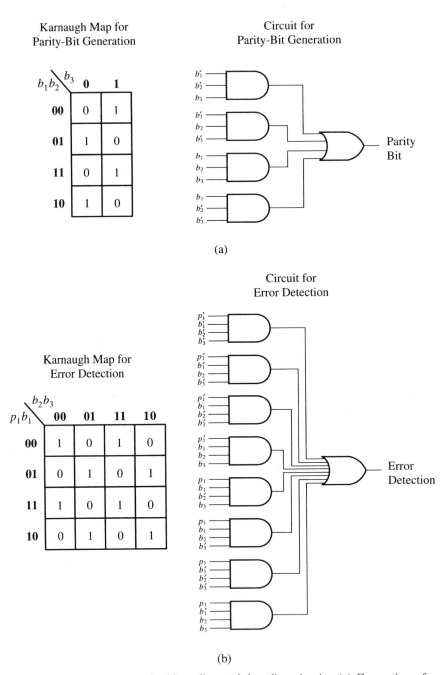

**Figure 2.1**  Elementary parity-bit coding and decoding circuits. (a) Generation of an even-parity bit for a 3-bit code word. (b) Detection of an error for an even-parity-bit code for a 3-bit code word.

The addition of the parity bit creates a set of legal and illegal words; thus we can detect an error if we check for legal or illegal words. In Fig. 2.1(b) the Karnaugh map displays ones for legal code words and zeroes for illegal code words. Again, there is no simplification since all the minterms are separated, so the error detector circuit can be composed by generating all the illegal word minterms (indicated by zeroes) in Fig. 2.1(b) using eight AND gates followed by an 8-input OR gate as shown in the figure. The circuits derived in Fig. 2.1 can be simplified by using exclusive or (EXOR) gates (as shown in the next section); however, we have demonstrated in Fig. 2.1 how check bits can be generated and how errors can be detected. Note that parity checking will detect errors that occur in either the message bits or the parity bit.

## 2.3 PARITY-BIT CODES

### 2.3.1 Applications

Three important applications of parity-bit error-checking codes are as follows:

1. The transmission of characters over telephone lines (or optical, microwave, radio, or satellite links). The best known application is the use of a modem to allow computers to communicate over telephone lines.
2. The transmission of data to and from electronic memory (memory read and write operations).
3. The exchange of data between units within a computer via various data and control buses.

Specific implementation details may differ among these three applications, but the basic concepts and circuitry are very similar. We will discuss the first application and use it as an illustration of the basic concepts.

### 2.3.2 Use of Exclusive OR Gates

This section will discuss how an additional bit can be added to a byte for error detection. It is common to represent alphanumeric characters in the input and output phases of computation by a single byte. The ASCII code is almost universally used. One technique uses the entire byte to represent $2^8 = 256$ possible characters (the extended character set that is used on IBM personal computers, containing some Greek letters, language accent marks, graphic characters, and so forth, as well as an additional ninth parity bit. The other approach limits the character set to 128, which can be expressed by seven bits, and uses the eighth bit for parity.

Suppose we wish to build a parity-bit generator and code checker for the case of seven message bits and one parity bit. Identifying the minterms will reveal a generalization of the checkerboard diagram similar to that given in the

**38** CODING TECHNIQUES

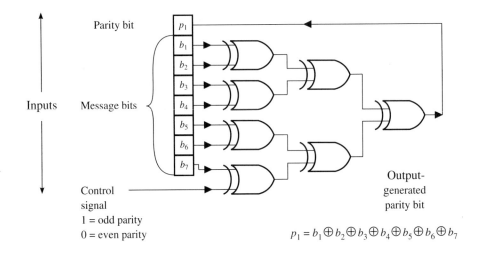

(a) Parity-Bit Encoder (generator)

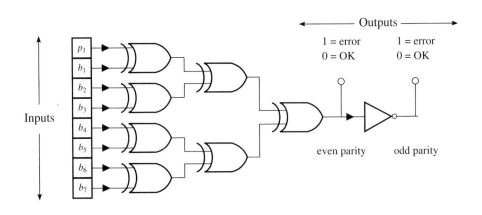

(b) Parity-Bit Decoder (checker)

**Figure 2.2** Parity-bit encoder and decoder for a transmitted byte: (a) A 7-bit parity encoder (generator); (b) an 8-bit parity decoder (checker).

Karnaugh maps of Fig. 2.1. Such checkerboard patterns indicate that EXOR gates can be used to simplify the circuit. A circuit using EXOR gates for parity-bit generation and for checking of an 8-bit byte is given in Fig. 2.2. Note that the circuit in Fig. 2.2(a) contains a control input that allows one to easily switch from even to odd parity. Similarly, the addition of the NOT gate (inverter) at the output of the checking circuit allows one to use either even or odd parity.

Most modems have these refinements, and a switch chooses either even or odd parity.

### 2.3.3 Reduction in Undetected Errors

The purpose of parity-bit checking is to detect errors. The extent to which such errors are detected is a measure of the success of the code, whereas the probability of not detecting an error, $P_{ue}$, is a measure of failure. In this section we analyze how parity-bit coding decreases $P_{ue}$. We include in this analysis the reliability of the parity-bit coding and decoding circuit by analyzing the reliability of a standard IC parity code generator/checker. We model the failure of the IC chip in a simple manner by assuming that it fails to detect errors, and we ignore the possibility that errors are detected when they are not present.

Let us consider the addition of a ninth parity bit to an 8-bit message byte. The parity bit adjusts the number of ones in the word to an even (odd) number and is computed by a parity-bit generator circuit that calculates the EXOR function of the 8 message bits. Similarly, an EXOR-detecting circuit is used to check for transmission errors. If 1, 3, 5, 7, or 9 errors are found in the received word, the parity is violated, and the checking circuit will detect an error. This can lead to several consequences, including "flagging" the error byte and retransmission of the byte until no errors are detected. The probability of interest is the probability of an undetected error, $P'_{ue}$, which is the probability of 2, 4, 6, or 8 errors, since these combinations do not violate the parity check. These probabilities can be calculated by simply using the binomial distribution (see Appendix A5.3). The probability of $r$ failures in $n$ occurrences with failure probability $q$ is given by the binomial probability $B(r:n,q)$. Specifically, $n=9$ (the number of bits) and $q=$ the probability of an error per transmitted bit; thus

General:

$$B(r:9,q) = \binom{9}{r} q^r (1-q)^{9-r} \tag{2.1}$$

Two errors:

$$B(2:9,q) = \binom{9}{2} q^2 (1-q)^{9-2} \tag{2.2}$$

Four errors:

$$B(4:9,q) = \binom{9}{4} q^4 (1-q)^{9-4} \tag{2.3}$$

and so on.

**40** CODING TECHNIQUES

For $q$, relatively small ($10^{-4}$), it is easy to see that Eq. (2.3) is much smaller than Eq. (2.2); thus only Eq. (2.2) needs to be considered (probabilities for $r = 4, 6,$ and 8 are negligible), and the probability of an undetected error with parity-bit coding becomes

$$P'_{ue} = B(2:9,q) = 36q^2(1-q)^7 \qquad (2.4)$$

We wish to compare this with the probabilty of an undetected error for an 8-bit transmission without any checking. With no checking, all errors are undetected; thus we must compute $B(1:8,q) + \cdots + B(8:8,q)$, but it is easier to compute

$$P_{ue} = 1 - P(0 \text{ errors}) = 1 - B(0:8,q) = 1 - \binom{8}{0} q^0 (1-q)^{8-0}$$
$$= 1 - (1-q)^8 \qquad (2.5)$$

Note that our convention is to use $P_{ue}$ for the case of no checking, and $P'_{ue}$ for the case of checking.

The ratio of Eqs. (2.5) and (2.4) yields the improvement ratio due to the parity-bit coding as follows:

$$P_{ue}/P'_{ue} = [1 - (1-q)^8]/[36q^2(1-q)^7] \qquad (2.6)$$

For small $q$ we can simplify Eq. (2.6) by replacing $(1 \pm q)^n$ by $1 \pm nq$ and $[1/(1-q)]$ by $1 + q$, which yields

$$P_{ue}/P'_{ue} = [2(1+7q)/9q] \qquad (2.7)$$

The parameter $q$, the probability of failure per bit transmitted, is quoted as $10^{-4}$ in Hill and Peterson [1981]. The failure probability $q$ was $10^{-5}$ or $10^{-6}$ in the 1960s and '70s; now, it may be as low as $10^{-7}$ for the best telephone lines [Rubin, 1990]. Equation (2.7) is evaluated for the range of $q$ values; the results appear in Table 2.3 and in Fig. 2.3.

The improvement ratio is quite significant, and the overhead—adding 1 parity bit out of 8 message bits—is only 12.5%, which is quite modest. This probably explains why a parity-bit code is so frequently used.

In the above analysis we assumed that the coder and decoder are perfect. We now examine the validity of that assumption by modeling the reliability of the coder and decoder. One could use a design similar to that of Fig. 2.2; however, it is more realistic to assume that we are using a commercial circuit device: the SN74180, a 9-bit odd/even parity generator/checker (see Texas Instruments [1988]), or the newer 74LS280 [Motorola, 1992]. The SN74180 has an equivalent circuit (see Fig. 2.4), which has 14 gates and inverters, whereas the pin-compatible 74LS280 with improved performance has 46 gates and inverters in

PARITY-BIT CODES  41

**TABLE 2.3  Evaluation of the Reduction in Undetected Errors from Parity-Bit Coding: Eq. (2.7)**

| Bit Error Probability, $q$ | Improvement Ratio: $P_{ue}/P'_{ue}$ |
|---|---|
| $10^{-4}$ | $2.223 \times 10^3$ |
| $10^{-5}$ | $2.222 \times 10^4$ |
| $10^{-6}$ | $2.222 \times 10^5$ |
| $10^{-7}$ | $2.222 \times 10^6$ |
| $10^{-8}$ | $2.222 \times 10^7$ |

its equivalent circuit. Current prices of the SN74180 and the similar 74LS280 ICs are about 10–75 cents each, depending on logic family and order quantity. We will use two such devices since the same chip can be used as a coder and a decoder (generator/checker). The logic diagram of this device is shown in Fig. 2.4.

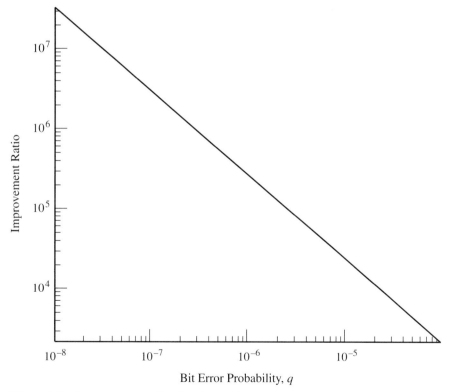

**Figure 2.3**  Improvement ratio of undetected error probability from parity-bit coding.

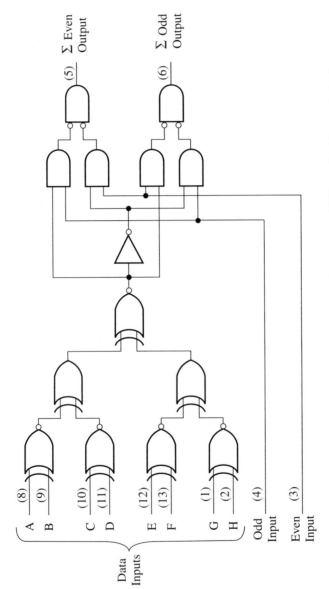

**Figure 2.4** Logic diagram for SN74180 [Texas Instruments, 1988, used with permission].

## 2.3.4 Effect of Coder–Decoder Failures

An approximate model for IC reliability is given in Appendix B3.3, Fig. B7. The model assumes the failure rate of an integrated circuit is proportional to the square root of the number of gates, $g$, in the equivalent logic model. Thus the failure rate per million hours is given as $\lambda_b = C(g)^{1/2}$, where $C$ was computed from 1985 IC failure-rate data as 0.004. We can use this model to estimate the failure rate and subsequently the reliability of an IC parity generator checker. In the equivalent gate model for the SN74180 given in Fig. 2.4, there are 5 EXNOR, 2 EXOR, 1 NOT, 4 AND, and 2 NOR gates. Note that the output gates (5) and (6) are NOR rather than OR gates. Sometimes for good and proper reasons integrated circuit designers use equivalent logic using different gates. Assuming the 2 EXOR and 5 EXNOR gates use about 1.5 times as many transistors to realize their function as the other gates, we consider them as equivalent to 10.5 gates. Thus we have 17.5 equivalent gates and $\lambda_b = 0.004(17.5)^{1/2}$ failures per million hours = $1.67 \times 10^{-8}$ failures per hour.

In formulating a reliability model for a parity-bit coder–decoder scheme, we must consider two modes of failure for the coded word: $A$, where the coder and decoder do not fail but the number of bit errors is an even number equal to 2 or more; and $B$, where the coder or decoder chip fails. We ignore chip failure modes, which sometimes give correct results. The probability of undetected error with the coding scheme is given by

$$P'_{ue} = P(A + B) = P(A) + P(B) \tag{2.8}$$

In Eq. (2.8), the chip failure rates are per hour; thus we write Eq. (2.8) as

$P'_{ue} = P[\text{no coder or decoder failure during 1 byte transmission}]$
$\quad \times P[\text{2 or more errors}]$
$\quad + P[\text{coder or decoder failure during 1 byte transmission}] \quad (2.9)$

If we let $B$ be the bit transmission rate per second, then the number of seconds to transmit a bit is $1/B$. Since a byte plus parity is 9 bits, it will take $9/B$ seconds to transmit and $9/3{,}600B$ hours to transmit the 9 bits.

If we assume a constant failure rate $\lambda_b$ for the coder and decoder, the reliability of a coder–decoder pair is $e^{-2\lambda_b t}$ and the probability of coder or decoder failure is $(1 - e^{-2\lambda_b t})$. This probability of 2 errors is given by Eq. (2.4); thus Eq. (2.9) becomes

$$P'_{ue} = e^{-2\lambda_b t} \times 36q^2(1-q)^7 + (1 - e^{-2\lambda_b t}) \tag{2.10}$$

where

$$t = 9/3{,}600B \tag{2.11}$$

**44** CODING TECHNIQUES

**TABLE 2.4 The Reduction in Undetected Errors from Parity-Rate Coding Including the Effect of Coder–Decoder Failures**

| Bit Error Probability $q$ | Improvement Ratio: $P_{ue}/P'_{ue}$ for Several Transmission Rates | | | |
|---|---|---|---|---|
| | 300 Bits/Sec | 1,200 Bits/Sec | 9,600 Bits/Sec | 56,000 Bits/Sec |
| $10^{-4}$ | $2.223 \times 10^3$ | $2.223 \times 10^3$ | $2.223 \times 10^3$ | $2.223 \times 10^3$ |
| $10^{-5}$ | $2.222 \times 10^4$ | $2.222 \times 10^4$ | $2.222 \times 10^4$ | $2.222 \times 10^4$ |
| $10^{-6}$ | $2.228 \times 10^5$ | $2.218 \times 10^5$ | $2.222 \times 10^5$ | $2.222 \times 10^5$ |
| $10^{-7}$ | $1.254 \times 10^6$ | $1.962 \times 10^6$ | $2.170 \times 10^6$ | $2.213 \times 10^6$ |
| $5 \times 10^{-8}$ | $1.087 \times 10^6$ | $2.507 \times 10^6$ | $4.053 \times 10^6$ | $4.372 \times 10^6$ |
| $10^{-8}$ | $2.841 \times 10^5$ | $1.093 \times 10^6$ | $6.505 \times 10^6$ | $1.577 \times 10^7$ |

The undetected error probability with no coding is given by Eq. (2.5) and is independent of time

$$P_{ue} = 1 - (1-q)^8 \qquad (2.12)$$

Clearly if the failure rate is small or the bit rate $B$ is large, $e^{-2\lambda_b t} \approx 1$, the failure probabilities of the coder–decoder chips are insignificant, and the ratio of Eq. (2.12) and Eq. (2.10) will reduce to Eq. (2.7) for high bit rates $B$. If we are using a parity code for memory bit checking, the bit rate will be essentially the memory cycle time if we assume that a long succession of memory operations and the effect of chip failures are negligible. However, in the case of parity-bit coding in a modem, the baud rate will be lower and chip failures can be significant, especially in the case where $q$ is small. The ratio of Eq. (2.12) to Eq. (2.10) is evaluated in Table 2.4 (and plotted in Fig. 2.5) for typical modem bit rates $B = 300, 1,200, 9,600$, and $56,000$. Note that the chip failure rate is insignificant for $q = 10^{-4}, 10^{-5}$, and $10^{-6}$; however, it does make a difference for $q = 10^{-7}$ and $10^{-8}$. If the bit rate $B$ is infinite, the effect of chip failure disappears, and we can view Table 2.3 as depicting this case.

## 2.4 HAMMING CODES

### 2.4.1 Introduction

In this section, we develop a class of codes created by Richard Hamming [1950], for whom they are named. These codes will employ $c$ check bits to detect more than a single error in a coded word, and if enough check bits are used, some of these errors can be corrected. The relationships among the number of check bits and the number of errors that can be detected and corrected are developed in the following section. It will not be surprising that the case in which $c = 1$ results in a code that can detect single errors but cannot correct errors; this is the parity-bit code that we had just discussed.

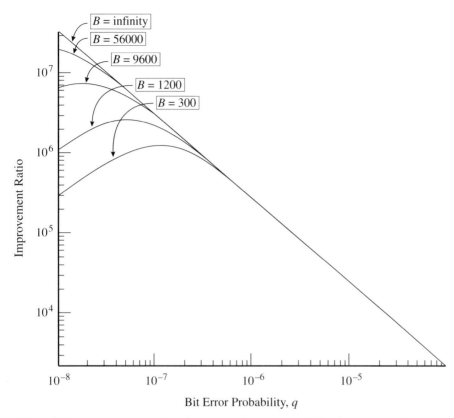

**Figure 2.5** Improvement ratio of undetected error probability from parity-bit coding (including the possibility of coder–decoder failure). $B$ is the transmission rate in bits per second.

### 2.4.2 Error-Detection and -Correction Capabilities

We defined the concept of Hamming distance of a code in the previous section. Now, we establish the error-detecting and -correcting abilities of a code based on its Hamming distance. The following results apply to *linear codes*, in which the difference and sum between any two code words (addition and subtraction of their binary representations) is also a code word. Most of this chapter will deal with linear codes. The following notations are used in this chapter:

$$d = \text{the Hamming distance of a code} \quad (2.13)$$
$$D = \text{the number of errors that a code can detect} \quad (2.14a)$$
$$C = \text{the number of errors that a code can correct} \quad (2.14b)$$
$$n = \text{the total number of bits in the coded word} \quad (2.15a)$$

$m$ = the number of message or information bits (2.15b)

$c$ = the number of check (parity) bits (2.15c)

where $d$, $D$, $C$, $n$, $m$, and $c$ are all integers $\geq 0$.

As we said previously, the model we will use is one in which the check bits are added to the message bits by the coder. The message is then "transmitted," and the decoder checks for any detectable errors. If there are enough check bits, and if the circuit is so designed, some of the errors are corrected. Initially, one can view the error-detection process as a check of each received word to see if the word belongs to the illegal set of words. Any set of errors that convert a legal code word into an illegal one are detected by this process, whereas errors that change a legal code word into another legal code word are not detected. To detect $D$ errors, the Hamming distance must be at least one larger than $D$.

$$d \geq D + 1 \qquad (2.16)$$

This relationship must be so because a single error in a code word produces a new word that is a distance of one from the transmitted word. However, if the code has a basic distance of one, this error results in a new word that belongs to the legal set of code words. Thus for this single error to be detectable, the code must have a basic distance of two so that the new word produced by the error does not belong to the legal set and therefore must correspond to the detectable illegal set. Similarly, we could argue that a code that can detect two errors must have a Hamming distance of three. By using induction, one establishes that Eq. (2.16) is true.

We now discuss the process of error correction. First, we note that to correct an error we must be able to detect that an error has occurred. Suppose we consider the parity-bit code of Table 2.2. From Eq. (2.16) we know that $d \geq 2$ for error detection; in fact, $d = 2$ for the parity-bit code, which means that we have a set of legal code words that are separated by a Hamming distance of at least two. A single bit error creates an illegal code word that is a distance of one from *more than 1* legal code word; thus we cannot correct the error by seeking the closest legal code word. For example, consider the legal code word 0000 in Table 2.2(b). Suppose that the last bit is changed to a one yielding 0001, which is the second illegal code word in Table 2.2(c). Unfortunately, the distance from that illegal word to each of the eight legal code words is 1, 1, 3, 1, 3, 1, 3, and 3 (respectively). Thus there is a four-way tie for the closest legal code word. Obviously we need a larger Hamming distance for error correction. Consider the number line representing the distance between any 2 legal code words for the case of $d = 3$ shown in Fig. 2.6(a). In this case, if there is 1 error, we move 1 unit to the right from word $a$ toward word $b$. We are still 2 units away from word $b$ and at least that far away from any other word, so we can recognize word $a$ as the closest and select it as the correct word. We can generalize this principle by examining Fig. 2.6(b). If there are $C$ errors to correct, we have moved a distance of $C$ away from code word $a$; to have this

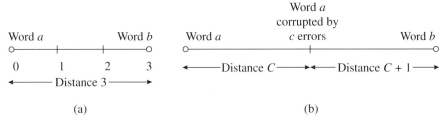

**Figure 2.6** Number lines representing the distances between two legal code words.

word closer than any other word, we must have at least a distance of $C + 1$ from the erroneous code word to the nearest other legal code word so we can correct the errors. This gives rise to the formula for the number of errors that can be corrected with a Hamming distance of $d$, as follows:

$$d \geq 2C + 1 \quad (2.17)$$

Inspecting Eqs. (2.16) and (2.17) shows that for the same value of $d$,

$$D \geq C \quad (2.18)$$

We can combine Eqs. (2.17) and (2.18) by rewriting Eq. (2.17) as

$$d \geq C + C + 1 \quad (2.19)$$

If we use the smallest value of $D$ from Eq. (2.18), that is, $D = C$, and substitute for one of the $C$s in Eq. (2.19), we obtain

$$d \geq D + C + 1 \quad (2.20)$$

which summarizes and combines Eqs. (2.16) to (2.18).

One can develop the entire class of Hamming codes by solving Eq. (2.20), remembering that $D \geq C$ and that $d$, $D$, and $C$ are integers $\geq 0$. For $d = 1$, $D = C = 0$—no code is possible; if $d = 2$, $D = 1$, $C = 0$—we have the parity bit code. The class of codes governed by Eq. (2.20) is given in Table 2.5.

The most popular codes are the parity code; the $d = 3$, $D = C = 1$ code—generally called a single error-correcting and single error-detecting (SECSED) code; and the $d = 4$, $D = 2$, $C = 1$ code—generally called a single error-correcting and double error-detecting (SECDED) code.

### 2.4.3 The Hamming SECSED Code

The Hamming SECSED code has a distance of 3, and corrects and detects 1 error. It can also be used as a double error-detecting code (DED).

Consider a Hamming SECSED code with 4 message bits ($b_1$, $b_2$, $b_3$, and $b_4$) and 3 check bits ($c_1$, $c_2$, and $c_3$) that are computed from the message bits by equations integral to the code design. Thus we are dealing with a 7-bit word. A brute

**48** CODING TECHNIQUES

**TABLE 2.5 Relationships Among $d$, $D$, and $C$**

| $d$ | $D$ | $C$ | Type of Code |
|---|---|---|---|
| 1 | 0 | 0 | No code possible |
| 2 | 1 | 0 | Parity bit |
| 3 | 1 | 1 | Single error detecting; single error correcting |
| 3 | 2 | 0 | Double error detecting; zero error correcting |
| 4 | 3 | 0 | Triple error detecting; zero error correcting |
| 4 | 2 | 1 | Double error detecting; single error correcting |
| 5 | 4 | 0 | Quadruple error detecting; zero error correcting |
| 5 | 3 | 1 | Triple error detecting; single error correcting |
| 5 | 2 | 2 | Double error detecting; double error correcting |
| 6 | 5 | 0 | Quintuple error detecting; zero error correcting |
| 6 | 4 | 1 | Quadruple error detecting; single error correcting |
| 6 | 3 | 2 | Triple error detecting; double error correcting |
| etc. | | | |

force detection–correction algorithm would be to compare the coded word in question with all the $2^7 = 128$ code words. No error is detected if the coded word matched any of the $2^4 = 16$ legal combinations of message bits. No detected errors means either that none have occurred or that too many errors have occurred (the code is not powerful enough to detect so many errors). If we detect an error, we compute the distance between the illegal code word and the 16 legal code words and effect error correction by choosing the code word that is closest. Of course, this can be done in one step by computing the distance between the coded word and all 16 legal code words. If one distance is 0, no errors are detected; otherwise the minimum distance points to the corrected word.

The information in Table 2.5 just tells us the possibilities in constructing a code; it does not tell us how to construct the code. Hamming [1950] devised a scheme for coding and decoding a SECSED code in his original work. Check bits are interspersed in the code word in bit positions that correspond to powers of 2. Word positions that are not occupied by check bits are filled with message bits. The length of the coded word is $n$ bits composed of $c$ check bits added to $m$ message bits. The common notation is to denote the code word (also called *binary word*, *binary block*, or *binary vector*) as $(n, m)$. As an example, consider a $(7, 4)$ code word. The 3 check bits and 4 message bits are located as shown in Table 2.6.

**TABLE 2.6 Bit Positions for Hamming SECSED ($d = 3$) Code**

| Bit positions | $x_1$ | $x_2$ | $x_3$ | $x_4$ | $x_5$ | $x_6$ | $x_7$ |
|---|---|---|---|---|---|---|---|
| Check bits | $c_1$ | $c_2$ | — | $c_3$ | — | — | — |
| Message bits | — | — | $b_1$ | — | $b_2$ | $b_3$ | $b_4$ |

**TABLE 2.7 Relationships Among $n$, $c$, and $m$ for a SECSED Hamming Code**

| Length, $n$ | Check Bits, $c$ | Message Bits, $m$ |
|---|---|---|
| 1 | 1 | 0 |
| 2 | 2 | 0 |
| 3 | 2 | 1 |
| 4 | 3 | 1 |
| 5 | 3 | 2 |
| 6 | 3 | 3 |
| 7 | 3 | 4 |
| 8 | 4 | 4 |
| 9 | 4 | 5 |
| 10 | 4 | 6 |
| 11 | 4 | 7 |
| 12 | 4 | 8 |
| 13 | 4 | 9 |
| 14 | 4 | 10 |
| 15 | 4 | 11 |
| 16 | 5 | 11 |
| etc. | | |

In the code shown, the 3 check bits are sufficient for codes with 1 to 4 message bits. If there were another message bit, it would occupy position $x_9$, and position $x_8$ would be occupied by a fourth check bit. In general, $c$ check bits will cover a maximum of $(2^c - 1)$ word bits or $2^c \geq n + 1$. Since $n = c + m$, we can write

$$2^c \geq [c + m + 1] \qquad (2.21)$$

where the notation $[c + m + 1]$ means the smallest integer value of $c$ that satisfies the relationship. One can solve Eq. (2.21) by assuming a value of $n$ and computing the number of message bits that the various values of $c$ can check. (See Table 2.7.)

If we examine the entry in Table 2.7 for a message that is 1 byte long, $m = 8$, we see that 4 check bits are needed and the total word length is 12 bits. Thus we can say that the ratio $c/m$ is a measure of the code overhead, which in this case is 50%. The overhead for common computer word lengths, $m$, is given in Table 2.8.

Clearly the overhead approaches 10% for long word lengths. Of course, one should remember that these codes are competing for efficiency with the parity-bit code, in which 1 check bit represents only a 1.6% overhead for a 64-bit word length.

We now return to our (7, 4) SECSED code example to explain how the check bits are generated. Hamming developed a much more ingenious and

## 50  CODING TECHNIQUES

**TABLE 2.8  Overhead for Various Word Lengths ($m$) for a Hamming SECSED Code**

| Code Length, $n$ | Word (Message) Length, $m$ | Number of Check Bits, $c$ | Overhead $(c/m) \times 100\%$ |
|---|---|---|---|
| 12 | 8 | 4 | 50 |
| 21 | 16 | 5 | 31 |
| 38 | 32 | 6 | 19 |
| 54 | 48 | 6 | 13 |
| 71 | 64 | 7 | 11 |

efficient design and method for detection and correction. The Hamming code positions for the check and message bits are given in Table 2.6, which yields the code word $c_1 c_2 b_1 c_3 b_2 b_3 b_4$. The check bits are calculated by computing the exclusive, or $\oplus$, of 3 appropriate message bits as shown in the following equations:

$$c_1 = b_1 \oplus b_2 \oplus b_4 \tag{2.22a}$$
$$c_2 = b_1 \oplus b_3 \oplus b_4 \tag{2.22b}$$
$$c_3 = b_2 \oplus b_3 \oplus b_4 \tag{2.22c}$$

Such a choice of check bits forms an obvious pattern if we write the 3 check equations below the word we are checking, as is shown in Table 2.9. Each parity bit and message bit present in Eqs. (2.22a–c) is indicated by a "1" in the respective rows (all other positions are 0). If we read down in each column, the last 3 bits are the binary number corresponding to the bit position in the word.

Clearly, the binary number pattern gives us a design procedure for constructing parity check equations for distance 3 codes of other word lengths. Reading across rows 3–5 of Table 2.9, we see that the check bit with a 1 is on the left side of the equation and all other bits appear as $\oplus$ on the right-hand side.

As an example, consider that the message bits $b_1 b_2 b_3 b_4$ are 1010, in which case the check bits are

**TABLE 2.9  Pattern of Parity Check Bits for a Hamming (7, 4) SECSED Code**

| Bit positions in word | $x_1$ | $x_2$ | $x_3$ | $x_4$ | $x_5$ | $x_6$ | $x_7$ |
|---|---|---|---|---|---|---|---|
| Code word | $c_1$ | $c_2$ | $b_1$ | $c_3$ | $b_2$ | $b_3$ | $b_4$ |
| Check bit $c_1$ | 1 | 0 | 1 | 0 | 1 | 0 | 1 |
| Check bit $c_2$ | 0 | 1 | 1 | 0 | 0 | 1 | 1 |
| Check bit $c_3$ | 0 | 0 | 0 | 1 | 1 | 1 | 1 |

$$c_1 = 1 \oplus 0 \oplus 0 = 1 \quad (2.23a)$$
$$c_2 = 1 \oplus 1 \oplus 0 = 0 \quad (2.23b)$$
$$c_3 = 0 \oplus 1 \oplus 0 = 1 \quad (2.23c)$$

and the code word is $c_1 c_2 b_1 c_3 b_2 b_3 b_4 = 1011010$.

To check the transmitted word, we recalculate the check bits using Eqs. (2.22a–c) and obtain $c'_1$, $c'_2$, and $c'_3$. The old and the new parity check bits are compared, and any disagreement indicates an error. Depending on which check bits disagree, we can determine which message bit is in error. Hamming devised an ingenious way to make this check, which we illustrate by example.

Suppose that bit 3 of the message we have been discussing changes from a "1" to a "0" because of a noise pulse. Our code word then becomes $c_1 c_2 b_1 c_3 b_2 b_3 b_4 = 1011000$. Then, application of Eqs. (2.22a–c) yields $c'_3$, $c'_2$, and $c'_1 = 110$ for the new check bits. Disagreement of the check bits in the message with the newly calculated check bits indicates that an error has been detected. To locate the error, we calculate error-address bits, $e_3 e_2 e_1$, as follows:

$$e_1 = c_1 \oplus c'_1 = 1 \oplus 1 = 0 \quad (2.24a)$$
$$e_2 = c_2 \oplus c'_2 = 0 \oplus 1 = 1 \quad (2.24b)$$
$$e_3 = c_3 \oplus c'_3 = 1 \oplus 0 = 1 \quad (2.24c)$$

The binary address of the error bit is given by $e_3 e_2 e_1$, which in our example is 110 or 6. Thus we have detected correctly that the sixth position, $b_3$, is in error. If the address of the error bit is 000, it indicates that no error has occurred; thus calculation of $e_3 e_2 e_1$ can serve as our means of error detection and correction. To correct a bit that is in error once we know its location, we replace the bit with its complement.

The generation and checking operations described above can be derived in terms of a parity code matrix (essentially the last three rows of Table 2.9), a column vector that is the coded word, and a row vector called the *syndrome*, which is $e_3 e_2 e_1$ that we called the binary address of the error bit. If no errors occur, the syndrome is zero. If a single error occurs, the syndrome gives the correct address of the erroneous bit. If a double error occurs, the syndrome is nonzero, indicating an error; however, the address of the erroneous bit is incorrect. In the case of triple errors, the syndrome is zero and the errors are not detected. For a further discussion of the matrix representation of Hamming codes, the reader is referred to Siewiorek [1992].

### 2.4.4 The Hamming SECDED Code

The SECDED code is a distance 4 code that can be viewed as a distance 3 code with one additional check bit. It can also be a triple error-detecting code (TED). It is easy to design such a code by first designing a SECSED code and

**TABLE 2.10** Interpretation of Syndrome for a Hamming (8, 4) SECDED Code

| $e_1$ | $e_2$ | $e_3$ | $e_4$ | Interpretation |
|---|---|---|---|---|
| 0 | 0 | 0 | 0 | No errors |
| $a_1$ | $a_2$ | $a_3$ | 1 | One error, $a_1 a_2 a_3$ |
| $a_1$ | $a_2$ | $a_3$ | 0 | Two errors, $a_1 a_2 a_3$, not 000 |
| 0 | 0 | 0 | 1 | Three errors |
| 0 | 0 | 0 | 0 | Four errors |

then adding an appended check bit, which is a parity bit over all the other message and check bits. An even-parity code is traditionally used; however, if the digital electronics generating the code word have a failure mode in which the chip is burned out and all bits are 0, it will not be detected by an even-parity scheme. Thus odd parity is preferred for such a case. We expand on the (7, 4) SECSED example of the previous section and affix an additional check bit ($c_4$) and an additional syndrome bit ($e_4$) to obtain a SECDED code.

$$c_4 = c_1 \oplus c_2 \oplus b_1 \oplus c_3 \oplus b_2 \oplus b_3 \oplus b_4 \qquad (2.25)$$

$$e_4 = c_4 \oplus c_4' \qquad (2.26)$$

The new coded word is $c_1 c_2 b_1 c_3 b_2 b_3 b_4 c_4$. The syndrome is interpreted as given in Table 2.10.

Table 2.8 can be modified for a SECDED code by adding 1 to the code length column and 1 to the check bits column. The overhead values become 63%, 38%, 22%, 15%, and 13%.

### 2.4.5 Reduction in Undetected Errors

The probability of an undetected error for a SECSED code depends on the error-correction philosophy. Either a nonzero syndrome can be viewed as a single error—and the error-correction circuitry is enabled—or it can be viewed as detection of a double error. Since the next section will treat uncorrected error probabilities, we assume in this section that the nonzero syndrome condition for a SECSED code means that we are detecting 1 or 2 errors. (Some people would call this simply a distance 3 double error-detecting, or DED, code.) In such a case, the error detection fails if 3 or more errors occur. We discuss these probability computations by using the example of a code for a 1-byte message, where $m = 8$ and $c = 4$ (see Table 2.8). If we assume that the dominant term in this computation is the probability of 3 errors, then we can see Eq. (2.1) and write

$$P_{ue}' = B(3:12) = 220 q^3 (1-q)^9 \qquad (2.27)$$

## HAMMING CODES

**TABLE 2.11  Evaluation of the Reduction in Undetected Errors for a Hamming SECSED Code: Eq. (2.25)**

| Bit Error Probability, $q$ | Improvement Ratio: $P_{ue}/P'_{ue}$ |
|---|---|
| $10^{-4}$ | $3.640 \times 10^{6}$ |
| $10^{-5}$ | $3.637 \times 10^{8}$ |
| $10^{-6}$ | $3.636 \times 10^{10}$ |
| $10^{-7}$ | $3.636 \times 10^{12}$ |
| $10^{-8}$ | $3.636 \times 10^{14}$ |

Following simplifications similar to those used to derive Eq. (2.7), the undetected error ratio becomes

$$P_{ue}/P'_{ue} = 2(1+9q)/55q^2 \qquad (2.28)$$

This ratio is evaluated in Table 2.11.

### 2.4.6  Effect of Coder–Decoder Failures

Clearly, the error improvement ratios in Table 2.11 are much larger than those in Table 2.3. We now must include the probability of the generator/checker circuitry failing. This should be a more significant effect than in the case of the parity-bit code for two reasons. First, the undetected error probabilities are much smaller with the SECSED code, and second, the generator/checker will be more complex. A practical circuit for checking a (7, 4) SECSED code is given in Wakerly [p. 298, 1990] and is reproduced in Fig. 2.7. For the reader who is not experienced in digital circuitry, some explanation is in order. The three 74LS280 ICs ($U_1$, $U_2$, and $U_3$) are similar to the SN74180 shown in Fig. 2.4. Substituting Eq. (2.22a) into Eq. (2.24a) shows that the syndrome bit $e_1$ is dependent on the $\oplus$ of $c_1$, $b_1$, $b_2$, and $b_4$, and from Table 2.6 we see that these are bit positions $x_1$, $x_3$, $x_5$, and $x_7$, which correspond to the inputs to $U_1$. Similarly, $U_2$ and $U_3$ compute $e_2$ and $e_3$. The decoder $U_4$ (see Appendix C6.3) activates one of its 8 outputs, which is the address of the error bit. The 8 output gates ($U_5$ and $U_6$) are exclusive or gates (see Appendix C; only 7 are used). The output of the $U_4$ selects the erroneous bit from the bus DU(1–7), complements it (performing a correction), and passes through the other 6 bits unchanged. Actually the outputs DU(1–7) are all complements of the desired values; however, this is simply corrected by a group of inverters at the output or inversion of the next stage of digital logic. For a check-bit generator, we can use three 74LS280 chips to generate $e_1$, $e_2$, and $e_3$.

We can compute the reliability of the generator/checker circuitry by again using the IC failure rate model of Section B3.3, $\lambda_b = 0.004\sqrt{g}$. We assume

**54** CODING TECHNIQUES

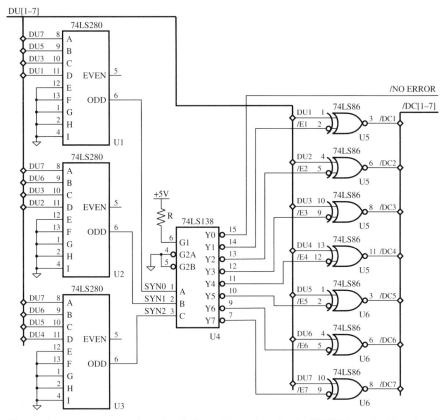

**Figure 2.7** Error-correcting circuit for a Hamming (7, 4) SECSED code [Reprinted by permission of Pearson Education, Inc., Upper Saddle River, NJ 07458; from Wakerly, 2000, p. 298].

that any failure in the IC causes system failure, so the reliability diagram is a series structure and the failure rates add. The computation is detailed in Table 2.12. (See also Fig. 2.7.)

Thus the failure rate for the coder plus decoder is $\lambda = 13.58 \times 10^{-8}$, which is about four times as large as that for the parity bit case ($2 \times 1.67 \times 10^{-8}$) that was calculated previously.

We now incorporate the possibility of generator/checker failure and how it affects the error-correction performance in the same manner as we did with the parity-bit code in Eqs. (2.8)–(2.11). From Table 2.8 we see that a 1-byte (8-bit) message requires 4 check bits; thus the SECSED code is (12, 8). The example developed in Table 2.12 and Fig. 2.7 was for a (7, 4) code, but we can easily modify these results for the (12, 8) code we have chosen to discuss. First, let us consider the code generator. The 74LS280 chips are designed to generate parity check bits for up to an 8-bit word, so they still suffice; however, we now

**TABLE 2.12 Computation of Failure Rates for a (7, 4) SECSED Hamming Generator/Checker Circuitry**

| IC | Function | Gates,[a] g | $\lambda_b = 0.004\sqrt{g} \times 10^{-6}$ | Number in Circuit | Failure Rate/hr |
|---|---|---|---|---|---|
| 74LS280 | Parity-bit generator | 17.5 | $1.67 \times 10^{-8}$ | 3 in generator | $5.01 \times 10^{-8}$ |
| 74LS280 | Parity-bit generator | 17.5 | $1.67 \times 10^{-8}$ | 3 in checker | $5.01 \times 10^{-8}$ |
| 74LS138 | Decoder | 16.0 | $1.60 \times 10^{-8}$ | 1 in checker | $1.60 \times 10^{-8}$ |
| 74LS86 | EXOR package | 6.0 | $9.80 \times 10^{-9}$ | 2 in checker | $1.96 \times 10^{-8}$ |
| | | | | Total | $13.58 \times 10^{-8}$ |

[a]Using 1.5 gates for each EXOR and ENOR gate.

**56**     CODING TECHNIQUES

need to generate 4 check bits, so a total of 4 will be required. In the case of the checker (see Fig. 2.7), we will also require four 74LS280 chips to generate the y-syndrome bits. Instead of a 3-to-8 decoder we will need a 4-to-16 decoder for the next stage, which can be implemented by using two 74LS138 chips and the appropriate connections at the enable inputs (G1, G2A, and G2B), as explained in Appendix C6.3. The output stage composed of 74LS86 chips will not be required if we are only considering error detection, since the nonerror output is sufficient for this. Thus we can modify Table 2.12 to compute the failure rate that is shown in Table 2.13. Note that one could argue that since we are only computing the error-detection probabilities, the decoders and output correction EXOR gates are not needed, and only an OR gate with the syndrome inputs is needed to detect a 0000 syndrome that indicates no errors.

Using the information in Table 2.13 and Eq. (2.27), we obtain an expression similar to Eq. (2.10), as follows:

$$P'_{ue} = e^{-\lambda t} 220 q^3 (1-q)^9 + (1 - e^{-\lambda t}) \qquad (2.29)$$

where $\lambda$ is $19.50 \times 10^{-8}$ failures per hour and $t$ is $12/3600B$.

We formulate the improvement ratio by dividing Eq. (2.29) by Eq. (2.12); the ratio is given in Table 2.14 and is plotted in Fig. 2.8. The data presented in Table 2.11 is also plotted in Fig. 2.8 and represents the line labeled $B \to \infty$, which represents the case for a nonfailing generator/checker.

### 2.4.7 How Coder–Decoder Failures Affect SECSED Codes

Because the Hamming SECSED code results in a lower value for undetected errors than the parity-bit code, the effect of chip failures is even more pronounced. Of course the coding is still a big improvement, but not as much as one would predict. In fact, by comparing Figs. 2.8 and 2.5 we see that for $B = 300$, the parity-bit scheme is superior to the SECSED scheme for values of $q$ less than about $2 \times 10^{-7}$; for $B = 1,200$, the parity-bit scheme is superior to the SECSED scheme for values of $q$ less than about $10^{-7}$. The general conclusion is that for more complex error detection schemes, one should evaluate the effects of generator/checker failures, since these may be of considerable importance for small values of $q$. (Chip-specific failure rates may be required.)

More generally, we should compute whether generator/checker failures significantly affect the code performance for the given values of $q$ and $B$. If such failures are significant, we can consider the following alternatives:

1. Consider a simpler coding scheme if $q$ is very small and $B$ is low.
2. Consider other coding schemes if they use simpler generator/checker circuitry.
3. Use other digital logic designs that utilize fewer but larger chips. Since the failure rate is proportional to $\sqrt{g}$, larger-scale integration improves reliability.

**TABLE 2.13 Computation of Failure Rates for a (12, 8) DED Hamming Generator/Checker Circuitry**

| IC | Function | Gates,[a] $g$ | $\lambda_b = 0.004\sqrt{g} \times 10^{-6}$ | Number in Circuit | Failure Rate/hr |
|---|---|---|---|---|---|
| 74LS280 | Parity-bit generator | 17.5 | $1.67 \times 10^{-8}$ | 4 in generator | $6.68 \times 10^{-8}$ |
| 74LS280 | Parity-bit generator | 17.5 | $1.67 \times 10^{-8}$ | 4 in checker | $6.68 \times 10^{-8}$ |
| 74LS138 | Decoder | 16.0 | $1.60 \times 10^{-8}$ | 2 in checker | $3.20 \times 10^{-8}$ |
| 74LS86 | EXOR package | 6.0 | $0.98 \times 10^{-8}$ | 3 in checker | $2.94 \times 10^{-8}$ |
| | | | | Total | $19.50 \times 10^{-8}$ |

[a]Using 1.5 gates for each EXOR and ENOR gate.

**TABLE 2.14  The Reduction in Undetected Errors from a Hamming (12, 8) DED Code Including the Effect of Coder–Decoder Failures**

| Bit Error Probability $q$ | Improvement Ratio: $P_{ue}/P'_{ue}$ for Several Transmission Rates | | | |
|---|---|---|---|---|
| | 300 Bits/Sec | 1,200 Bits/Sec | 9,600 Bits/Sec | 56,000 Bits/Sec |
| $10^{-4}$ | $3.608 \times 10^6$ | $3.629 \times 10^6$ | $3.637 \times 10^6$ | $3.638 \times 10^6$ |
| $10^{-5}$ | $3.88 \times 10^7$ | $1.176 \times 10^8$ | $2.883 \times 10^8$ | $3.480 \times 10^8$ |
| $10^{-6}$ | $4.34 \times 10^6$ | $1.738 \times 10^7$ | $1.386 \times 10^8$ | $7.939 \times 10^8$ |
| $10^{-7}$ | $4.35 \times 10^5$ | $1.739 \times 10^6$ | $1.391 \times 10^7$ | $8.116 \times 10^7$ |
| $10^{-8}$ | $4.35 \times 10^4$ | $1.739 \times 10^5$ | $1.391 \times 10^6$ | $8.116 \times 10^6$ |

4. Seek to lower IC failure rates via improved derating, burn-in, use of high reliability ICs, and so forth.
5. Seek fault-tolerant or redundant schemes for code generator and code checker circuitry.

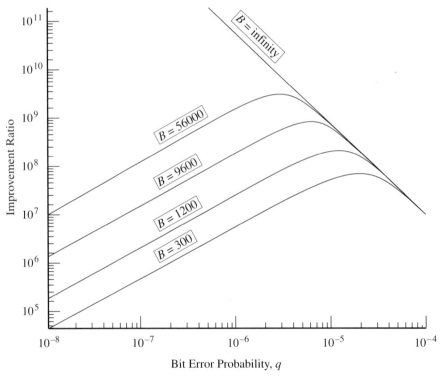

**Figure 2.8**  Improvement ratio of undetected error probability from a SECSED code, including the possibility of coder–decoder failure. $B$ is the transmission rate in bits per second.

## 2.5 ERROR-DETECTION AND RETRANSMISSION CODES

### 2.5.1 Introduction

We have discussed both error detection and correction in the previous sections of this chapter. However, performance metrics (the probabilities of undetected errors) have been discussed only for error detection. In this section, we introduce metrics for evaluating the error-correction performance of various codes. In discussing the applications for parity and Hamming codes, we have focused on information transmission as a typical application. Clearly, the implementations and metrics we have developed apply equally well to memory scheme protection, cache checking, bus-transmission checks, and so forth. Thus, when we again use a data-transmission data application to discuss error correction, the results will also apply to the other application.

The Hamming error-correcting codes provide a direct means of error correction; however, if our transmission channel allows communication in both directions (bidirectional), there is another possibility. If we detect an error, we can send control signals back to the source to ask for retransmission of the erroneous byte, work, or code block. In general, the appropriate measure of error correction is the reliability (probability of no error).

### 2.5.2 Reliability of a SECSED Code

To discuss the reliability of transmission, we again focus on 1 transmitted byte and compute the reliability with and without error correction. The reliability of a single transmitted byte without any error correction is just the probability of no errors occurring, which was calculated as the second term in Eq. (2.5).

$$R = (1 - q)^8 \qquad (2.30)$$

In the case of a SECSED code (12, 8), single errors are corrected; thus the reliability is given by

$$R = P(\text{no errors} + 1 \text{ error}) \qquad (2.31)$$

and since these are mutually exclusive events,

$$R = P(\text{no errors}) + P(1 \text{ error}) \qquad (2.32)$$

the binomial distribution yields

$$R' = (1 - q)^{12} + 12q(1 - q)^{11} = (1 - q)^{11}(1 + 11q) \qquad (2.33)$$

Clearly, $R' \geq R$; however, for small values of $q$, both are very close to 1, and it is easier to compare the unreliability $U = 1 - R$. Thus a measure of the improvement of a SECSED code is given by

**60** CODING TECHNIQUES

**TABLE 2.15** Evaluation of the Reduction in Unreliability for a Hamming SECSED Code: Eq. (2.35)

| Bit Error Probability, $q$ | Improvement Ratio: $\dfrac{1-U}{1-U'}$ |
|---|---|
| $10^{-4}$ | $6.61 \times 10^2$ |
| $10^{-5}$ | $6.61 \times 10^3$ |
| $10^{-6}$ | $6.61 \times 10^4$ |
| $10^{-7}$ | $6.61 \times 10^5$ |
| $10^{-8}$ | $6.61 \times 10^6$ |

$$(1 - U)/(1 - U') = [1 - (1 - q)^8]/[1 - (1 - q)^{11}(1 + 11q)] \qquad (2.34)$$

and approximating this for small $q$ yields

$$(1 - U)/(1 - U') = 8/121q \qquad (2.35)$$

which is evaluated for typical values of $q$ in Table 2.15.

The foregoing evaluations neglected the probability of IC generator and checker failure. However, the analysis can be broadened to include these effects as was done in the preceding sections.

### 2.5.3 Reliability of a Retransmitted Code

If it is possible to retransmit a code block after an error has been detected, one can improve the reliability of the transmission. In such a case, the reliability expression becomes

$$R' = P(\text{no error} + \text{detected error and no error on retransmisson}) \qquad (2.36)$$

and since these are mutually exclusive events and independent events,

$$R' = P(\text{no error}) + P(\text{detected error}) \times P(\text{no error on retransmission}) \qquad (2.37)$$

Since the error probabilities on initial transmission and on retransmission are the same, we obtain

$$R' = P(\text{no error})[1 + P(\text{detected error})] \qquad (2.38)$$

For the case of a parity-bit code, we transmit 9 bits; the probability of detecting an error is approximately the probability of 1 error. Substitution in Eq. (2.38) yields

## ERROR-DETECTION AND RETRANSMISSION CODES

$$R' = (1-q)^9[1 + 9q(1-q)^8] \tag{2.39}$$

Comparing the ratio of unreliabilities yields

$$(1-U)/(1-U') = [1 - (1-q)^8]/[1 - [(1-q)^9[1 + 9q(1-q)^8]]] \tag{2.40}$$

and simplification for small $q$ yields

$$(1-U)/(1-U') = 8q/[9q^2 - 828q^3] \tag{2.41}$$

Similarly, we can use a Hamming distance 3 code (12, 8) to detect up to 2 errors and retransmit. In this case, the probability of detecting an error is approximately the probability of 1 or 2 errors. Substitution in Eq. (2.38) yields

$$R' = (1-q)^{12}[1 + (12q(1-q)^{11} + 66q^2(1-q)^{10})] \tag{2.42}$$

and the unreliability ratio becomes

$$(1-U)/(1-U') = [1 - (1-q)^8]/[1 - [(1-q)^{12}[1 + (12q(1-q)^{11} + 66q^2(1-q)^{10}]]] \tag{2.43}$$

and simplification for small $q$ yields

$$(1-U)/(1-U') = 8q/[78q^2 - 66q^3] \tag{2.44}$$

Equations (2.41) and (2.44) are evaluated in Table 2.16 for typical values of $q$. Comparison of Tables 2.15 and 2.16 shows that both retransmit schemes are superior to the error correction of a SECSED code, and that the parity-bit retransmit scheme is the best. However, retransmit has at least a 100% overhead penalty, and Table 2.8 shows typical SECSED overheads of 11–50%.

**TABLE 2.16  Evaluation of the Improvement in Reliability by Code Retransmission for Parity and Hamming $d = 3$ Code**

| Bit Error Probability, $q$ | Parity-Bit Retransmission $(1-U)/(1-U')$: Eq. (2.41) | Hamming $d = 3$ Retransmission $(1-U)/(1-U')$: Eq. (2.44) |
|---|---|---|
| $10^{-4}$ | $8.97 \times 10^3$ | $1.026 \times 10^3$ |
| $10^{-5}$ | $8.90 \times 10^4$ | $1.026 \times 10^4$ |
| $10^{-6}$ | $8.89 \times 10^5$ | $1.026 \times 10^5$ |
| $10^{-7}$ | $8.89 \times 10^6$ | $1.026 \times 10^6$ |
| $10^{-8}$ | $8.89 \times 10^7$ | $1.026 \times 10^7$ |

**62**  CODING TECHNIQUES

The foregoing evaluations neglected the probability of IC generator and checker failure as well as the circuitry involved in controlling retransmission. However, the analysis can be broadened to include these effects, and a more detailed comparison can be made.

## 2.6 BURST ERROR-CORRECTION CODES

### 2.6.1 Introduction

The codes previously discussed have all been based on the assumption that the probability that bit $b_i$ is corrupted by an error is largely independent of whether bit $b_{i-1}$ is correct or is in error. Furthermore, the probability of a single bit error, $q$, is relatively small; thus the probability of more than one error in a word is quite small. In the case of a burst error, the probability that bit $b_i$ is corrupted by an error is much larger if bit $b_{i-1}$ is incorrect than if bit $b_{i-1}$ is correct. In other words, the errors commonly come in bursts rather than singly. One class of applications that are subject to burst errors are rotational magnetic and optical storage devices (e.g., music CDs, CD-ROMs, and hard and floppy disk drives). Magnetic tape used for pictures, sound, or data is also affected by burst errors.

Examples of the patterns of typical burst errors are given in the four 12-bit messages ($m_1$–$m_4$) shown in the forthcoming equations. The common notation is used where $b$ represents a correct message bit and $x$ represents an erroneous message bit. (For the purpose of identification, assume that the bits are numbered 1–12 from left to right.)

$$m_1 = bbbxxbxbbbbb \qquad (2.45a)$$
$$m_2 = bxbxxbbbbbbb \qquad (2.45b)$$
$$m_3 = bbbbxbxbbbbb \qquad (2.45c)$$
$$m_4 = bxxbbbbbbbbb \qquad (2.45d)$$

Messages 1 and 2 each have 3 errors that extend over 4 bits (e.g., in $m_1$ the error bits are in positions 4, 5, and 7); we would refer to them as bursts of length 4. In message 3, the burst is of length 3; in message 4, the burst is of length 2. In general, we call the burst length $t$. The burst length is really a matter of definition; for example, one could interpret messages 1 and 2 as 2 bursts—one of length 1 and one of length 2. In practice, this causes no confusion, for $t$ is a parameter of a burst code and is fixed in the initial design of the code. Thus if $t$ is chosen as length 4, all 4 of the messages would have 1 burst. If $t$ is chosen as length 3, messages 1 and 2 would have two bursts, and messages 3 and 4 would have 1 burst.

Most burst error codes are more complex than the Hamming codes that were just discussed; thus the remainder of this chapter will present a succinct introduction to the basis of such codes and will briefly introduce one of the most popular burst codes: the Reed–Solomon code [Golumb, 1986].

### 2.6.2 Error Detection

We begin by giving an example of a burst error-detection code [Arazi, 1988]. Consider a 12-bit-long code word (also called a *code block* or *code vector*, $V$), which includes both message and check bits as follows:

$$V = (x_1 x_2 x_3 x_4 x_5 x_6 x_7 x_8 x_9 x_{10} x_{11} x_{12}) \tag{2.46}$$

Let us choose to deal with bursts of length $t = 4$. Equations for calculating the check bits in terms of the message bits can be developed by writing a set of equations in which the bits are separated by $t$ positions. Thus for $t = 4$, each equation contains every fourth bit.

$$x_1 \oplus x_5 \oplus x_9 = 0 \tag{2.47a}$$
$$x_2 \oplus x_6 \oplus x_{10} = 0 \tag{2.47b}$$
$$x_3 \oplus x_7 \oplus x_{11} = 0 \tag{2.47c}$$
$$x_4 \oplus x_8 \oplus x_{12} = 0 \tag{2.47d}$$

Each bit appears in only one equation. Assume there is either 0 or only 1 burst in the code vector (multiple bursts in a single word are excluded). Thus each time there is 1 erroneous bit, one of the four equations will equal 1 rather than 0, indicating a single error. To illustrate this, suppose $x_2$ is an error bit. Since we are assuming a burst length of 4 and at most 1 burst per code vector, the only other *possible* erroneous bits are $x_3$, $x_4$, and $x_5$. (At this point, we don't know if 0, 1, 2, or 3 errors occur in bits 3–5.) Examining Eq. (2.47b), we see that it is not possible for $x_6$ or $x_{10}$ to be erroneous bits, so it is not possible for 2 errors to cancel out in evaluating Eq. (2.47b). In fact, if we analyze the set of Eqs. (2.47a–d), we see that the number of nonzero equations in the set is equal to the number of bit errors in the burst.

Since there are 4 check equations, we need 4 check bits; any set of 4 bits in the vector can be chosen as check bits, provided that 1 bit is chosen from each equation (2.47a–d). For clarity, it probably makes sense to choose the 4 check bits as the first or last 4 bits in the vector; such a choice in any type of code is referred to as a *systematic code*. Suppose we choose the first 4 bits. We then obtain a (12, 8) systematic burst code of length 4, where $c_i$ stands for a check bit and $b_i$ a message bit.

$$V = (c_1 c_2 c_3 c_4 b_1 b_2 b_3 b_4 b_5 b_6 b_7 b_8) \tag{2.48}$$

A moment's reflection shows that we have now maneuvered Eqs. (2.47a–d) so that with $c$s and $b$s substituted for the $x$s, we obtain

$$c_1 \oplus b_1 \oplus b_5 = 0 \quad (2.49a)$$
$$c_2 \oplus b_2 \oplus b_6 = 0 \quad (2.49b)$$
$$c_3 \oplus b_3 \oplus b_7 = 0 \quad (2.49c)$$
$$c_4 \oplus b_4 \oplus b_8 = 0 \quad (2.49d)$$

which can be used to compute the check bits. These equations are therefore the basis of the check-bit generator, which can be done with 74180 or 74280 IC chips.

The same set of equations form the basis of the error-checking circuitry. Based on the fact that the number of nonzero equations in the set of Eqs. (2.47a–d) is equal to the number of bit errors in the burst, we can modify Eqs. (2.47a–c) so that they explicitly yield bits of a syndrome vector, $e_1 e_2 e_3 e_4$.

$$e_1 = x_1 \oplus x_5 \oplus x_9 \quad (2.50a)$$
$$e_2 = x_2 \oplus x_6 \oplus x_{10} \quad (2.50b)$$
$$e_3 = x_3 \oplus x_7 \oplus x_{11} \quad (2.50c)$$
$$e_4 = x_4 \oplus x_8 \oplus x_{12} \quad (2.50d)$$

The nonerror condition occurs when all the syndrome bits are 0. In general, the number of errors detected is the arithmetic sum: $e_1 + e_2 + e_3 + e_4$. Note that because we originally chose $t = 4$ in this design, no more than 4 errors can be detected. Again, the checker can be done with 74180 or 74280 IC chips. Alternatively, one can use individual gates. To generate the check bits, 4 EXOR gates are sufficient; 8 EXOR gates and an output OR gate are sufficient for error checking (cf. Fig. 2.2). However, if one wishes to determine how many errors have occurred, the output OR gate in the checker can be replaced by a few half-adders or full-adders to compute the arithmetic sum: $e_1 + e_2 + e_3 + e_4$.

We can now state some properties of burst codes that were illustrated by the above discussion. The reader is referred to the references for proof [Arazi, 1988].

**Properties of Burst Codes**

1. For a burst length of $t$, $t$ check bits are needed for error detection. (Note: this is independent of the message length $m$.)
2. For $m$ message bits and a burst length of $t$, the code word length $n = m + t$.
3. There are $t$ check-bit equations:
   (a) The first check-bit equation starts with bit 1 and contains all the bits that are $t + 1, 2t + 1, \ldots kt + 1$ (where $kt + 1 \leq n$).
   (b) The second check-bit equation starts with bit 2 and contains all the bits that are $t + 2, 2t + 2, \ldots kt + 2$ (where $kt + 2 \leq n$).
   ............................................................
   (t) The $t'$th check-bit equation starts with bit $t$ and contains all the bits that are $2t, 3t, \ldots kt$ (where $kt \leq n$).

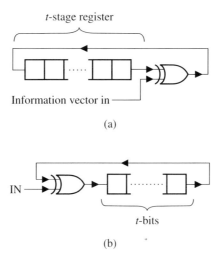

**Figure 2.9** Burst error-detection circuitry using an LFSR: (a) encoder; (b) decoder. [Reprinted by permission of MIT Press, Cambridge, MA 02142; from Arazi, 1988, p. 108.]

4. The EXOR of all the bits in $3a$ should $= 0$ and similarly for properties $3b, \ldots 3t$.
5. The word length $n$ need not be an integer multiple of $t$, but for practicality, we assume that it is. If necessary, the word can be padded with additional dummy bits to achieve this.
6. Generation and checking for a burst error code (as well as other codes) can be realized by a linear feedback shift register (LFSR). (See Fig. 2.9.)
7. In general, the LFSR has a delay of $t \times$ the shift time.
8. The generating and checking for a burst error code can be realized by an EXOR tree circuit (cf. Fig. 2.2), in which the number of stages is $\leq \log_2(t)$ and the delay is $\leq \log_2(t) \times$ the EXOR gate-switching time.

These properties are explored further in the problems at the end of this chapter. To summarize, in this section we have developed the basic equations for burst error-detection codes and have shown that the check-bit generator and checker circuitry can be implemented with EXOR trees, parity-bit chips, or LFSRs. In general, the LFSR implementation requires less hardware, but the delay time is linear in the burst length $t$. In the case of EXOR trees, there is more hardware needed; however, the time delay is less, for it increases proportionally to the log of $t$. In either case, for the modest size $t = 4$ or 5, the differences in time delay and hardware are not that significant. Both designs should be attempted, and a choice should be made.

The case of burst error *correction* is more difficult. It is discussed in the next section.

## 2.6.3 Error Correction

We now state some additional properties of burst codes that will lead us to an error-correction procedure. In general, these are properties associated with a shifting of the error syndrome of a burst code and an ancient theorem of number theory related to the mod function. The theorem from number theory is called the *Chinese Remainder Theorem* [Rosen, 1991, p. 134] and was first given as a puzzle by the first-century Chinese mathematician Sun-Tsu. It will turn out that the method of error correction will depend on first locating a region in the code word of $t$ consecutive bits that contains the start of the error burst, followed by pinpointing which of these $t$ bits is the start of the burst. The methodology is illustrated by applying the principles to the example given in Eq. (2.46). For a development of the theory and proofs, the reader is referred to Arazi [1988] and Rosen [1991].

The error syndrome can be viewed as a cyclic shift of the burst error pattern. For example, if we assume a single burst and $t = 4$, then substitution of error pattern for $x_1 x_2 x_3 x_4$ into Eqs. (2.50a–d) will yield a particular syndrome pattern. To compute what the syndrome would be, we note that if $x_1 x_2 x_3 x_4 = $ **bbbb**, all the bits are correct and the syndrome must be 0000. If bit 1 is in error (either changed from a correct 1 to an erroneous 0 or from a correct 0 to an erroneous 1), then Eq. (4.50a) will yield a 1 for $e_1$ (since there is only 1 burst, bits $x_5$–$x_{12}$ must be all valid $b$s). Suppose the error pattern is $x_1 x_2 x_3 x_4 = $ **xbxx**, then all other bits in the 12-bit vector are $b$ and substitution into Eqs. (2.50a–d) yields

$$e_1 = \mathbf{x} \oplus x_5 \oplus x_9 = 1 \tag{2.51a}$$
$$e_2 = \mathbf{b} \oplus x_6 \oplus x_{10} = 0 \tag{2.51b}$$
$$e_3 = \mathbf{x} \oplus x_7 \oplus x_{11} = 1 \tag{2.51c}$$
$$e_4 = \mathbf{x} \oplus x_8 \oplus x_{12} = 1 \tag{2.51d}$$

which is a syndrome pattern $e_1 e_2 e_3 e_4 = 1011$. Similarly, error pattern $x_4 x_5 x_6 x_7 = $ **xbxx**, where all other bits are $b$, yields syndrome equations as follows:

$$e_1 = x_1 \oplus \mathbf{b} \oplus x_9 = 0 \tag{2.52a}$$
$$e_2 = x_2 \oplus \mathbf{x} \oplus x_{10} = 1 \tag{2.52b}$$
$$e_3 = x_3 \oplus \mathbf{x} \oplus x_{11} = 1 \tag{2.52c}$$
$$e_4 = \mathbf{x} \oplus x_8 \oplus x_{12} = 1 \tag{2.52d}$$

which is a syndrome pattern $e_1 e_2 e_3 e_4 = 0111$. We can view 0111 as a pattern that can be transformed into 1011 by cyclic-shifting *left* (end-around-rotation left) three times. We will show in the following material that the same syndrome is obtained by shifting the code vector *right* four times.

We begin marking a burst error pattern with the first erroneous bit in the word; thus burst error patterns always start with an **x**. Since the burst is $t$ bits long, the syndrome equations (2.50a–d) include bits that differ by $t$ positions. Therefore, if we shift the burst error pattern in the code vector by $t$ positions to the right, the burst error pattern generates the same syndrome. There can be at most $u$ placements of the burst pattern in a code vector that results in the same syndrome; if the code vector is $n$ bits long, $u$ is the largest integer such that $ut \leq n$. Without loss of generality, we can always pad the message bits with dummy bits such that $ut = n$. We define the *mod* function $x$ mod $y$ as the remainder that is obtained when we divide the integer $x$ by the integer $y$. Thus, if $ut = n$, we can then say that $n$ mod $u = 0$. These relationships will soon be used to devise an algorithm for burst error correction.

The location of the start of the burst error pattern in a word is related to the amount of shift (end-around and cyclic) of the pattern that is observed in the syndrome. We can illustrate this relationship by using the burst pattern **xbxx** as an example, where **xbxx** is denoted by 1011: meaning incorrect, correct, incorrect, incorrect. In Table 2.17, we illustrate the relationship between the start of the error burst and the rotational shift (end-around shift) in the detected error syndrome. We begin by renumbering the code vector, Eq. (2.46), so it starts with bit 0:

$$V = (x_0 x_1 x_2 x_3 x_4 x_5 x_6 x_7 x_8 x_9 x_{10} x_{11}) \tag{2.53}$$

A study of Table 2.17 shows that the number of syndrome shifts is related to the bit number by (bit number) mod 4. For example, if the burst starts with bit no. 3, we have 3 mod 4 (which is 3), so the syndrome is the error pattern shifted 3 places to the right. If we want to recover the syndrome, we shift 3 places to the left. In the case of a burst starting with bit no. 4, 4 mod 4 is 0, so the syndrome pattern and the burst pattern agree.

Thus, if we know the position in the code word at which the burst starts (defined as $x$), and if the burst length is $t$, then we can obtain the burst pattern by shifting the syndrome $x$ mod $t$ places to the left. Knowing the starting position of the burst ($x$) and the burst pattern, we can correct any erroneous bits. Thus our task is now to find $x$.

The procedure for solving for $x$ depends on the Chinese Remainder Theorem, a previously mentioned mathematical theorem in number theory. This theorem states that if $p$ and $q$ are relatively prime numbers (meaning their only common factor is 1), and if $0 \leq x \leq (pq-1)$, then knowledge of $x$ mod $p$ and $x$ mod $q$ allows us to solve for $x$. We already have one equation: $x$ mod $t$; to generate another equation, we define $u = 2t-1$ and calculate $x$ from $x$ mod $u$ [Arazi, 1988]. Note that $t$ and $2t-1$ are relatively prime since if a number divides $t$, it also divides $2t$ but not $2t-1$. Also, we must show that $0 \leq x \leq (tu-1)$; however, we already showed that $tu \leq n$. Substitution yields $0 \leq x \leq (n-1)$, which must be true since the latest bit position to start a burst error ($x$) for a burst of length $t$ is $n - t < n - 1$.

The above relationships show that it is possible to solve for the beginning

**TABLE 2.17  Relationship Between Start of the Error Burst and the Syndrome for the 12-Bit-Long Code Given in Eq. (2.49)**

| Burst Start Bit No. | Code Vector Positions | | | | | | | | | | | | Error Syndrome | | | | To Recover Syndrome Shift |
|---|---|---|---|---|---|---|---|---|---|---|---|---|---|---|---|---|---|
| | 0 | 1 | 2 | 3 | 4 | 5 | 6 | 7 | 8 | 9 | 10 | 11 | $e_1$ | $e_2$ | $e_3$ | $e_4$ | |
| 0 | x | b | x | x | b | b | b | b | b | b | b | b | 1 | 0 | 1 | 1 | 0 |
| 1 | b | x | b | x | x | b | b | b | b | b | b | b | 1 | 1 | 0 | 1 | 1 left |
| 2 | b | b | x | b | x | x | b | b | b | b | b | b | 1 | 1 | 1 | 0 | 2 left |
| 3 | b | b | b | x | b | x | x | b | b | b | b | b | 0 | 1 | 1 | 1 | 3 left |
| 4 | b | b | b | b | x | b | x | x | b | b | b | b | 1 | 0 | 0 | 1 | 0 |
| 5 | b | b | b | b | b | x | b | x | x | b | b | b | 1 | 1 | 1 | 1 | 1 left |
| 6 | b | b | b | b | b | b | x | b | x | x | b | b | 1 | 1 | 1 | 0 | 2 left |
| 7 | b | b | b | b | b | b | b | x | b | x | x | b | 0 | 1 | 1 | 1 | 3 left |
| 8 | b | b | b | b | b | b | b | b | x | b | x | x | 1 | 0 | 1 | 1 | 0 |

of the burst error $x$ and the burst error pattern. Given this information, by simply complementing the incorrect bits, error correction is performed. The remainder of this section details how we set up equations to calculate the check bits (generator) and to calculate the burst pattern and location (checker); this is done by means of an illustrative example. One circuit implementation using shift registers is discussed as well.

The number of check bits is equal to $u + t$, and since $u = 2c - 1$ and $n = ut$, the number of message bits is determined. We formulate check bit equations in a manner analogous to that used in error checking.

The following example illustrates how the two sets of check bits are generated, how one formulates and solves for $x \bmod u$ and $x \bmod t$ to solve for $x$, and how the burst error pattern is determined. In our example, we let $t = 3$ and calculate $u = 2t - 1 = 2 \times 3 - 1 = 5$. In this case, the word length $n = u \times t = 5 \times 3 = 15$. The code vector is given by

$$V = (x_0 x_1 x_2 x_3 x_4 x_5 x_6 x_7 x_8 x_9 x_{10} x_{11} x_{12} x_{13} x_{14}) \tag{2.54}$$

The $t + u$ check equations are generated from a set of $u$ equations that form the auxiliary syndrome. For our example, the $u = 5$ auxiliary syndrome equations are:

$$s_0 = x_0 \oplus x_5 \oplus x_{10} \tag{2.55a}$$
$$s_1 = x_1 \oplus x_6 \oplus x_{11} \tag{2.55b}$$
$$s_2 = x_2 \oplus x_7 \oplus x_{12} \tag{2.55c}$$
$$s_3 = x_3 \oplus x_8 \oplus x_{13} \tag{2.55d}$$
$$s_4 = x_4 \oplus x_9 \oplus x_{14} \tag{2.55e}$$

and the set of $t = 3$ equations that form the syndrome are

$$e_1 = x_0 \oplus x_3 \oplus x_6 \oplus x_9 \oplus x_{12} \tag{2.56a}$$
$$e_2 = x_1 \oplus x_4 \oplus x_7 \oplus x_{10} \oplus x_{13} \tag{2.56b}$$
$$e_3 = x_2 \oplus x_5 \oplus x_8 \oplus x_{11} \oplus x_{14} \tag{2.56c}$$

If we want a systematic code, we can place the 8 check bits at the beginning or the end of the word. Let us assume that they go at the end ($x_7$–$x_{14}$) and that these check bits $c_0$–$c_7$ are calculated from Eqs. (2.55a–e) and (2.56a–c). The first 7 bits ($x_0$–$x_6$) are message bits, and the transmitted word is

$$V = (b_0 b_1 b_2 b_3 b_4 b_5 b_6 c_0 c_1 c_2 c_3 c_4 c_5 c_6 c_7) \tag{2.57}$$

As an example, let us assume that the message bits $b_0$–$b_6$ are 1011010. Substitution of these values in Eqs. (2.55a–e) and (2.56a–c) *that must initially be 0* yields a set of equations that can be solved for the values of $c_0$–$c_7$. One can show by substitution that the values $c_0$–$c_7$ = 10000010 satisfy the equations.

# 70   CODING TECHNIQUES

(Shortly, we will describe code generation circuitry that can solve for the check bits in a straightforward manner.) Thus the transmitted word is

$$V_t = (b_0 b_1 b_2 b_3 b_4 b_5 b_6) = 1011010 \quad \text{for the message part} \quad (2.58a)$$
$$V_t = (c_0 c_1 c_2 c_3 c_4 c_5 c_6 c_7) = 10000010 \quad \text{for the check part} \quad (2.58b)$$

Let us assume that the received word is

$$V_r = (101101000100010) \tag{2.59}$$

We now begin the error-recovery procedure by calculating the auxiliary syndrome by substitution of Eq. (2.59) in Eqs. (2.55a–e) yielding

$$s_0 = 1 \oplus 1 \oplus 0 = 0 \tag{2.60a}$$
$$s_1 = 0 \oplus 0 \oplus 0 = 0 \tag{2.60b}$$
$$s_2 = 1 \oplus 0 \oplus 1 = 1 \tag{2.60c}$$
$$s_3 = 1 \oplus 0 \oplus 0 = 0 \tag{2.60d}$$
$$s_4 = 0 \oplus 1 \oplus 0 = 1 \tag{2.60e}$$

The fact that the auxiliary syndrome is not all 0's indicates that 1 or more errors have occurred. In fact, since two equations are nonzero, there are two errors. Furthermore, it can be shown that the burst error pattern associated with the auxiliary syndrome must always start with an $x$ and all bits $> t$ must be valid bits. Thus, the burst error pattern (since $t = 3$) must be $x??bb = 1??00$. This means the auxiliary syndrome pattern should start with a 1 and end in two 0's. The unique solution is that the auxiliary syndrome pattern must be shifted to the left two places yielding 10100 so that the first bit is 1 and the last two bits are 0. In addition, we deduce that the real syndrome (and the burst pattern) is 101. Similarly, Eqs. (2.56a–c) yield

$$e_0 = 1 \oplus 1 \oplus 0 \oplus 1 \oplus 0 = 1 \tag{2.61a}$$
$$e_1 = 0 \oplus 0 \oplus 0 \oplus 0 \oplus 1 = 1 \tag{2.61b}$$
$$e_2 = 1 \oplus 1 \oplus 0 \oplus 0 \oplus 0 = 0 \tag{2.61c}$$

Thus, to get the known syndrome—found from Eqs. (2.61a–c)—to be 101, we must shift the real syndrome left one place. Based on these shift results, our two mod equations become

$$\text{for } u: \quad x \bmod u = x \bmod 5 = 2 \tag{2.62a}$$
$$\text{for } t: \quad x \bmod t = x \bmod 3 = 1 \tag{2.62b}$$

We now know the burst pattern 101 and have two equations (2.62a, b) that can be solved for the start of the burst pattern given by $x$. Substitution of trial values into Eq. (2.62a) yields $x = 2$, which satisfies (2.62a) but not (2.62b). The

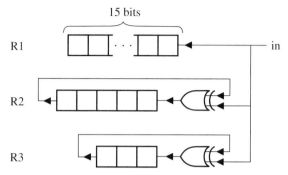

**Figure 2.10** Basic error decoder for $u = 5$ and $t = 3$ burst code based on three shift registers. (Additional circuitry is needed for a complete decoder.) The input (IN) is a train of shift pulses. [Reprinted by permission of MIT Press, Cambridge, MA 02142; from Arazi, 1988, p. 123.]

next value that satisfies Eq. (2.62a) is $x = 7$, and since this value also satisfies Eq. (2.62b), it is a solution. We conclude that the burst error started at position $x = 7$ (the eighth bit, since the count starts with 0) and that is was $xbx$, so the eighth and tenth bits must be complemented. Thus the received and corrected versions of the code vector are

$$V_r = (101101000100010) \qquad (2.63a)$$

$$\Updownarrow$$

$$V_c = (101101010000010) \qquad (2.63b)$$

Note that Eqs. (2.63a, b) agrees with Eqs. (2.58a, b).

One practical decoder implementation for the $u = 5$ and $t = 3$ code discussed above is based on three shift registers (R1, R2, and R3) shown in Fig. 2.10. Such a circuit is said to employ linear feedback shift registers (LFSR).

Initially, R1 is loaded with the received code vector, R2 is loaded with the auxiliary syndrome calculated from EXOR trees or parity-bit chips that implement Eqs. (2.60a–e), and R3 is loaded with the syndrome calculated from EXOR trees or parity-bit chips that implement Eqs. (2.61a–c). Using our previous example, R1 is loaded with Eqs. (2.58a, b), R2 with 00101, and R3 with 110. R2 and R3 are shifted left until the left 3 bits of R2 agree with R3, and the leftmost bit is a 1. A count of the number of left shifts yields the start position of the burst error ($x$), and the contents of R3 is the burst pattern. Circuitry to complement the appropriate bits results in error correction. In the circuit shown, when the error pattern is recovered in R3, R1 has the burst error in the left 3 bits of the register. If correction is to be performed by shifting, the leftmost 3 bits in R1 and R3 can be EXORed and restored in R1. This would assume that the bits shifted out of R1 go to a storage register or are circulated back to R1 and, after error detection, the bits in the repaired word are shifted to their proper position. For more details, see Arazi [1988].

## 72  CODING TECHNIQUES

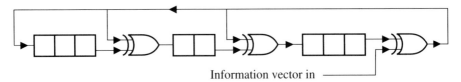

Information vector in

**Figure 2.11** Basic encoder circuit for $u = 5$ and $t = 3$ burst code based on three shift registers. (Additional circuitry is needed for a complete decoder.) The input (IN) is the information vector (message). [Reprinted by permission of MIT Press, Cambridge, MA 02142; from Arazi, 1988, p. 125.]

One can also generate the check bits (encoder) by using LFSRs. One such circuit for our code example is given in Fig. 2.11. For more details, see Arazi [1988].

## 2.7 REED–SOLOMON CODES

### 2.7.1 Introduction

One technique to mitigate against burst errors is to simply interleave data so that a burst does not affect more than a few consecutive data bits at a time. A more efficient approach is to use codes that are designed to detect and correct burst errors. One of the most popular types of error-correcting codes is the Reed–Solomon (RS) code. This code is useful for correcting both random and burst errors, but it is especially popular in burst error situations and is often used with other codes in a convolutional code (see Section 2.8).

### 2.7.2 Block Structure

The RS code is a block-type code and operates on multiple rather than individual bits. Data is processed in a batch called a block instead of continuously. Each block is composed of $n$ symbols, each of which has $m$ bits. The block length $n = 2^m - 1$ symbols. A message is $k$ symbols long, and $n-k$ additional check symbols are added to allow error correction of up to $t$ error symbols. Block length and symbol sizes can be adjusted to accommodate a wide range of message sizes. For an RS code, one can show that

$$(n - k) = 2t \qquad \text{for } n-k \text{ even} \qquad (2.64a)$$
$$(n - k) = 2t + 1 \qquad \text{for } n-k \text{ odd} \qquad (2.64b)$$
$$\text{minimum distance} = d_{\min} = 2t + 1 \text{ symbols} \qquad (2.64c)$$

As a typical example [AHA Applications Note], we will assume $n = 255$ and $m = 8$ (a symbol is 1 byte long). Thus from Eq. (2.64a), if we wish to correct up to 10 errors, then $t = 10$ and $(n - k) = 20$. We therefore have 235 message symbols and 20 check symbols. The code rate (efficiency) of the code is given by $k/n$, which is $(235/255) = 0.92$ or 92%.

### 2.7.3 Interleaving

Interleaving is a technique that can be used with RS and other block codes to improve performance. Individual bits are shifted to spread them over several code blocks. The effect is to spread out long bursts so that error correction can occur even for code bursts that are longer than $t$ bits. After the message is received, the bits are deinterleaved.

### 2.7.4 Improvement from the RS Code

We can calculate the improvement from the RS code in a manner similar to that which was used in the Hamming code. Now, the $P_{ue}$ is the probability of an undetected error in a code block and $P_{se}$ is the probability of a symbol error. Since the code can correct up to $t$ errors, the block error probability is that of having more than $t$ symbol errors in a block, which can be written as

$$P_{ue} = 1 - \sum_{i=0}^{t} \binom{n}{i} (P_{se})^i (1 - P_{se})^{n-i} \tag{2.65}$$

If we didn't have an RS code, any error in a code block would be uncorrectable, and the probability is given as

$$P_{ue} = 1 - (1 - P_{se})^n \tag{2.66}$$

One can plot a set of curves to illustrate the error-correcting performance of the code. A graph of Eq. (2.65) appears in Fig. 2.12 for the example in our discussion. Figure 2.12 is similar to Figs. 2.5 and 2.8 except that the $x$-axis is plotted in opposite order and the $y$-axis has not been normalized by dividing by Eq. (2.66). Reading from the curve, we see for the case where $t = 5$ and $P_{se} = 10^{-3}$:

$$P_{ue} = 3 \times 10^{-7} \tag{2.67}$$

### 2.7.5 Effect of RS Coder–Decoder Failures

We can use Eqs. (2.8) and (2.9) to evaluate the effect of coder–decoder failures. However, instead of computing per byte of transmission, we compute per block of transmission. Thus, by analogy with Eqs. (2.10) and (2.11), for our example we have

$$P_{ue} = e^{-2\lambda_b t} \times 3 \times 10^{-7} + (1 - e^{-2\lambda_b t}) \tag{2.68}$$

where

$$t = 8 \times 255/3{,}600B \tag{2.69}$$

**74** CODING TECHNIQUES

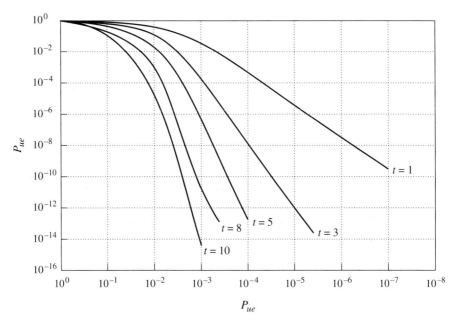

**Figure 2.12** Probability of an uncorrected error in a block of 255 1-byte symbols with 235 message symbols, 20 check symbols, and an error-correcting capacity of up to 10 errors versus the probability of symbol error [AHA Applications Note, used with permission].

We can compute when $P_{ue}$ is composed of equal values for code failures and chip failures by equating the first and second terms of Eq. (2.68). Substituting a typical value of $B = 19{,}200$, we find that this occurs when the chip failure rate is equal to about $5.04 \times 10^{-3}$ failures per hour. Using our model, the chip failure rate $= 0.004\sqrt{g} \; 10^{-6}$, which is equivalent to $g = 1.6 \times 10^{12}$—a very unlikely value. However, if we assume that $P_{se} = 10^{-4}$, then from Fig. 2.12 we see that $P_{ue} = 3 \times 10^{-13}$ and for $B = 19{,}200$ that the effects are equal if the chip failure is equal to about $5.08 \times 10^{-9}$. Substitution into our chip failure rate model shows that this occurs when $g \approx 2$. Thus coder–decoder failures predominate for the second case.

Another approach to investigating the impact of chip failures is to use manufacturers' data on RS coder–decoder failures. Some data exists [AHA Reliability Report, 1995] that is derived from accelerated tests. To collect enough failure data for low-failure-rate components, an accelerated life test—the Arrhenius Relationship—is used to scale back the failure rates to normal operating temperatures (70–85°C). The resulting failure rates range from 50 to $700 \times 10^{-9}$ failures per hour, which certainly exceeds the just-calculated significant failure rate threshold of $5.08 \times 10^{-9}$, which was the value calculated for 19,200 baud and a block error of $10^{-4}$. (Note: using the gate model, we calculate $\lambda =$

$700 \times 10^{-9}$ as equivalent to about 30,000 gates.) Clearly we conclude that the chip failures will predominate for some common ranges of the system parameters.

## 2.8 OTHER CODES

There are many other types of error codes. We will briefly discuss the special features of the more popular codes and refer the reader to the references for additional details.

1. *Burst error codes.* All the foregoing codes assume that errors occur infrequently and are independent, generally corrupting a single bit or a few bits in a word. In some cases, errors may occur in bursts. If a study of the failure modes of the device or medium we wish to protect by our coding indicates that errors occur in bursts to affect all the bits in a word, other coding techniques are required. The reader should consult the references for works that discuss Binary Block codes, $m$-out-of-$n$ codes, Berger codes, Cyclic codes, and Reed–Solomon codes [Pradhan, 1986, 1993].
   *BCH codes.* This is a code that was independently discovered by Bose, Chaudhury, and Hocquenghem. (Reed–Solomon codes are a subclass of BCH codes.) These codes can be viewed as extensions of Hamming codes, which are easier to design and implement for a large number of correctable errors.
   *Concatenated codes.* This refers to the process of lumping more than one code word together to reduce the overhead—generally less for long code words (cf., Table 2.8). Disadvantages include higher error probability (since check bits cover several words), more complexity and depth, and a delay for associated decoding trees.
   *Convolutional codes.* Sometimes, codes are "nested"; for example, information can be coded by an inner code, and the resulting alphabet of legal code words can be treated as a "symbol" subjected to an outer code. An example might be the use of a Hamming SECSED code as the inner code word and a Reed–Solomon code as an outer code scheme.
   *Check sum.* The sum of all the numbers in a block of words is added, modulo 2, and the block and the sum are transmitted. The words in the received block are added again and the check sum is recomputed and checked with the transmitted sum. This is an error-detecting code.
   *Duplication.* One can always transmit the result twice and check the two copies. Although this may be inefficient, it is the only technique in some cases: for example, if we wish to check logical operations, AND, OR, and NAND.
   *Fire code.* An interleaved code for burst errors. The similar Reed–

Solomon code is now more popular since it is somewhat more efficient.

*Hamming codes.* Other codes in the family use more error-correcting and -detecting bits, thereby achieving higher levels of fault tolerance.

*IC chip parity.* Can be one bit per word, one bit per byte, or interlaced parity where $b$ bits are watched over by $i$ check bits. Thus each check bit "watches over" $b/i$ bits.

*Interleaving.* One approach to dealing with burst codes is to disassemble codes into a number of words, then reassemble them so that one bit is chosen from each word. For example, one could take 8 bytes and interleave (also called interlace) the bits so that a new byte is constructed from all the first bits of the original 8 bytes, another is constructed from all the second bits, and so forth. In this example, as long as the burst length is less than 8 bits and we have only one burst per 8 bytes, we are guaranteed that each new word can contain at most one error.

*Residue* m *codes.* This is used for checking certain arithmetic operations, such as addition, multiplication, and shifting. One computes the code bits (residue, $R$) that are concatenated ($|$, i.e., appended) to the message $N$ to from $N|R$. The residue is the remainder left when $N/m$. After transmission or computation, the new message bits $N'$ are divided by $m$ to form $R'$. Disagreement of $R$ and $R'$ indicates an error.

*Viterby decoding.* A decoding algorithm for error correction of a Reed–Solomon or other convolutional code based on enumerating all the legal code words and choosing the one closest to the received words. For medium-sized search spaces, an organized search resembling a branching tree was devised by Viterbi in 1967; it is often used to shorten the search. Forney recognized in 1968 that such trees are repetitive, so he devised an improvement that led to a diagram looking like a "lattice" used for supporting plants and trees.

## REFERENCES

AHA Reliability Report No. 4011. Reed–Solomon Coder/Decoder. Advanced Hardware Architectures, Inc., Pullman, WA, 1995.

AHA Applications Note. Primer: Reed–Solomon Error Correction Codes (ECC). Advanced Hardware Architectures, Pullman, WA, Inc.

Arazi, B. *A Commonsense Approach to the Theory of Error Correcting Codes.* MIT Press, Cambridge, MA, 1988.

Forney, G. D. Jr. *Concatenated Codes.* MIT Press Research Monograph, no. 37. MIT Press, Cambridge, MA, 1966.

Golomb, S. W. Optical Disk Error Correction. *Byte* (May 1986): 203–210.

Gravano, S. *Introduction to Error Control Codes.* Oxford University Press, New York, 2000.

Hamming, R. W. Error Detecting and Correcting Codes. *Bell System Technical Journal* 29 (April 1950): 147–160.

Houghton, A. D. *The Engineer's Error Coding Handbook*. Chapman and Hall, New York, 1997.

Johnson, B. W. *Design and Analysis of Fault Tolerant Digital Systems*. Addison-Wesley, Reading, MA, 1989.

Johnson, B. W. *Design and Analysis of Fault Tolerant Digital Systems*, 2d ed. Addison-Wesley, Reading, MA, 1994.

Jones, G. A., and J. M. Jones. *Information and Coding Theory*. Springer-Verlag, New York, 2000.

Lala, P. K. *Fault Tolerant and Fault Testable Hardware Design*. Prentice-Hall, Englewood Cliffs, NJ, 1985.

Lala, P. K. *Self-Checking and Fault-Tolerant Digital Design*. Academic Press, San Diego, CA, 2000.

Lee, C. *Error-Control Block Codes for Communications Engineers*. Telecommunication Library, Artech House, Norwood, MA, 2000.

Peterson, W. W. *Error-Correcting Codes*. MIT Press (Cambridge, MA) and Wiley (New York), 1961.

Peterson, W. W., and E. J. Weldon Jr. *Error Correcting Codes*, 2d ed. MIT Press, Cambridge, MA, 1972.

Pless, V. *Introduction to the Theory of Error-Correcting Codes*. Wiley, New York, 1998.

Pradhan, D. K. *Fault-Tolerant Computing Theory and Technique*, vol. I. Prentice-Hall, Englewood Cliffs, NJ, 1986.

Pradhan, D. K. *Fault-Tolerant Computing Theory and Technique*, vol. II. Prentice-Hall, Englewood Cliffs, NJ, 1993.

Rao, T. R. N., and E. Fujiwara. *Error-Control Coding for Computer Systems*. Prentice-Hall, Englewood Cliffs, NJ, 1989.

Shooman, M. L. *Probabilistic Reliability: An Engineering Approach*. McGraw-Hill, New York, 1968.

Shooman, M. L. *Probabilistic Reliability: An Engineering Approach*, 2d ed. Krieger, Melbourne, FL, 1990.

Shooman, M. L. The Reliability of Error-Correcting Code Implementations. *Proceedings Annual Reliability and Maintainability Symposium*, Las Vegas, NV, January 22–25, 1996.

Shooman, M. L., and F. A. Cassara. The Reliability of Error-Correcting Codes on Wireless Information Networks. *International Journal of Reliability, Quality, and Safety Engineering*, special issue on Reliability of Wireless Communication Systems, 1996.

Siewiorek, D. P., and F. S. Swarz. *The Theory and Practice of Reliable System Design*. The Digital Press, Bedford, MA, 1982.

Siewiorek, D. P., and R. S. Swarz. *Reliable Computer Systems Design and Evaluation*, 2d ed. The Digital Press, Bedford, MA, 1992.

Siewiorek, D. P., and R. S. Swarz. *Reliable Computer Systems Design and Evaluation*, 3d ed. A. K. Peters, www.akpeters.com, 1998.

Spencer, J. L. The Highs and Lows of Reliability Predictions. *Proceedings Annual Reliability and Maintainability Symposium*, 1986. IEEE, New York, NY, pp. 156–162.

Stapper, C. H. et al. High-Reliability Fault-Tolerant 16-M Bit Memory Chip. *Proceedings Annual Reliability and Maintainability Symposium*, January 1991. IEEE, New York, NY, pp. 48–56.

Taylor, L. *Air Travel How Safe Is It?* BSP Professional Books, Cambridge, MA, 1988.

Texas Instruments, *TTL Logic Data Book*. 1988, pp. 2-597–2-599.

Wakerly, J. F. *Digital Design Principles and Practice.* Prentice-Hall, Englewood Cliffs, NJ, 1994.

Wakerly, J. F. *Digital Design Principles and Practice*, 3d. ed. Prentice-Hall, Englewood Cliffs, NJ, 2000.

Wells, R. B. *Applied Coding and Information Theory for Engineers.* Prentice-Hall, Englewood Cliffs, NJ, 1998.

Wicker, S. B., and V. K. Bhargava. *Reed–Solomon Codes and their Applications.* IEEE Press, New York, 1994.

Wiggert, D. *Codes for Error Control and Synchronization.* Communications and Defense Library, Artech House, Norwood, MA, 1998.

Wolf, J. J., M. L. Shooman, and R. R. Boorstyn. Algebraic Coding and Digital Redundancy. *IEEE Transactions on Reliability* R-18, 3 (August 1969): 91–107.

## PROBLEMS

**2.1.** Find a recent edition of *Jane's all the World's Aircraft* in a technical or public library. Examine the data given in Table 2.1 for soft failures for the 6 aircraft types listed. From the book, determine the approximate number of electronic systems (aircraft avionics) for each of the aircraft that are computer-controlled (digital rather than analog). You may have to do some intelligent estimation to determine this number. One section in the book gives information on the avionics systems installed. Also, it may help to know that the U.S. companies (all mergers) that provide most of the avionics systems are Bendix/King/Allied, Sperry/Honeywell, and Collins/Rockwell. (Hint: You may have to visit the Web sites of the aircraft manufacturers or the avionics suppliers for more details.

  **(a)** Plot the number of soft fails per aircraft versus the number of avionics systems on board. Comment on the results.

  **(b)** It would be better to plot soft fails per aircraft versus the number of words of main memory for the avionics systems on board. Do you have any ideas on how you could obtain such data?

**2.2.** Compute the minimum code distance for all the code words given in Table 2.2.

  **(a)** Compute for column (a) and comment.

  **(b)** Compute for column (b) and comment.

  **(c)** Compute for column (c) and comment.

**2.3.** Repeat the parity-bit coder and decoder designs given in Fig. 2.1 for an 8-bit word with 7 message bits and 1 parity bit. Does this approach to design of a coder and decoder present any difficulties?

**2.4.** Compare the design of problem 2.3 with that given in Fig. 2.2 on the basis of ease of design, complexity, practicality, delay time (assume all gates have a delay of $D$), and number of gates.

**2.5.** Compare the results of problem 2.4 with the circuit of Fig. 2.4.

**2.6.** Compute the binomial probabilities $B(r:8,q)$ for $r = 1$ to 8.
  (a) Does the sum of these probabilities check with Eq. (2.5)?
  (b) Show for what values of $q$ the term $B(1:8,q)$ dominates all the error-occurrence probabilities.

**2.7.** Find a copy of the latest military failure-rate manual (MIL-HDBK-217) and plot the data on Fig. B7 of Appendix B. Does it agree? Can you find any other IC failure-rate information? (Hint: The telecommunication industry and the various national telephone companies maintain large failure-rate databases. Also, the *Annual Reliability and Maintainability Symposium* from the IEEE regularly publishes papers with failure-rate data.) Does this data agree with the other results? What advances have been made in the last decade or so?

**2.8.** Assume that a 10% reduction in the probability of undetected error from coder and decoder failures is acceptable (see Sec. 2.3.4).
  (a) Compute the value of $B$ at which a 10% reduction occurs for fixed values of $q$.
  (b) Plot the results of part (a) and interpret.

**2.9.** Check the results given in Table 2.5. How is the distance $d$ related to the number of check bits? Explain.

**2.10.** Check the values given in Tables 2.7 and 2.8.

**2.11.** The Hamming SECSED code with 4 message bits and 3 check bits is used in the text as an example (Section 2.4.3). It was stated that we could use a brute force technique of checking all the legal or illegal code words for error detection, as was done for the parity-bit code in Fig. 2.1.
  (a) List all the legal and illegal code words for this example and show that the code distance is 3.
  (b) Design an error-detector circuit using minimized two-level logic (cf. Fig. 2.1).

**2.12.** Design a check bit generating circuit for problem 2.11 using Eqs. (2.22a–c) and EXOR gates.

**2.13.** One technique for error correction is to pick the nearest code word as the correct word once an error has been detected.

**80**  CODING TECHNIQUES

- (a) Devise a software algorithm that can be used to program a microprocessor to perform such error correction.
- (b) Devise a hardware design that performs the error correction by choosing the closest word.
- (c) Compare complexity and speed of designs (a) and (b).

**2.14.** An error-correcting circuit for a Hamming (7, 4) SECSED is given in Fig. 2.7. How would you generate the check bits that are defined in Eqs. (2.22a–c)? Is there a better way than that suggested in problem 2.12?

**2.15.** Compare the designs of problems 2.11, 2.12, and 2.13 with Hamming's technique in problem 2.14.

**2.16.** Give a complete design for the code generator and checker for a Hamming (12, 8) SECSED code following the approach of Fig. 2.7.

**2.17.** Repeat problem 2.16 for a SECDED code.

**2.18.** Repeat problem 2.8 for the design referred to in Table 2.14.

**2.19.** Retransmission as described in Section 2.5 tends to decrease the effective baud rate ($B$) of the transmission. Compare the unreliability and the effective baud rate for the following designs:
- (a) Transmit each word twice and retransmit when they disagree.
- (b) Transmit each word three times and use the majority of the three values to determine the output.
- (c) Use a parity-bit code and only retransmit when the code detects an error.
- (d) Use a Hamming SECSED code and only retransmit when the code detects an error.

**2.20.** Add the probabilities of generator and checker failure for the reliability examples given in Section 2.5.3.

**2.21.** Assume we are dealing with a burst code design for error detection with a word length of 12 bits and a maximum burst length of 4, as noted in Eqs. (2.46)–(2.50). Assume the code vector $V(x_1, x_2, \ldots, x_{12}) = V(c_1 c_2 c_3 c_4 10100011)$.
- (a) Compute $c_1 c_2 c_3 c_4$.
- (b) Assume no errors and show how the syndrome works.
- (c) Assume one error in bit $c_2$ and show how the syndrome works.
- (d) Assume one error in bit $x_9$; then show how the syndrome works.
- (e) Assume two errors in bits $x_8$ and $x_9$; then show how the syndrome works.
- (f) Assume three errors in bits $x_8$, $x_9$, and $x_{10}$; then show how the syndrome works.

(g) Assume four errors in bits $x_7$, $x_8$, $x_9$, and $x_{10}$; then show how the syndrome works.

(h) Assume five errors in bits $x_7$, $x_8$, $x_9$, $x_{10}$, and $x_{11}$; then show how the syndrome fails.

(i) Repeat the preceding computations using a different set of four equations to calculate the check bits.

**2.22.** Draw a circuit for generating the check bits, the syndrome vector, and the error-detection output for the burst error-detecting code example of Section 2.6.2.

(a) Use parallel computation and use EXOR gates.

(b) Use serial computation and a linear feedback shift register.

**2.23.** Compute the probability of undetected error for the code of problem 2.22 and compare with the probability of undetected error for the case of no error detection. Assume perfect hardware.

**2.24.** Repeat problem 2.23 assuming that the hardware is imperfect.

(a) Assume a model as in Section 2.3.4 and 2.4.5.

(b) Plot the results as in Figs. 2.5 and 2.8.

**2.25.** Repeat problem 2.22 for the burst error-detecting code in Section 2.6.3.

**2.26.** Repeat problem 2.23 for the burst error-detecting code in Section 2.6.3.

**2.27.** Repeat problem 2.24 for the burst error-detecting code in Section 2.6.3.

**2.28.** Analyze the design of Fig. 2.4 and show that it is equivalent to Fig. 2.2. Also, explain how it can be used as a generator and checker.

**2.29.** Explain in detail the operation of the error-correcting circuit given in Fig. 2.7.

**2.30.** Design a check bit generator circuit for the SECDED code example in Section 2.4.4.

**2.31.** Design an error-correcting circuit for the SECDED code example in Section 2.4.4.

**2.32.** Explain how a distance 3 code can be implemented as a double error-detecting code (DED). Give the circuit for the generator and checker.

**2.33.** Explain how a distance 4 code can be implemented as a triple error-detecting code (TED). Give the circuit for the generator and checker.

**2.34.** Construct a table showing the relationship between the burst length $t$, the auxiliary check bits $u$, the total number of check bits, the number of message bits, and the length of the code word. Use a tabular format similar to Table 2.7.

**82**  CODING TECHNIQUES

**2.35.** Show for the $u = 5$ and $t = 3$ code example given in Section 2.6.3 that after $x$ shifts, the leftmost bits of R2 and R3 in Fig. 2.10 agree.

**2.36.** Show a complete circuit for error correction that includes Fig. 2.10 in addition to a counter, a decoder, a bit-complementing circuit, and a corrected word storage register, as well as control logic.

**2.37.** Show a complete circuit for error correction that includes Fig. 2.10 in addition to a counter, an EXOR-complementing circuit, and a corrected word storage register, as well as control logic.

**2.38.** Show a complete circuit for error correction that includes Fig. 2.10 in addition to a counter, an EXOR-complementing circuit, and a circulating register for R1 to contain the corrected word, as well as control logic.

**2.39.** Explain how the circuit of Fig. 2.11 acts as a coder. Input the message bits; then show what is generated and which bits correspond to the auxiliary syndrome and which ones correspond to the real syndrome.

**2.40.** What additional circuitry is needed (if any) to supplement Fig. 2.11 to produce a coder. Explain.

**2.41.** Using Fig. 2.12 for the Reed–Solomon code, plot a graph similar to Fig. 2.8.

# 3

# REDUNDANCY, SPARES, AND REPAIRS

## 3.1 INTRODUCTION

This chapter deals with a variety of techniques for improving system reliability and availability. Underlying all these techniques is the basic concept of redundancy, providing alternate paths to allow the system to continue operation even when some components fail. Alternate paths can be provided by parallel components (or systems). The parallel elements can all be continuously operated, in which case all elements are powered up and the term *parallel redundancy* or *hot standby* is often used. It is also possible to provide one element that is powered up (on-line) along with additional elements that are powered down (standby), which are powered up and switched into use, either automatically or manually, when the on-line element fails. This technique is called *standby redundancy* or *cold redundancy*. These techniques have all been known for many years; however, with the advent of modern computer-controlled digital systems, a rich variety of ways to implement these approaches is available. Sometimes, system engineers use the general term *redundancy management* to refer to this body of techniques. In a way, the ultimate cold redundancy technique is the use of spares or repairs to renew the system. At this level of thinking, a spare and a repair are the same thing—except the repair takes longer to be effected. In either case for a system with a single element, we must be able to tolerate some system downtime to effect the replacement or repair. The situation is somewhat different if we have a system with two hot or cold standby elements combined with spares or repairs. In such a case, once one of the redundant elements fails *and we detect the failure*, we can replace or repair the failed element while the system continues to operate; as long as the

## 84   REDUNDANCY, SPARES, AND REPAIRS

replacement or repair takes place before the operating element fails, the system never goes down. The only way the system goes down is for the remaining element(s) to fail before the replacement or repair is completed.

This chapter deals with conventional techniques of improving system or component reliability, such as the following:

1. Improving the manufacturing or design process to significantly lower the system or component failure rate. Sometimes innovative engineering does not increase cost, but in general, improved reliability requires higher cost or increases in weight or volume. In most cases, however, the gains in reliability and decreases in life-cycle costs justify the expenditures.

2. Parallel redundancy, where one or more extra components are operating and waiting to take over in case of a failure of the primary system. In the case of two computers and, say, two disk memories, synchronization of the primary and the extra systems may be a bit complex.

3. A standby system is like parallel redundancy; however, power is off in the extra system so that it cannot fail while in standby. Sometimes the sensing of primary system failure and switching over to the standby system is complex.

4. Often the use of replacement components or repairs in conjunction with parallel or standby systems increases reliability by another substantial factor. Essentially, once the primary system fails, it is a race to fix or replace it before the extra system(s) fails. Since the repair rate is generally much higher than the failure rate, the repair almost always wins the race, and reliability is greatly increased.

Because fault-tolerant systems generally have very low failure rates, it is hard and expensive to obtain failure data from tests. Thus second-order factors, such as common mode and dependent failures, may become more important than they usually are.

The reader will need to use the concepts of probability in Appendix A, Sections A1–A6.3 and those of reliability in Appendix B3 for this chapter. Markov modeling will appear later in the chapter; thus the principles of the Markov model given in Appendices A8 and B6 will be used. The reader who is unfamiliar with this material or needs review should consult these sections.

If we are dealing with large complex systems, as is often the case, it is expedient to divide the overall problem into a number of smaller subproblems (the "divide and conquer" strategy). An approximate and very useful approach to such a strategy is the method of apportionment discussed in the next section.

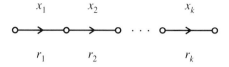

**Figure 3.1** A system model composed of $k$ major subsystems, all of which are necessary for system success.

## 3.2 APPORTIONMENT

One might conceive system design as an optimization problem in which one has a budget of resources (dollars, pounds, cubic feet, watts, etc.), and the goal is to achieve the highest reliability within the constraints of the available budget. Such an approach is discussed in Chapter 7; however, we need to use some of the simple approaches to optimization as a structure for comparison of the various methods discussed in this chapter. Also, in a truly large system, there are too many possible combinations of approach; a top–down design philosophy is therefore useful to decompose the problem into simpler subproblems. The technique of apportionment serves well as a "divide and conquer" strategy to break down a large problem.

Apportionment techniques generally assume that the highest level—the overall system—can be divided into 5–10 major subsystems, all of which must work for the system to work. Thus we have a series structure as shown in Fig. 3.1.

We denote $x_1$ as the event success of element (subsystem) 1, $x_1'$ is the event failure of element 1, $P(x_1) = 1 - P(x_1')$ is the probability of success (the reliability, $r_1$). The system reliability is given by

$$R_s = P(x_1 \cap x_2 \cdots \cap x_k) \tag{3.1a}$$

and if we use the more common engineering notation, this equation becomes

$$R_s = P(x_1 x_2 \cdots x_k) \tag{3.1b}$$

If we assume that all the elements are independent, Eq. (3.1a) becomes

$$R_s = \prod_{i=1}^{k} r_i \tag{3.2}$$

To illustrate the approach, let us assume that the goal is to achieve a system reliability equal to or greater than the system goal, $R_0$, within the cost budget, $c_0$. We let the single constraint be cost, and the total cost, $c$, is given by the sum of the individual component costs, $c_i$.

$$c = \sum_{i=1}^{k} c_i \tag{3.3}$$

We assume that the system reliability given by Eq. (3.2) is below the system specification or goal, and that the designer must improve the reliability of the system. We further assume that the maximum allowable system cost, $c_0$, is generally sufficiently greater than $c$ so that the system reliability can be improved to meet its reliability goal, $R_s \geq R_0$; otherwise, the goal cannot be reached, and the best solution is the one with the highest reliability within the allowable cost constraint.

Assume that we have a method for obtaining optimal solutions and, in the case where more than one solution exceeds the reliability goal within the cost constraint, that it is useful to display a number of "good" solutions. The designer may choose to just meet the reliability goal with one of the suboptimal solutions and save some money. Alternatively, there may be secondary factors that favor a good suboptimal solution. Lastly, a single optimum value does not give much insight into how the solution changes if some of the cost or reliability values assumed as parameters are somewhat in error. A family of solutions and some sensitivity studies may reveal a good suboptimal solution that is less sensitive to parameter changes than the true optimum.

A simple approach to solving this problem is to assume an equal apportionment of all the elements $r_i = r_1$ to achieve $R_0$ will be a good starting place. Thus Eq. (3.2) becomes

$$R_0 = \prod_{i=1}^{k} r_i = (r_1)^k \tag{3.4}$$

and solving for $r_1$ yields

$$r_1 = (R_0)^{1/k} \tag{3.5}$$

Thus we have a simple approximate solution for the problem of how to apportion the subsystem reliability goals based on the overall system goal. More details of such optimization techniques appear in Chapter 7.

## 3.3 SYSTEM VERSUS COMPONENT REDUNDANCY

There are many ways to implement redundancy. In Shooman [1990, Section 6.6.1], three different designs for a redundant auto-braking system are compared: a split system, which presently is used on American autos either front/rear or LR–RF/RR–LF diagonals; two complete systems; or redundant components (e.g., parallel lines). Other applications suggest different possibilities. Two redundancy techniques that are easily classified and studied are component and system redundancy. In fact, one can prove that component redundancy is superior to system redundancy in a wide variety of situations.

Consider the three systems shown in Fig. 3.2. The reliability expression for system (a) is

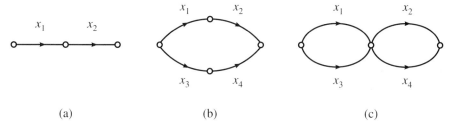

**Figure 3.2** Comparison of three different systems: (a) single system, (b) unit redundancy, and (c) component redundancy.

$$R_a(p) = P(x_1)P(x_2) = p^2 \tag{3.6}$$

where both $x_1$ and $x_2$ are independent and identical and $P(x_1) = P(x_2) = p$. The reliability expression for system (b) is given simply by

$$R_b(p) = P(x_1 x_2 + x_3 x_4) \tag{3.7a}$$

For independent identical units (IIU) with reliability of $p$,

$$R_b(p) = 2R_a - R_a^2 = p^2(2 - p^2) \tag{3.7b}$$

In the case of system (c), one can combine each component pair in parallel to obtain

$$R_b(p) = P(x_1 + x_3)P(x_2 + x_4) \tag{3.8a}$$

Assuming IIU, we obtain

$$R_c(p) = p^2(2 - p)^2 \tag{3.8b}$$

To compare Eqs. (3.8b) and (3.7b), we use the ratio

$$\frac{R_c(p)}{R_b(p)} = \frac{p^2(2-p)^2}{p^2(2-p^2)} = \frac{(2-p)^2}{(2-p^2)} \tag{3.9}$$

Algebraic manipulation yields

$$\frac{R_c(p)}{R_b(p)} = \frac{(2-p)^2}{(2-p^2)} = \frac{4 - 4p + p^2}{2 - p^2} = \frac{(2-p^2) + 2(1-p)^2}{2 - p^2} = 1 + \frac{2(1-p)^2}{2 - p^2}$$

$$\tag{3.10}$$

Because $0 < p < 1$, the term $2 - p^2 > 0$, and $R_c(p)/R_b(p) \geq 1$; thus component redundancy is superior to system redundancy for this structure. (Of course, they are equal at the extremes when $p = 0$ or $p = 1$.)

We can extend these chain structures into an $n$-element series structure, two parallel $n$-element system-redundant structures, and a series of $n$ structures of two parallel elements. In this case, Eq. (3.9) becomes

$$\frac{R_c(p)}{R_b(p)} = \frac{(2-p)^n}{(2-p^n)} \qquad (3.11)$$

Roberts [1964, p. 260] proves by induction that this ratio is always greater than 1 and that component redundancy is superior regardless of the number of elements $n$.

The superiority of component redundancy over system redundancy also holds true for nonidentical elements; an algebraic proof is given in Shooman [1990, p. 282].

A simpler proof of the foregoing principle can be formulated by considering the system tie-sets. Clearly, in Fig. 3.2(b), the tie-sets are $x_1 x_2$ and $x_3 x_4$, whereas in Fig. 3.2(c), the tie-sets are $x_1 x_2$, $x_3 x_4$, $x_1 x_4$, and $x_3 x_2$. Since the system reliability is the probability of the union of the tie-sets, and since system (c) has the same two tie-sets as system (b) as well as two additional ones, the component redundancy configuration has a larger reliability than the unit redundancy configuration. It is easy to see that this tie-set proof can be extended to the general case.

The specific result can be broadened to include a large number of structures. As an example, consider the system of Fig. 3.3(a) that can be viewed as a simple series structure if the parallel combination of $x_1$ and $x_2$ is replaced by an equivalent branch that we will call $x_5$. Then $x_5$, $x_3$, and $x_4$ form a simple chain structure, and component redundancy, as shown in Fig. 3.3(b), is clearly superior. Many complex configurations can be examined in a similar manner. Unit and component redundancy are compared graphically in Fig. 3.4.

Another interesting case in which one can compare component and unit

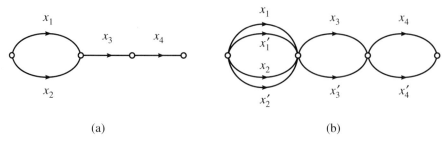

**Figure 3.3** Component redundancy: (a) original system and (b) redundant system.

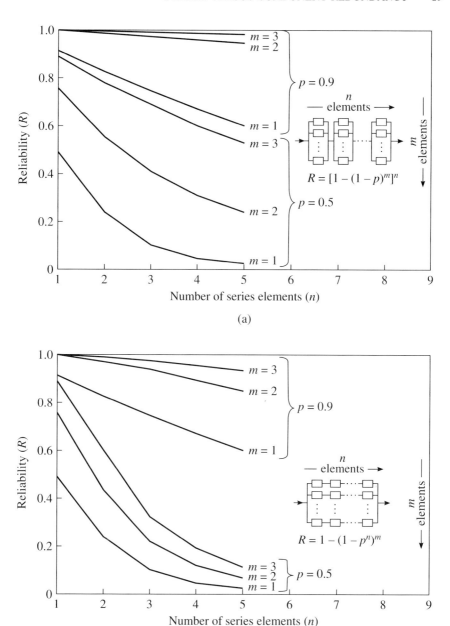

**Figure 3.4** Redundancy comparison: (a) component redundancy and (b) unit redundancy. [Adapted from Figs. 7.10 and 7.11, *Reliability Engineering*, ARINC Research Corporation, used with permission, Prentice-Hall, Englewood Cliffs, NJ, 1964.]

**90**  REDUNDANCY, SPARES, AND REPAIRS

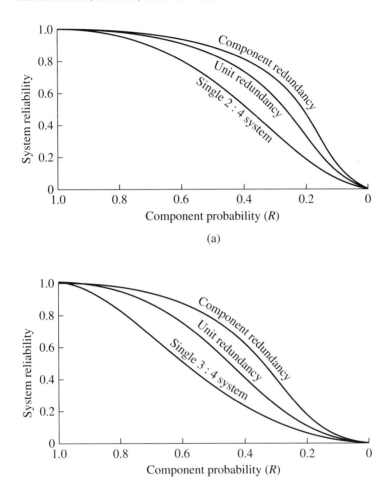

**Figure 3.5** Comparison of component and unit redundancy for $r$-out-of-$n$ systems: (a) a 2-out-of-4 system and (b) a 3-out-of-4 system.

redundancy is in an $r$-out-of-$n$ system (the system succeeds if $r$-out-of-$n$ components succeed). Immediately, one can see that for $r = n$, the structure is a series system, and the previous result applies. If $r = 1$, the structure reduces to $n$ parallel elements, and component and unit redundancy are identical. The interesting cases are then $2 \leq r < n$. The results for 2-out-of-4 and 3-out-of-4 systems are plotted in Fig. 3.5. Again, component redundancy is superior. The superiority of component over unit redundancy in an $r$-out-of-$n$ system is easily proven by considering the system tie-sets.

All the above analysis applies to two-state systems. Different results are obtained for multistate models; see Shooman [1990, p. 286].

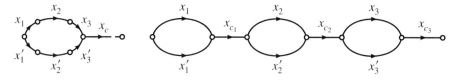

(a) System redundancy (one coupler)

(b) Component redundancy (three couplers)

**Figure 3.6** Comparison of system and component redundancy, including coupling.

In a practical case, implementing redundancy is a bit more complex than indicated in the reliability graphs used in the preceding analyses. A simple example illustrates the issues involved. We all know that public address systems consisting of microphones, connectors and cables, amplifiers, and speakers are notoriously unreliable. Using our principle that component redundancy is better, we should have two microphones that are connected to a switching box, and we should have two connecting cables from the switching box to dual inputs to amplifier 1 or 2 that can be selected from a front panel switch, and we select one of two speakers, each with dual wires from each of the amplifiers. We now have added the reliability of the switches in *series* with the parallel components, which lowers the reliability a bit; however, the net result should be a gain. Suppose we carry component redundancy to the extreme by trying to parallel the resistors, capacitors, and transistors in the amplifier. In most cases, it is far from simple to merely parallel the components. Thus how low a level of redundancy is feasible is a decision that must be left to the system designer.

We can study the required circuitry needed to allow redundancy; we will call such circuitry or components *couplers*. Assume, for example, that we have a system composed of three components and wish to include the effects of coupling in studying system versus component reliability by using the model shown in Fig. 3.6. (Note that the prime notation is used to represent a "companion" element, not a logical complement.) For the model in Fig. 3.6(a), the reliability expression becomes

$$R_a = P(x_1 x_2 x_3 + x_1' x_2' x_3') P(x_c) \tag{3.12}$$

and if we have IIU and $P(x_c) = Kp(x_c) = Kp$,

$$R_a = (2p^3 - p^6) Kp \tag{3.13}$$

Similarly, for Fig. 3.6(b) we have

$$R_b = P(x_1 + x_1') P(x_2 + x_2') P(x_3 + x_3') P(x_{c1}) P(x_{c2}) P(x_{c3}) \tag{3.14}$$

and if we have IIU and $P(x_{c1}) = P(x_{c2}) = P(x_{c3}) = Kp$,

$$R_b = (2p - p^2)^3 K^3 p^3 \tag{3.15}$$

We now wish to explore for what value of $K$ Eqs. (3.13) and (3.15) are equal:

$$(2p^3 - p^6)Kp = (2p - p^2)^3 K^3 p^3 \tag{3.16a}$$

Solving for $K$ yields

$$K^2 = \frac{(2p^3 - p^6)}{(2p - p^2)^3 p^2} \tag{3.16b}$$

If $p = 0.9$, substitution in Eq. (3.16) yields $K = 1.085778501$, and the coupling reliability $Kp$ becomes 0.9772006509. The easiest way to interpret this result is to say that if the component failure probability $1 - p$ is 0.1, then component and system reliability are equal if the coupler failure probability is 0.0228. In other words, if the coupler failure probability is less than 22.8% of the component failure probability, component redundancy is superior. Clearly, coupler reliability will probably be significant in practical situations.

Most reliability models deal with two element states—good and bad; however, in some cases, there are more distinct states. The classical case is a diode, which has three states: good, failed-open, and failed-shorted. There are also analogous elements, such as leaking and blocked hydraulic lines. (One could contemplate even more than three states; for example, in the case of a diode, the two "hard"-failure states could be augmented by an "intermittent" short-failure state.) For a treatment of redundancy for such three-state elements, see Shooman [1990, p. 286].

## 3.4 APPROXIMATE RELIABILITY FUNCTIONS

Most system reliability expressions simplify to sums and differences of various exponential functions once the expressions for the hazard functions are substituted. Such functions may be hard to interpret; often a simple computer program and a graph are needed for interpretation. Notwithstanding the case of computer computations, it is still often advantageous to have techniques that yield approximate analytical expressions.

### 3.4.1 Exponential Expansions

A general and very useful approximation technique commonly used in many branches of engineering is the *truncated series expansion*. In reliability work, terms of the form $e^{-z}$ occur time and again; the expressions can be simplified by series expansion of the exponential function. The Maclaurin series expansion of $e^{-Z}$ about $Z = 0$ can be written as follows:

$$e^{-Z} = 1 - Z + \frac{Z^2}{2!} - \frac{Z^3}{3!} + \cdots + \frac{(-Z)^n}{n!} + \cdots \qquad (3.17)$$

We can also write the series in $n$ terms and a remainder term [Thomas, 1965, p. 791], which accounts for all the terms after $(-Z)^n/n!$

$$e^{-Z} = 1 - Z + \frac{Z^2}{2!} - \frac{Z^3}{3!} + \cdots + \frac{(-Z)^n}{n!} + R_n(Z) \qquad (3.18)$$

where

$$R_n(Z) = (-1)^{n+1} \int_0^Z \frac{(Z - \xi)^n}{n!} e^{-\xi} \, d\xi \qquad (3.19)$$

We can therefore approximate $e^{-Z}$ by $n$ terms of the series and use $R_n(Z)$ to approximate the remainder. In general, we use only two or three terms of the series, since in the high-reliability region $e^{-Z} \sim 1$, $Z$ is small, and the high-order terms $Z^n$ in the series expansion becomes insignificant. For example, the reliability of two parallel elements is given by

$$(2e^{-Z}) + (-e^{-2Z}) = \left(2 - 2Z + \frac{2Z^2}{2!} - \frac{2Z^3}{3!} + \cdots + \frac{2(-Z)^n}{n!} + \cdots\right)$$

$$+ \left(-1 + 2Z - \frac{(2Z)^2}{2!} + \frac{(2Z)^3}{3!} - \cdots - \frac{(2Z)^n}{n!} + \cdots\right)$$

$$= 1 - Z^2 + Z^3 - \frac{7}{12} Z^4 + \frac{1}{4} Z^5 - \cdots + \qquad (3.20)$$

Two- and three-term approximations to Eqs. (3.17) and (3.20) are compared with the complete expressions in Fig. 3.7(a) and (b). Note that the two-term approximation is a "pessimistic" one, whereas the three-term expression is slightly "optimistic"; inclusion of additional terms will give a sequence of alternate upper and lower bounds. In Shooman [1990, p. 217], it is shown that the magnitude of the $n$th term is an upper bound on the error term, $R_n(Z)$, in an $n$-term approximation.

If the system being modeled involves repair, generally a Markov model is used, and oftentimes Laplace transforms are used to solve the Markov equations. In Section B8.3, a simplified technique for finding the series expansion of a reliability function—cf. Eq. (3.20)—directly from a Laplace transform is discussed.

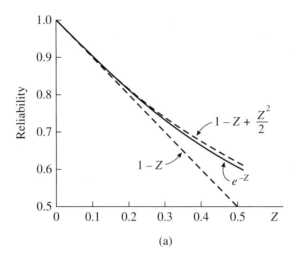

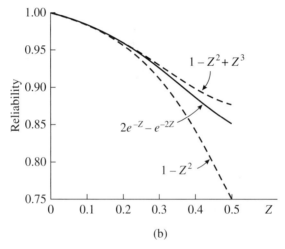

**Figure 3.7** Comparison of exact and approximate reliability functions: (a) single unit and (b) two parallel units.

### 3.4.2 System Hazard Function

Sometimes it is useful to compute and study the system hazard function (failure rate). For example, suppose that a system consists of two series elements, $x_2 x_3$, in parallel with a third, $x_1$. Thus, the system has two "success paths": it succeeds if $x_1$ works or if $x_2$ and $x_3$ both work. If all elements have identical constant hazards, $\lambda$, the reliability function is given by

$$R(t) = P(x_1 + x_2 x_3) = e^{-\lambda t} + e^{-2\lambda t} - e^{-3\lambda t} \qquad (3.21)$$

From Appendix B, we see that $z(t)$ is given by the density function divided by the reliability function, which can be written as the negative of the time derivative of the reliability function divided by the reliability function.

$$z(t) = \frac{f(t)}{R(t)} = -\frac{\dot{R}(t)}{R(t)} = \frac{\lambda(1 + 2e^{-\lambda t} - 3e^{-2\lambda t})}{1 + e^{-\lambda t} - e^{-2\lambda t}} \qquad (3.22)$$

Expanding $z(t)$ in a Taylor series,

$$z(t) = 1 + \lambda t - 3\lambda^2 t^2/2 + \cdots \qquad (3.23)$$

We can use such approximations to compare the equivalent hazard of various systems.

### 3.4.3 Mean Time to Failure

In the last section, it was shown that reliabiilty calculations become very complicated in a large system when there are many components and a diverse reliability structure. Not only was the reliability expression difficult to write down in such a case, but computation was lengthy, and interpretation of the individual component contributions was not easy. One method of simplifying the situation is to ask for less detailed information about the system. A useful figure of merit for a system is the mean time to failure (MTTF).

As was derived in Eq. (B51) of Appendix B, the MTTF is the expected value of the time to failure. The standard formula for the expected value involves the integral of $tf(t)$; however, this can be expressed in terms of the reliability function.

$$\text{MTTF} = \int_0^\infty R(t)\,dt \qquad (3.24)$$

We can use this expression to compute the MTTF for various configurations. For a series reliability configuration of $n$ elements in which each of the elements has a failure rate $z_i(t)$ and $Z(t) = \int z(t)\,dt$, one can write the reliability expression as

$$R(t) = \exp\left[-\sum_{i=1}^{n} Z_i(t)\right] \qquad (3.25a)$$

and the MTTF is given by

$$\text{MTTF} = \int_0^\infty \left\{\exp\left[-\sum_{i=1}^{n} Z_i(t)\right]\right\} dt \qquad (3.25b)$$

**96** REDUNDANCY, SPARES, AND REPAIRS

If the series system has components with more than one type of hazard model, the integral in Eq. (3.25b) is difficult to evaluate in closed form but can always be done using a series approximation for the exponential integrand; see Shooman [1990, p. 20].

Different equations hold for a parallel system. For two parallel elements, the reliability expression is written as $R(t) = e^{-Z_1(t)} + e^{-Z_2(t)} - e^{[-Z_1(t)+Z_2(t)]}$. If both system components have a constant-hazard rate, and we apply Eq. (3.24) to each term in the reliability expression,

$$\text{MTTF} = \frac{1}{\lambda_1} + \frac{1}{\lambda_2} - \frac{1}{\lambda_1 + \lambda_2} \tag{3.26}$$

In the general case of $n$ parallel elements with constant-hazard rate, the expression becomes

$$\text{MTTF} = \left(\frac{1}{\lambda_1} + \frac{1}{\lambda_2} + \cdots + \frac{1}{\lambda_n}\right) - \left(\frac{1}{\lambda_1 + \lambda_2} + \frac{1}{\lambda_1 + \lambda_3} + \cdots + \frac{1}{\lambda_i + \lambda_j}\right)$$
$$+ \left(\frac{1}{\lambda_1 + \lambda_2 + \lambda_3} + \frac{1}{\lambda_1 + \lambda_2 + \lambda_4} + \cdots + \frac{1}{\lambda_i + \lambda_j + \lambda_k}\right)$$
$$- \cdots + (-1)^{n+1} \frac{1}{\sum_{i=1}^{n} \lambda_i} \tag{3.27}$$

If the $n$ units are identical—that is, $\lambda_1 = \lambda_2 = \cdots = \lambda_n = \lambda$—then Eq. (3.27) becomes

$$\text{MTTF} = \left[\frac{\binom{n}{1}}{1} - \frac{\binom{n}{2}}{2} + \frac{\binom{n}{3}}{3} - \cdots + (-1)^{n+1} \frac{\binom{n}{n}}{n}\right]^{1/\lambda} = \frac{1}{\lambda} \sum_{i=1}^{n} \frac{1}{i} \tag{3.28a}$$

The preceding series is called the harmonic series; the summation form is given in Jolley [1961, p. 26, Eq. (200)] or Courant [1951, pp. 380]. This series occurs in number theory, and a series expansion is attributed to the famous mathematician Euler; the constant in the expansion (0.577) is called Euler's constant [Jolley, 1961, p. 14, Eq. (70)].

$$\frac{1}{\lambda} \sum_{i=1}^{n} \frac{1}{i} = \frac{1}{\lambda}\left[0.577 + \ln n + \frac{1}{2n} - \frac{1}{12n(n+1)} \cdots\right] \tag{3.28b}$$

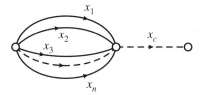

**Figure 3.8** Parallel reliability configuration of $n$ elements and a coupling device $x_c$.

## 3.5 PARALLEL REDUNDANCY

### 3.5.1 Independent Failures

One classical approach to improving reliability is to provide a number of elements in the system, any one of which can perform the necessary function. If a system of $n$ elements can function properly when only one of the elements is good, a parallel configuration is indicated. (A parallel configuration of $n$ items is shown in Fig. 3.8.) The reliability expression for a parallel system may be expressed in terms of the probability of success of each component or, more conveniently, in terms of the probability of failure (coupling devices ignored).

$$R(t) = P(x_1 + x_2 + \cdots + x_n) = 1 - P(\bar{x}_1 \bar{x}_2 \cdots \bar{x}_n) \tag{3.29}$$

In the case of constant-hazard components, $P_f = P(\bar{x}_i) = 1 - e^{-\lambda_i t}$, and Eq. (3.29) becomes

$$R(t) = 1 - \left[ \prod_{i=1}^{n} (1 - e^{-\lambda_i t}) \right] \tag{3.30}$$

In the case of linearly increasing hazard, the expression becomes

$$R(t) = 1 - \left[ \prod_{i=1}^{n} (1 - e^{-K_i t^2/2}) \right] \tag{3.31}$$

We recall that in the example of Fig. 3.6(a), we introduced the notion that a coupling device is needed. Thus, in the general case, the system reliability function is

$$R(t) = \left\{ 1 - \left[ \prod_{i=1}^{n} (1 - e^{-Z_i(t)}) \right] \right\} P(x_c) \tag{3.32}$$

If we have IIU with constant-failure rates, then Eq. (3.32) becomes

**98** REDUNDANCY, SPARES, AND REPAIRS

$$R(t) = [1 - (1 - e^{-\lambda t})^n]e^{-\lambda_c t} \tag{3.33a}$$

where $\lambda$ is the element failure rate and $\lambda_c$ is the coupler failure rate. Assuming $\lambda_c t < \lambda t \ll 1$, we can simplify Eq. (3.33) by approximating $e^{-\lambda_c t}$ and $e^{-\lambda t}$ by the first two terms in the expansion—cf. Eq. (3.17)—yielding $(1 - e^{-\lambda t}) \approx \lambda t$, $e^{-\lambda_c t} \approx 1 - \lambda_c t$. Substituting these approximations into Eq. (3.33a),

$$R(t) \approx [1 - (\lambda t)^n](1 - \lambda_c t) \tag{3.33b}$$

Neglecting the last term in Eq. (3.33b), we have

$$R(t) \approx 1 - \lambda_c t - (\lambda t)^n \tag{3.34}$$

Clearly, the coupling term in Eq. (3.34) must be small or it becomes the dominant portion of the probability of failure. We can obtain an "upper limit" for $\lambda_c$ if we equate the second and third terms in Eq. (3.34) (the probabilities of coupler failure and parallel system failure) yielding

$$\frac{\lambda_c}{\lambda} < (\lambda t)^{n-1} \tag{3.35}$$

For the case of $n = 3$ and a comparison at $\lambda t = 0.1$, we see that $\lambda_c/\lambda < 0.01$. Thus the failure rate of the coupling device must be less than $1/100$ that of the element. In this example, if $\lambda_c = 0.01\lambda$, then the coupling system probability of failure is equal to the parallel system probability of failure. This is a limiting factor in the application of parallel reliability and is, unfortunately, sometimes neglected in design and analysis. In many practical cases, the reliability of the several elements in parallel is so close to unity that the reliability of the coupling element dominates.

If we examine Eq. (3.34) and assume that $\lambda_c \approx 0$, we see that the number of parallel elements $n$ affects the curvature of $R(t)$ versus $t$. In general, the more parallelism in a reliability block diagram, the less the initial slope of the reliability curve. The converse is true with more series elements. As an example, compare the reliability functions for the three reliability graphs in Fig. 3.9 that are plotted in Fig. 3.10.

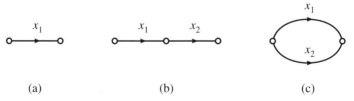

**Figure 3.9** Three reliability structures: (a) single element, (b) two series elements, and (c) two parallel elements.

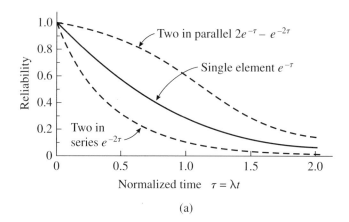

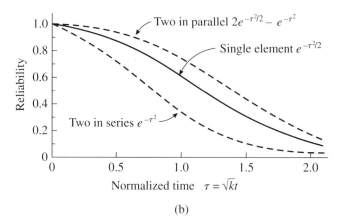

**Figure 3.10** Comparison of reliability functions: (a) constant-hazard elements and (b) linearly increasing hazard elements.

### 3.5.2 Dependent and Common Mode Effects

There are two additional effects that must be discussed in analyzing a parallel system: that of *common mode* (common cause) failures and that of *dependent failures*. A common mode failure is one that affects all the elements in a redundant system. The term was popularized when the first reliability and risk analyses of nuclear reactors were performed in the 1970s [McCormick, 1981, Chapter 12]. To protect against core melt, reactors have two emergency core-cooling systems. One important failure scenario—that of an earthquake—is likely to rupture the piping on both cooling systems.

Another example of common mode activity occurred early in the space program. During the reentry of a *Gemini* spacecraft, one of the two guidance computers failed, and a few minutes later the second computer failed. Fortunately, the astronauts had an additional backup procedure. Based on rehearsed procedures and precomputations, the Ground Control advised the astronauts to maneuver the spacecraft, to align the horizon with one of a set of horizontal scribe marks on the windows, and to rotate the spacecraft so that the Sun was aligned with one set of vertical scribe marks. The Ground Control then gave the astronauts a countdown to retro-rocket ignition and a second countdown to rocket cutoff. The spacecraft splashed into the ocean—closer to the recovery ship than in any previous computer-controlled reentry. Subsequent analysis showed that the temperature inside the two computers was much higher than expected and that the diodes in the separate power supply of each computer had burned out. From this example, we learn several lessons:

1. The designers provided two computers for redundancy.
2. Correctly, two separate power supplies were provided, one for each computer, to avoid a common power-supply failure mode.
3. An unexpectedly high ambient temperature caused identical failues in the diodes, resulting in a common mode failure.
4. Fortunately, there was a third redundant mode that depended on a completely different mechanism, the scribe marks, and visual alignment. When parallel elements are purposely chosen to involve devices with different failure mechanisms to avoid common mode failures, the term *diversity* is used.

In terms of analysis, common mode failures behave much like failures of a coupling mechanism that was studied previously. In fact, we can use Eq. (3.33) to analyze the effect if we use $\lambda_c$ to represent the sum of coupling and common mode failure rates. (A fortuitous choice of subscript!)

Another effect to consider in parallel systems is the effect of dependent failures. Suppose we wish to use two parallel satellite channels for reliable communication, and the probability of each channel failure is 0.01. For a single channel, the reliability would be 0.99; for two parallel channels, $c_1$ and $c_2$, we would have

$$R = P(c_1 + c_2) = 1 - P(\bar{c}_1 \bar{c}_2) \tag{3.36}$$

Expanding the last term in Eq. (3.36) yields

$$R = 1 - P(\bar{c}_1 \bar{c}_2) = 1 - P(\bar{c}_1)P(\bar{c}_2|\bar{c}_1) \tag{3.37}$$

If the failures of both channels, $c_1$ and $c_2$, are independent, Eq. (3.37) yields $R = 1 - 0.01 \times 0.01 = 0.9999$. However, suppose that one-quarter of satellite transmission failures are due to atmospheric interference that would affect both channels. In this case, $P(\bar{c}_2|\bar{c}_1)$ is 0.25, and Eq. (3.37) yields $R = 1 - 0.01 \times 0.25 = 0.9975$. Thus for a single channel, the probability of failure is

0.01; with two independent parallel channels, it is 0.0001, but for dependent channels, it is 0.0025. This means that dependency has reduced the expected 100-fold reduction in failure probabilities to a reduction by only a factor of 4. In general, a modeling of dependent failures requires some knowledge of the failure mechanisms that result in dependent modes.

The above analysis has explored many factors that must be considered in analyzing parallel systems: coupling failures, common mode failures, and dependent failures. Clearly, only simple models were used in each case. More complex models may be formulated by using Markov process models—to be discussed in Section 3.7, where we analyze standby redundancy.

## 3.6 AN $r$-OUT-OF-$n$ STRUCTURE

Another simple structure that serves as a useful model for many reliability problems is an $r$-out-of-$n$ structure. Such a model represents a system of $n$ components in which $r$ of the $n$ items must be good for the system to succeed. (Of course, $r$ is less than $n$.) An example of an $r$-out-of-$n$ structure is a fiber-optic cable, which has a capacity of $n$ circuits. If the application requires $r$ channels of the transmission, this is an $r$-out-of-$n$ system $(r:n)$. If the capacity of the cable $n$ exceeds $r$ by a significant amount, this represents a form of parallel redundancy. We are of course assuming that if a circuit fails it can be switched to one of the $n-r$ "extra circuits."

We may formulate a structural model for an $r$-out-of-$n$ system, but it is simpler to use the binomial distribution if applicable. The binomial distribution can be used only when the $n$ components are independent and identical. If the components differ or are dependent, the structural-model approach must be used. Success of exactly $r$-out-of-$n$ identical and independent items is given by

$$B(r:n) = \binom{n}{r} p^r (1-p)^{n-r} \qquad (3.38)$$

where $r:n$ stands for $r$ out of $n$, and the success of at least $r$-out-of-$n$ items is given by

$$P_s = \sum_{k=r}^{n} B(k:n) \qquad (3.39)$$

For constant-hazard components, Eq. (3.38) becomes

$$R(t) = \sum_{k=r}^{n} \binom{n}{k} e^{-k\lambda t}(1 - e^{-\lambda t})^{n-k} \qquad (3.40)$$

**102**  REDUNDANCY, SPARES, AND REPAIRS

Similarly, for linearly increasing or Weibull components, the reliability functions are

$$R(t) = \sum_{k=r}^{n} \binom{n}{k} e^{-kKt^2/2}(1 - e^{-Kt^2/2})^{n-k} \quad (3.41\text{a})$$

and

$$R(t) = \sum_{k=r}^{n} \binom{n}{k} e^{-kKt^{m+1}/(m+1)}(1 - e^{-Kt^{m+1}/(m+1)})^{n-k} \quad (3.41\text{b})$$

Clearly, Eqs. (3.39)–(3.41) can be studied and evaluated by a parametric computer study. In many cases, it is useful to approximate the result, although numerical evaluation via a computer program is not difficult. For an $r$-out-of-$n$ structure of identical components, the exact reliability expression is given by Eq. (3.38). As is well known, we can approximate the binomial distribution by the Poisson or normal distributions, depending on the values of $n$ and $p$ (see Shooman, 1990, Sections 2.5.6 and 2.6.8). Interestingly, we can also develop similar approximations for the case in which the $n$ parameters are not identical.

The Poisson approximation to the binomial holds for $p \leq 0.05$ and $n \geq 20$, which represents the low-reliability region. If we are interested in the high-reliability region, we switch to failure probabilities, requiring $q = 1 - p \leq 0.05$ and $n \geq 20$. Since we are assuming different components, we define average probabilities of success and failure $\bar{p}$ and $\bar{q}$ as

$$\bar{p} = \frac{1}{n} \sum_{i=1}^{n} p_i = 1 - \bar{q} = 1 - \frac{1}{n} \sum_{i=1}^{n} (1 - p_i) \quad (3.42)$$

Thus, for the high-reliability region, we compute the probability of $n-r$ or fewer failures as

$$R(t) = \sum_{k=0}^{n-r} \frac{(n\bar{q})^k e^{-n\bar{q}}}{k!} \quad (3.43)$$

and for the low-reliability region, we compute the probability of $r$ or more successes as

$$R(t) = \sum_{k=r}^{n} \frac{(n\bar{p})^k e^{-n\bar{p}}}{k!} \quad (3.44)$$

Equations (3.43) and (3.44) avoid a great deal of algebra in dealing with nonidentical $r$-out-of-$n$ components. The question of accuracy is somewhat difficult to answer since it depends on the system structure and the range of values of $p$ that make up $\bar{p}$. For example, if the values of $q$ vary only over a 2 : 1 range, and if $\bar{q} \le 0.05$ and $n \ge 20$, intuition tells us that we should obtain reasonably accurate results. Clearly, modern computer power makes explicit enumeration of Eqs. (3.39)–(3.41) a simple procedure, and Eqs. (3.43) and (3.44) are useful mainly as simplified analytical expressions that provide a check on computations. [Note that Eqs. (3.43) and (3.44) also hold true for IIU with $\bar{p} = p$ and $\bar{q} = q$.]

We can appreciate the power of an $r:n$ design by considering the following example. Suppose we have a fiber-optic cable with 20 channels (strands) and a system that requires all 20 channels for success. (For simplicity of the discussion, assume that the associated electronics will not fail.) Suppose the probability of failure of each channel within the cable is $q = 0.0005$ and $p = 0.9995$. Since all 20 channels are needed for success, the reliability of a 20-channel cable will be $R_{20} = (0.9995)^{20} = 0.990047$. Another option is to use two parallel 20-channel cables and associated electronics switch from cable A to cable B whenever there is any failure in cable A. The reliability of such an ordinary parallel system of two 20-channel cables is given by $R_{2/20} = 2(0.990047) - (0.990047)^2 = 0.9999009$. Another design option is to include extra channels in the single cable beyond the 20 that are needed—in such a case, we have an $r:n$ system. Suppose we approach the design in a trial-and-error fashion. We begin by trying $n = 21$ channels, in which case we have

$$R_{21} = B(21:21) + B(20:21) = p^{21}q^0 + 21p^{20}q$$
$$= (0.9995)^{21} + 21(0.9995)^{20}(0.0005) = 0.98755223 + 0.010395497$$
$$= 0.999947831 \tag{3.45}$$

Thus $R_{21}$ exceeds the design with two 20-channel cables. Clearly, all the designs require some electronic steering (couplers) for the choice of channels, and the coupler reliability should be included in a detailed comparison. Of course, one should worry about common mode failures, which could completely change the foregoing results. Construction damage—that is, line-severing by a contractor's excavating maching (backhoe)—is a significant failure mode for in-soil fiber-optic telephone lines.

As a check on Eq. (3.45), we compute the approximation Eq. (3.43) for $n = 21$, $r = 20$.

$$R(t) = \sum_{k=0}^{1} \frac{(n\bar{q})^k e^{-n\bar{q}}}{k!} = (1 + nq)e^{-nq} = [1 + 21(0.0005)]e^{-22 \times 0.0005}$$
$$= 0.999831687 \tag{3.46}$$

These values are summarized in Table 3.1.

**TABLE 3.1 Comparison of Design for Fiber-Optic Cable Example**

| System | Reliability, $R$ | Unreliability, $(1 - R)$ |
|---|---|---|
| Single 20-channel cable | 0.990047 | 0.00995 |
| Two 20-channel cables in parallel | 0.9999009 | 0.000099 |
| A 21-channel cable (exact) | 0.999948 | 0.000052 |
| A 21-channel cable (approx.) | 0.99983 | 0.00017 |

Essentially, the efficiency of the $r:n$ system is because the redundancy is applied at a lower level. In practice, a 24- or 25-channel cable would probably be used, since a large portion of the cable cost would arise from the land used and the laying of the cable. Therefore, the increased cost of including four or five extra channels would be "money well spent," since several channels could fail and be locked out before the cable failed. If we were discussing the number of channels in a satellite communications system, the major cost would be the launch; the economics of including a few extra channels would be similar.

## 3.7 STANDBY SYSTEMS

### 3.7.1 Introduction

Suppose we consider two components, $x_1$ and $x_1'$, in parallel. For discussion purposes, we can think of $x_1$ as the primary system and $x_1'$ as the backup; however, the systems are identical and could be interchanged. In an ordinary parallel system, both $x_1$ and $x_1'$ begin operation at time $t = 0$, and both can fail. If $t_1$ is the time to failure of $x_1$, and $t_2$ is the time to failure of $x_2$, then the time to system failure is the maximum value of $(t_1, t_2)$. An improvement would be to energize the primary system $x_1$ and have backup system $x_1'$ unenergized so that it cannot fail. Assume that we can immediately detect the failure of $x_1$ and can energize $x_1'$ so that it becomes the active element. Such a configuration is called a standby system, $x_1$ is called the *on-line* system, and $x_1'$ the *standby* system. Sometimes an ordinary parallel system is called a "hot" standby, and a standby system is called a "cold" standby. The time to system failure for a standby system is given by $t = t_1 + t_2$. Clearly, $t_1 + t_2 > \max(t_1, t_2)$, and a standby system is superior to a parallel system. The "coupler" element in a standby system is more complex than in a parallel system, requiring a more detailed analysis.

One can take a number of different approaches to deriving the equations for a standby system. One is to determine the probability distribution of $t = t_1 + t_2$, given the distributions of $t_1$ and $t_2$ [Papoulis, 1965, pp. 193–194]. Another approach is to develop a more general system of probability equations known

**TABLE 3.2  States for a Parallel System**

$s_0 = x_1 x_2 =$ Both components good.
$s_1 = x_1 \bar{x}_2 = x_1$, good; $x_2$, failed.
$s_2 = \bar{x}_1 x_2 = x_1$, failed; $x_2$, good.
$s_3 = \bar{x}_1 \bar{x}_2 =$ Both components failed.

---

as Markov models. This approach is developed in Appendix B and will be used later in this chapter to describe repairable systems.

In the next section, we take a slightly simpler approach: we develop two difference equations, solve them, and by means of a limiting process develop the needed probabilities. In reality, we are developing a simplified Markov model without going through some of the formalism.

### 3.7.2  Success Probabilities for a Standby System

One can characterize an ordinary parallel system with components $x_1$ and $x_2$ by the four states given in Table 3.2. If we assume that the standby component in a standby system won't fail until energized, then the three states given in Table 3.3 describe the system. The probability that element $x$ fails in time interval $\Delta t$ is given by the product of the failure rate $\lambda$ (failures per hour) and $\Delta t$. Similarly, the probability of no failure in this interval is $(1 - \lambda \Delta t)$. We can summarize this information by the probabilistic state model (probabilistic graph, Markov model) shown in Fig. 3.11.

The probability that the system makes a transition from state $s_0$ to state $s_1$ in time $\Delta t$ is given by $\lambda_1 \Delta t$, and the transition probability for staying in state $s_0$ is $(1 - \lambda_1 \Delta t)$. Similar expressions are shown in the figure for staying in state $s_1$ or making a transition to state $s_2$. The probabilities of being in the various system states at time $t = t + \Delta t$ are governed by the following difference equations:

$$P_{s_0}(t + \Delta t) = (1 - \lambda_1 \Delta t) P_{s_0}(t), \tag{3.47a}$$

$$P_{s_1}(t + \Delta t) = \lambda_1 \Delta t P_{s_0}(t) + (1 - \lambda_2 \Delta t) P_{s_1}(t) \tag{3.47b}$$

$$P_{s_2}(t + \Delta t) = \lambda_2 \Delta t P_{s_1}(t) + (1) P_{s_2}(t) \tag{3.47c}$$

We can rewrite Eq. (3.47) as

**TABLE 3.3  States for a Standby System**

$s_0 = x_1 x_2 =$ On-line and standby components good.
$s_1 = \bar{x}_1 x_2 =$ On-line failed and standby component good.
$s_2 = \bar{x}_1 \bar{x}_2 =$ On-line and standby components failed.

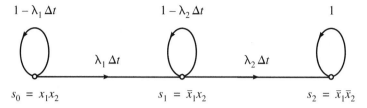

**Figure 3.11** A probabilistic state model for a standby system.

$$P_{s_0}(t + \Delta t) - P_{s_0}(t) = -\lambda_1 \Delta t P_{s_0}(t) \tag{3.48a}$$

$$\frac{P_{s_0}(t + \Delta t) - P_{s_0}(t)}{\Delta t} = -\lambda_1 P_{s_0}(t) \tag{3.48b}$$

Taking the limit of the left-hand side of Eq. (3.48b) as $\Delta t \to 0$ yields the time derivative, and the equation becomes

$$\frac{dP_{s_0}(t)}{dt} + \lambda_1 P_{s_0} = 0 \tag{3.49}$$

This is a linear, first-order, homogeneous differential equation and is known to have the solution $P_{s_0} = Ae^{-\lambda_1 t}$. To verify that this is a solution, we substitute into Eq. (3.49) and obtain

$$-\lambda_1 A e^{-\lambda_1 t} + \lambda_1 A e^{-\lambda_1 t} = 0$$

The value of $A$ is determined from the initial condition. If we start with a good system, $P_{s_0}(t = 0) = 1$; thus $A = 1$ and

$$P_{s_0} = e^{-\lambda_1 t} \tag{3.50}$$

In a similar manner, we can rewrite Eq. (3.47b) and take the limit obtaining

$$\frac{dP_{s_1}(t)}{dt} + \lambda P_{s_1}(t) = \lambda_1 P_{s_0} \tag{3.51}$$

This equation has the solution

$$P_{s_1}(t) = B_1 e^{-\lambda_1 t} + B_2 e^{-\lambda_2 t} \tag{3.52}$$

Substitution of Eq. (3.52) into Eq. (3.51) yields a group of exponential terms that reduces to

$$[\lambda_2 B_1 - \lambda_1 B_1 - \lambda_1]e^{-\lambda_1 t} = 0 \tag{3.53}$$

and solving for $B_1$ yields

$$B_1 = \frac{\lambda_1}{\lambda_2 - \lambda_1} \tag{3.54}$$

We can obtain the other constant by substituting the initial condition $P_{s_1}(t=0) = 0$, and solving for $B_2$ yields

$$B_2 = -B_1 = \frac{\lambda_1}{\lambda_1 - \lambda_2} \tag{3.55}$$

The complete solution is

$$P_{s_1}(t) = \frac{\lambda_1}{\lambda_2 - \lambda_1}[e^{-\lambda_1 t} - e^{-\lambda_2 t}] \tag{3.56}$$

Note that the system is successful if we are in state 0 or state 1 (state 2 is a failure). Thus the reliability is given by

$$R(t) = P_{s_0}(t) + P_{s_1}(t) \tag{3.57}$$

Equation (3.57) yields the reliability expression for a standby system where the on-line and the standby components have two different failure rates. In the more general case, both the on-line and standby components have the same failure rate, and we have a small difficulty since Eq. (3.56) becomes 0/0. The standard approach in such cases is to use l'Hospital's rule from calculus. The procedure is to take the derivative of the numerator and the denominator separately with respect to $\lambda_2$; then to take the limit as $\lambda_2 \to \lambda_1$. This results in the expression for the reliability of a standby system with two identical on-line and standby components:

$$R(t) = e^{-\lambda t} + \lambda t e^{-\lambda t} \tag{3.58}$$

A few general comments are appropriate at this point.

1. The solution given in Eq. (3.58) can be recognized as the first two terms in the Poisson distribution, the probability of zero occurrences in time $t$ plus the probability of one occurrence in time $t$ hours, where $\lambda$ is the occurrence rate per hour. Since the "exposure time" for the standby component does not start until the on-line element has failed, the occurrences are a sequence in time that follows the Poisson distribution.
2. The model in Fig. 3.11 could have been extended to the right to incorporate a very large number of components and states. The general solution of such a model would have yielded the Poisson distribution.

**108** REDUNDANCY, SPARES, AND REPAIRS

3. A model could have been constructed composed of four states: ($x_1 x_2$, $x_1 \bar{x}_2$, $\bar{x}_1 x_2$, $\bar{x}_1 \bar{x}_2$). Solution of this model would yield the probability expressions for a parallel system. However, solution of a parallel system via a Markov model is seldom done except for tutorial purposes because the direct methods of Section 3.5 are simpler.
4. Generalization of a probabilistic graph, the resulting differential equations, the solution process, and the summing of appropriate probabilities leads to a generalized Markov model. This is further illustrated in the next section on repair.
5. In Section 3.8.2 and Chapter 4, we study the formulation of Markov models using a more general algorithm to derive the equations, and we use Laplace transforms to solve the equations.

### 3.7.3 Comparison of Parallel and Standby Systems

It is assumed that the reader has studied the material in Sections A8 and B6 that cover Markov models. We now compare the reliability of parallel and standby systems in this section. Standby systems are inherently superior to parallel systems; however, much of this superiority depends on the reliability of the standby switch. Also, the reliability of the coupler in a parallel system must also be considered in the comparison. The reliability of the standby system with an imperfect switch will require a more complex Markov model than that developed in the previous section, and such a model is discussed below.

The switch in a standby system must perform three functions:

1. It must have some sort of decision element or algorithm that is capable of sensing improper operation.
2. The switch must then remove the signal input from the on-line unit and apply it to the standby unit, and it must also switch the output as well.
3. If the element is an active one, the power must be transferred from the on-line to the standby element (see Fig. 3.12). In some cases, the input and output signals can be permanently connected to the two elements; only the power needs to be switched.

Often the decision unit and the input (and output) switch can be incorporated into one unit: either an analog circuit or a digital logic circuit or processor algorithm. Generally, the power switch would be some sort of relay or electronic switch, or it could be a mechanical device in the case of a mechanical, hydraulic, or pneumatic system. The specific implementation will vary with the application and the ingenuity of the designer.

The reliability expression for a two-element standby system with constant hazards and a perfect switch was given in Eqs. (3.50), (3.56), and (3.57) and for identical elements in Eq. (3.58). We now introduce the possibility that the switch is imperfect.

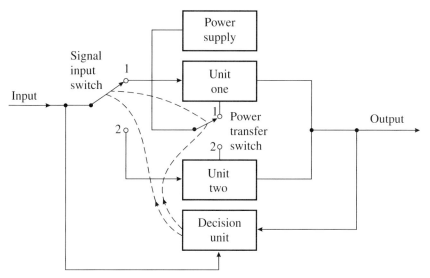

**Figure 3.12**  A standby system in which input and power switching are shown.

We begin with a simple model for the switch where we assume that any failure of the switch is a failure of the system, even in the case where both the on-line and the standby components are good. This is a conservative model that is easy to formulate. If we assume that the switch failures are independent of the on-line and standby component failures and that the switch has a constant failure rate $\lambda_s$, then Eq. (3.58) holds. Thus we obtain

$$R_1(t) = e^{-\lambda_s t}(e^{-\lambda t} + \lambda t e^{-\lambda t}) \qquad (3.59)$$

Clearly, the switch reliability multiplies the reliability of the standby system and degrades the system reliability. We can evaluate how significant the switch reliability problem is by comparing it with an ordinary parallel system. A comparison of Eqs. (3.59) and (3.30) (for $n = 2$ and identical failure rates) is given in Fig. 3.13. Note that when the switch failure rate is only 10% of the component failure rates ($\lambda_s = 0.1\lambda$), the degradation is only minor, especially in the high-reliability region of most interest: ($1 \geq R(t) \geq 0.9$). The standby system degrades to about the same reliability as the parallel system when the switch failure rate is about half the component failure rate.

A simple way to improve the switch reliability model is to assume that the switch failure mode is such that it only fails to switch from on-line to standby when the on-line element fails (it never switches erroneously when the on-line element is good). In such a case, the probability of no failures is a good state and the probability of one failure and no switch failure is also a good state, that is, the switch reliability only multiplies the second term in Eq. (3.58). In such a case, the reliability expression becomes

**110**   REDUNDANCY, SPARES, AND REPAIRS

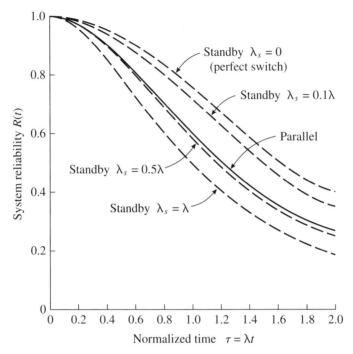

**Figure 3.13** A comparison of a two-element ordinary parallel system with a two-element standby system with imperfect switch reliability.

$$R_2(t) = e^{-\lambda t} + \lambda t e^{-\lambda t} e^{-\lambda_s t} \tag{3.60}$$

Clearly, this is less conservative and a more realistic switch model than the previous one.

One can construct even more complex failure models for the switch in a standby system [Shooman, 1990, Section 6.9].

1. Switch failure modes where the switching occurs even when the on-line element is good or where the switch jitters between elements can be included.
2. The failure rate of $n$ nonidentical standby elements was first derived by Bazovsky [1961, p. 117]; this can be shown as related to the gamma distribution and to approach the normal distribution for large $n$ [Shooman, 1990].
3. For $n$ identical standby elements, the system succeeds if there are $n-1$ or fewer failures, and the probabilities are given by the Poisson distribution that leads to the expression

$$R(t) = e^{-\lambda t} \sum_{i=0}^{n-1} \frac{(\lambda t)^i}{i!} \qquad (3.61)$$

## 3.8 REPAIRABLE SYSTEMS

### 3.8.1 Introduction

Repair or replacement can be viewed as the same process, that is, replacement of a failed component with a spare is just a fast repair. A complete description of the repair process takes into account several steps: (a) detection that a failure has occurred; (b) diagnosis or localization of the cause of the failure; (c) the delay for replacement or repair, which includes the logistic delay in waiting for a replacement component or part to arrive; and (d) test and/or recalibration of the system. In this section, we concentrate on modeling the basics of repair and will not decompose the repair process into a finer model that details all of these substates.

The decomposition of a repair process into substates results in a non-constant-repair rate (see Shooman [1990, pp. 348–350]). In fact, there is evidence that some repair processes lead to lognormal repair distributions or other nonconstant-repair distributions. One can show that a number of distributions (e.g., lognormal, Weibull, gamma, Erlang) can be used to model a repair process [Muth, 1967, Chapter 3]. Some software for modeling system availability permits nonconstant-failure and -repair rates. Only in special cases is such detailed data available, and constant-repair rates are commonly used. In fact, it is not clear how much difference there is in compiling the steady-state availability for constant- and nonconstant-repair rates [Shooman, 1990, Eq. (6.106) ff.]. For a general discussion of repair modeling, see Ascher [1984].

In general, repair improves two different measures of system performance: the reliability and the availability. We begin our discussion by considering a single computer and the following two different types of computer systems: an air traffic control system and a file server that provides electronic mail and network access to a group of users. Since there is only a single system, a failure of the computer represents a system failure, and repair will not affect the system *reliability* function. The *availability* of the system is a measure of how much of the operating time the system is up. In the case of the air traffic control system, the fact that the system may occasionally be down for short time periods while repair or replacement goes on may not be tolerable, whereas in the case of the file server, a small amount of downtime may be acceptable. Thus a computation of both the reliability and the availability of the system is required; however, for some critical applications, the most important measure is the reliability. If we say the basic system is composed of two computers in parallel or standby, then the problem changes. In either case, the system can tolerate one computer failure and stay up. It then becomes a race to see if the failed element can be repaired and restored before the remaining element fails. The system only goes down in the rare event that the second component fails before the repair or replacement is completed.

In the following sections, we will model a two-element parallel and a two-element standby system with repair and will comment on the improvements in reliability and availability due to repair. To facilitate the solutions of the ensuing Markov models, some simple features of the Laplace transform method will be employed. It is assumed that the reader is familiar with Laplace transforms or will have already read the brief introduction to Laplace transform methods given in Appendix B, Section B8. We begin our discussion by developing a general Markov model for two elements with repair.

### 3.8.2 Reliability of a Two-Element System with Repair

The benefits of repair in improving system reliability are easy to illustrate in a two-element system, which is the simplest system used in high-reliability fault-tolerant situations. Repair improves both a hot standby and a cold standby system. In fact, we can use the same Markov model to describe both situations if we appropriately modify the transition probabilities. A Markov model for two parallel or standby systems with repair is given in Fig. 3.14. The transition rate from state $s_0$ to $s_1$ is given by $2\lambda$ in the case of an ordinary parallel system because two elements are operating and either one can fail. In the case of a standby system, the transition is given by $\lambda$ since only one component is powered and only that one can fail (for this model, we ignore the possibility that the standby system can fail). The transition rate from state $s_1$ to $s_0$ represents the repair process. If only one repairman is present (the usual case), then this transition is governed by the constant repair rate $\mu$. In a rare case, more than one repairman will be present, and if all work cooperatively, the repair rate is $>\mu$. In some circumstances, there will be only a shared repairman among a number of equipments, in which case the repair rate is $<\mu$.

In many cases, study of the repair statistics shows a nonexponential distribution (the exponential distribution is the one corresponding to a constant transition rate)—specifically, the lognormal distribution [Ascher, 1984; Shooman, 1990, pp. 348–350]. However, much of the benefits of repair are illustrated by

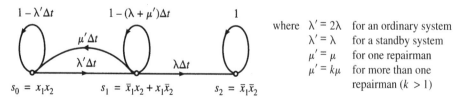

**Figure 3.14** A Markov reliability model for two identical parallel elements and $k$ repairmen.

the constant transition rate repair model. The Markov equations corresponding to Fig. 3.14 can be written by utilizing a simple algorithm:

1. The terms with 1 and $\Delta t$ in the Markov graph are deleted.
2. A first-order Markov differential equation is written for each node where the left-hand side of the equation is the first-order time derivative of the probability of being in that state at time $t$.
3. The right-hand side of each equation is a sum of probability terms for each branch that enters the node in question. The coefficient of each probability term is the transition probability for the entering branch.

We will illustrate the use of these steps in formulating the Markov of Fig. 3.14.

$$\frac{dP_{s_0}(t)}{dt} = -\lambda' P_{s_0}(t) + \mu' P_{s_1}(t) \tag{3.62a}$$

$$\frac{dP_{s_1}(t)}{dt} = \lambda' P_{s_0}(t) - (\lambda + \mu') P_{s_1}(t) \tag{3.62b}$$

$$\frac{dP_{s_2}(t)}{dt} = \lambda' P_{s_1}(t) \tag{3.62c}$$

Assuming that both systems are initially good, the initial conditions are

$$P_{s_0}(0) = 1, \qquad P_{s_1}(0) = P_{s_2}(0) = 0$$

One great advantage of the Laplace transform method is that it deals simply with initial conditions. Another is that it transforms differential equations in the time domain into a set of algebraic equations in the Laplace transform domain (often called the frequency domain), which are written in terms of the Laplace operator $s$.

To transform the set of equations (3.62a–c) into the Laplace domain, we utilize transform theorem 2 (which incorporates initial conditions) from Table B7 of Appendix B, yielding

$$sP_{s_0}(s) - 1 = -\lambda' P_{s_0}(s) + \mu' P_{s_1}(s) \tag{3.63a}$$
$$sP_{s_1}(s) - 0 = \lambda' P_{s_0}(s) - (\lambda + \mu') P_{s_1}(s) \tag{3.63b}$$
$$sP_{s_2}(s) - 0 = \lambda P_{s_1}(s) \tag{3.63c}$$

Writing these equations in a more symmetric form yields

**114** REDUNDANCY, SPARES, AND REPAIRS

$$(s + \lambda')P_{s_0}(s) - \mu'P_{s_1}(s) = 1 \tag{3.64a}$$

$$-\lambda'P_{s_0}(s) + (s + \mu' + \lambda)P_{s_1}(s) = 0 \tag{3.64b}$$

$$-\lambda P_{s_1}(s) + sP_{s_2}(s) = 0 \tag{3.64c}$$

Clearly, Eqs. (3.64a–c) lead to a matrix formulation if desired. However, we can simply solve these equations using Cramer's rule since they are now algebraic equations.

$$P_{s_0}(s) = \frac{(s + \lambda + \mu')}{[s^2 + (\lambda + \lambda' + \mu')s + \lambda\lambda']} \tag{3.65a}$$

$$P_{s_1}(s) = \frac{\lambda'}{[s^2 + (\lambda + \lambda' + \mu')s + \lambda\lambda']} \tag{3.65b}$$

$$P_{s_2}(s) = \frac{\lambda\lambda'}{s[s^2 + (\lambda + \lambda' + \mu')s + \lambda\lambda']} \tag{3.65c}$$

We must now invert these equations—transform them from the frequency domain to the time domain—to find the desired time solutions. There are several alternatives at this point. One can apply transform No. 10 from Table B6 of Appendix B to Eqs. (3.65a, b) to obtain the solution as a sum of two exponentials, or one can use a partial fraction expansion as illustrated in Eq. (B104) of the appendix. An algebraic solution of these equations using partial fractions appears in Shooman [1990, pp. 341–342], and further solution and plotting of these equations is covered in the problems at the end of this chapter as well as in Appendix B8. One can, however, make a simple comparison of the effects of repair by computing the MTTF for the various models.

### 3.8.3 MTTF for Various Systems with Repair

Rather than compute the complete reliabiity function of the several systems we wish to compare, we can simplify the analysis by comparing the MTTF for these systems. Furthermore, the MTTF is given by an integral of the reliability function, and by using Laplace theory we can show [Section B8.2, Eqs. (B105)–(B106)] that the MTTF is just given by the limit of the Laplace transform expression as $s \to 0$.

For the model of Fig. 3.14, the reliability expression is the sum of the first two-state probabilities; thus, the MTTF is the limit of the sum of Eqs. (3.65a, b) as $s \to 0$, which yields

$$\text{MTTF} = \frac{\lambda + \mu' + \lambda'}{(\lambda\lambda')} \tag{3.66}$$

**TABLE 3.4 Comparison of MTTF for Several Systems**

| Element | Formula | For $\lambda = 1$, $\mu = 10$ |
|---|---|---|
| Single element | $1/\lambda$ | 1.0 |
| Two parallel elements—no repair | $1.5/\lambda$ | 1.5 |
| Two standby elements—no repair | $2/\lambda$ | 2.0 |
| Two parallel elements—with repair | $(3\lambda + \mu)/2\lambda^2$ | 6.5 |
| Two standby elements—with repair | $(2\lambda + \mu)/\lambda^2$ | 12.0 |

We substitute the various values of $\lambda'$ shown in Fig. 3.14 in the expression; since we are assuming a single repairman, $\mu' = \mu$. The MTTF for several systems is compared in Table 3.4. Note how repair strongly increases the MTTF of the last two systems in the table. For large $\mu/\lambda$ ratios, which are common in practice, the MTTF of the last two systems approaches $0.5\mu/\lambda^2$ and $\mu/\lambda^2$.

### 3.8.4 The Effect of Coverage on System Reliability

In Fig. 3.12, we portrayed a fairly complex block diagram for a standby system. We have already modeled the possibility of imperfection in the switching mechanism. In this section, we develop a model for imperfections in the decision unit that detects failures and switches from the on-line system to the standby system. In some cases, even in the $n$-ordinary parallel system (hot standby), it is not possible to have both systems fully connected, and a decision unit and switch are needed. Another way of describing this phenomenon is to say that the decision unit cannot detect 100% of all the on-line unit failures; it only "covers" (detects) the fraction $c$ ($0 < c < 1$) of all the possible failures. (The formulation of this concept is generally attributed to Bouricius, Carter, and Schneider [1969].) The problem is that if the decision unit does not detect a failure of the on-line unit, input and output remain connected to the failed on-line element. The result is a system failure, because although the standby unit is good, there is no indication that it must be switched into use. We can formulate a Markov model in Fig. 3.15, which allows us to evaluate the effect of coverage. (Compare with the model of Fig. 3.14.) In fact, we can use Fig. 3.15 to model the effects of coverage on either a hot or cold standby system. Note that the symbol $D$ stands for the decision unit correctly detecting a failure in the on-line unit, and the symbol $\overline{D}$ means that the decision unit has not been able to (failed to) detect a failure in the on-line unit. Also, a new arc has been added in the figure from the good state $s_0$ to the failed state $s_2$ for modeling the failure of the decision unit to "cover" a failure of the on-line element.

The Markov equations for Fig. 3.15 become the following:

# REDUNDANCY, SPARES, AND REPAIRS

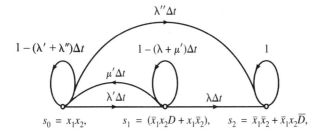

where  $\lambda' = 2c\lambda$ for an ordinary parallel system
$\lambda'' = 2(1-c)\lambda$ for an ordinary parallel system
$\lambda' = c\lambda$ for a standby system
$\lambda'' = (1-c)\lambda$ for a standby system
$\mu' = \mu$ for one repairman

**Figure 3.15** A Markov reliability model for two identical, parallel elements, $k$ repairmen, and coverage effects.

$$sP_{s_0}(s) - 1 = -(\lambda' + \lambda'')P_{s_0}(s) + \mu' P_{s_1}(s) \qquad (3.67a)$$
$$sP_{s_1}(s) - 0 = \lambda' P_{s_0}(s) - (\lambda + \mu')P_{s_1}(s) \qquad (3.67b)$$
$$sP_{s_2}(s) - 0 = \lambda'' P_{s_0}(s) + \lambda P_{s_1}(s) \qquad (3.67c)$$

Compare the preceding equations with Eqs. (3.63a–c) and (3.64a–c). Writing these equations in a more symmetric form yields

$$(s + \lambda' + \lambda'')P_{s_0}(s) - \mu' P_{s_1}(s) = 1 \qquad (3.68a)$$
$$-\lambda' P_{s_0}(s) + (s + \mu' + \lambda)P_{s_1}(s) = 0 \qquad (3.68b)$$
$$-\lambda'' P_{s_0}(s) - \lambda P_{s_1}(s) + sP_{s_2}(s) = 0 \qquad (3.68c)$$

The solution of these equations yields

$$P_{s_0}(s) = \frac{(s + \lambda + \mu')}{s^2 + (\lambda + \lambda' + \lambda'' + \mu')s + (\lambda\lambda' + \lambda''\mu' + \lambda\lambda'')} \qquad (3.69a)$$

$$P_{s_1}(s) = \frac{\lambda'}{s^2 + (\lambda + \lambda' + \lambda'' + \mu')s + (\lambda\lambda' + \lambda''\mu' + \lambda\lambda'')} \qquad (3.69b)$$

$$P_{s_2}(s) = \frac{\lambda''s + \lambda\lambda' + \mu'\lambda'' + \lambda\lambda''}{s[s^2 + (\lambda + \lambda' + \lambda'' + \mu')s + (\lambda\lambda' + \lambda''\mu' + \lambda\lambda'')]} \qquad (3.69c)$$

For the model of Fig. 3.15, the reliability expression is the sum of the first two-state probabilities; thus the MTTF is the limit of the sum of Eqs. (3.69a, b) as $s \to 0$, which yields

**TABLE 3.5 Comparison of MTTF for Several Systems**

| Element | Formula | For $\lambda = 1$, $\mu = 10$, $c = 1$ | For $\lambda = 1$, $\mu = 10$, $c = 0.95$ | For $\lambda = 1$, $\mu = 10$, $c = 0.90$ |
|---|---|---|---|---|
| Single element | $1/\lambda$ | 1.0 | — | — |
| Two parallel elements—no repair: $[\mu' = 0, \lambda' = 2c\lambda, \lambda'' = 2(1-c)\lambda]$ | $(0.5 + c)/\lambda$ | 1.5 | 1.45 | 1.40 |
| Two standby elements—no repair: $[\mu' = 0, \lambda' = c\lambda, \lambda'' = (1-c)\lambda]$ | $(1 + c)/\lambda$ | 2.0 | 1.95 | 1.90 |
| Two parallel elements—with repair: $[\mu' = \mu, \lambda' = 2c\lambda, \lambda'' = 2(1-c)\lambda]$ | $\dfrac{(1+2c)\lambda + \mu}{2\lambda[\lambda + (1-c)\mu]}$ | 6.5 | 4.3 | 3.2 |
| Two standby elements—with repair: $[\mu' = \mu, \lambda' = c\lambda, \lambda'' = (1-c)\lambda]$ | $\dfrac{(1+c)\lambda + \mu}{\lambda[\lambda + (1-c)\mu]}$ | 12.0 | 7.97 | 5.95 |

$$\text{MTTF} = \frac{\lambda + \mu' + \lambda'}{(\lambda\lambda' + \lambda''\mu' + \lambda\lambda'')} \tag{3.70}$$

When $c = 1$, $\lambda'' = 0$, and we see that Eq. (3.70) reduces to Eq. (3.66). The effect of coverage on the MTTF is evaluated in Table 3.5 by making appropriate substitutions for $\lambda'$, $\lambda''$, and $\mu'$. Notice what a strong effect the coverage factor has on the MTTF of the systems with repair. For two parallel and two standby systems, $c = 0.90$—more than half the MTTF. Practical values for $c$ are hard to find in the literature and are dependent on design. Sieworek [1992, p. 288] comments, "a typical diagnostic program, for example, may detect only 80–90% of possible faults." Bell [1978, p. 91] states that static testing of PDP-11 computers at the end of the manufacturing process was able to find 95% of faults, such as solder shorts, open-circuit etch connections, dead components, and incorrectly valued resistors. Toy [1987, p. 20] states, "realistic coverages range between 95% and 99%." Clearly, the value of $c$ should be a major concern in the design of repairable systems.

A more detailed treatment of coverage can be found in the literature. See Bouricius and Carter [1969, 1971]; Dugan [1989, 1996]; Kaufman and Johnson [1998]; and Pecht [1995].

### 3.8.5 Availability Models

In some systems, it is tolerable to have a small amount of downtime as the system is rapidly repaired and restored. In such a case, we allow repair out

# 118 REDUNDANCY, SPARES, AND REPAIRS

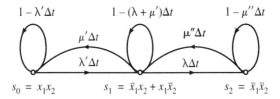

where  $\lambda' = 2\lambda$ for an ordinary system  
$\lambda' = \lambda$ for a standby system  
$\mu' = \mu$ for one repairman  
$\mu' = k_1\mu$ for more than one repairman ($k_1 > 1$)

$\mu'' = \mu$ for one repairman  
$\mu'' = 2\mu$ for two repairmen  
$\mu'' = k_2\mu$ for more than one repairman ($k_2 > 1$)

**Figure 3.16** Markov availability graph for two identical parallel elements.

of the system down state, and the model of Fig. 3.16 is obtained. Note that Fig. 3.14 and Fig. 3.16 only differ in the repair branch from state $s_2$ to state $s_1$. Using the same techniques that we used above, one can show that the equations for this model become

$$(s + \lambda')P_{s_0}(s) - \mu'P_{s_1}(s) = 1 \qquad (3.71\text{a})$$
$$-\lambda'P_{s_0}(s) + (s + \mu' + \lambda)P_{s_1}(s) - \mu''P_{s_2}(s) = 0 \qquad (3.71\text{b})$$
$$-\lambda P_{s_1}(s) + (s + \mu'')P_{s_2}(s) = 0 \qquad (3.71\text{c})$$

See Shooman [1990, Section 6.10] for more information.

The solution follows the same procedure as before. In this case, the sum of the probabilities for states 0 and 1 is not the reliability function but the availability function: $A(t)$. In most cases, $A(t)$ does not go to 0 as $t \to \infty$, as is true with the $R(t)$ function. $A(t)$ starts at 1 and, for well-designed systems, decays to a steady-state value close to 1. Thus a lower bound on the availability function is the steady-state value. A simple means for solving for the steady-state value is to formulate the differential equations for the Markov model and set all the time derivatives to 0. The set of equations now becomes an algebraic set of equations; however, the set is not independent. We obtain an independent set of equations by replacing any of these equations by the equation—the sum of all the state probabilities = 1. The algebraic solution for the steady-state availability is often used in practice. An even simpler procedure for computing the steady-state availability is to apply the final value theorem to the transformed expression for $A(s)$. This method is used in Section 4.9.2.

This chapter and Chapter 4 are linked in many ways. The technique of voting reliability joins parallel and standby system reliability as the three most common techniques for fault tolerance. Also, the analytical techniques involving Markov models are used in both chapters. In Chapter 4, a comparison is made of the reliability and availability of parallel, standby, and voting systems; in addition, some of the Markov modeling begun in this chapter is extended in Chapter 4 for the purpose of this comparison. The following chapter also has a more extensive discussion of the many shortcuts provided by Laplace transforms.

## 3.9 RAID SYSTEMS RELIABILITY

### 3.9.1 Introduction

The reliability techniques discussed in Chapter 2 involved coding to detect and correct errors in data streams. In this chapter, various parallel and standby techniques have been introduced that significantly increase the reliability of various systems and components. This section will discuss a newly developed technology for constructing computer secondary-storage systems that utilize the techniques of both Chapters 2 and 3 for the design of reliable, compact, high-performance storage systems. The generic term for such memory system technology is *redundant disk arrays* [Gibson, 1992]; however, it was soon changed to redundant array of inexpensive disks (RAID), and as technology evolved so that the quality and capacity of small disks rapidly increased, the word "inexpensive" was replaced by "independent." The term "array," when used in this context, means a collection of many disks organized in a specific fashion to improve speed of data transfer and reliability. As the RAID technology evolved, cache techniques (the use of small, very high-speed memories to accelerate processing by temporarily retaining items expected to be needed again soon) were added to the mix. Many varieties of RAID have been developed and more will probably emerge in the future. The RAID systems that employ cache techniques for speed improvement are sometimes called cached array of inexpensive disks (CAID) [Buzen, 1993]. The technology is driven by the variety of techniques available for connecting multiple disks, as well as various coding techniques, alternative read-and-write techniques, and the flexibility in organization to "tune" the architecture of the RAID system to match various user needs.

Prior to 1990, the dominant technology for secondary storage was a group of very large disks, typically 5–15, in a cabinet the size of a clothes washer. Buzen [1993] uses the term single large expensive disk (SLED) to refer to this technology. RAID technology utilizes a large number, typically 50–100, of small disks the size of those used in a typical personal computer. Each disk drive is assumed to have one actuator to position reads or writes, and large and small drives are assumed to have the same I/O read- or write-time. The bandwidth (BW) of such a disk is the reciprocal of the read-time. If data is broken into "chunks" and read (written) in parallel chunks to each of the $n$ small disks in a RAID array, the effective BW increases. There is some "overhead" in implementing such a parallel read-write scheme, however, in the limit:

$$\text{effective bandwidth} \to n\text{BW} \tag{3.72}$$

Thus, one possible beneficial effect of a RAID configuration in which many disks are written in parallel is a large increase in the BW.

If the RAID configuration depends on all the disks working, then the reliability of so many disks is lower than a smaller number of large disks. If the failure rate of each of the $n$ disks is denoted by $\lambda = 1/\text{MTTF}$, then the failure rate and MTTF of $n$ disks is given by

$$\text{effective failure rate} = n\lambda = 1/\text{effective MTTF} = n/\text{MTTF} \tag{3.73}$$

The failure rate is $n$ times as large and the MTTF is $n$ times smaller. If data is stored in "chunks" over many disks so that the write operation occurs in parallel for increased BW, the reliability of the block of data decreases significantly as per Eq. (3.73). Writing data in a distributed manner over a group of disks is called striping or interleaving. The size of the "chunk" is a design parameter in striping. To increase the reliability of a striped array, one can use redundant disks and/or error-detecting and -correcting codes for "chunks" of data of various sizes. We have purposely used the nonspecific term "chunk" because one of the design choices, which will soon be discussed, is the size of the "chunk" and how "the chunk" is distributed across various disks.

The various trade-offs among objectives and architectural approaches have changed over the decade (1990–2000) in which RAID storage systems were developed. At the beginning, small disks had modest capacity, longer access and transfer times, higher cost, and lower reliability. The improvements in all these parameters have had major effects on design.

The designers of RAID systems utilize various techniques of redundancy and error-correcting codes to raise the reliability of the RAID sysem [Buzen, 1993]. The early literature defined six levels of RAID [Patterson, 1987, 1988; Gibson, 1992], and most manufacturers followed these levels as guidelines in describing their products. However, as variations and options developed, classification became difficult, and some marketing managers took the classification system to mean that a higher level of RAID meant a better system. Thus, one vendor whose system included features of RAID 2 and RAID 5 decided to call his product RAID 10, claiming the levels multiplied! [Friedman, 1996.] Situations such as these led to the creation of the RAID Advisory Board, which serves as an industry standards body to define seven (and possibly more) levels of RAID [RAID, 1995; Massaglia, 1997]. The basic levels of RAID are given in Table 3.6, and the reader is cautioned to remember that because the various levels of RAID are to differentiate architectural approach, an increase in level does not necessarily correspond to an increase in BW or reliability. Complexity, however, does probably increase as the RAID level increases.

**TABLE 3.6 Summary Comparison of RAID Technologies**

| Level | Common Name | Features |
|---|---|---|
| 0 | No RAID or JBOD ("just a bunch of disks"). | No redundancy; thus, many claim that to consider this RAID is a misnomer. A Level 0 system could have a striped array and even a cache for speed improvement. There is, however, decreased reliability compared to a single disk if striping is employed, and the BW is increased. |
| 1 | Mirrored disks (duplexing, shadowing). | Two physical disk drives store identical copies of the data, one on each drive. This concept may be generalized to $n$ drives with $n$ identical copies or to $k$ sets of pairs with identical data. It is a simple scheme with high reliability and speed improvement, but there is high cost. |
| 2 | Hamming error-correcting code with bit-level interleaving. | Hamming SECSED (SECDED) code is computed on the data blocks and is striped across the disks. It is not often used in practice. |
| 3 | Parity-bit code at the bit level. | A parity-bit code is applied at the bit level and the parity bits are stored on a separate parity disk. Since parity bits are calculated for each strip, and strips appear on different disks, error detection is possible with a simple parity code. The parity disk must be accessed on all reads; generally, the disk spindles are synchronized. |
| 4 | Parity-bit code at the block level. | A parity-bit code is applied at the block level, and the parity bits are stored on a separate parity disk. |
| 5 | Parity-bit code at the sector level. | A parity-bit code is applied at the sector level and the parity information is distributed across the disks. |
| 6 | Parity-bit code at the bit level applied in two ways to provide correction when two disks fail. | Parity is computed in two different independent manners so that the array can recover from two disk failures. |

*Source:* [*The RAIDbook*, 1995].

## 3.9.2 RAID Level 0

This level was introduced as a means of classifying techniques that utilize a disk array and striping to improve the BW; however, no redundancy is included, and the reliability decreases. Equations (3.72) and (3.73) describe these basic effects. The BW of the array has increased over individual disks, but the reliability has decreased. Since high reliability is generally required in the disk storage system, this level would rarely be used except for special applications.

## 3.9.3 RAID Level 1

The use of mirrored disks is an obvious way to improve reliability; if the two disks are written in parallel, the BW is increased. If the data is striped across the two disks, the parallel reading of a transaction can increase the BW by a factor of 2. However, the second (backup) copy of the transaction must be written, and if there is a continuous transaction stream, the duplicate data copy requirement reduces the BW by a factor of 2, resulting in no change in the BW. However, if transactions occur in bursts with delays between bursts, the primary copy can be written at twice the BW during the burst, and the backup copy can be performed during the pauses between bursts. Thus the doubling of BW can be realized under those circumstances. Since memory systems represent 40%–60% of the cost of computer systems [Gibson, 1992, pp. 50–51], the use of mirrored disks greatly increases the cost of a computer system. Also, since the reliability is that of a parallel system, the reliability function is given by Eq. (3.8) and the MTTF by Eq. (3.26) for constant disk failure rates. If both disks are identical and have the same failure rates, the MTTF of the mirrored disks becomes 1.5 times greater than that of a single disk system. The Tandem line of "Nonstop" computers (discussed in Section 3.10.1) are essentially mirrored disks with the addition of duplicate computers, disk controllers, and I/O buses. *The RAIDbook* [1995, p. 45] calls the Tandem configuration a *fully redundant system*.

RAID systems of Level 2 and higher all have at least one hot spare disk. When a disk error is detected via an error-detecting code or other form of built-in disk monitoring, the disk system takes the remaining stored and redundant information and reconstructs a valid copy on the hot disk, which is switched-in, instead of the failed disk. Sometime later during maintenance, the failed disk is repaired or replaced. The differences among the following RAID levels are determined by the means of error detection, the size of the chunk that has associated error checking, and the pattern of striping.

## 3.9.4 RAID Level 2

This level of RAID introduces Hamming error-correcting codes similar to those discussed in Chapter 2 to detect and correct data errors. The error-correcting codes are added to the "chunks" of data and striped across the disks. In general, this level of RAID employs a SECSED code or a SECDED code such as one described in Chapter 2. The code is applied to data blocks, and the disk spindles must be synchronized. One can roughly compare the reliability of this scheme with a Level 1 system. For the Level 1 RAID system to fail, both disks must fail, and the probability of failure is

$$P_{f1} = q^2 \tag{3.74}$$

For a Level 2 system to fail, one of the two disks must fail that has a probability of $2q$, and the Hamming code must fail to detect an error. The example used in the *The RAIDbook* [1995] to discuss a Level 2 system is for ten data disks and four check disks, representing a 40% cost overhead for redundancy compared with a 100% overhead for a Level 1 system. In Chapter 2, we computed the probability of undetected error for eight data bits and four check bits in Eq. (2.27) and shall use these results to estimate the probability of failure of a typical Level 2 system. For this example,

$$P_{f2} = (2q) \times [220q^3(1-q)^9] \tag{3.75}$$

Clearly, for very small $q$, the Level 2 system has a smaller probability of failure. The two equations—(3.74) and (3.75)—are approximately equal for $q = 0.064$, at which level the probability of failure is $0.064^2 = 0.00041$.

To appreciate how this level would apply to a typical disk, let us assume that the MTTF for a typical disk is 300,000 hours. Assuming a constant failure-rate model, $\lambda = 1/300,000 = 3.3 \times 10^{-6}$. The associated probability of failure for a single disk would be $1 - \exp(-3.3 \times 10^{-6} t)$, and setting this expression to 0.064 shows that a single disk reaches this probability of failure at about 20,000 hours. Since a year is roughly 10,000 hours (8,766), a mirrored disk system would be superior for a few years of operation. A detailed reliability comparison would require a prior design of a Level 2 system with the appropriate number of disks, choice of chunk level (bit, byte, block, etc.), inclusion of a swapping disk, disk striping, and other details.

Detailed design of a Level 2 system such a disk system leads to nonstandard disks, significantly raising the cost of the system, and the technique is seldom used in practice.

### 3.9.5 RAID Levels 3, 4, and 5

In Chapter 2, we discussed the fact that a single parity bit is an inexpensive and fairly effective way of significantly increasing reliability. Levels 3, 4, and 5 apply such a parity-bit code to different size data "chunks" in various ways to increase the reliability of a disk array at a lower cost than a mirrored disk. We will model the reliability of a Level 3 system as an example. A disk can fail in

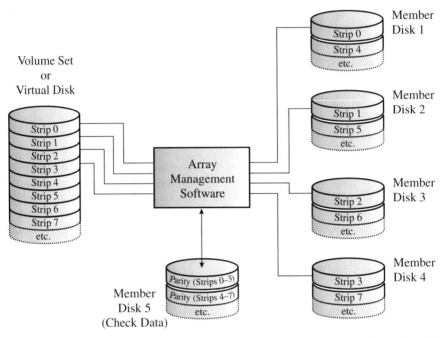

**Figure 3.17** A common mapping for a RAID Level 3 array [adapted from Fig. 48, *The RAIDbook*, 1995].

several ways: two are a surface failure (where stored bits are corrupted) and an actuator, head, or spindle failure (where the entire disk does not work—total failure). We assume that disk optimization software that periodically locks out bad bits on a disk generally protects against local surface failures, and the main category of failures requiring fault tolerance are total failures.

Normally, a single parity bit will provide an error-detecting but not an error-correcting code; however, the availability of parity checks for more than one group of strips provides error-correcting ability. Consider the typical example shown in Fig. 3.17. The parity disk computes a parity copy for strips (0–3) and (4–7) using the EXOR function:

$$P(0\text{–}3) = \text{strip } 0 \oplus \text{strip } 1 \oplus \text{strip } 2 \oplus \text{strip } 3 \quad (3.76)$$

$$P(4\text{–}7) = \text{strip } 4 \oplus \text{strip } 5 \oplus \text{strip } 6 \oplus \text{strip } 7 \quad (3.77)$$

Assume that there is a failure of disk 2, corrupting the data on strip 1 and strip 5. To regenerate the data on strip 1, we compute the EXOR of $P(0\text{–}3)$ along with strip 0, strip 2, and strip 3 that are on unfailed disks 5, 1, 3, 4.

$$\text{REGEN}(1) = P(0\text{–}3) \oplus \text{strip } 0 \oplus \text{strip } 2 \oplus \text{strip } 3 \quad (3.78a)$$

and substitution of Eq. (3.76) into Eq. (3.78a) yields

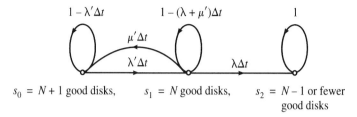

**Figure 3.18** A Markov model for $N + 1$ disks protected by a single parity disk.

$$\text{REGEN}(1) = (\text{strip } 0 \oplus \text{strip } 1 \oplus \text{strip } 2 \oplus \text{strip } 3)$$
$$\oplus (\text{strip } 0 \oplus \text{strip } 2 \oplus \text{strip } 3) \quad (3.78b)$$

Since strip $0 \oplus$ strip $0 = 0$, and similarly for strip 2 and strip 3, Eq. (3.78b) results in the regeneration of strip 1.

$$\text{REGEN}(1) = \text{strip } 1 \quad (3.78c)$$

The conclusion is that we can regenerate the information on strip 1, which was on the catastrophically failed disk 2 from the other unfailed disks. Clearly one could recover the other data for strip 5, which is also on failed disk 2 in a similar manner. These recovery procedures generalize to other Level 3, 4, and 5 recovery procedures. Allocating data to strips is called *striping*.

A Level 3 system has $N$ data disks that store the system data and one parity disk that stores the error-detection data for a total of $N + 1$ disks. The system succeeds if there are zero failures or one disk failure, since the damaged strips can be regenerated (repaired) using the above procedures. A Markov model for such operation is shown in Fig. 3.18. The solution follows the same path as that of Fig 3.14, and the same solution can be used if we set $\lambda' = (N + 1)\lambda$, $\lambda = N\lambda$, and $\mu' = \mu$. Substitution of these values into Eqs. (3.65a, b) and adding these probabilities yields the reliability function. Substitution into Eq. (3.66) yields the MTTF:

$$\text{MTTF} = [N\lambda + \mu + (N + 1)\lambda]/[N\lambda(N + 1)\lambda] \quad (3.79a)$$
$$\text{MTTF} = [(2N + 1)\lambda + \mu]/[N(N + 1)\lambda^2] \quad (3.79b)$$

These equations check with the model given in Gibson [1992, pp. 137–139]. In most cases, $\mu \gg \lambda$, and the MTTF expression given in Eq. (3.79b) becomes $\text{MTTF} = \mu/[N(N + 1)\lambda^2]$. If the recovery time were 1 hour, $N = 4$ as in the design of Fig. 3.17, and $\lambda = 1/300{,}000$ as previously assumed, then MTTF $= 4.5 \times 10^9$. Clearly, the recovery built into this example makes the loss of data very improbable. A comprehensive analysis would include the investigation of other possible modes of failure, common mode failures, and so forth. If one wishes to compute the availability of a RAID Level 3 system, a model similar to that given in Fig. 3.16 can be used.

## 3.9.6 RAID Level 6

There are several choices for establishing two independent parity checks. One approach is a horizontal–vertical parity approach. A parity bit for a string is computed in two different ways. Several rows of strings are written, from which a set of horizontal parity bits are computed for each row and a set of vertical parity bits are computed for each column. Actually, this description is just one approach to Level 6; any technique that independently computes two parity bits is classified as Level 6 (e.g., applying parity to two sets of bits, using two different algorithms for computing parity, and Reed–Solomon codes). For more comprehensive analysis of RAID systems, see Gibson [1992]. A comparison of the various RAID levels was given in Table 3.6, on page 121.

## 3.10 TYPICAL COMMERCIAL FAULT-TOLERANT SYSTEMS: TANDEM AND STRATUS

### 3.10.1 Tandem Systems

In the 1980s, Tandem captured a significant share of the business market with its "NonStop" computer systems. The name was a great asset, since it captured the aspirations of many customers in the on-line transaction processing market who required high reliability, such as banks, airlines, and financial institutions. Since 1997, Tandem Computers has been owned by the Compaq Computer Corporation, and it still stresses fault-tolerant computer systems. A 1999 survey estimates that 66% of credit card transactions, 95% of securities transactions, and 80% of automated teller machine transactions are processed by Tandem computers (now called NonStop Himalaya computers). "As late as 1985 it was estimated that a conventional, well-managed, transaction-processing system failed about once every two weeks for about an hour" [Siewiorek, 1992, p. 586]. Since there are 168 hours in a week, substitution into the basic steady-state equation for availability Eq. (B95a) yields an availability of 0.997. (Remember that $\lambda = 1/\text{MTTF}$ and $\mu = 1/\text{MTTR}$ for constant failure and repair rates.) To appreciate how mediocre such an availability is for a high-reliability system, let us consider the availability of an automobile. Typically an auto may require one repair per year (sometimes referred to as nonscheduled maintenance to eliminate inclusion of scheduled maintenance, such as oil changes, tire replacements, and new spark plugs), which takes one day (drop-off to pickup time). The repair rate becomes 1 per day; the failure rate, 1/365 per day. Substitution into Eq. (B95a) yields a steady-state availability of 0.99726—nearly identical to our computer computation. Clearly, a highly reliable computer system should have a much better availability than a car! Tandem's original goal was to build a system with an MTTF of 100 years! There was clearly much to do to improve the availability in terms of increasing the MTTF, decreasing the MTTR, and structuring a system configuration with greatly increased reliability and availability. Suppose one chooses a goal of 1 hour for repair. This may be realistic for repairs such as board-swapping, but suppose the replacement part is not available? If we assume that 1 hour represents 90% of the repairs but that 10% of the repairs require a replacement part that is unavailable and must be obtained by overnight mail (24 hours), the weighted repair time is then $(0.9 \times 1 + 0.1 \times 24) = 3.3$ hours. Clearly, the MTTR will depend on the distribution of failure modes, the stock of spare parts on hand, and the efficiency of ordering procedures for spare parts that must be ordered from the manufacturer. If one were to achieve an MTTF of 100 years and an MTTR of 3.3 hours, the availability given by Eq. (B95) would be an impressive 0.999996.

The design objectives of Tandem computers were the following [Anderson, 1985]:

- No single hardware failure should stop the system.
- Hardware elements should be maintainable with the on-line system.
- Database integrity should be ensured.
- The system should be modularly extensible without incurring application software changes.

The last objective, extensibility of the system without a software change, played a role in Tandem's success. The software allowed the system to grow by adding new pairs of Tandem computers while the operation continued. Many of Tandem's competitors required that the system be brought down for system expansion, that new software and hardware be installed, and that the expanded system be regenerated.

The original Tandem system was a combination of hardware and software fault tolerance. (The author thanks Dr. Alan P. Wood of Compaq Corporation for his help in clarifying the Tandem architecture and providing details for this section [Wood, 2001].) Each major hardware subsystem (CPUs, disks, power supplies, controllers, and so forth) was (and still is) implemented with parallel units continuously operating (hot redundancy). A diagram depicting the Tandem architecture is shown in Fig. 3.19. The architecture supports $N$ processors in which $N$ is at an even number between 2 and 16.

The Tandem processor subsystem uses hardware fault detection and software fault tolerance to recover from processor failures. The Tandem operating system called Guardian creates and manages *heartbeat signals*, saying "I'm alive," which each processor sends to all the other processors every second. If a processor has not received a heartbeat signal from another processor within two seconds, each operating processor enters a system state called *regroup*. The regroup algorithm determines the hardware element(s) that has failed (which could be a processor or the communications between a group of processors, or it could be multiple failures) and also determines which system resources are still available, avoiding bisection of the system, called the *split-brain* condition, in which communications are lost between two processor groups and each group tries to continue on its own. At the end of the regroup, each processor knows the available system resources.

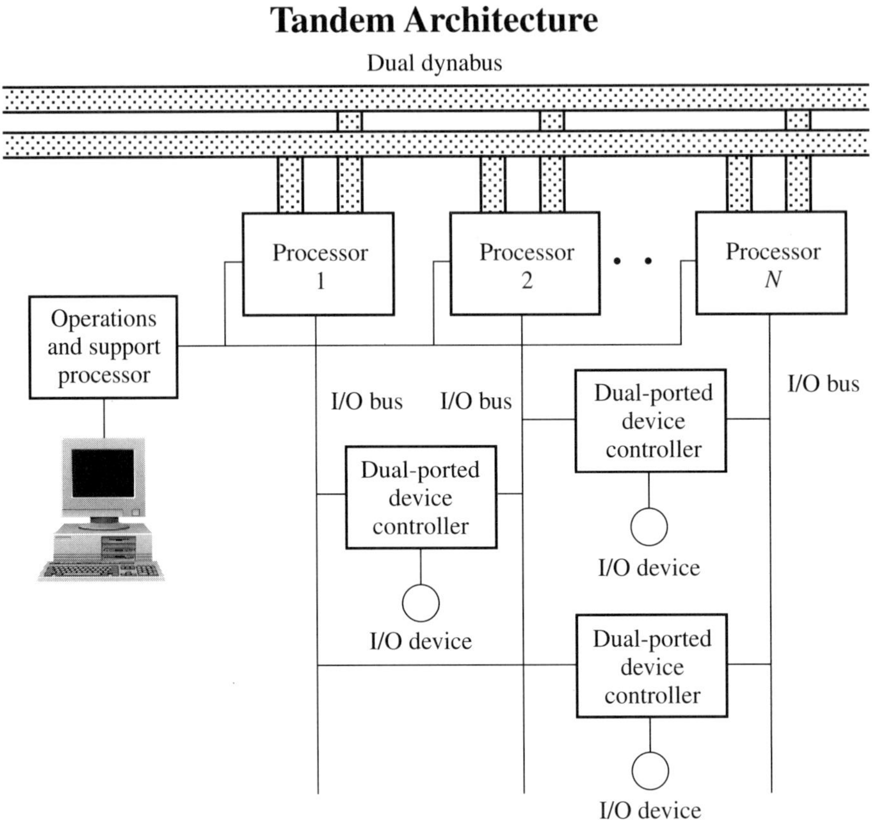

**Tandem Architecture**
Dual dynabus

**Figure 3.19** Basic architecture of a Tandem NonStop computer system. [Reprinted with permission of Compaq Computer Corporation.]

The original Tandem systems used custom microprocessors and checking logic to detect hardware faults. If a hardware fault was detected, the processor would stop sending output (including the heartbeat signal), causing the remaining processors to regroup. Software fault tolerance is implemented via process pairs using the Tandem Guardian operating system. A process pair consists of a primary and a backup process running in separate processors. If the primary process fails because of a software defect or processor hardware failure, the backup process assumes all the duties of the primary process. While the primary process is running, it sends checkpoint messages to the backup process for ensuring that the backup process has all the process state information it needs to assume responsibility in case of a failure. When a processor failure is detected, the backup processes for all the processes that were running in that processor take over, using the process state from the last checkpoint and reexecuting any operations that were pending at the time of the failure.

Since checkpointing requires very little processing, the "backup" processor is actually the primary processor for many tasks. In other words, all Tandem processors spend most of their time processing transactions; only a small fraction of their time is spent doing backup processing to protect against a failure.

In the Tandem system, hardware fault tolerance consists of multiple processors performing the same operations and determining the correct output by using either comparative or self-checking logic. The redundant processors serve as standbys for the primary processor and do not perform additional useful work. If a single processor fails, a redundant processor continues to operate, which prevents an outage. The process pairs in the Tandem system provide software fault tolerance and, like hardware fault tolerance, provide the ability to recover from single hardware failures. Unlike hardware fault tolerance, however, they protect against transient software failures because the backup process reexecutes an operation rather than simultaneously performing the same operation.

The K-series NonStop Himalaya computers released by Tandem in 1992 operate under the same basic principles as the original machines. However, they use commercial microprocessors instead of custom-designed microprocessors. Since commercial microprocessors do not have the custom fault-detection capabilities of custom-designed microprocessors, Tandem had to develop a new architecture to ensure data integrity. Each NonStop Himalaya processor contains two microprocessor chips. These microprocessors are lock-stepped—that is, they run exactly the same instruction stream. The output from the two microprocessors is compared; if it should ever differ, the processor output is frozen within a few nanoseconds so that the corrupted data cannot propagate. The output comparison provides the processor fault detection. The takeover is still managed by process pairs using the Tandem operating system, which is now called the NonStop Kernel.

The S-series NonStop Himalaya servers released in 1997 provided new architectural features. The processor and I/O buses were replaced with a network architecture called ServerNet (see Fig. 3.20). The network architecture allows any device controller to serve as the backup for any other device controller. ServerNet incorporates a number of data integrity and fault-isolation features, such as a 32-bit cyclic redundancy check (CRC) [Siewiorek, 1992, pp. 120–123], on all data packets and automatic low-level link error detection. It also provides the interconnect for NonStop Himalaya servers to move beyond the 16-processor node limit using an architecture called ServerNet Clusters. Another feature of NonStop Himalaya servers is that all hardware replacements and reconfigurations can be done without interrupting system operations. The database can be reconfigured and some software patches can be installed without interrupting system operations as well.

The S-series line incorporates many additional fault-tolerant features. The power and cooling systems are redundant and derated so that a single power supply or fan has sufficient capability to power or cool an entire cabinet. The speed of the remaining fans automatically increases to maintain cooling if any fan should fail. Temperature and voltage levels at key points are continuously mon-

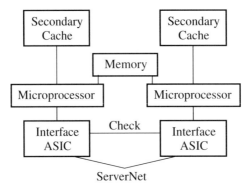

**Figure 3.20** S-Series NonStop Himalaya architecture. (Supplied courtesy of Wood [2001].)

itored, and alarms are sounded whenever the levels exceed safe thresholds. Battery backup is provided to continue operation through any short-duration power outages (up to 30 seconds) and to preserve the contents of memory to provide a fast restart from outages shorter than 2 hours. (If it is necessary to protect against longer power outages, the common solution for high-availability systems is to provide a power supply with backup storage batteries plus DC–AC converters and diesel generators to recharge the batteries. The superior procedure is to have autostart generators, which automatically start when a power outage is detected; however, they must be tested—perhaps once a week—to see if they will start.) All controllers are redundant and dual-ported to serve the primary and secondary connection paths. Each hardware and software module is self-checking and halts immediately instead of permitting an error to propagate—a concept known as the fail-fast design, which makes it possible to determine the source of errors and correct them. NonStop systems incorporate state-of-the-art memory-detection and -correction codes to correct single-bit errors, detect double-bit errors, and detect "nibble" errors (3 or 4 bits in a row). Tandem has modified the memory vendor's error-correcting code (ECC) to include address bits, which helps avoid the reading from or writing to the wrong block of memory. Active techniques are used to check for latent faults. A background memory "sniffer" checks the entire memory every few hours.

System data is protected in many ways. The multiple data paths provided for fault tolerance are alternately used to ensure correct operation. Data on all the buses is parity-protected, and parity errors cause immediate interrupts to trigger error recovery. Disk-driver software provides an end-to-end checksum that is appended to a standard 512-byte disk sector. For structured data, such as SQL files, an additional end-to-end checksum (called a *block checksum*) encodes data values, the physical location of the data, and transaction information. These checksums protect against corrupted data values, partial writes, and misplaced or misaligned data. NonStop systems can use the NonStop remote duplicate database facility (NonStop RDF) to help recover from disasters such as earthquakes, hurricanes, fires, and floods. NonStop RDF sends database updates to a remote site up to thousands of miles away. If a disaster occurs, the remote system takes over within a few minutes without losing any transactions. NonStop Himalaya servers are even "coffee fault-tolerant," meaning the air vents are on the sides to protect against coffee spills on top of the processor cabinet (or, more likely, if the sprinkler system in the computer room is triggered). One would hope that Tandem has also thought about protection against failure modes caused by inadvertant operator errors. Tandem plans to use the alpha microprocessor sometime in the future.

To analyze the Tandem fault-tolerant system, one would formulate a Markov model and proceed as was done previously in this chapter (but for more detail, consult Chapter 4). One must also anticipate the possibilities of errors of commission and omission in generating and detecting the heartbeat signals. This could be modeled by a coverage factor representing the fraction of processor faults that the heartbeat signal would diagnose. (This basic approach is explored in the problems at the end of this chapter.) In Chapter 4, the availability formulas are derived for a parallel system to compare with the availability of a voting system [see Eq. (4.48) and Table 4.9]. Typical computations at the end of Section 4.9.2 for a parallel system apply to the Tandem system. A complete analysis would require the use of a Markov modeling program and multiple models that include more detail and fault-tolerant features.

The original Guardian operating system was responsible for creating, destroying, and monitoring processes, reporting on the failure or restoration of processors, and handling the conventional functions of operating systems in addition to multiprogramming system functions and I/O handling. The early Guardian system required the user to exactingly program the checkpointing, the record locking, and other functions. Thus expert programmers were needed for these tasks, which were often slow in addition to exacting. To avoid such problems, Tandem developed two simpler software systems: the terminal control program (TCP) called Pathway, which provided users with a control program having screen-handling modules written in a higher level (COBOL-like) language to issue checkpoints and dealt with process management and processor failure; and the transaction-monitoring facility (TMF) program, which dealt with the consistency and recoverability of the database and provided concurrence control. The new Himalaya software greatly simplifies such programming, and it provides options to increase throughput. It also supports Tuxedo, Corba, and Java to allow users to write to industry-standard interfaces and still get the benefits of fault tolerance. For further details, see Anderson [1985], Baker [1995], Siewiorek [1992, p. 586], Wood [1995], and the Tandem Web site: [http://himalaya.compaq.com]. Also, see the discussion in Chapter 5, Section 5.10.

### 3.10.2 Stratus Systems

The Stratus line of continuous processing systems is designed to provide uninterrupted operation without loss of data and performance degradation, as well

**132**   REDUNDANCY, SPARES, AND REPAIRS

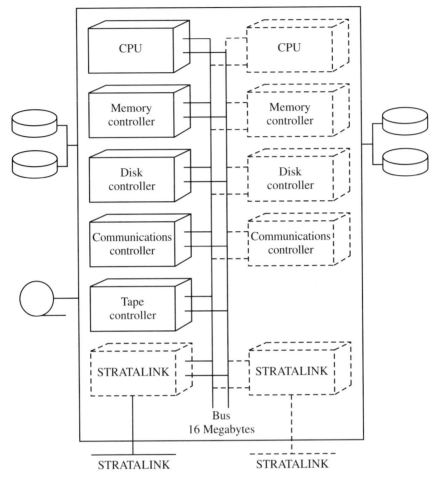

**Figure 3.21**  Basic Stratus architecture. [Reprinted with permission of Stratus Computer.]

as without special application programming. In 1999, Stratus was acquired by Investcorp, but it continues its operation as Stratus Computers. Stratus's customers include major credit card companies, 4 of the 6 U.S. regional securities exchanges, the largest stock exchange in Asia, 15 of the world's 20 top banks, 9-1-1 emergency services, and others. (The author thanks Larry Sherman of Stratus Computers for providing additional information about Stratus.) The Stratus system uses the basic architecture shown in Fig. 3.21. Comparison with the Tandem system architecture shown in Fig. 3.19 shows that both systems have duplicated CPUs, I/O and memory controllers, disk controllers, communication controllers, and high-speed buses. In addition, power supplies and other buses are duplicated.

The Stratus lockstepped microprocessor architecture appears similar to the Tandem architecture described in the previous section, but fault tolerance is achieved through different mechanisms. The Stratus architecture is hardware fault-tolerant, with four microprocessors (all running the same instruction stream) configured as redundant pairs of physical processors. Processor failure is detected by a microprocessor miscompare, and the redundant processor (pair of microprocessors) continues processing with no takeover time. The Tandem architecture is software fault-tolerant; although failure of a processor is also detected by a microprocessor miscomparison, takeover is managed by software requiring a few seconds' delay.

To summarize the comparison, the Tandem system is more complex, higher in cost, and aimed at the upper end of the market. The Stratus system, on the other hand, is more simple, lower in cost, and competes in the middle and lower end portion of the market.

Each major Stratus circuit board has self-checking hardware that continuously monitors operation, and if the checks fail, the circuit board removes itself from service. In addition, each CPU board has two or more CPUs that process the same data, and the outputs are compared at each clock cycle. If the comparison fails, the CPU board removes itself from service and its twin board continues processing without stop. Stratus calls the architecture with two CPUs being checked *pair and spare*, and claims that its architecture is superior in detecting transient errors, is lower in cost, and does not require intensive programming. Tandem points out that software fault tolerance also protects against software faults (90% of all software faults are transient); note, however, that there is the small possibility of missed or imagined software errors. The Stratus approach requires a dedicated backup processor, whereas the Tandem system can use the backup processor in a two-processor configuration to do "useful work" before a failure occurs.

For a further description of the pair-and-spare architecture, consider logical processor $A$ and $B$. As previously discussed in the case of Tandem, logical processor $A$ is composed of lockstepped microprocesors $A_1$ and $A_2$ and logical processor $B$ is composed of lockstepped microprocessors $B_1$ and $B_2$. Processors $A_1$ and $A_2$ compare outputs and will lock out processor $A$ if there is disagreement. A similar comparison is made for processor $B$, as lockout of processor $B$ occurs if processors $B_1$ and $B_2$ disagree. The basic mode of failure is if there is a failure of one processor from logical $A$ and one processor from logical $B$. The outputs of logical processors $A$ and $B$ are not further checked and are ORED on the output bus. Thus, if a very rare failure mode occurs where both processors $A_1$ and $A_2$ fail in the same manner and if both have the same wrong output, the comparitor would be fooled, the faulty output of logical processor $A$ would be ORED with the correct output of logical processor $B$, and wrong results would appear on the output bus. Because of symmetry, identical failures of $B_1$ and $B_2$ would also pass the comparitor and corrupt the output. Although these two failure modes would be rare, they should be included and evaluated in a detailed analysis.

Recovery of partially completed transactions is performed by software using the Stratus virtual operating system (VOS) and the transaction protection facility (TPF). The latest Stratus servers also support Microsoft Windows 2000 operating systems. The Stratus Continuum 400 systems are based on the Hewlett-Packard (HP) PA-RISC microprocessor family and run a version of the HP-UX operating system.

The system can be expanded vertically by adding more processor boards or horizontally via the StrataLINK. The StrataLINK will connect modules within a building or miles away if extenders are used. Networking allows distributed processing at remote distances under control of the VOS: one module could run a program, another could acess a file, and a third could print the results. To shorten repair time, a board taken out of service is self-tested to determine if it is a transient or permanent fault. In the former case, the system automatically returns the board to service. In the case of a permanent failure, however, the customer assistance center can immediately ship replacement parts or aid in the diagnosis of problems by means of a secured, built-in communications link.

Stratus claims that its systems have about five minutes of downtime per year. One can relate this statistic to availability if we start with Eq. (4.53), which was derived for a single element; however, in this case the element is a system. Repair rates are related to the amount of downtime in an interval and failure rates to the amount of uptime in an interval. For convenience, we let the interval be one year and denote the average uptime by $\overline{U}$ and the average downtime by $\overline{D}$. The repair rate, in repairs per year, is the reciprocal of the years per repair, which is the downtime per year; thus, $\mu = 1/\overline{D}$. Similar reasoning leads to a relationship for the failure rate, $\lambda = 1/\overline{U}$. Substituting the above expressions for $\lambda$ and $\mu$ into Eq. (B95a) yields (also see Section 1.3.4):

$$A_{ss} = \frac{\mu}{\mu + \lambda} = \frac{\frac{1}{\overline{D}}}{\frac{1}{\overline{D}} + \frac{1}{\overline{U}}} = \frac{\overline{U}}{\overline{U} + \overline{D}} \qquad (3.80)$$

Since a year contains 8,766 hours, and 5 minutes of downtime is 5/60 of an hour, we can substitute in Eq. (3.80) and obtain

$$A_{ss} = \frac{8,766 - \frac{5}{60}}{8,766} = 0.9999905 \qquad (3.81)$$

Stratus calls this result a "five-nines availability." The quoted value is slightly less than the Bell Labs' ESS No. 1A goal of 2 hours downtime in 40 years (which yields an availability of 0.9999943) and is equivalent to 3 minutes of downtime per year (see Section 1.3.5). Of course, it is easier to compare the unavailability, $\overline{A} = 1 - A$, of such highly reliable systems. Thus ESS No. 1 had an unavailability goal of $57 \times 10^{-7}$, and Stratus claims that it achieves an unavailability of $95 \times 10^{-7}$, which is (5/3) larger. The availability formulation given in Eq. (3.80) is often used to estimate availability based on measured up- and downtimes. For more details on the derivation, see Shooman [1990, pp. 358–359].

To analyze such a system, one would formulate a Markov model and proceed as was done in this chapter and also in Chapter 4. One must also anticipate the possibilities of errors of commission and omission in the hardware comparisons of the various processors. This could be modeled by a coverage factor representing the fraction of processor faults that go undetected by the comparison logic. This basic approach is explored in the problems at the end of this chapter.

Considerable effort must be expended during the design of a high-availability computer system to decrease the mean time between repairs and increase the mean time between failures. Stratus provides a number of diagnostic LEDs (light-emitting diodes) to aid in diagnosis and repair. The status of various subsystems is indicated by green, red, and sometimes amber lights (there may also be flashing red lights). Also, considerable emphasis is given to the power supply. Manufacturers of high-reliability equipment know that the power supply of a computer system is sometimes an overlooked feature but that it is of great importance. During the late 1960s, the SABRE airlines reservation system was one of the first large-scale multilocation transaction systems. During the early stages of operation, many of the system outages were caused by power supply problems [Shooman, 1983, p. 502]. As was previously stated, power supplies for such large critical installations as air traffic control and nuclear plant control are dual systems with a local power company as a primary supply backed up by storage batteries with DC–AC converters and diesel generators as a third line of defense. Small details must be attended to, such as running the diesel generators for a few minutes a week to ensure that they will start in an emergency. The Stratus power supply system contains three or four power supply units as well as backup batteries and battery-temperature monitoring. The batteries have sufficient load capacity to power the system for up to four minutes, which is sufficient for one minute of operation during a power fluctuation plus time for safe shutdown, or four consecutive outages of less than one minute without time to recharge the batteries. Clearly, long power outages will bring down the system unless there are backup batteries and generators. High battery temperature and low battery voltage are monitored. To increase the MTTF of the fan system (and to reduce acoustic noise), fans are normally run at two-thirds speed, and in the case of overtemperature, failures, or other warning conditions, they increase to full speed to enhance cooling.

For more details on Stratus systems, see Anderson [1985], Siewiorek [1992, p. 648], and the Stratus Web site: [http://www.stratus.com].

### 3.10.3 Clusters

In general, the term *cluster* refers to a group of off-the-shelf computers organized by software to serve a specific purpose requiring very large computing power or high availability, fault tolerance, and on-line repairability. We are of course interested in the latter application of clustering; however, we should first cite two historic achievements of clusters designed for the former application class [Hennessy, 1998, pp. 734–736].

- In 1997, the IBM SP2 computer, a cluster of 32 IBM nodes similar to the RS/6000 workstation with added hardware accelerators for chessboard evaluation, beat the then-reigning world chess champion Gary Kasparov in a human–machine contest.
- A cluster of 100 Sun UltraSPARC computers at the University of California–Berkeley, connected by 160 MB/sec Myrinet switches, set two world records: (a), 8.6 gigabytes of data stored on disk was sorted in 1 minute; and (b), a 40-bit DES key encrypted message was cracked in 3.5 hours.

Fault-tolerant applications of clusters involve a different architecture. The simplest scheme is to have two computers: one that is processing on-line and the other that is operating in standby. If the operating system senses a failure of the on-line computer, a recovery procedure is started to bring the second computer on line. Unfortunately, such an architecture results in downtime during the recovery period, which may be either satisfactory or unsatisfactory depending on the application. For a university-computing center, downtime is acceptable as long as it is minimal, but even a small amount of downtime would be inadequate for electronic funds transfer. A superior procedure is to have facilities in the operating system that allow transfer from the on-line to the standby computer without the system going down and without the loss of information. The Tandem system can be considered a cluster, and some of the VAX clusters in the 1980s were very popular.

As an example, we will discuss the hardware and Solaris operating-system features used by a Sun cluster [www.sun.com, 2000]. Some of the incorporated fault-tolerant features are the following:

- Error-correcting codes are used on all memories and caches.
- RAID controllers.
- Redundant power supplies and cooling fans, each with overcapacity.
- The system can lock out bad components during operation or when the server is rebooted.
- The Solaris 8 operating system has error-capture capabilities, and more such capabilities will be included in future releases.
- The Solaris 8 operating system provides recovery with a reboot, though outages occur.
- The Sun Cluster 2.2 software, which is an add-on to the Solaris system, will handle up to four nodes, providing networking and fiber-channel inter-

connections as well as some form of nonstop processing when failures occur.
- The Sun Cluster 3.0 software, released in 2000, will improve on Sun Cluster 2.2 by increasing the number of nodes and simplifying the software.

It seems that the Sun Cluster software is now beginning to develop fault-tolerant features that have been available for many years in the Tandem systems. For a comprehensive discussion of clusters, see Pfister [1995].

## REFERENCES

Advanced Computer and Networks Corporation. White Paper on RAID (http://www.raid-storage.com/aboutacnc.html), 1997.

Anderson, T. *Resilient Computing Systems*. Wiley, New York, 1985.

ARINC Research Corporation. *Reliability Engineering*. Prentice-Hall, Englewood Cliffs, NJ, 1964.

Ascher, H., and H. Feingold. *Repairable Systems Reliability*. Marcel Dekker, New York, 1984.

Baker, W. A Flexible ServerNet-Based Fault-Tolerant Architecture. *Proceedings of the 25th International Symposium on Fault-Tolerant Computing*, 1995. IEEE, New York, NY.

Bazovsky, I. *Reliability Theory and Practice*. Prentice-Hall, Englewood Cliffs, NJ, 1961.

Berlot, A. et al. Unavailability of a Repairable System with One or Two Replacement Options. *Proceedings Annual Reliability and Maintainability Symposium*, 2000. IEEE, New York, NY, pp. 51–57.

Bouricius, W. G., W. C. Carter, and P. R. Schneider. Reliability Modeling Techniques for Self-Repairing Computer Systems. *Proceedings of 24th National Conference of the ACM*, 1969. ACM, pp. 295–309.

Bouricius, W. G., W. C. Carter, and P. R. Schneider. Reliability Modeling Techniques and Trade-Off Studies for Self-Repairing Computers. IBM RC2378, 1969.

Bouricius, W. G. et al. Reliability Modeling for Fault-Tolerant Computers. *IEEE Transactions on Computers* C-20 (November 1971): 1306–1311.

Buzen, J. P., and A. W. Shum. RAID, CAID, and Virtual Disks: I/O Performance at the Crossroads. Computer Measurement Group (CMG), 1993, pp. 658–667.

Coit, D. W., and J. R. English. System Reliability Modeling Considering the Dependence of Component Environmental Influences. *Proceedings Annual Reliability and Maintainability Symposium*, 1999. IEEE, New York, NY, pp. 214–218.

Courant, R. *Differential and Integral Calculus*, vol. I. Interscience Publishers, New York, 1957.

Dugan, J. B., and K. S. Trivedi. Coverage Modeling for Dependability Analysis of Fault-Tolerant Systems. *IEEE Transactions on Computers* 38, 6 (1989): 775–787.

Dugan, J. B. "Software System Analysis Using Fault Trees." In *Handbook of Software Reliability Engineering*, M. R. Lyu (ed.). McGraw-Hill, New York, 1996, ch. 15.

Elks, C. R., J. B. Dugan, and B. W. Johnson. Reliability Analysis of Hard Real-Time Systems in the Presence of Controller Malfunctions. *Proceedings Annual Reliability and Maintainability Symposium*, 2000. IEEE, New York, NY, pp. 58–64.

Elrath, J. G. et al. Reliability Management and Engineering in a Commercial Computer Environment [Tandem]. *Proceedings Annual Reliability and Maintainability Symposium*, 1999. IEEE, New York, NY, pp. 323–329.

Flynn, M. J. *Computer Architecture Pipelined and Parallel Processor Design.* Jones and Bartlett Publishers, Boston, 1995.

Friedman, M. B. Raid Keeps Going and Going and ... from its Conception as a Small, Simple, Inexpensive Array of Redundant Magnetic Disks, RAID Has Grown into a Sophisticated Technology. *IEEE Spectrum* (April 1996): pp. 73–79.

Gibson, G. A. *Redundant Disk Arrays: Reliable, Parallel Secondary Storage.* MIT Press, Cambridge, MA, 1992.

Hennessy, J. L., and D. A. Patterson. *Computer Organization and Design The Hardware/Software Interface.* Morgan Kaufman, San Francisco, 1998.

Huang, J., and M. J. Zuo. Multi-State $k$-out-of-$n$ System Model and its Applications. *Proceedings Annual Reliability and Maintainability Symposium*, 2000. IEEE, New York, NY, pp. 264–268.

Jolley, L. B. W. *Summation of Series.* Dover Publications, New York, 1961.

Kaufman, L. M., and B. W. Johnson. The Importance of Fault Detection Coverage in Safety Critical Systems. *Proceedings of the Twenty-Sixth Water Reactor Safety Information Meeting.* NUCREG/CP-0166, vol. 2, October 1998, pp. 5–28.

Kaufman, L. M., S. Bhide, and B. W. Johnson. Modeling of Common-Mode Failures in Digital Embedded Systems. *Proceedings Annual Reliability and Maintainability Symposium*, 2000. IEEE, New York, NY, pp. 350–357.

Massiglia, P. (ed.). *The Raidbook: A Storage Systems Technology*, 6th ed. (www.peer-to-peer.com), 1997.

McCormick, N. J. *Reliability and Risk Analysis.* Academic Press, New York, 1981.

Muth, E. J. *Stochastic Theory of Repairable Systems.* Ph.D. dissertation, Polytechnic Institute of Brooklyn, New York, June 1967.

Osaki, S. *Stochastic System Reliability Modeling.* World Scientific, Philadelphia, 1985.

Papoulis, A. *Probability, Random Variables, and Stochastic Processes.* McGraw-Hill, New York, 1965.

Paterson, D., R. Katz, and G. Gibson. A Case for Redundant Arrays of Inexpensive Disks (RAID). UCB/CSD 87/391, University of California Technical Report, Berkeley, CA, December 1987. [Also published in *Proceedings of the 1988 ACM Conference on Management of Data (SIGMOD)*, Chicago, IL, June 1988, pp. 109–116.]

Pecht, M. G. (ed.). *Product Reliability, Maintainability, and Supportability Handbook.* CRC Pub. Co (www.crcpub.com), 1995.

Pfister, G. *In Search of Clusters.* Prentice-Hall, Englewood Cliffs, NJ, 1995.

RAID Advisory Board (RAB). *The RAIDbook A Source Book for Disk Array Technology*, 5th ed. The *RAID* Advisory Board, 13 Marie Lane, St. Peter, MN, September 1995.

Roberts, N. H. *Mathematical Methods in Reliability Engineering*. McGraw-Hill, New York, 1964.

Sherman, L. Stratus Computers private communication, January 2001. See also the Stratus Web site for papers written by this author.

Shooman, M. L. *Probabilistic Reliability: An Engineering Approach*. McGraw-Hill, New York, 1968.

Shooman, M. L. *Probabilistic Reliability: An Engineering Approach*, 2d ed. Krieger, Melbourne, FL, 1990.

Siewiorek, D. P., and R. S. Swarz. *The Theory and Practice of Reliable System Design*. The Digital Press, Bedford, MA, 1982.

Siewiorek, D. P., and R. S. Swarz. *Reliable Computer Systems Design and Evaluation*, 2d ed. The Digital Press, Bedford, MA, 1992.

Siewiorek, D. P. and R. S. Swarz. *Reliable Computer Systems Design and Evaluation*, 3d ed. A. K. Peters, www.akpeters.com, 1998.

Stratus Web site: http://www.stratus.com.

Tandem Web site: http://himalaya.compaq.com.

Thomas, G. B. *Calculus and Analytic Geometry*, 3d ed. Addison-Wesley, Reading, MA, 1965.

Toy, W. N. Dual Versus Triplication Reliability Estimates. *AT&T Technical Journal* (November/December 1987): p. 15.

Wood, A. P. Predicting Client/Server Availability. *IEEE Computer Magazine* 28, 4 (April 1995).

Wood, A. P. Compaq Computers (Tandem Division Reliability Engineering) private communication, January 2001.

www.sun.com/software/white-papers: A Developer's Perspective on Sun Solaris Operating Environment, Reliability, Availability, Serviceability. D. H. Brown Associates, Port Chester, NY, February 2000.

## PROBLEMS

**3.1.** Assume that a system consists of five series elements. Each of the elements has the same reliability $p$, and the system goal is $R_s = 0.9$. Find $p$.

**3.2.** Assume that a system consists of five series elements. Three of the elements have the same reliability $p$, and two have known reliabilities of 0.95 and 0.97. The system goal is $R_s = 0.9$. Find $p$.

**3.3.** Assume that a system consists of five series elements. The initial reliability of all the elements is 0.9, each costing $1,000. All components must be improved so that they have a lower failure rate for the system to meet its goal of $R_s = 0.9$. Suppose that for three of the elements, each 50% reduction in failure probability adds $200 to the element cost; for the other two components, each 50% reduction in failure probability adds $300 to the element cost. Find the lowest cost system that meets the system goal of $R_s = 0.9$.

**3.4.** Would it be cheaper to use component redundancy for some or all of the elements in problem 3.3? Explain. Give the lowest cost system design.

**3.5.** Compute the reliability of the system given in problem 3.1, assuming that one is to use
  (a) System reliability for all elements.
  (b) Component reliability for all elements.
  (c) Component reliability for selected elements.

**3.6.** Compute the reliability of the system given in problem 3.2, assuming that one is to use
  (a) System reliability for all elements.
  (b) Component reliability for all elements.
  (c) Component reliability for selected elements.

**3.7.** Verify the curves for $m = 3$ for Fig. 3.4.

**3.8.** Verify the curves for Fig. 3.5.

**3.9.** Plot the system reliability versus $K$ ($0 < K < 2$) for Eqs. (3.13) and (3.15).

**3.10.** Verify that Eq. (3.16) leads to the solution $Kp = 0.9772$ for $p = 0.9$.

**3.11.** Find the solution for problem 3.10 corresponding to $p = 0.95$.

**3.12.** Use the approximate exponential expansion method discussed in Section 3.4.1 to compute an approximate reliability expression for the systems shown in Figs. 3.3(a) and 3.3(b). Use these expressions to compare the reliability of the two configurations.

**3.13.** Repeat problem 3.12 for the systems of Fig. 3.6(a) and 3.6(b). Are you able to verify the result given in problem 3.10 using these equations? Explain.

**3.14.** Compute the system hazard function as discussed in Section 3.4.2 for the systems of Fig. 3.3(a) and Fig. 3.3(b). Do these expressions allow you to compare the reliability of the two configurations?

**3.15.** Repeat problem 3.14 for the systems of Fig. 3.6(a) and 3.6(b). Are you able to verify the result given in problem 3.10 using these equations? Explain.

**3.16.** The mean time to failure, MTTF, is defined as the mean (expected value, first moment) of the time to failure distribution [density function $f(t)$]. Thus, the basic definition is

$$\text{MTTF} = \int_{t=0}^{\infty} t f(t) \, dt$$

Using integration by parts, show that this expression reduces to Eq. (3.24).

**3.17.** Compute the MTTF for Fig. 3.2(a)–(c) and compare.

**3.18.** Compute the MTTF for [Note: All elements have idential constant failure values]
   (a) Fig. 3.3(a) and (b).
   (b) Fig. 3.6(a) and (b).
   (c) Fig. 3.8.
   (d) Eq. (3.40).

**3.19.** Sometimes a component may have more than one failure state. For example, consider a diode that has 3 states: good, $x_1$; failed as an open circuit, $\bar{x}_o$; failed as a short circuit, $\bar{x}_s$;
   (a) Make an RBD model.
   (b) Write the reliability equation for a single diode.
   (c) Write the reliability equation for two diodes in series.
   (d) Write the reliability equation for two diodes in parallel.
   (e) If the $P(x_1) = 0.9$, $P(\bar{x}_o) = 0.07$, $P(\bar{x}_s) = 0.03$, calculate the reliability for parts (b), (c), and (d).

**3.20.** Suppose that in problem 3.19 you had only made a two-state model—diode either good or bad, $P(x_g) = 0.9$, $P(\bar{x}_b) = 0.1$. Would the reliabilities of the three systems have been the same? Explain.

**3.21.** A mechanical component, such as a valve, can have two modes of failure: leaking and blocked. Can we treat this with a three-state model as we did in problem 3.19? Explain.

**3.22.** It is generally difficult to set up a reliability model for a system with common mode failures. Oftentimes, making a three-state model will help. Suppose $x_1$ denotes element 1 that is good, $\bar{x}_c$ denotes element 1 that has failed in a common mode, and $\bar{x}_i$ denotes element 1 that has failed in an independent mode. Set up reliability models and equations for a single element, two series elements, and two parallel elements based on the one success and two failures modes. Given the probabilities $P(x_1) = 0.9$, $P(\bar{x}_c) = 0.03$, $P(\bar{x}_i) = 0.07$, evaluate the reliabilities of the three systems.

**3.23.** Suppose we made a two-state model for problem 3.22 in which the element was either good or bad, $P(x_1) = 0.9$, $P(\bar{x}_1) = 0.10$. Would the reliabilities of the single element, two in series, and two in parallel be the same as computed in problem 3.22?

**3.24.** Show that the sum of Eqs. (3.65a–c) is unity in the time domain. Is this result correct? Explain why.

**3.25.** Make a model of a standby system with one on-line element and two standby elements, all with identical failure rates. Formulate the Markov model, write the equations, and solve for the reliability.

**3.26.** Compute the MTTF for problem 3.25.

**3.27.** Extend the model of Fig. 3.11 to $n$ states. If all the transition probabilities are equal, show that the state probabilities follow the Poisson distribution. (This is one way of deriving the Poisson distribution.). Hint: use of Laplace transforms helps in the derivation.

**3.28.** Compute the MTTF for problem 3.27.

**3.29.** Compute the reliability of a two-element standby system with unequal on-line failure rates for the two components. Modify Fig. 3.11.

**3.30.** Compute the MTTF for problem 3.29.

**3.31.** Compute the reliability of a two-element standby system with equal on-line failure rates and a nonzero standby failure rate.

**3.32.** Compute the MTTF for problem 3.31.

**3.33.** Verify Fig. 3.13.

**3.34.** Plot a figure similar to Fig. 3.13, where Eq. (3.60) replaces Eq. (3.58). Under what conditions are the parallel and standby systems now approximately equal? Compare with Fig. 3.13 and comment.

**3.35.** Reformulate the Markov model of Fig. 3.14 for two nonidentical parallel elements with one repairman; then write the equations and solve for the reliability.

**3.36.** Compute the MTTF for problem 3.35.

**3.37.** Reformulate the Markov model of Fig. 3.14 for an online and a standby element with one repairman and a nonzero standby failure rate. Write the equations and solve for the reliability.

**3.38.** Compute the MTTF for problem 3.37.

**3.39.** Compute the reliability of a two-element standby system with unequal on-line failure rates for the two components. Include coverage. Modify Fig. 3.11 and Fig. 3.15.

**3.40.** Compute the MTTF for problem 3.39.

**3.41.** Compute the reliability of a two-element standby system with equal on-line and a nonzero standby failure rate. Include coverage.

**3.42.** Compute the MTTF for problem 3.1.

**3.43.** Plot a figure similar to Fig. 3.13 where we compare the effect of coverage (rather than an imperfect switch) in reducing the reliability of a standby system. For what value of coverage are the parallel and standby systems approximately equal? Compare with Fig. 3.13 and comment.

**3.44.** Reformulate the Markov model of Fig. 3.14 for two nonidentical parallel elements with one repairman; then write the equations and solve for the reliability. Include coverage.

**3.45.** Compute the MTTF for problem 3.44.

**3.46.** Reformulate the Markov model of Fig. 3.14 for two identical parallel elements with one repairman and a nonzero standby failure rate. Write the equations and solve for the reliability. Include coverage.

**3.47.** Compute the MTTF for problem 3.46.

(In the following problems, you may wish to use a program that solves differential equations or Laplace transform equations algebraically or numerically: Maple, Mathcad, and so forth. See Appendix D.)

**3.48.** Compute the availability of a single element with repair. Draw the Markov model and show that the availability becomes

$$A(t) = \frac{\mu}{\lambda + \mu} + \frac{\lambda}{\lambda + \mu} e^{-(\lambda + \mu)t}$$

Plot this availability function for $\mu = 10\lambda$, $\mu = 100\lambda$, and $\mu = 1,000\lambda$.

**3.49.** If we apply the MTTF formula to the $A(t)$ function, what quantity do we get? Compute for problem 3.48 and explain.

**3.50.** Show how we can get the steady-state value of $A(t)$ for problem 3.48,

$$A(t \to \infty) = \frac{\mu}{\lambda + \mu}$$

in the following two ways:
(a) Set the time derivatives equal to zero in the Markov equations and and combine with the equation that states that the sum of all the probabilities is unity.
(b) Use the Laplace transform final value theorem.

**3.51.** Solve the model of Fig. 3.16 for one repairman, an ordinary parallel system, and values of $\mu = 10\lambda$, $\mu = 100\lambda$, and $\mu = 1,000\lambda$. Plot the results.

**3.52.** Find the steady-state value of $A(t \to \infty)$ for problem 3.51.

**3.53.** Solve the model of Fig. 3.16 for one repairman, a standby system, and values of $\mu = 10\lambda$, $\mu = 100\lambda$, and $\mu = 1,000\lambda$. Plot the results.

**3.54.** Find the steady-state value of $A(t \to \infty)$ for problem 3.53.

**3.55.** Solve the model of Fig. 3.16 augmented to include coverage for one repairman, an ordinary parallel system, and values of $\mu = 10\lambda$, $\mu = 100\lambda$, $\mu = 1,000\lambda$, $c = 0.95$, and $c = 0.90$. Plot the results.

**3.56.** Find the steady-state value of $A(t \to \infty)$ for problem 3.55.

**3.57.** Solve the model of Fig. 3.16 augmented to include coverage for one repairman, a standby system, and values of $\mu = 10\lambda$, $\mu = 100\lambda$, $\mu = 1,000\lambda$, $c = 0.95$, and $c = 0.90$. Plot the results.

**3.58.** Find the steady-state value of $A(t \to \infty)$ for problem 3.57.

**3.59.** Show by induction that Eq. (3.11) is always greater than unity.

**3.60.** Derive Eqs. (3.22) and (3.23).

**3.61.** Derive Eqs. (3.27) and (3.28).

**3.62.** Consider the effect of common mode failures on the computation of Eq. (3.45). How large would the probability of common mode failures have to be to negate the advantage of a 20:21 system?

**3.63.** Formulate a Markov model for a Tandem computer system. Include the possibilities of errors of commission and omission in generating the heartbeat signal—a coverage factor representing the fraction of processor faults that the heartbeat signal would diagnose. Discuss, but do not solve.

**3.64.** Formulate a Markov model for a Stratus computer system. Include the possibilities of errors of commission and omission in the hardware comparison of the various processors. This could be modeled by a coverage factor representing the fraction of processor faults that go undetected by the comparison logic. Discuss, but do not solve.

**3.65.** Compare the models of problems 3.63 and 3.64. What factors will determine which system has a higher availability?

**3.66.** Determine what fault-tolerant features are supported by the latest release of the Sun operating system.

**3.67.** Model the reliability of the system described in problem 3.66.

**3.68.** Model the availability of the system described in problem 3.66.

**3.69.** Search the Web to see if the Digital Equipment Corporation's popular VAX computer clusters are still being produced by Digital now that they are owned by Compaq. (Note: Tandem is also owned by Compaq.) If so, compare with the Sun cluster system.

# 4
# *N*-MODULAR REDUNDANCY

## 4.1 INTRODUCTION

In the previous chapter, parallel and standby systems were discussed as means of introducing redundancy and ways to improve system reliability. After the concepts were introduced, we saw that one of the complicating design features was that of the coupler in a parallel system and that of the decision unit and switch in a standby system. These complications are present in the design of analog systems as well as digital systems. However, a technique known as *voting redundancy* eliminates some of these problems by taking advantage of the digital nature of the output of digital elements. The concept is simple to explain if we view the output of a digital circuit as a string of bits. Without loss of generality, we can view the output as a parallel byte (8 bits long). (The concept generalizes to serial or parallel outputs $n$ bits long.) Assume that we apply the same input to two identical digital elements and compare the outputs. If each bit agrees, then either they are both working properly (likely) or they have both failed in an identical manner (unlikely). Using the concepts of coding theory, we can describe this as an error-detection, not an error-correction, method. If we detect a difference between the two outputs, then there is an error, although we cannot tell which element is in error. Suppose we add a third element and compare all three. If all three outputs agree bitwise, then either all three are working properly (most likely) or all three have failed in the same manner (most unlikely). If two of the element outputs (say, one and three) agree, then most likely element two has failed and we can rely on the output of elements one and three. Thus with three elements, we are able to correct one error. If two errors have occurred, it is very possible that they will fail in the same manner, and the comparison will agree (vote along) with the majority. The bitwise comparison of the outputs (which are 1s or 0s) can be done easily with simple digital logic. The next section references some early works that led to the development of this concept, now called *N-modular redundancy*.

This chapter and Chapter 3 are linked in many ways. For example, the technique of voting reliability joins the parallel and standby system reliability of the previous chapter as the three most common techniques for fault tolerance. (Also, the analytical techniques involving binomial probabilities and Markov models are used in both chapters.) Thus many of the analyses in this chapter that are aimed at comparing the three techniques constitute a continuation of the analyses that were begun in the previous chapter.

The reader not familiar with the binomial distribution discussed in Sections A5.3 and B2.4 or the concepts of Markov modeling in Sections A8 and B7 should read the material in these appendix sections first. Also, the introductory material on digital logic in Appendix C is used in this chapter for discussing voter circuitry.

## 4.2 THE HISTORY OF *N*-MODULAR REDUNDANCY

The history of majority voting begins with the work of some of the most illustrious mathematicians of the 20th century, as outlined by Pierce [1965, pp. 2–7]. There were underlying currents of thought (linked together by theoreticians) that focused on the following:

1. How to use automata theory (logic gates and state machines) to model digital circuit and digital computer operation.
2. A model of the human nervous system based on an interconnection of logic elements.
3. A means of making reliable computing machines from unreliable components.

The third topic was driven by the maintenance problems of the early computers related to relay and vacuum tube failures. A study of the Univac computer that was undertaken by Bell and Newell [1971, pp. 157–169] yields insight into these problems. The first Univac system passed its acceptance tests and was put into operation by the Bureau of the Census in March 1951. This machine was designed to operate 24 hours per day, 7 days per week (168 hours), except for approximately 32 hours of regularly scheduled preventative maintenance per week. Thus the availability would be 136/168 (81%) if there were no failures. In the 7-month period from June to December 1951, the computer experienced about 22 hours of nonscheduled engineering time (repair time due to failures), which reduced availability to 114/168 (68%). Some of the stated causes of troubles were uniservo failures, noise, long time constants, and tube failures occurring at a rate of about 2 per week. It is therefore clear that reliability was a compelling issue.

Moore and Shannon of Bell Labs in a classic article [1956] developed methods for making reliable relay circuits by various series and parallel connections of relay contacts. (The relay was the active element of its time in the switching networks of the telephone company as well as many elevator control systems and many early computers built at Bell Labs starting in 1937. See Randell [1975, Chapter VI] and Shooman [1990, pp. 310–320] for more information.) The classic paper on majority logic was written by John von Neuman (published in the work of Moore and Shannon [1956]), who developed the basic idea of majority voting into a sophisticated scheme with many *NAND* elements in parallel. Each input to the *NAND* element is supplied by a bundle of $N$ identical inputs, and the $2N$ inputs are cross-coupled so that each *NAND* element has one input from each bundle. One of von Neuman's elements was called a *restoring organ*, since erroneous data that entered at the input was compared with the correct input data, producing the correct output and restoring the data.

## 4.3 TRIPLE MODULAR REDUNDANCY

### 4.3.1 Introduction

The basic modular redundancy circuit is triple modular redundancy (often called TMR). The system shown in Fig. 4.1 consists of three parallel digital circuits—$A$, $B$, and $C$—all with the same input. The outputs of the three circuits are compared by the voter, which sides with the majority and gives the majority opinion as the system output. If all three circuits are operating properly, all outputs agree; thus the system output is correct. However, if one element has failed so that it has produced an incorrect output, the voter chooses the output of the two good elements as the system output because they both agree; thus the system output is correct. If two elements have failed, the voter agrees with the majority (the two that have failed); thus the system output is incorrect. The system output is also incorrect if all three circuits have failed. All the foregoing conclusions assume that a circuit fault is such that it always yields the complement of the correct input. A slightly different failure model is often used that assumes the digital circuit to have a fault that makes it *stuck-at-one* (s-a-1) or *stuck-at-zero* (s-a-0). Assuming that rapidly changing signals are exciting the circuit, a failure occurs within fractions of a microsecond of the fault occurrence regardless of the failure model assumed. Therefore, for reliability purposes, the two models are essentially equivalent; however, the error-rate computation differs from that discussed in Section 4.3.3. For further discussion of fault models, see Siewiorek [1982, pp. 17; 105–107] and [1992, pp. 22; 32; 35; 37; 357; 804].

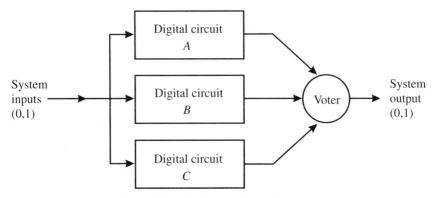

**Figure 4.1** Triple modular redundancy.

### 4.3.2 System Reliability

To apply TMR, all circuits—A, B, and C—must have equivalent logic and must have the same truth tables. In most cases, they are three replications of the same design and are identical. Using this assumption, and assuming that the *voter does not fail*, the system reliability is given by

$$R = P(A \cdot B + A \cdot C + B \cdot C) \tag{4.1}$$

If all the digital circuits are independent and identical with probability of success $p$, then this equation can be rewritten as follows in terms of the binomial theorem.

$$\begin{aligned} R &= B(3:3) + B(2:3) \\ &= \binom{3}{3} p^3 (1-p)^0 + \binom{3}{2} p^2 (1-p)^1 \\ &= 3p^2 - 2p^3 = p^2(3-2p) \end{aligned} \tag{4.2}$$

This is, of course, the reliability expression for a two-out-of-three system. The assumption that the digital elements fail so that they produce the complement of the correct input may not be valid. (It is, however, a worst-case type of result and should yield a lower bound, i.e., a pessimistic answer.)

### 4.3.3 System Error Rate

The probability model derived in the previous secton enabled us to compute the system reliability, that is, the probability of no failures. In many problems, this is the primary measure of interest; however, there are also a number of applications in which another approach is important. In a digital communications system, for example, we are interested not only in the probability that the system makes no errors but also in the error rate. In other words, we assume that errors from temporary equipment malfunction or noise are not catastrophic if they occur only rarely, and we wish to compute the probability of such occurrence. Similarly, in digital computer processing of non-safety-critical data, we could occasionally tolerate an error without shutting down the operation for repair. A third, less clear-cut example is that of an inertial guidance computer for a rocket. At every computation cycle, the computer generates a course change and directs the missile control system accordingly. An error in one computation will direct the missile off course. If the error is large, the time between computations moderately long, the missile control system and dynamics quick to respond, and the flight near its end, the target may be missed, from which a catastrophic failure occurs. If these factors are reversed, however, a small error will temporarily steer the missile off course, much as a wind gust does. As long as the error has cleared in one or two computation cycles, the missile will rapidly return to its proper course. A model for computing transmission-error probabilities is discussed below.

To construct the type of failure model discussed previously, we assume that one good state and two failed states exist:

$A_1$ = element $A$ gives a one output regardless of input (stuck-at-one, or s-a-1)

$\overline{A}_0$ = element $A$ gives a zero output regardless of input (stuck-at-zero, or s-a-0)

To work with this three-state model, we shall change our definition of reliability to "the probability that the digital circuit gives the correct output to any given input." Thus, for the circuits of Fig. 4.1, if the correct output is to be a one, the probability expression is

$$P_1 = 1 - P(\overline{A}_0\overline{B}_0 + \overline{A}_0\overline{C}_0 + \overline{B}_0\overline{C}_0) \tag{4.3a}$$

Equation (4.3a) states that the probability of correctly indicating a one output is given by unity minus the probability of two or more "zero failures." Similarly, the probability of correctly indicating zero output is given by Eq. (4.3b):

$$P_0 = 1 - P(\overline{A}_1\overline{B}_1 + \overline{A}_1\overline{C}_1 + \overline{B}_1\overline{C}_1) \tag{4.3b}$$

If we assume that a one output and a zero output have equal probability of occurrence, 1/2, on any particular transmisson, then the system reliability is the average of Eqs. (4.3a) and (4.3b). If we let

$$P(A) = P(B) = P(C) = p \tag{4.4a}$$
$$P(\overline{A}_1) = P(\overline{B}_1) = P(\overline{C}_1) = q_1 \tag{4.4b}$$
$$P(\overline{A}_0) = P(\overline{B}_0) = P(\overline{C}_0) = q_0 \tag{4.4c}$$

and assume that all states and all elements fail independently, keeping in mind that the expansion of the second term in Eq. (4.3a) has seven terms, then substitution of Eqs. (4.4a–c) in Eq. (4.3a) yields the following equations:

$$P_1 = 1 - P(\overline{A}_0\overline{B}_0) - P(\overline{A}_0\overline{C}_0) - P(\overline{B}_0\overline{C}_0) + 2P(\overline{A}_0\overline{B}_0\overline{C}_0) \quad (4.5a)$$
$$= 1 - 3q_0^2 + 2q_0^3 \quad (4.5b)$$

Similarly, Eq. (4.3b) becomes

$$P_0 = 1 - P(\overline{A}_1\overline{B}_1) - P(\overline{A}_1\overline{C}_1) - P(\overline{B}_1\overline{C}_1) + 2P(\overline{A}_1\overline{B}_1\overline{C}_1) \quad (4.6a)$$
$$= 1 - 3q_1^2 + 2q_1^3 \quad (4.6b)$$

Averaging Eq. (4.5a) and Eq. (4.6a) gives

$$P = \frac{P_0 + P_1}{2} \quad (4.7a)$$

$$= -\frac{1}{2}(3q_0^2 + 3q_1^2 - 2q_0^3 - 2q_1^3) \quad (4.7b)$$

To compare Eq. (4.7b) with Eq. (4.2), we choose the same probability for both failure modes $q_0 = q_1 = q$; therefore, $p + q_0 + q_1 = p + q + q = 1$, and $q = (1 - p)/2$. Substitution in Eq. (4.7b) yields

$$P = \frac{1}{2} + \frac{3}{4}p - \frac{1}{4}p^3 \quad (4.8)$$

The two probabilities, Eq. (4.2) and Eq. (4.8), are compared in Fig. 4.2.

To interpret the results, it is assumed that the digital circuit in Fig. 4.1 is turned on at $t = 0$ and that initially the probability of each digital circuit being successful is $p = 1.00$. Thus both the reliability and probability of successful transmission are unity. If after 1 year of continuous operation $p$ drops to 0.750, the system reliability becomes 0.844; however, the probability that any one message is successfully transmitted is 0.957. To put the result another way, if 1,000 such digital circuits were operated for 1 year, on average 156 would not be operating properly at that time. However, the mistakes made by these machines would amount to 43 mistakes per 1,000 on the average. Thus, for the entire group, the error rate would be 4.3% after 1 year.

### 4.3.4 TMR Options

Systems with $N$-modular redundancy can be designed to behave in different ways in practice [Toy, 1987; Arsenault, 1980, p. 137]. Let us examine in more detail the way a TMR system works. As previously described, the TMR sys-

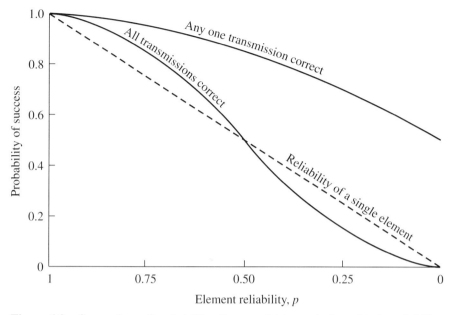

**Figure 4.2** Comparison of probability of successful transmission with the reliability.

tem functions properly if there are no system failures or one system failure. The reliability expression was previously derived in terms of the probability of element success, $p$, as

$$R = 3p^2 - 2p^3 \tag{4.9}$$

If we assume a constant-failure rate $\lambda$, then each component has a reliability $p = e^{-\lambda t}$, and substitution into Eq. (4.9) yields

$$R(t) = 3e^{-2\lambda t} - 2e^{-3\lambda t} \tag{4.10}$$

We can compute the MTTF for this system by integrating the reliability function, which yields

$$\text{MTTF} = \frac{3}{2\lambda} - \frac{2}{3\lambda} = \frac{5}{6\lambda} \tag{4.11}$$

Toy calls this a TMR *3–2 system* because the system succeeds if 3 or 2 units are good. Thus when a second failure occurs, the voter does not know which of the systems has failed and cannot determine which is the good system.

In some cases, additional information is available by such means as observation (from a human operator or an automated system) of the two remaining units after the first failure occurs. For agreement in the event of failure, if one of the two remaining units has behaved strangely or erratically, the "strange" system would be locked out (i.e., disconnected) and the other unit would be assumed to operate properly. In such a case, the TMR system really becomes a 1 : 3 system with a voter, which Toy calls a TMR 3–2–1 system. Equation (4.9) will change, and we must add the binomial probability of 1 : 3 to the equation, that is, $B(1:3) = 3p(1-p)^2$, yielding

$$R = 3p^2 - 2p^3 + 3p(1-p)^2 = p^3 - 3p^2 + 3p \qquad (4.12a)$$

Substitution of $p = e^{-\lambda t}$ gives

$$R(t) = e^{-3\lambda t} - 3e^{-2\lambda t} + 3e^{-\lambda t} \qquad (4.12b)$$

and an MTTF calculation yields

$$\text{MTTF} = \frac{1}{3\lambda} - \frac{3}{2\lambda} + \frac{3}{\lambda} = \frac{11}{6\lambda} \qquad (4.13)$$

If we compare these results with those given in Table 3.4, we see that on the basis of MTTF, the TMR 3–2 system is slightly worse than a system with two standby elements. However, if we make a series expansion of the two functions and compare them in the high-reliability region, the TMR 3–2 system is superior. In the case of the TMR 3–2–1 system, it has an MTTF that is nearly the same as two standby elements. Again, a series expansion of the two functions and comparison in the high-reliability region is instructive.

For a single element, the truncated expansion of the reliability function $e^{-\lambda t}$ is

$$R_s \cong 1 - \lambda t \qquad (4.14)$$

For a TMR 3–2 system, the truncated expansion of the reliability function, Eq. (4.9), is

$$R_{\text{TMR}}(3\text{--}2) = e^{-2\lambda t}(3 - 2e^{-\lambda t}) \cong [1 - 2\lambda t + (2\lambda t)^2/2] \\ \cdot [3 - 2(1 - \lambda t + (\lambda t)^2/2)] \cong 1 - 3(\lambda t)^2 \qquad (4.15)$$

For a TMR 3–2–1 system, the truncated expansion of the reliability function, Eq. (4.12b), is

$$R_{\text{TMR}}(3\text{--}2\text{--}1) = e^{-3\lambda t} - 3e^{-2\lambda t} + 3e^{-\lambda t} \cong [1 - 3\lambda t + (3\lambda t)^2/2 - (3\lambda t)^3/6] \\ - 3[1 - 2\lambda t + (2\lambda t)^2/2 - (2\lambda t)^3/6] \\ + 3[1 - \lambda t + (\lambda t)^2/2 - (\lambda t)^3/6] = 1 - \lambda^3 t^3 \qquad (4.16)$$

Equations (4.14), (4.15), and (4.16) are plotted in Fig. 4.3 showing the superiority of the TMR systems in the high-reliability region. Note that the TMR(3–2) system reliability decreases to about the same value as a single

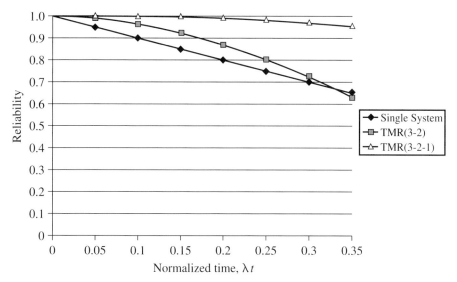

**Figure 4.3** Comparison of the reliability functions of a single system, a TMR 3–2 system, and a TMR 3–2–1 system in the high-reliability region.

element when $\lambda t$ increases from about 0.3 to 0.35. Thus, the TMR is of most use for $\lambda t < 0.2$, whereas TMR (3–2–1) is of greater benefit and provides a considerably higher reliability for $\lambda t < 0.5$.

For further comparisons of MTTF and reliability for $N$-modular systems, refer to the problems at the end of the chapter.

## 4.4 N-MODULAR REDUNDANCY

### 4.4.1 Introduction

The preceding section introduced TMR as a majority voting scheme for improving the reliability of digital systems and components. Of course, this is the most common implementation of majority logic because of the increased cost of replicating systems. However, with the reduction in cost of digital systems from integrated circuit advances, it is practical to discuss $N$-version voting or, as it is now more popularly called, $N$-modular redundancy. In general, $N$ is an odd integer; however, if we have additional information on which systems are malfunctioning and also the ability to lock out malfunctioning systems, it is feasible to let $N$ be an even integer. (Compare advanced voting techniques in Section 4.11 and the Space Shuttle control system example in Section 5.9.3.)

The reader should note there is a pitfall to be skirted if we contemplate the design of, say, a 5-level majority logic circuit on a chip. If the five digital circuits plus the voter are all on the same chip, and if only input and output signals are accessible, there would be no way to test the chip, for which reason additional best outputs would be needed. This subject is discussed further in Sections 4.6.2 and 4.7.4.

In addition, if we contemplate using $N$-modular redundancy for a digital system composed of the three subsystems $A$, $B$, and $C$, the question arises: Do we use $N$-modular redundancy on three systems ($A_1B_1C_1$, $A_2B_2C_2$, and $A_3B_3C_3$) with one voter, or do we apply voting on a lower level, with one voter comparing $A_1A_2A_3$, a second comparing $B_1B_2B_3$, and a third comparing $C_1C_2C_3$? If we apply the principles of Section 3.3, we will expect that voting on a component level is superior and that the reliability of the voter must be considered. This section explores such models.

### 4.4.2 System Voting

A general treatment of $N$-modular redundancy was developed in the 1960s [Knox-Seith, 1953; Pierce, 1961]. If one considers a system of $2n + 1$ voters (note that this is an odd number), parallel digital elements, and a single perfect voter, the reliability expression is given by

$$R = \sum_{i=n+1}^{2n+1} B(i : 2n+1) = \sum_{i=n+1}^{2n+1} \binom{2n+1}{i} p^i (1-p)^{2n+1-i} \qquad (4.17)$$

The preceding expression is plotted in Fig. 4.4 for the case of one, three, five, and nine elements, assuming $p = e^{-\lambda t}$. Note that as $n \to \infty$, the MTTF of the system $\to 0.69/\lambda$. The limiting behavior of Eq. (4.17) as $n \to \infty$ is discussed in Shooman [1990, p. 302]; the reliability function approaches the three straight lines shown in Fig. 4.4. Further study of this figure reveals another important principle—$N$-modular redundancy is only superior to a single system in the high-reliability region. To be more specific, $N$-modular redundancy is superior to a single element for $\lambda t < 0.69$; thus, in system design, one must carefully evaluate the values of reliability obtained over the range $0 < t <$ maximum mission time for various values of $n$ and $\lambda$.

Note that in the foregoing analysis, we assumed a perfect voter, that is, one with a reliability equal to unity. Shortly, we will discard this assumption and assign a more realistic reliability to voting elements. However, before we investigate the effect of the voter, it is germane to study the benefits of partitioning the original system into subsystems and using voting techniques on the subsystem level.

### 4.4.3 Subsystem Level Voting

Assume that a digital system is composed of $m$ series subsystems, each having a constant-failure rate $\lambda$, and that voting is to be applied at the subsystem level. The majority voting circuit is shown in Fig. 4.5. Since this configuration is composed of just the $m$-independent series groups of the same configuration

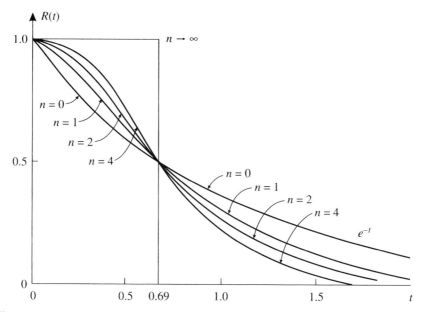

**Figure 4.4** Reliability of a majority voter containing $2n + 1$ circuits. (Adapted from Knox-Seith [1963, p. 12].) (Note that graph is normalized for $\lambda = 1$.)

as previously considered, the reliability is simply given by Eq. (4.17) raised to the $m$th power.

$$R = \left[ \sum_{i=n+1}^{2n+1} \binom{2n+1}{i} p_{ss}^i (1 - p_{ss})^{2n+1-i} \right]^m \quad (4.18)$$

where $p_{ss}$ is the subsystem reliability.

The subsystem reliability $p_{ss}$ is, of course, not equal to a fixed value of $p$; it instead decays in time. In fact, if we assume that all subsystems are identical and have constant-hazard and -failure rates, and if the system failure rate if $\lambda$, the subsystem failure rate would be $\lambda/n$, and $p_{ss} = e^{-\lambda t/m}$. Substitution of the time-dependent expression ($p_{ss} = e^{-\lambda t/m}$) into Eq. (4.18) yields the time-dependent expression for $R(t)$.

Numerical computations of the system reliability functions for several values of $m$ and $n$ appear in Fig. 4.6. Knox-Seith [1963] notes that as $n \to \infty$, the MTTF $\approx 0.7m/\lambda$. This is a direct consequence of the limiting behavior of Eq. (4.17), as was discussed previously.

To use Eq. (4.18) in design, one chooses values of $n$ and $m$ that yield a value of $R$, which meets the design goals. If there is a choice of values ($n$, $m$) that yield the desired reliability, one would choose the pair that represents the lowest cost system. The subject of optimizing voter systems is discussed further in Chapter 7.

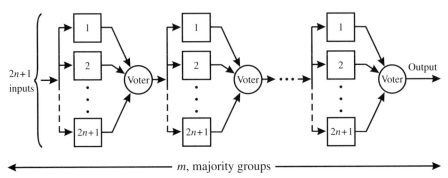

**Figure 4.5** Component redundancy and majority voting.

## 4.5 IMPERFECT VOTERS

### 4.5.1 Limitations on Voter Reliability

One of the main reasons for using a voter to implement redundancy in a digital circuit or system is the ease with which a comparison is made of the digital signals. In this section, we consider an imperfect voter and compute the effect that voter failure will have on the system reliability. (The reader should compare the following analysis with the analogous effect of coupler reliability in the discussion of parallel redundancy in Section 3.5.)

In the analysis presented so far in this chapter, we have assumed that the voter itself cannot fail. This is, of course, untrue; in fact, intuition tells us that if the voter is poor, its unreliability will wipe out the gains of the redundancy scheme. Returning to the example of Fig. 4.1, the digital circuit reliability will be called $p_c$, and the voter reliability will be called $p_v$. The system reliability formerly given by Eq. (4.2) must be modified to yield

$$R = p_v(3p_c^2 - 2p_c^3) = p_v p_c^2(3 - 2p_c) \qquad (4.19)$$

To achieve an overall gain, the voting scheme with the imperfect voter must be better than a single element, and

$$R > p_c \quad \text{or} \quad \frac{R}{p_c} > 1 \qquad (4.20)$$

Obviously, this requires that

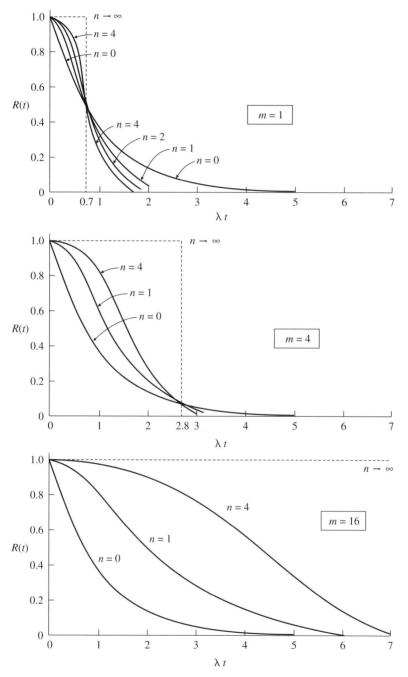

**Figure 4.6** Reliability for a system with $m$ majority vote takers and $(2n+1)m$ circuits. (Adapted from Knox-Seith [1963, p. 19].)

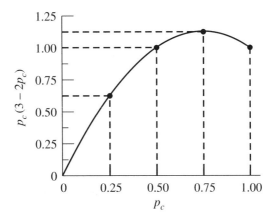

**Figure 4.7** Plot of function $p_c(3 - 2p_c)$ versus $p_c$.

$$\frac{R}{p_c} = p_v p_c(3 - 2p_c) > 1 \qquad (4.21)$$

The minimum value of $p_v$ for reliability improvement can be computed by setting $p_v p_c(3 - 2p_c) = 1$. A plot of $p_c(3 - 2p_c)$ is given in Fig. 4.7. One can obtain information on the values of $p_v$ that allow improvement over a single circuit by studying this equation. To begin with, we know that since $p_v$ is a probability, $0 < p_v < 1$. Furthermore, a study of Fig. 4.3 (lower curve) and Fig. 4.4 (note that $e^{-0.69} = 0.5$) reminds us that N-modular redundancy is only beneficial if $0 < p_c < 1$. Examining Fig. 4.7, we see that the minimum value of $p_v$ will be obtained when the expression $p_c(3-2p_c) = 3p_c - 2p_c^2$. Differentiating with respect to $p_c$ and equating to zero yields $p_c = 3/4$, which agrees with Fig. 4.7. Substituting this value of $p_c$ into $[p_v p_c(3 - 2p_c) = 1]$ yields $p_v = 8/9 = 0.889$, which is the reciprocal of the maximum of Fig. 4.7. (For additional details concerning voter reliability, see Siewiorek [1992, pp. 140–141].) This result has been generalized by Grisamone [1963] for N-voter redundancy, and the results are shown in Table 4.1. This table provides lower bounds on voter reliability that are useful during design; however, most voters have a much higher reliability. The main objective is to make $p_v$ close enough to unity by using reliable components, by derating, and by exercising conservative design so that the voter reliability has only a negligible effect on the value of $R$ given in Eq. (4.19).

### 4.5.2 Use of Redundant Voters

In some cases, it is not possible to devise individual voters that have a high enough reliability to meet the requirements of an ultrareliable system. Since the voter reliability multiplies the N-modular redundancy reliability, as illustrated in Eq. (4.19), the system reliability can never exceed that of the voter. If voting

**TABLE 4.1 Minimum Voter Reliability**

| Number of redundant circuits, $2n + 1$ | 3 | 5 | 7 | 9 | 11 | ∞ |
|---|---|---|---|---|---|---|
| Minimum voter reliability, $p_v$ | 0.889 | 0.837 | 0.807 | 0.789 | 0.777 | 0.75 |

is done at the component level, as shown in Fig. 4.5, the situation is even worse: the reliability function in Eq. (4.18) is multiplied by $p_v^m$, which can significantly lower the reliability of the $N$-modular redundancy scheme. In such cases, one should consider the possibility of using redundant voters.

The standard TMR configuration including redundant voters is shown in Fig. 4.8. Note that Fig. 4.8 depicts a system composed of $n$ subsystems with a triple of subsystems $A$, $B$, and $C$ and a triple of voters $V$, $V'$, $V''$. Also, in the last stage of voting, only a single voter can be employed. One interesting property of the circuit in Fig. 4.8 is that errors do not propagate more than one stage. If we assume that subsystems $A_1$, $B_1$, and $C_1$ are all operating properly and that their outputs should be one, then the outputs of the triplicated voters $V_1$ should also all be one. Say that one circuit, $B_1$, has failed, yielding a zero output; then, each of the three voters $V_1$, $V_1'$, $V_1''$ will agree with the majority ($A_1 = C_1 = 1$) and have a unity output, and the single error does not show up at the output of any voter. In the case of voter failure, say that voter $V_1''$ fails and yields an erroneous output of zero. Circuits $A_2$ and $B_2$ will have the correct inputs and outputs, and $C_2$ will have an incorrect output since it has an incorrect input. However, the next stage of voters will have two correct inputs from $A_2$ and $B_2$, and these will outvote the erroneous output from $V_1''$; thus, voters $V_2$, $V_2'$, and $V_2''$ will all have the correct output. One can say that single circuit errors do not propagate at all and that single voter errors only propagate for one stage.

The reliability expressions for the system of Fig. 4.8 and other similar arrangements are more complex and depend on which of the following assumptions (or combination of assumptions) is true:

1. All circuits $A_i$, $B_i$, and $C_i$ and voters $V_i$ are independent circuits or independent integrated circuit chips.
2. All circuits $A_i$, $B_i$, and $C_i$ are independent circuits or independent integrated circuit chips, and voters $V_i$, $V_i'$, and $V_i''$ are all on the same chip.

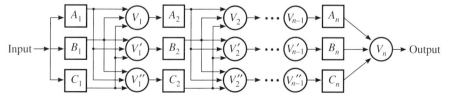

**Figure 4.8** A TMR circuit with redundant voters.

3. All voters $V_i$, $V_i'$, and $V_i''$ are independent circuits or independent integrated circuit chips, and circuits $A_i$, $B_i$, and $C_i$ are all on the same chip.
4. All circuits $A_i$, $B_i$, and $C_i$ are all on the same chip, and voters $V_i$, $V_i'$, and $V_i''$ are all on the same chip.
5. All circuits $A_i$, $B_i$, and $C_i$ and voters $V_i$, $V_i'$, and $V_i''$ are on one large chip.

Reliability expressions for some of these different assumptions are developed in the problems at the end of this chapter.

### 4.5.3 Modeling Limitations

The emphasis of this book up to this point has been on analytical models for predicting the reliability of various digital systems. Although this viewpoint will also prevail for the remainder of the text, there are limitations. This section will briefly discuss a few situations that limit the accuracy of analytical models.

The following situations can be viewed as effects that are difficult to model analytically, that lead to pessimistic results from analytical models, and that represent cases in which the methods of Appendix D would be warranted.

1. Some of the failures in digital (and analog) systems are transient in nature [compare the rationale behind adaptive voting; see Eq. (4.63)]. A transient failure only occurs over a brief period of time or following certain triggering events. Thus the equipment may or may not be operating at any point in time. The analysis associated with the upper curve in Fig. 4.2 took such effects into account.
2. Sometimes, the resulting output of a TMR circuit is correct even if there are two failures. Suppose that all three circuits compute one bit, that unit two is good, unit one has failed s-a-1, and that unit three has failed s-a-0. If the correct output should be a one, then the good unit produces a one output that votes along with the failed unit one, producing a correct voter output. Similarly, if zero were the correct output, unit three would vote with the good unit, producing a correct voter output.
3. Suppose that the circuit in question produces a 4-bit binary word and that circuit one is working properly and produces the 4-bit word 0110. If the first bit of circuit two is bad, we obtain 1110; if the last bit of circuit three is bad, we obtain 0111. Thus, if we vote on the three complete words, then no two agree, but if we vote on the outputs one bit at a time, we get the correct results for all bits.

The more complex fault-tolerant computer programs discussed in Appendix D allow many of these features, as well as other, more complex issues, to be modeled.

**TABLE 4.2  A Truth Table for a Three-Input Majority Voter**

| Inputs | | | Outputs | |
|---|---|---|---|---|
| $x_1$ | $x_2$ | $x_3$ | $f_v(x_1x_2x_3)$ | |
| 0 | 0 | 0 | 0 | Two |
| 0 | 0 | 1 | 0 | or |
| 0 | 1 | 0 | 0 | three |
| 1 | 0 | 0 | 0 | zeroes |
| 1 | 1 | 0 | 1 | Two |
| 1 | 0 | 1 | 1 | or |
| 0 | 1 | 1 | 1 | three |
| 1 | 1 | 1 | 1 | ones |

## 4.6 VOTER LOGIC

### 4.6.1 Voting

It is useful to discuss the structure of a majority logic voter. This allows the designer to appreciate the complexity of a voter and to judge when majority voter techniques are appropriate. The structure of a voter is easy to realize in terms of logic gates and also through the use of other digital logic-design techniques [Shiva, 1988; Wakerly, 1994]. The basic logic function for a TMR voter is based on the Truth Table given in Table 4.2, which leads to the simple Karnaugh map shown in Table 4.3.

A direct approach to designing a majority voter is to include a term for all the minterms in Table 4.2, that is, the last four rows corresponding to an output of one. The logic circuit would require three three-input AND gates, a three-input OR gate, and three inverters (NOT gates) for each bit.

$$f_v(x_1x_2x_3) = x_1x_2\bar{x}_3 + x_1\bar{x}_2x_3 + \bar{x}_1x_2x_3 \qquad (4.22)$$

**TABLE 4.3  Karnaugh Map for a TMR Voter**

| $x_1$ \ $x_2x_3$ | 00 | 01 | 11 | 10 |
|---|---|---|---|---|
| **0** | 0 | 0 | 1 | 0 |
| **1** | 0 | 1 | 1 | 1 |

## TABLE 4.4  Minterm Simplification for Table 4.3

| $x_1$ \ $x_2x_3$ | 00 | 01 | 11 | 10 |
|---|---|---|---|---|
| 0 | 0 | 0 | (1) | 0 |
| 1 | 0 | (1) | (1) | (1) |

The minterm simplification for the TMR voter is shown in Table 4.4 and yields the logic function given in Eq. (4.23). The result of the simplification yields a voter logic function, as follows:

$$f_v(x_1 x_2 x_3) = x_1 x_2 + x_1 x_3 + x_2 x_3 \quad (4.23)$$

Such a circuit is easy to realize with basic logic gates as shown in Fig. 4.9(a), where three AND gates plus one OR gate is used, and in Fig. 4.9(b), where four

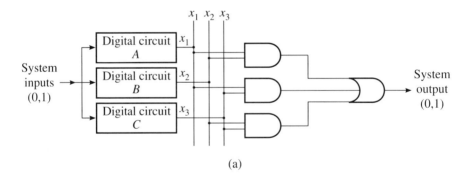

(a)

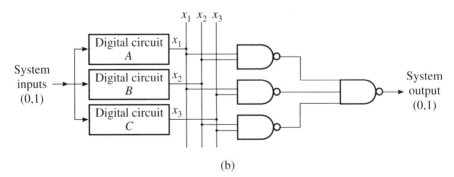

(b)

**Figure 4.9**  Two circuit realizations of a TMR voter. (a) A voter constructed from AND/OR gates; and (b) a voter constructed from NAND gates.

NAND gates are used. The voter in Fig. 4.9(b) can be seen as equivalent to that in Fig. 4.9(a) if one examines the output and applies DeMorgan's theorem:

$$f_v(x_1x_2x_3) = \overline{(\overline{x_1x_2}) \cdot (\overline{x_1x_3}) \cdot (\overline{x_2x_3})} = x_1x_2 + x_1x_3 + x_2x_3 \quad (4.24)$$

### 4.6.2 Voting and Error Detection

There are many reasons why it is important to know which circuit has failed when $N$-modular redundancy is employed, such as the following:

1. If a panel with light-emitting diodes (LEDs) indicates circuit failures, the operator has a warning about which circuits are operative and can initiate replacement or repair of the failed circuit. This eliminates much of the need for off-line testing.
2. The operator can take the failure information into account in making a decision.
3. The operator can automatically lock out a failed circuit.
4. If spare circuits are available, they can be powered up and switched in to replace a failed component.

If one compares the voter inputs the first time that a circuit disagrees with the majority, a failed warning can be initiated along with any automatic action. We can illustrate this by deriving the logic circuits that would be obtained for a TMR system. If we let $f_v(x_1x_2x_3)$ represent the voter output as before and $f_{e1}(x_1x_2x_3), f_{e2}(x_1x_2x_3)$, and $f_{e3}(x_1x_2x_3)$ represent the signals that indicate errors in circuits one, two, and three, respectively, then the truth table shown in Table 4.5 holds.

A simple logic realization of these 4 outputs using NAND gates is shown in

**TABLE 4.5 Truth Table for a TMR Voter Including Error-Detection Outputs**

| Inputs | | | Outputs | | | |
|---|---|---|---|---|---|---|
| $x_1$ | $x_2$ | $x_3$ | $f_v$ | $f_{e1}$ | $f_{e2}$ | $f_{e3}$ |
| 0 | 0 | 0 | 0 | 0 | 0 | 0 |
| 0 | 0 | 1 | 0 | 0 | 0 | 1 |
| 0 | 1 | 0 | 0 | 0 | 1 | 0 |
| 0 | 1 | 1 | 1 | 1 | 0 | 0 |
| 1 | 0 | 0 | 0 | 1 | 0 | 0 |
| 1 | 0 | 1 | 1 | 0 | 1 | 0 |
| 1 | 1 | 0 | 1 | 0 | 0 | 1 |
| 1 | 1 | 1 | 1 | 0 | 0 | 0 |

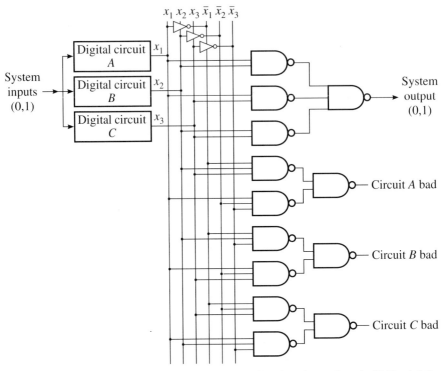

**Figure 4.10** Circuit that realizes the four switching functions given in Table 4.5 for a TMR majority voter and error detector.

Fig. 4.10. The reader should realize that this circuit, with 13 NAND gates and 3 inverters, is only for a single bit output. For a 32-bit computer word, the circuit will have 96 inverters and 416 NAND gates. In Appendix B, Fig. B7, we show that the integrated circuit failure rate, $\lambda$, is *roughly* proportional to the square root of the number of gates, $\lambda \sim \sqrt{g}$, and for our example, $\lambda \sim \sqrt{512} = 22.6$. If we assume that the circuit on which we are voting should have 10 times the failure rate of the voter, the circuit would have 51,076 or about 50,000 gates. The implication of this computation is clear: One should not employ voters to improve the reliability of small circuits because the voter reliability may wipe out most of the intended improvement. Clearly, it would also be wise to consult an experienced logic circuit designer to see if the 512-gate circuit just discussed could be simplified by using other technology, semicustom gate circuits, available microelectronic chips, and so forth.

The circuit given in Fig. 4.10 could also be used to solve the chip test problem mentioned in Section 4.4.1. If the entire circuit of Fig. 4.10 were on a single IC, the outputs "circuit *A, B, C* bad" would allow initial testing and subsequent monitoring of the IC.

## 4.7 N-MODULAR REDUNDANCY WITH REPAIR

### 4.7.1 Introduction

In Chapter 3, we argued that as long as the operating system possesses redundancy, the addition of repair raises the reliability. One might ask at the outset why N-modular redundancy should be used with repair when ordinary parallel or standby redundancy with repair is very effective in achieving highly reliable and available systems. The answer to this question involves the coupling device reliability that was explored in Chapter 3. To be specific, suppose that we wish to compare the reliability of two parallel systems with that of a TMR system. Both systems fail if two of the elements fail, but in the TMR case, there are three systems that could fail; thus the probability of failure is higher. However, in general, the coupler in a parallel system will be more complex than a TMR voter, so a comparison of the two designs requires a detailed evaluation of coupler versus voter reliability. Analysis of TMR system reliability and availability can be found in Siewiorek [1992, p. 335] and in Toy [1987].

### 4.7.2 Reliability Computations

One might expect that it would be most efficient to seek a general solution for the reliability and availability of a system with N-modular redundancy and repair, then specify that $N = 3$ for a TMR system, $N = 5$ for 5-level voting, and so on. A moment's thought, however, suggests quite a different approach. The conventional solution for the reliability and availability of a system with repair involves making a Markov model and solving it much as was done in Chapter 3. In the process, the Laplace transform was computed, and a partial fraction expansion was used to find the individual exponential terms in the solution. For the case of repair, in general the repair rates couple the $n$ states, and solution of the set of $n$ first-order differential equations leads to the solution of an $n$th-order differential equation. If one applies Laplace transform theory, solution of the $n$th-order differential equation is "transformed into" a simpler sequence of steps. However, one step involves the solution for the roots of an $n$th-order polynomial.

Unfortunately, *closed-form* solutions exist only for first- through fourth-order polynomials, and solution procedures for cubic and quadratic polynomials are lengthy and seldom used. We learned in high-school algebra the formula for the roots of a quadratic equation (polynomial). A somewhat more complex solution exists for the solution of a cubic, which is listed in various handbooks [Iyanaga, p. 1396], and also for a fourth-order equation [Iyanaga, p. 1396].

A brief historical note about the origin of closed-form solutions is of interest. The formula for the third-order equation is generally attributed to Giordamo Cardano (also known as Jerome Cardan) [Cardano, 1545; Cardan, 1963]; however, he obtained the solution from Nicolo Tartaglia, and apparently it was discovered by Scipio Ferreo in circa 1505 [Hall, 1957, pp. 480–481]. Ludovico Ferrari, a pupil of Cardan, developed the formula for the fourth-order equation.

Neils Henrik Abel developed a proof that no closed-form solution exists for $n \geq 5$ [Iyanaga, p. 1].

The conclusion from the foregoing information on polynomial roots is that we should start with TMR and other simpler systems if we wish to use algebraic solutions. Numerical solutions are always possible for higher-order equations, and the mathematical software discussed in Appendix D expedites such an approach; however, the insight of an analytical solution is generally lacking. Another approach is to use simplifications and approximations such as those discussed in Appendix B (Sections B8.2 and B8.3). We will use the tried and true three-step engineering approach:

1. Represent the main features of the system by a low-order model that is amenable to closed-form solution.
2. Add further effects one at a time that complicate the model; study the effect (if necessary, use simplifying assumptions and approximations or numerical results computed over a range of parameters).
3. Put all the effects into a comprehensive model and solve numerically.

Our development begins by studying the reliability and availability of a TMR system, assuming that the design is truly TMR or that we are using a TMR model as step one in our solution approach.

### 4.7.3 TMR Reliability

***Markov Model.*** We begin the analysis of voting systems with repair by analyzing the reliability of a TMR system. The Markov reliability diagram for a TMR system composed of a voter, $V$, and three digital subsystems $x_1$, $x_2$, and $x_3$ is given in Fig. 4.11. It is assumed that the $x$s are identical and have the same failure rate, $\lambda$, and that the voter does not fail.

If we compare Fig. 4.11 with the model given in Fig. 3.14 of Chapter 3, we see that they are essentially the same, only with different parameter values (transition rates). There are three states in both models: repair occurs from state $s_1$ to $s_0$, and state $s_2$ is an absorbing state. (Actually, a complete model for Fig. 4.11 would have a fourth state, $s_3$, which is reached by an additional failure from state $s_2$. However, we have included both states in state $s_2$ since either two or three failures both represent system failure. As a rule, it is almost always easier to use a Markov model with fewer states even if one or more of the states represent combined states. State $s_2$ is actually a combined state, also known as a merged state, and a complete discussion of the rules for merging appears in Shooman [1990, p. 529]. One could decompose the third state in Fig. 4.11 into $s_2 = \bar{x}_1 \bar{x}_2 x_3 + \bar{x}_1 x_2 \bar{x}_3 + x_1 \bar{x}_2 \bar{x}_3$ and $s_3 = \bar{x}_1 \bar{x}_2 \bar{x}_3$ by reformulating the model as a more complex four-state model. However, the four-state model is not needed to solve for the upstate probabilities $P_{s_0}$ and $P_{s_1}$. Thus the simpler three-state model of Fig. 4.11 will be used.)

# N-MODULAR REDUNDANCY WITH REPAIR

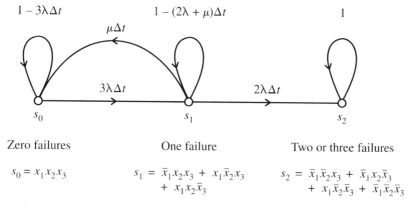

**Figure 4.11** A Markov reliability model for a TMR system with repair.

In the TMR model of Fig. 4.11, there are three ways to experience a single failure from $s_0$ to $s_1$ and two ways for failures to move the system state from $s_1$ to $s_2$. Figure 3.14 of Chapter 3 uses failure rates of $\lambda'$ and $\lambda$ in the model; by substituting appropriate values, the model could hold for two parallel elements or for one on-line and one standby element. One can save repeating a lot of analysis and solution by realizing that the solution given in Eqs. (3.62)–(3.66) will also hold for the model of Fig. 4.11 if we let $\lambda' = 3\lambda$ (three ways to go from state $s_1$ to state $s_2$); $\lambda = 2\lambda$ (two ways to go from state $s_2$ to state $s_3$); and $\mu' = \mu$ (single repairman in both cases). Substituting these values in Eqs. (3.65) yields

$$P_{s_0}(s) = \frac{s + 2\lambda + \mu}{s^2 + (5\lambda + \mu)s + 6\lambda^2} \tag{4.25a}$$

$$P_{s_1}(s) = \frac{3\lambda}{s^2 + (5\lambda + \mu)s + 6\lambda^2} \tag{4.25b}$$

$$P_{s_2}(s) = \frac{6\lambda}{s[s^2 + (5\lambda + \mu)s + 6\lambda^2]} \tag{4.25c}$$

Note that as a check, we sum Eqs. (4.25a–c) and obtain the value $1/s$, which is the transform of unity. Thus the three equations sum to 1, as they should.

One can add the equations for $P_{s_0}$ and $P_{s_1}$ to obtain the reliability of a TMR system with repair in the transform domain.

$$R_{\text{TMR}}(s) = \frac{s + 5\lambda + \mu}{s^2 + (5\lambda + \mu)s + 6\lambda^2} \tag{4.26a}$$

For simplicity let $\lambda = \mu = 1$. The denominator polynomial factors into $(s + 1.268)$ and $(s + 4.732)$, and partial fraction expansion yields

$$R_{\text{TMR}}(s) = \frac{-0.3661}{s + 4.732} + \frac{1.3661}{s + 1.268} \tag{4.26b}$$

Using transform #4 in Table B6 in Appendix B, we obtain the time function:

$$R_{\text{TMR}} = 1.3661e^{1.268t} - 0.3661e^{-4.7632t} \tag{4.26c}$$

For $t = 1$, $R_{\text{TMR}} = 0.3841$. This is an improvement over no repair, which yields $R_{\text{TMR}} = 0.3064$. Larger values of $\mu$, which are common, yield more improvement. which of course agrees with the result previously computed [see Eq. (4.2)].

***Initial Behavior.*** The complete solution for the reliability of a TMR system with repair is given in Eq. (4.26c). It is useful to practice with the simplifying effects of initial behavior, final behavior, and MTTF solutions on this simple problem before they are applied later in this chapter to more complex models where the simplification is needed. One can evaluate the effects of repair on the initial behavior of the TMR system simply by using the transform for $t^n$, which is discussed in Appendix B, Section B8.3. We begin with Eq. (4.26a), where division of the denominator into the numerator using polynomial long division yields for the first three terms:

$$R_{\text{TMR}}(s) = \frac{1}{s} - \frac{6\lambda^2}{s^3} + \frac{6\lambda^2(5\lambda + \mu)}{s^4} - \cdots \tag{4.27a}$$

Using inverse transform no. 5 of Table B6 of Appendix B yields

$$\mathcal{L}\left\{\frac{1}{(n-1)!} t^{n-1} e^{-at}\right\} = \frac{1}{(s+a)^n} \tag{4.27b}$$

Setting $a = 0$ yields

$$\mathcal{L}\left\{\frac{1}{(n-1)!} t^{n-1}\right\} = \frac{1}{(s)^n} \tag{4.27c}$$

Using the transform in Eq. (4.27c) converts Eq. (4.27a) into the time function, which is a three-term polynomial in $t$ (the first three terms in the Taylor series expansion of the time function).

$$R_{\text{TMR}}(t) = 1 - 3\lambda^2 t^2 + \lambda^2(5\lambda + \mu)t^3 \cdots \tag{4.27d}$$

We previously studied the first two terms in the Taylor series expansion of the TMR reliability expansion in Eq. (4.15). In Eq. (4.27d), we have a three-term solution, and one can compare Eqs. (4.15) and (4.27b) by calculating an additional third term in the expansion of Eq. (4.15). The expansions in Eq. (4.15) are augmented by including the cubic terms in the expansions of the bracketed terms, that is, $-4\lambda^3 t^3/3$ in the first bracket and $+\lambda^3 t^3/3$ in the second bracket. Carrying out the algebra adds a third term, and Eq. (4.15) becomes expanded as follows:

$$R_{\text{TMR}}(3-2) = 1 - 3\lambda^2 t^2 + 5\lambda^3 t^3 \qquad (4.27e)$$

Thus the first three terms of Eq. (4.15) and Eq. (4.27d) are identical for the case of no repair, $\mu = 0$. Equation (4.27d) is larger (closer to unity) than the expanded version of Eq. (4.15) because of the additional term $+\lambda^2 \mu t^3$ that is significant for large values of repair rate; we therefore see that repair improves the reliability. However, we note that repair only affects the cubic term in Eq. (4.27d) and not the quadratic term. Thus, for *very* small $t$, repair does not affect the initial behavior; however, from the above solution, we can see that it is beneficial for small and modest size $t$.

A numerical example will illustrate the improvement in initial reliability due to repair. Let $\mu = 10\lambda$; then the third term in Eq. (4.27d) becomes $+15\lambda^3 t^3$ rather than $+5\lambda^3 t^3$ with no repair. One can evaluate the increase due to $\mu = 10\lambda$ at one point in time by letting $t = 0.1/\lambda$. At this point in time, the TMR reliability without repair is equal to 0.975; with repair, it is 0.985. Further comparisons of the effects of repair appear in the problems at the end of the chapter.

The approximate analysis of this section led to a useful evaluation of the effects of repair through the computation of the power series expansion of the time function for the model with repair. This approximate result avoids the need to factor the denominator polynomial in the Laplace transform solution, which was found to be a stumbling block in obtaining a complete closed solution for higher-order systems. The next section will discuss the mean time to failure (MTTF) as another approximate solution that also avoids polynomial factoring.

***Mean Time to Failure.*** As we saw in the preceding chapter, the computation of MTTF greatly simplifies the analysis, but it is not without pitfalls. The MTTF computes the "area under the reliability curve" (see also Section 3.8.3). Thus, for a single element with a reliability function of $e^{-\lambda t}$, the area under the curve yields $1/\lambda$; however, the MTTF calculation for the TMR system given in Eq. (4.11) yields a value of $5/6\lambda$. This implies that a single element is better than TMR, but we know that TMR has a higher reliability than a single element (see also Siewiorek [1992, p. 294]). The explanation of this apparent contradiction is simple if we examine the $n = 0$ and $n = 1$ curves in Fig. 4.4. In the region of primary interest, $0 < \lambda t < 0.69$, TMR *is* superior to a single element, but in the region $0.69 < \lambda t < \infty$ (not a region of primary interest), the single element has a superior reliability. Thus, in computing the integral between $t = 0$ and $t = \infty$, the long tail controls the result. The lesson is that we should not trust an MTTF comparison without further study unless there is a significant superiority or unless the two reliability functions have the same shape. Clearly, if the two functions have the same shape, then a comparison of the MTTF values should be definitive. Graphing of reliability functions in the high-reliability region should always be included in an analysis, especially with the ready availability, power, and ease provided by software on a modern PC. One can also easily integrate the functions in question by using an analysis program to compute MTTF.

We now apply the simple method given in Appendix B, Section B8.2 to evaluate the MTTF by letting $s$ approach zero in the Laplace transform of the reliability function—Eq. (4.26a). The result is

$$\text{MTTF} = \frac{5 + \mu/\lambda}{6\lambda} \tag{4.28}$$

To evaluate the effect of repair, let $\mu = 10\lambda$. The MTTF without repair increases from $5/6\lambda$ to $16/6\lambda$—a threefold improvement.

***Final Behavior.*** The Laplace transform has a simple theorem that allows us to easily calculate the final value of a time function based on its transform. (See Appendix B, Table B7, Theorem 7.) The final-value theorem states that the value of the time function $f(t)$ as $t \to \infty$ is given by $sF(s)$ (the transform multiplied by $s$) as $s \to 0$. Applying this to Eq. (4.26a), we obtain

$$\lim_{s \to 0} \{sR_{\text{TMR}}\} = \lim_{s \to 0} \frac{s(s + 5\lambda + \mu)}{s^2 + (5\lambda + \mu)s + 6\lambda^2} = 0 \tag{4.29}$$

A little thought shows that this is the correct result since all reliability functions go to zero as time increases. However, when we study the availability function later in this chapter, we will see that the final value of the availability is nonzero. This value is an important measure of system behavior.

### 4.7.4 N-Modular Reliability

Having explored the analysis of the reliability of a TMR system with repair, it would be useful to develop general expressions for the reliability, MTTF, and initial behavior for $N$-modular systems. This task is difficult and probably unnecessary since most practical systems have 3- or 5-level majority voting. (An intermediate system with 4-level voting used by NASA in the Space Shuttle will be discussed later in this chapter.) The main focus of this section will therefore be the analysis.

***Markov Model.*** We begin the analysis of 5-level modular reliability with

# N-MODULAR REDUNDANCY WITH REPAIR

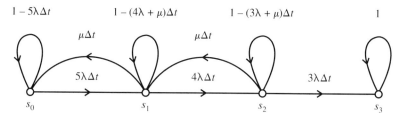

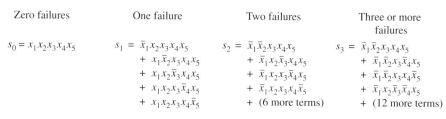

**Figure 4.12** A Markov reliability model for a 5-level majority voting system with repair.

repair by formulating the Markov model given in Fig. 4.12. We follow the same approach used to formulate the Markov model given in Fig. 4.11. There are, however, additional states. (Actually, there is one additional state that lumps together three other states.)

The Markov time-domain differential equations are written in a manner analogous to that used in developing Eqs. (3.62a–c). The notation $\dot{P}_s = dP_s/dt$ is used for convenience, and the following equations are obtained:

$$\dot{P}_{s_0}(t) = -5\lambda P_{s_0}(t) + \mu P_{s_1}(t) \quad (4.30\text{a})$$
$$\dot{P}_{s_1}(t) = 5\lambda P_{s_0}(t) - (4\lambda + \mu)P_{s_1}(t) + \mu P_{s_2}(t) \quad (4.30\text{b})$$
$$\dot{P}_{s_2}(t) = 4\lambda P_{s_1}(t) - (3\lambda + \mu)P_{s_2}(t) \quad (4.30\text{c})$$
$$\dot{P}_{s_3}(t) = 3\lambda P_{s_2}(t) \quad (4.30\text{d})$$

Taking the Laplace transform of the preceding equations and incorporating the initial conditions $P_{s_0}(0) = 1$, $P_{s_1}(0) = P_{s_2}(0) = P_{s_3}(0) = 0$ leads to the transformed equations as follows:

$$(s + 5\lambda)P_{s_0}(s) - \mu P_{s_1}(s) = 1 \quad (4.31\text{a})$$
$$-5\lambda P_{s_0}(s) + (s + 4\lambda + \mu)P_{s_1}(s) - \mu P_{s_2}(s) = 0 \quad (4.31\text{b})$$
$$-4\lambda P_{s_1}(s) + (s + 3\lambda + \mu)P_{s_2}(s) = 0 \quad (4.31\text{c})$$
$$-3\lambda P_{s_2}(s) + sP_{s_3}(s) = 0 \quad (4.31\text{d})$$

Equations (4.31a–d) can be solved by a variety of means for the probabilities $P_{s_0}(t)$, $P_{s_1}(t)$, $P_{s_2}(t)$, and $P_{s_3}(t)$. One technique based on Cramer's rule is to formulate a set of determinants associated with the equations. Each of the probabilities becomes a ratio of two of the determinants: a numerator determinant divided by a denominator determinant. The denominator determinant is the same for each ratio; it is generally denoted by $\Delta$ and is the determinant of the coefficients of the equations. (One can develop the form of these equations in a more elaborate fashion using matrix theory; see Shooman [1990, pp. 239–243].) A brief inspection of Eqs. (4.31a–d) shows that the first three are uncoupled from the last and can be solved separately, simplifying the algebra (this will always be true in a Markov model with repair when the last state is an absorbing one). Thus, for the first three equations,

$$\Delta = \begin{vmatrix} s+5\lambda & -\mu & 0 \\ -5\lambda & s+4\lambda+\mu & -\mu \\ 0 & -4\lambda & s+3\lambda+\mu \end{vmatrix} \quad (4.32)$$

The numerator determinants in the solution are similar to the denominator determinants; however, one column is replaced by the right-hand side of the Eqs. (4.31a–d); that is,

$$\Delta_1 = \begin{vmatrix} 1 & -\mu & 0 \\ 0 & s+4\lambda+\mu & -\mu \\ 0 & -4\lambda & s+3\lambda+\mu \end{vmatrix} \quad (4.33a)$$

$$\Delta_2 = \begin{vmatrix} s+5\lambda & 1 & 0 \\ -5\lambda & 0 & -\mu \\ 0 & 0 & s+3\lambda+\mu \end{vmatrix} \quad (4.33b)$$

$$\Delta_3 = \begin{vmatrix} s+5\lambda & -\mu & 1 \\ -5\lambda & s+4\lambda+\mu & 0 \\ 0 & -4\lambda & 0 \end{vmatrix} \quad (4.33c)$$

In terms of this group of determinants, the probabilities are

$$P_{s_0}(s) = \frac{\Delta_1}{\Delta} \quad (4.34a)$$

$$P_{s_1}(s) = \frac{\Delta_2}{\Delta} \quad (4.34b)$$

$$P_{s_2}(s) = \frac{\Delta_3}{\Delta} \quad (4.34c)$$

The reliability of the 5-level modular redundancy system is given by

$$R_{5\,\mathrm{MR}}(t) = P_{s_0}(t) + P_{s_1}(t) + P_{s_2}(t) \quad (4.35)$$

Expansion of the denominator determinant yields the following polynomial:

$$\Delta = s^3 + (12\lambda + 2\mu)s^2 + (47\lambda^2 + 8\lambda\mu + \mu^2)s + 60\lambda^3 \quad (4.36a)$$

Similarly, expanding the other determinants yields the following polynomials:

$$\Delta_1 = s^2 + (7\lambda + 2\mu)s + 12\lambda^2 + 3\lambda\mu + \mu^2 \quad (4.36b)$$
$$\Delta_2 = 5\lambda(s + 3\lambda + \mu) \quad (4.36c)$$
$$\Delta_3 = 20\lambda^2 \quad (4.36d)$$

Substitution in Eqs. (4.34a–c) and (4.35) yields the transform of the reliability function:

$$R_{5\,\mathrm{MR}}(s) = \frac{s^2 + (12\lambda + 2\mu)s + 47\lambda^2 + 8\lambda\mu + \mu^2}{s^3 + (12\lambda + 2\mu)s^2 + (47\lambda^2 + 8\lambda\mu + \mu^2)s + 60\lambda^3} \quad (4.37)$$

As a check, we compute the probability of being in the fourth state $P_{s_3}(s)$ from Eq. (4.31d) as

$$P_{s_3}(s) = \frac{3\lambda P_{s_2}(s)}{s} = \frac{60\lambda^3}{s\Delta} \quad (4.38)$$

Adding Eq. (4.37) to Eq. (4.38) and performing some algebraic manipulation yields $1/s$, which is the transform of unity. Thus the sum of all the state probabilities adds to unity as it should and the results check.

***Initial Behavior.*** As in the preceding section, we can model the initial behavior by expanding the transform Eq. (4.37) into a series in inverse powers of $s$ using polynomial division. The division yields

$$R_{5\,\mathrm{MR}}(s) = \frac{1}{2} - \frac{60\lambda^3}{s^4} + \frac{60\lambda^3(12\lambda + 2\mu)}{s^5} - \cdots \quad (4.39a)$$

Applying the inverse transform of Eq. (4.27c) yields

$$R_{5\,\mathrm{MR}}(s) = 1 - 10\lambda^3 t^3 + 2.5\lambda^3(12\lambda + 2\mu)t^4 \cdots \quad (4.39b)$$

We can compare the gain due to 5-level modular redundancy with repair to that of TMR with repair by letting $\mu = 10\lambda$ and $t = 0.1/\lambda$, as in Section 4.7.3, which gives a reliability of 0.998. Without repair, the reliability would be 0.993. These values should be compared with the TMR reliability without repair, which is equal to 0.975, and TMR with repair, which is 0.985. Since it is difficult to compare reliabilities close to unity, we can focus on the unreliabilities with repair. The 5-level voting has an unreliability of 0.002; the TMR, 0.015. Thus, the change in voting from 3-level to 5-level has reduced the unre-

**TABLE 4.6  Comparison of the MTTF for Several Voting and Parallel Systems with Repair**

| System | MTTF Equation | $\mu = 0$ | $\mu = 10\lambda$ | $\mu = 100\lambda$ |
|---|---|---|---|---|
| TMR with repair | $\dfrac{5 + \dfrac{\mu}{\lambda}}{6\lambda}$ | $\dfrac{0.83}{\lambda}$ | $\dfrac{2.5}{\lambda}$ | $\dfrac{17.5}{\lambda}$ |
| 5MR with repair | $\dfrac{47 + 8\dfrac{\mu}{\lambda} + \left(\dfrac{\mu}{\lambda}\right)^2}{60\lambda}$ | $\dfrac{0.78}{\lambda}$ | $\dfrac{3.78}{\lambda}$ | $\dfrac{180.78}{\lambda}$ |
| Two parallel | $\dfrac{3\lambda + \mu}{2\lambda^2}$ | $\dfrac{1.5}{\lambda}$ | $\dfrac{6.5}{\lambda}$ | $\dfrac{51.5}{\lambda}$ |
| Two standby | $\dfrac{2\lambda + \mu}{\lambda^2}$ | $\dfrac{2}{\lambda}$ | $\dfrac{12}{\lambda}$ | $\dfrac{102}{\lambda}$ |

liability by a factor of 7.5. Further comparisons of the effects of repair appear in the problems at the end of this chapter.

***Mean Time to Failure Comparison.*** The MTTF for 5-level voting is easily computed by letting $s$ approach 0 in the transform equation, which yields

$$\text{MTTF}_{5\,\text{MR}} = \frac{47\lambda^2 + 8\lambda\mu + \mu^2}{60\lambda^3} \qquad (4.40)$$

This MTTF is compared with some other systems in Table 4.6. The table shows, as expected, that 5MR is superior to TMR when repair is present. Note that two parallel or two standby elements appear more reliable. Once reduction in reliability due to the reliability of the coupler and coverage is included and compared with the reduction due to the reliability of the voter, this advantage may disappear.

***Initial Behavior Comparison.*** The initial behavior of the systems given in Table 4.6 is compared in Table 4.7 using Eqs. (4.27d) and (4.39b) for TMR and 5MR systems. For the case of two ordinary parallel and two standby systems, we must derive the initial behavior equation by adding Eqs. (3.65a) and (3.65b) to obtain the transform of the reliability function that holds for both parallel and standby systems.

$$R(s) = P_{s_0}(s) + P_{s_1}(s) = \frac{s + \lambda + \lambda' + \mu'}{s^2 + (\lambda + \lambda' + \mu')s + \lambda\lambda'} \qquad (4.41)$$

For an ordinary parallel system, $\lambda' = 2\lambda$ and $\mu' = \mu$, and substitution into Eq. (4.41), long division of the denominator into the numerator, and inversion of

**TABLE 4.7 Comparison of the Initial Behavior for Several Voting and Parallel Systems with Repair**

| System | Initial Reliability Equation, $\mu = 10\lambda$ | Value of $t$ at which $R = 0.999$ |
|---|---|---|
| TMR with repair | $1 - 3(\lambda t)^2 + 15(\lambda t)^3$ | $\dfrac{0.0192}{\lambda}$ |
| 5MR with repair | $1 - 10(\lambda t)^3 + 80(\lambda t)^4$ | $\dfrac{0.057}{\lambda}$ |
| Two parallel | $1 - (\lambda t)^2 + 4.33(\lambda t)^3$ | $\dfrac{0.034}{\lambda}$ |
| Two standby | $1 - 0.5(\lambda t)^2 + 2(\lambda t)^3$ | $\dfrac{0.045}{\lambda}$ |

the transform (as was done previously) yields

$$R_{\text{parallel}}(t) = 1 - (\lambda t)^2 + \lambda^2(3\lambda + \mu)t^3/3 \tag{4.42a}$$

For a standby system, $\lambda' = \lambda$ and $\mu' = \mu$, and substitution into Eq. (4.41), long division, and inversion of the transform yields

$$R_{\text{standby}}(t) = 1 - (\lambda t)^2/2 + \lambda^2(2\lambda + \mu)t^3/6 \tag{4.42b}$$

Equations (4.42a) and (4.42b) appear in Table 4.7 along with Eqs. (4.27d) and (4.39b), where $\mu = 10\lambda$ has been substituted.

Table 4.7 shows that the length of time the reliability takes to decay from 1 to 0.999, which makes it clearly a high-reliability region. For the TMR system, the duration is $t = 0.0192\lambda$; for the 5-level voting system, $t = 0.057\lambda$. Thus the 5-level system represents an increase of nearly 3 over the 3-level system. One can better appreciate these numerical values if typical values are substituted for $\lambda$. The length of a year is 8,766 hours, which is often approximated as 10,000 hours. A high-reliability computer may have an MTTF($1/\lambda$) of about 10 years, or approximately 100,000 hours. Substituting this value for $t$ shows that the reliability of a TMR system with a repair rate of 10 times the failure rate will have a reliability exceeding 0.999 for about 1,920 hours. Similarly, a 5-level voting system will have a reliability exceeding 0.999 for about 5,700 hours. In the case of the parallel and standby systems, the high-reliability region is longer than in a TMR system, but is less than in a 5-level voter system.

*Higher-Level Voting.* One could extend the above analysis to cover higher-level voting systems; for example, 7-level and 9-level voting. Even though it is easy to replicate many different copies of a logic circuit on a chip at low cost, one seldom goes beyond the 3-level or 5-level voting system, although the foregoing methods could be used to solve for the reliability of such higher-level systems.

If one fabricates a very large scale integrated circuit (VLSI) with many circuits and a voter, an interesting question arises. There is a yield problem with complex chips caused by imperfections. With so much redundancy, how can one be sure that the chip does not contain such imperfections that a 5-level voter system with imperfections is really equivalent to a 4- or 3-level voter system? In fact, a 5-level voter system with two failed circuits is actually inferior to a 3-level voter. One more failure in the former will result in three failed and two good circuits, and the voter believes the failed three. In the case of a 3-level voter, a single failure will still leave the remaining two good circuits in control. The solution is to provide internal test inputs on an IC voter system so that the components of the system can be tested. This means that extra pins on the chip must be dedicated to test points. The extra outputs in Fig. 4.10 could provide these test points, as was discussed in Section 4.6.2.

The next section discusses the effect of voter reliability on $N$-modular redundancy. Note that we have not discussed the effects of coverage in a TMR system. In general, the simple nature of a voter catches almost all failures, and coverage is not significant in modeling the system.

## 4.8 $N$-MODULAR REDUNDANCY WITH REPAIR AND IMPERFECT VOTERS

### 4.8.1 Introduction

The analysis of the preceding section did not include two imperfections in a voting system: the reliability of the voter itself and also the concept of coverage. In the case of parallel and standby systems, which were treated in Chapter 3, coverage made a considerable difference in the reliability. The circuit that detected failures of the active system and switched to the standby (hot or cold) element in a parallel or standby system is reasonably complex and will have a significant failure rate. Furthermore, it will have the problem that it cannot detect all faults and will sometimes fail to switch when it should or switch when it should not. In the case of a voter, the concept and the resulting circuit is much simpler. Thus one might be justified in assuming that the voter does not have a coverage problem and so reduce our evaluation to the reliability of a voter and how it affects the system reliability. This can then be contrasted with the reliability of a coupler and a parallel system (introduced in Section 3.5).

### 4.8.2 Voter Reliability

We begin our discussion of voter reliability by considering the reliability of a TMR system as shown in Fig. 4.1 and the reliability expression given in

Eq. (4.19). In Section 4.5, we asked how small the voter reliability, $p_v$, can be so that the gains of TMR still exceed the reliability of a single circuit. The analysis was given in Eqs. (3.34) and (3.35). Now, we perform a similar analysis for a TMR system with an imperfect voter. The computation proceeds from a consideration of Eq. (4.19). If the voter were perfect, $p_v = 1$, then the reliability would be computed as

$$R_{TMR} = 3p_c^2 - 2p_c^3 \qquad (4.43a)$$

If we include an imperfect voter, this expression becomes

$$R_{TMR} = 3p_v p_c^2 - 2p_v p_c^3 = p_v(3p_c^2 - 2p_c^3) \qquad (4.43b)$$

If we assume constant-failure rates for the voter and the circuits in the TMR configuration, then for the voter we have $p_v = e^{-\lambda_v t}$, and for the TMR circuits, $p = e^{-\lambda t}$. If we use a three-term approximation for the exponential and substitute into Eq. (4.43b), one obtains an expression for the initial reliability, as follows:

$$R_{TMR} = \left(1 - \lambda_v t + \frac{(\lambda_v t)^2}{2!} - \frac{(\lambda_v t)^3}{3!}\right) \times \left[3\left(1 - 2\lambda_v t + \frac{(2\lambda t)^2}{2!} - \frac{(2\lambda t)^3}{3!}\right)\right.$$
$$\left. - 2\left(1 - 3\lambda t + \frac{(3\lambda t)^2}{2!} - \frac{(3\lambda_v t)^3}{3!}\right)\right] \qquad (4.44a)$$

Expanding the preceding equation and retaining only the first four terms yields

$$R_{TMR} = 1 - \lambda_v t + \frac{(\lambda_v t)^2}{2} - 3(\lambda t)^2 \qquad (4.44b)$$

Furthermore, we are mainly interested in the cases where $\lambda_v < \lambda$; thus we can omit the third term (which is a second-order term in $\lambda_v$) and obtain

$$R_{TMR} = 1 - \lambda_v t - 3(\lambda t)^2 \qquad (4.44c)$$

If we want the effect of the voter to be negligible, we let $\lambda_v t < 3(\lambda t)^2$,

$$\frac{\lambda_v}{\lambda} < 3\lambda t \qquad (4.45)$$

One can compare this result with that given in Eq. (3.35) for two parallel systems by setting $n = 2$, yielding

$$\frac{\lambda_c}{\lambda} < \lambda t \qquad (3.35)$$

The approximate result is that the coupler must have a failure rate three times smaller than that of the voter for the same decrease in reliability.

One can examine the effect of repair on the above results by examining Eq. (4.27d) and Eq. (4.42). In both cases, the effect of the repair rate does not appear until the cubic term is encountered. The above comparisons only involved the linear and quadratic terms, so the effect of repair would only become apparent if the repair rate were very large and the time interval of interest were extended.

### 4.8.3 Comparison of TMR, Parallel, and Standby Systems

Another advantage of voter reliability over parallel and standby reliability is that there is a straightforward scheme for implementing voter redundancy (e.g., Fig. 4.8). Of course, one can also make redundant couplers for parallel or standby systems, but they may be more complex than redundant voters.

It is easy to make a simple model for Fig. 4.8. Assume that the voters fail so that their outputs are stuck-at-zero or stuck-at-one and that voter failures do not corrupt the outputs of the circuits that feed the voters (e.g., $A_1$, $B_1$, and $C_1$). Assume just a single stage ($A_1$, $B_1$, and $C_1$) and a single redundant voter system ($V_1$, $V'_1$, and $V''_1$). The voter works if two or three of the three voters work. Thus this is the same formula for TMR systems, and the reliability of the system becomes

$$R_{\text{TMR}} \times R_{\text{voter}} = (3p_c^2 - 2p_c^3) \times (3p_v^2 - 2p_v^3) \qquad (4.46)$$

It is easy to evaluate the advantages of redundant voters. Assume that $p_c = 0.9$ and that the voter is 10 times as reliable: $(1 - p_c) = 0.1$, $(1 - p_v) = 0.01$, and $p_v = 0.99$. With a single voter, $R = 0.99[3(0.9)^2 - 2(0.9)^3] = 0.99 \times 0.972 = 0.962$. In the case of a redundant voter, we have $[3(0.99)^2 - 2(0.99)^3] \times [3(0.9)^2 - 2(0.9)^3] = 0.999702 \times 0.972 = 0.9717$. The redundant voter is thus significant; if the voter is less reliable, voter redundancy is even more effective. Assume that $p_v = 0.95$; for a single voter, $R = 0.95\,[3(0.9)^2 - 2(0.9)^3] = 0.95 \times 0.972 = 0.923$. In the case of a redundant voter, we have $[3(0.95)^2 - 2(0.95)^3] \times [3(0.9)^2 - 2(0.9)^3] = 0.99275 \times 0.972 = 0.964953$.

The foregoing calculations and discussions were performed for a TMR circuit with a single voter or redundant voters. It is possible to extend these computations to the subsystem level for a system such as that depicted in Fig. 4.8. In addition, one can repair a failed component of a redundant voter; thus one can use the analysis techniques previously derived for TMR and 5MR systems where the systems and voters can both be repaired. However, repair of voters really begs a larger question: How will we modularize the system architecture?

Assume one is going to design the system architecture with redundant voters and voting at a subsystem level. If the voters are to be placed on a single chip along with the circuits, then there is no separate repair of a voter system—only repair of the circuit and voter subsystem. The alternative is to make a separate chip for the $N$ circuits and a separate chip for the redundant voter. The proper strategy to choose depends on whether there will be scheduled downtime for the system during which testing and replacement can occur and also whether the chips have sufficient test points. No general conclusion can be reached; the system architecture should be critiqued with these issues in mind.

## 4.9 AVAILABILITY OF $N$-MODULAR REDUNDANCY WITH REPAIR AND IMPERFECT VOTERS

### 4.9.1 Introduction

When repair is present in a system, it is often possible for the system to fail and be down for a short period of time without serious operational effects. Suppose a computer used for electronic funds transfers is down for a short period of time. This is not catastrophic if the system is designed so that it can tolerate brief outages and perform the funds transfers at a later time period. If the system is designed to be self-diagnostic, and if a technician and a replacement plug in boards are both available, the machine can be restored quickly to operational status. For such systems, availability is a useful measure of system performance, as with reliability, and is the probability that the system is up at any point in time. It can be measured during operation by recording the downtimes and operating times for several failure and repair cycles. The availability is given by the ratio of the sum of the uptimes for the system divided by the sum of the uptimes and the downtimes. (Formally, this ratio becomes the availability in the limit as the system operating time approaches infinity.) The availability $A(t)$ is the probability that the system is up at time $t$, which can be written as a sum of probabilities:

$$A(t) = P(\text{no failures}) + P(\text{one failure + one repair})$$
$$+ P(\text{two failures + two repairs})$$
$$+ \cdots + P(n \text{ failures} + n \text{ repairs}) + \cdots \quad (4.47)$$

Availability is always higher than reliability, since the first term in Eq. (4.47) is the reliability and all the other terms are positive numbers. Note that only the first few terms in Eq. (4.47) are significant for a moderate time interval and higher-order terms become negligible. Thus one could evaluate availability analytically by computing the terms in Eq. (4.47); however, the use of the Markov model simplifies such a computation.

# 180  N-MODULAR REDUNDANCY

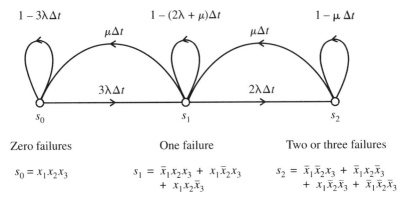

**Figure 4.13**  A Markov availability model for a TMR system with repair.

## 4.9.2  Markov Availability Models

A brief introduction to availability models appeared in Section 3.8.5; such computations will continue to be used in this section, and availabilities for TMR systems, parallel systems, and standby systems will be computed and compared. As in the previous section, we will make use of the fact that the Markov availability model given in Fig. 3.16 will hold with minor modifications (see Fig. 4.13). In Fig. 3.16, the value of $\lambda'$ is either one or two times $\lambda$, but in the case of TMR, it is three times $\lambda$. For the second transmission between $s_1$ and $s_2$ for the TMR system, there are two possibilities of failure; thus the transition rate is $2\lambda$. Since there is only one repairman, the repair rate is $\mu$.

A set of Markov equations can be written that will hold for two in parallel and two in standby, as well as for TMR. The algorithm used in the preceding chapter will be employed. The terms 1 and $\Delta t$ are deleted from Fig. 4.13. The time derivative of the probability of being in state $s_0$ is set equal to the "flows" from the other nodes; for example, $-\lambda' P_{s_0}(t)$ is from the self-loop and $\mu' P_{s_1}(t)$ is from the repair branch. Applying the algorithm to the other nodes and using algebraic manipulation yields the following:

$$\dot{P}_{s_0}(t) + \lambda' P_{s_0}(t) = \mu' P_{s_1}(t) \tag{4.48a}$$

$$\dot{P}_{s_1}(t) + (\lambda + \mu') P_{s_1}(t) = \lambda' P_{s_0}(t) + \mu'' P_{s_2}(t) \tag{4.48b}$$

$$\dot{P}_{s_2}(t) + \mu'' P_{s_2}(t) = \lambda P_{s_1}(t) \tag{4.48c}$$

$$P_{s_0}(0) = 1 \quad P_{s_1}(0) = P_{s_2}(0) = 0 \tag{4.48d}$$

The appropriate values of parameters for this set of equations is given in Table 4.8. A complete solution of these equations is given in Shooman [1990, pp. 344–347]. We will use the Laplace transform theorems previously introduced to simplify the solution.

The Laplace transforms of Eqs. (4.48a–d) become

## AVAILABILITY OF N-MODULAR REDUNDANCY WITH REPAIR

**TABLE 4.8** Parameters of Eqs. (4.48a–d) for Various Systems

| System | $\lambda$ | $\lambda'$ | $\mu'$ | $\mu''$ |
|---|---|---|---|---|
| Two in parallel | $\lambda$ | $2\lambda$ | $\mu$ | $\mu$ |
| Two standby | $\lambda$ | $\lambda$ | $\mu$ | $\mu$ |
| TMR | $2\lambda$ | $3\lambda$ | $\mu$ | $\mu$ |

$$(s+\lambda')P_{s_0}(s) \quad -\mu'P_{s_1}(s) \quad\quad\quad\quad = 1 \quad (4.49a)$$
$$-\lambda'P_{s_0}(s) + s(s+\lambda+\mu')P_{s_1}(s) \quad -\mu''P_{s_2}(s) = 0 \quad (4.49b)$$
$$-\lambda P_{s_1}(s) + (s+\mu'')P_{s_2}(s) = 0 \quad (4.49c)$$

In the case of a system composed of two in parallel, two in standby, or TMR, the system is up if it is in state $s_0$ or state $s_1$. The availability is thus the sum of the probabilities of being in one of these two states. If one uses Cramer's rule or a similar technique to solve Eqs. (4.49a–c), one obtains a ratio of polynomials in $s$ for the availability:

$$A(s) = P_{s_0}(s) + P_{s_1}(s) = \frac{s^2 + (\lambda+\lambda'+\mu'+\mu'')s + (\lambda'\mu''+\mu'\mu'')}{s[s^2 + (\lambda+\lambda'+\mu'+\mu'')s + (\lambda\lambda'+\lambda'\mu''+\mu'\mu'')]}$$

(4.50)

Before we begin applying the various Laplace transform theorems to this availability function, we should discuss the nature of availability and what sort of analysis is needed. In general, availability always starts at 1 because the system is always assumed to be up at $t = 0$. Examination of Eq. (4.47) shows that initially near $t = 0$, the availability is just the reliability function that of course starts at 1. Gradually, the next term $P$(one failure and one repair) becomes significant in the availability equation; as time progresses, other terms in the series contribute. Although the overall effect based on the summation of these many terms is hard to understand, we note that they generally lead to a slow decay of the availability function to some steady-state value that is reasonably close to 1. Thus the initial behavior of the availability function is not as important as that of the reliability function. In addition, the MTTF is not always a significant measure of system behavior. The one measure of interest is the final value of the availability function. If the availability function for a particular system has an initial value of unity at $t = 0$ and decays slowly to a steady-state value close to unity, this system must always have a high value of availability, in which case the final value is a lower bound on the availability. Examining Table B7 in Appendix B, Section B8.1, we see that the final value and initial value theorems both depend on the limit of $sF(s)$ [in our case, $sA(s)$] as $s$ approaches 0 and $\infty$. The initial value is when $s$ approaches $\infty$. Examination of Eq. (4.50) shows that multiplication of $A(s)$ by $s$ results in a cancellation of

**TABLE 4.9 Comparison of the Steady-State Availability, Eq. (4.50) for Various Systems**

| System | Eq. (4.50) | $\mu = \lambda$ | $\mu = 10\lambda$ | $\mu = 100\lambda$ |
|---|---|---|---|---|
| Two in parallel | $\dfrac{\mu(2\lambda + \mu)}{2\lambda^2 + 2\lambda\mu + \mu^2}$ | 0.6 | 0.984 | 0.9998 |
| Two standby | $\dfrac{\mu(\lambda + \mu)}{\lambda^2 + \lambda\mu + \mu^2}$ | 0.667 | 0.991 | 0.9999 |
| TMR | $\dfrac{\mu(3\lambda + \mu)}{6\lambda^2 + 3\lambda\mu + \mu^2}$ | 0.4 | 0.956 | 0.9994 |

the multiplying $s$ term in the denominator. As $s$ approaches infinity, both the numerator and denominator polynomials approach $s^2$; thus the ratio approaches 1, as it should. However, to find the final value, we let $s$ approach zero and obtain the ratio of the two constant terms given in Eq. (4.51).

$$A(\text{steady state}) = \frac{(\lambda'\mu'' + \mu'\mu'')}{(\lambda\lambda' + \lambda'\mu'' + \mu'\mu'')} \quad (4.51)$$

The values of the parameters given in Table 4.8 are substituted in this equation, and the steady-state availabilities are compared for the three systems noted in Table 4.9.

Clearly, the Laplace transform has been of great help in solving for steady-state availability and is superior to the simplified time-domain method: (a) let all time derivatives equal 0; (b) delete one of the resulting algebraic equations; (c) add the equation's sum of all probabilities to equal 1; and (d) solve (see Section B7.5).

Table 4.9 shows that the steady-state availability of two elements in standby exceeds that of two parallel items by a small amount, and they both exceed the TMR system by a greater margin. In most systems, the repair rate is much higher than the failure, so the results of the last column in the table are probably the most realistic. Note that these steady-state availabilities depend only on the ratio $\mu/\lambda$. Before one concludes that the small advantages of one system over another in the table are significant, the following factors should be investigated:

- It is assumed that a standby element cannot fail when it is in standby. This is not always true, since batteries discharge in standby, corrosion can occur, insulation can break down, etc., all of which may significantly change the comparison.
- The reliability of the coupling device in a standby or parallel system is more complex than the voter reliability in a TMR circuit. These effects on availability may be significant.
- Repair in any of these systems is predicated on knowing when a system

AVAILABILITY OF *N*-MODULAR REDUNDANCY WITH REPAIR   183

has failed. In the case of TMR, we gave a simple logic circuit that would detect which element has failed. The equivalent detection circuit in the case of a parallel or standby system is more complex and may have poorer coverage.

Some of these effects are treated in the problems at the end of this chapter. It is likely, however, that the detailed design of comparative systems must be modeled to make a comprehensive comparison.

A simple numerical example will show the power of increasing system availability using parallel and standby system configurations. In Section 3.10.1, typical failure and repair information for a circa-1985 transaction-processing system was quoted. The time between failures of once every two weeks translates into a failure rate $\lambda = 1/(2 \times 168) = 2.98 \times 10^{-3}$ failures/hour, and the time to repair of one hour becomes a repair rate $\mu = 1$ repair/hour. These values were shown to yield a steady-state availability of 0.997—a poor value for what should be a highly reliable system. If we assume that the computer system architecture will be configured as a parallel system or a standby system, we can use the formulas of Table 4.9 to compute the expected increase in availability. For an ordinary parallel system, the steady-state availability would be 0.999982; for a standby system, it would be 0.9999911. Both translate into unavailability values $\bar{A} = 1 - A$ of $1.8 \times 10^{-5}$ and $8.9 \times 10^{-6}$. The unavailability of the single system would of course be $3 \times 10^{-3}$. The steady-state availability of the Stratus system was discussed in Section 3.10.2 and, based on claimed downtime, was computed as 0.9999905, which is equivalent to an unavailability of $95 \times 10^{-7}$. In Section 3.10.1, the Tandem unavailability, based on hypothetical goals, was $4 \times 10^{-6}$. Comparison of these four unavailability values yields the following: (a) for a single system, $3,000 \times 10^{-6}$; (b) for a parallel system, $18 \times 10^{-6}$; (c) for a standby system, $8.9 \times 10^{-6}$; (d) for a Stratus system, $9.5 \times 10^{-6}$; and (e) for a Tandem system, $4 \times 10^{-6}$. Also compare the Bell Labs' ESS switching system unavailability goals and demonstrated availability of $5.7 \times 10^{-6}$ and $3.8 \times 10^{-6}$. (See Table 1.4.) Of course, more definitive data or complete models are needed for detailed comparisons.

### 4.9.3 Decoupled Availability Models

A simplified technique can be used to compute the steady-state value of availability for parallel and TMR systems. Availability computations really involve the evaluation of certain conditional probabilities. Since conditional probabilities are difficult to deal with, we introduced the Markov model computation technique. There is a case in which the dependent probabilities become independent and the computations simplify. We will introduce this case by focusing on the availability of two parallel elements.

Assume that we wish to compute the steady-state availability of two parallel elements, $A$ and $B$. The reliability is the probability of no system failures in interval 0 to $t$, which is the probability that either $A$ or $B$ is good,

$P(A_g + B_g) = P(A_g) + P(B_g) - P(A_g B_g)$. The subscript "g" means that the element is good, that is, has not failed. Similarly, the availability is the probability that the system is up at time $t$, which is the probability that either $A$ or $B$ is up, $P(A_{up} + B_{up}) = P(A_{up}) + P(B_{up}) = P(A_{up} B_{up})$. The subscript "up" means that the element is up, that is, is working at time $t$. The product terms in each of the above expressions, $P(A_g B_g) = P(A_g) P(B_g | A_g)$ and $P(A_{up} B_{up}) = P(A_{up}) P(B_{up} | A_{up})$ are the conditional probabilities discussed previously. If there are two repairmen—one assigned to component $A$ and one assigned to component $B$—the events $(B_g | A_g)$ and $(B_{up} | A_{up})$ become *decoupled*, that is, the events are independent. The coupling (dependence) comes from the repairmen. If there is only one repairman and element $A$ is down and being repaired, then if element $B$ fails, it will take longer to restore $B$ to operation; the repairman must first finish fixing $A$ before working on $B$. In the case of individual repairmen, there is no wait for repair of the second element if two items have failed because each has its own assigned repairman. In the case of such decoupling, the dependent probabilities become independent and $P(B_g | A_g) = P(B_g)$ and $P(B_{up} | A_{up}) = P(B_{up})$. This represents considerable simplification; it means that one can compute $P(B_g)$, $P(A_g)$, $P(B_{up})$, and $P(A_{up})$ separately and substitute into the reliability or availability equation to achieve a simple solution. Before we apply this technique and illustrate the simplicity of the solution, we should comment that because of the high cost, it is unlikely that there will be two separate repairmen. However, if the repair rate is much larger than the failure rate, $\mu \gg \lambda$, the decoupled case is approached. This is true since repairs are relatively fast and there is only a small probability that a failed element $A$ will still be under repair when element $B$ fails. For a more complete discussion of this decoupled approximation, consult Shooman [1990, pp. 521–529].

To illustrate the use of this approximation, we calculate the steady-state availability of two parallel elements. In the steady state,

$$A(\text{steady state}) = P(A_{ss}) + P(B_{ss}) - P(A_{ss})P(B_{ss}) \quad (4.52)$$

The steady-state availability for a single element is given by

$$A_{ss} = \frac{\mu}{\lambda + \mu} \quad (4.53)$$

One can verify this formula by reading the derivation in Appendix B, Sections B7.3 and B7.4, or by examining Fig. 3.16. We can reduce Fig. 3.16 to a single element model by setting $\lambda = 0$ to remove state $s_2$ and letting $\lambda' = \lambda$ and $\mu' = \mu$. Solving Eqs. (3.71a, b) for $P_{s_0}(t)$ and applying the final value theorem (multiply by $s$ and let $s$ approach 0) also yields Eq. (4.53). If $A$ and $B$ have identical failure and repair rates, substitution of Eq. (4.53) into Eq. (4.52) for both $A_{ss}$ and $B_{ss}$ yields

$$A_{ss} = \frac{2\mu}{\lambda + \mu} - \left(\frac{\mu}{\lambda + \mu}\right)^2 = \frac{\mu(2\lambda + \mu)}{(\lambda + \mu)^2} \quad (4.54)$$

If we compare this result with the exact one in Table 4.9, we see that the numerator is the same and the denominator differs only by a coefficient of two in the $\lambda^2$ term. Furthermore, since we are assuming that $\mu \gg \lambda$, the difference is very small.

We can repeat this simplification technique for a TMR system. The TMR reliability equation is given by Eq. (4.2), and modification for computing the availability yields

$$A(\text{steady state}) = [P(A_{ss})]^2[3 - P(A_{ss})] \quad (4.55)$$

Substitution of Eq. (4.53) into Eq. (4.55) gives

$$A(\text{steady state}) = \left(\frac{\mu}{\lambda + \mu}\right)^2 \left(3 - \frac{2\mu}{\lambda + \mu}\right) = \left(\frac{\mu}{\lambda + \mu}\right)^2 \left(\frac{3\lambda + \mu}{\lambda + \mu}\right) \quad (4.56)$$

There is no obvious comparison between Eq. (4.56) and the exact TMR availability expression in Table 4.9. However, numerical comparison will show that the formulas yield nearly equivalent results.

The development of approximate expressions for a standby system requires some preliminary work. The Poisson distribution (Appendix A, Section A5.4) describes the probabilities of success and failure in a standby system. The system succeeds if there are no failures or one failure; thus the reliability expression is computed from the Poisson distribution as

$$R(\text{standby}) = P(0 \text{ failures}) + P(1 \text{ failure}) = e^{-\lambda t} + \lambda t e^{-\lambda t} \quad (4.57)$$

If we wish to transform this equation in terms of the probability of success $p$ of a single element, we obtain $p = e^{-\lambda t}$ and $\lambda t = -\ln p$. (See also Shooman [1990, p. 147].) Substitution into Eq. (4.57) yields

$$R(\text{standby}) = p(1 - \ln p) \quad (4.58)$$

Finally, substitution in Eq. (4.58) of the steady-state availability from Eq. (4.53) yields an approximate expression for the availability of a standby system as follows:

$$A(\text{steady state}) = \left[\frac{\mu}{\lambda + \mu}\right]\left[1 - \ln\left(\frac{\mu}{\lambda + \mu}\right)\right] \quad (4.59)$$

Comparing Eq. (4.59) with the exact expression in Table 4.9 is difficult because of the different forms of the equations. The exact and approximate expressions are compared numerically in Table 4.10. Clearly, the approximations are close to the exact values. The best way to compare availability numbers, since they are all so close to unity, is to compare the differences with the unavailability $1 - A$. Thus, in Table 4.10, the difference in the results for the parallel system is $(0.99990197 - 0.99980396)/(1 - 0.99980396) = 0.49995$, or about 50%. Similarly, for the standby system, the difference in the results is $(0.999950823 - 0.999901)/(1 - 0.999901) = 0.50326$, which is also 50%. For the TMR system, the difference in the results is $(0.999707852 - 0.999417815)/(1 - 0.999417815) = 0.498819$—again, 50%. The reader will note that these results are good approximations, all approximations yield a slightly higher result than the exact value, and all are satisfactory for preliminary calculations. It is recommended that an exact computation be made once a design is chosen; however, these approximations are always useful in checking more exact results obtained from analysis or a computer program.

The foregoing approximations are frequently used in industry. However, it is important to check their accuracy. The first reference known to the author of such approximations appears in Calabro [1962, pp. 136–139].

## 4.10 MICROCODE-LEVEL REDUNDANCY

One can employ redundancy at the microcode level in a computer. Microcode consists of the elementary instructions that control the CPU or microprocessor—the heart of modern computers. Microinstructions perform such elementary operations as the addition of two numbers, the complement of a number, and shift left or right operations. When one structures the microcode of the computing chip, more than one algorithm can often be used to realize a particular operation. If several equivalent algorithms can be written, each one can serve the same purpose as the independent circuits in the $N$-modular redundancy. If the algorithms are processed in parallel, there is no reduction in computing speed except for the time to perform a voting algorithm. Of course, if all the algorithms use some of the same elements, and if those elements are faulty, the computations are not independent. One of the earliest works on microinstruction redundancy is Miller [1967].

## 4.11 ADVANCED VOTING TECHNIQUES

The voting techniques described so far in this chapter have all followed a simple majority voting logic. Many other techniques have been proposed, some of which have been implemented. This section introduces a number of these techniques.

### 4.11.1 Voting with Lockout

When $N$-modular redundancy is employed and $N$ is greater than three, additional considerations emerge. Let us consider a 4-level majority voter as an

TABLE 4.10 Comparison of the Exact and Approximate Steady-State Availability Equations for Various Systems

| System | Exact, Eq. (4.50) | Approximate, Eqs. (4.54), (4.56), and (4.59) | Exact, $\mu = 100\lambda$ | Approximate, $\mu = 100\lambda$ |
|---|---|---|---|---|
| Two in parallel | $\dfrac{\mu(2\lambda + \mu)}{2\lambda^2 + 2\lambda\mu + \mu^2}$ | $\dfrac{\mu(2\lambda + \mu)}{(\lambda + \mu)^2}$ | 0.99980396 | 0.99990197 |
| Two standby | $\dfrac{\mu(\lambda + \mu)}{\lambda^2 + \lambda\mu + \mu^2}$ | $\left(\dfrac{\mu}{\lambda + \mu}\right)\left[1 - \ln\left(\dfrac{\mu}{\lambda + \mu}\right)\right]$ | 0.999901 | 0.999950823 |
| TMR | $\dfrac{\mu(3\lambda + \mu)}{6\lambda^2 + 3\lambda\mu + \mu^2}$ | $\left(\dfrac{\mu}{\lambda + \mu}\right)^2 \left(\dfrac{3\lambda + \mu}{\lambda + \mu}\right)$ | 0.9994417815 | 0.999707852 |

example. (This is essentially the same architecture that is embedded into the Space Shuttle's primary flight control system—discussed in Chapter 5 as an example of software redundancy and shown in Fig. 5.19. However, if we focus on the first four computers in the primary flight control system, we have an example of 4-level voting with lockout. The backup flight control system serves as an additional level of redundancy; it will be discussed in Chapter 5.)

The question arises of what to do with a failed system when $N$ is greater than three. To provide a more detailed discussion, we introduce the fact that failures can be permanent as well as transient. Suppose that hardware $B$ in Fig. 5.19 experiences a failure and we know that it is permanent. There is no reason to leave it in the circuit if we have a way to remove it. The reasoning is that if there is a second failure, there is a possibility that the two failed elements will agree and the two good elements will agree, creating a standoff. Clearly, this can be avoided if the first element is disconnected (locked out) from the comparison. In the Space Shuttle control system, this is done by an astronaut who has access to onboard computer diagnostic information and also by consultation with Ground Control, which has access to telemetered data on the control system. The switch shown at the output of each computer in Fig. 5.19 is activated by an astronaut after appropriate deliberation and can be reversed at any time. NASA refers to this system as fail-safe–fail-operational, meaning that the system can experience two failures, can disconnect the two failed computers, and can have two remaining operating computers connected in a comparison arrangement. The flight rules that NASA uses to decide on safe modes of shuttle operation would rule on whether the shuttle must terminate a mission if only two valid computers in the primary system remain. In any event, there would clearly be an emergency situation in which the shuttle is still in orbit and one of the two remaining computers fails. If other tests could determine which computer gives valid information, then the system could continue with a single computer. One such test would be to switch out one of the computers and see if the vehicle is still stable and handles properly. The computers could then be swapped, and stability and control can be observed for the second computer. If such a test identifies the failed computer, the system is still operating with one good computer. Clearly, with Ground Control and an astronaut dealing with an emergency, there is the possibility of switching back in a previously disconnected computer in the hope that the old failure was only a transient problem that no longer exists. Many of these cases are analyzed and compared in the following paragraphs.

If we consider that the lockout works perfectly, the system will succeed if there are 0, 1, or 2 failures. The probability computation is simple using the binomial distribution.

$$\begin{aligned} R(2:4) &= B(4:4) + B(3:4) + B(2:4) \\ &= [p^4] + [4p^3 - 4p^4] + [6p^2 - 12p^3 + 6p^4] \\ &= 3p^4 - 8p^3 + 6p^2 \end{aligned} \qquad (4.60)$$

ADVANCED VOTING TECHNIQUES   189

**TABLE 4.11  Comparison of Reliabilities for Various Voting Systems**

| Single Element | TMR Voting | Two-out-of-Four | One-out-of-Four |
|---|---|---|---|
| $p$ | $p^2(3-2p)$ | $p^2(3p^2 - 8p + 6)$ | $p(4p^2 - p^3 - 6p + 4)$ |
| 1 | 1 | 1 | 1 |
| 0.8 | 0.896 | 0.9728 | 0.9984 |
| 0.6 | 0.648 | 0.8208 | 0.9744 |
| 0.4 | 0.352 | 0.5248 | 0.8704 |
| 0.2 | 0.104 | 0.1808 | 0.5904 |
| 0 | 0 | 0 | 0 |

The reliability will be higher if we can detect and isolate a third failure. To compute the reliability, we start with Eq. (4.60) and add the binomial probability $B(1:3) = (-p^4 + 4p^3 - 6p^2 + 4p)$. The result is given in the following equation:

$$R(1:4) = R(2:4) + B(1:4)$$
$$= -p^4 + 4p^3 - 6p^2 + 4p \quad (4.61)$$

Note that deriving Eqs. (4.60) and (4.61) involves some algebra, and a simple check on the result can help detect *some* common errors. We know that if every element in a system has failed, $p = 0$ and the reliability must be 0 regardless of the system configuration. Thus, one necessary but not sufficient check is to substitute $p = 0$ in the reliability polynomial and see if the reliability is 0. Clearly both Eqs. (4.60) and (4.61) satisfy this requirement. Similarly, we can check to see that the reliability is 1 when $p = 1$. Again, both equations also satisfy this necessary check. Equations (4.60) and (4.61) are compared with a TMR system Eq. (4.43a) and a single element in Table 4.11 and Fig. 4.14. Note that the TMR voter is poorer than a single element for $p < 0.5$ but better than a single element for $p > 0.5$.

### 4.11.2  Adjudicator Algorithms

A comprehensive discussion of various voting techniques appears in McAllister and Vouk [1996]. The authors frame the discussion of voting based on software redundancy—the use of two or more independently developed versions of the same software. In this book, *N*-version software is discussed in Sections 5.9.2 and 5.9.3. The more advanced voting techniques will be discussed in this section since most apply to both hardware and software.

McAllister and Vouk [1996] introduce a more general term for the voter element: an *adjudicator*, the underlying logic of which is the *adjudicator algorithm*. The adjudicator algorithm for majority voting (*N*-modular redundancy) is simply $n + 1$ or more agreements out of $N = 2n + 1$ elements (see also Section 4.4), where $n$ is an integer greater than 0 (it is commonly 1 or 2). This algorithm is formulated for an odd number of elements. If we wish to

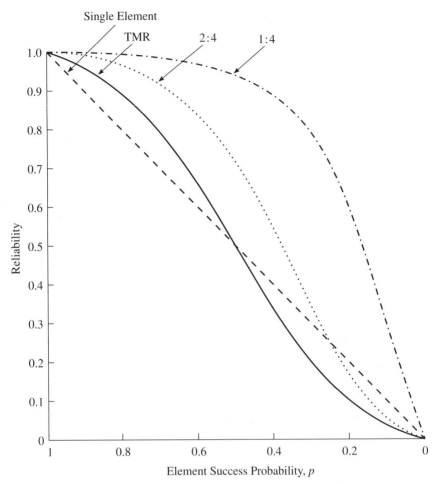

**Figure 4.14** Reliability comparison of the three voter circuits given in Table 4.11.

also include even values of $N$, we can describe the algorithm as an $m$-out-of-$N$ voter, with $N$ taking on any integer value equal to or larger than 3. The algorithm represents agreement if $m$ or more element outputs agree and $m$ is the integer, which is the ceiling function of $(N + 1)/2$ written as $m \geq \lceil (N + 1)/2 \rceil$. The ceiling function, $\lceil x \rceil$, is the smallest integer that is greater than or equal to $x$ (e.g., the roundup function).

### 4.11.3 Consensus Voting

If there is a sizable number of elements that process in parallel (hardware or software), then a number of agreement situations arise. The majority vote may fail, yet there may be agreement among some of the elements. An adjudication algorithm can be defined for the consensus case, which is more complex than majority voting. Again, $N$ is the number of parallel elements ($N > 1$) and $k$ is the largest number of element outputs that agree. The symbol $O_k$ denotes the set of $k$-element outputs that agree. In some cases, there can be more than one set of agreements, resulting in $O_{k_i}$, and the adjudication must choose between the multiple agreements. A flow chart is given in Fig. 4.15 that is based on the consensus voting algorithm in McAllister and Vouk [1996, p. 578].

If $k = 1$, there are obviously ties in the consensus algorithm. A similar situation ensues if $k > 1$, but because there is more then one group with the same value of $k$, a tie-breaking algorithm must be used. One such algorithm is a random choice among the ties; another is to test the elements for correct operation, which in terms of software version consensus is called acceptance testing of the software. Initially, such testing may seem better suited to software than to hardware; in reality, however, such is not the case because hardware testing has been used in the past. The Carousel Inertial Navigation System used on the early Boeing 747 and other aircraft had three stable platforms, three computers, and a redundancy management system that performed majority voting. One means of checking the validity of any of the computers was to submit a stored problem for solution and to check the results with a stored solution. The purpose was to help diagnose computer malfunctions and lock a defective computer out of the system. Also during the time when adaptive flight control systems were in use, some designs used test signals mixed with the normal control signals. By comparing the input test signals and the output response, one could measure the parameters of the aircraft (the coefficients of the governing differential equations) and dynamically adjust the feedback signals for best control.

### 4.11.4 Test and Switch Techniques

The discussion in the previous section established the fact that hardware testing is possible in certain circumstances. Assuming that such testing has a high probability of determining success or failure of an element and that two or more elements are present, we can operate with element one alone as long as it tests valid. When a failure of element one is detected, we can switch to element two, etc. The logic of such a system differs little from that of the standby system shown in Fig. 3.12 for the case of two elements, but the detailed implementation of test and switch may differ somewhat from the standby system. When these concepts are applied to software, the adjudication algorithm becomes an *acceptance test*. The switch to an earlier state of the process before failure was detected and the substitution of a second version of the software is called *rollback and recovery*, but the overall philosophy is generally referred to as the *recovery block* technique.

### 4.11.5 Pairwise Comparison

We assume that the number of elements is divisible by two, that is, $N = 2n$, where $n$ is an integer greater than one. The outputs of modules are compared

# 192 N-MODULAR REDUNDANCY

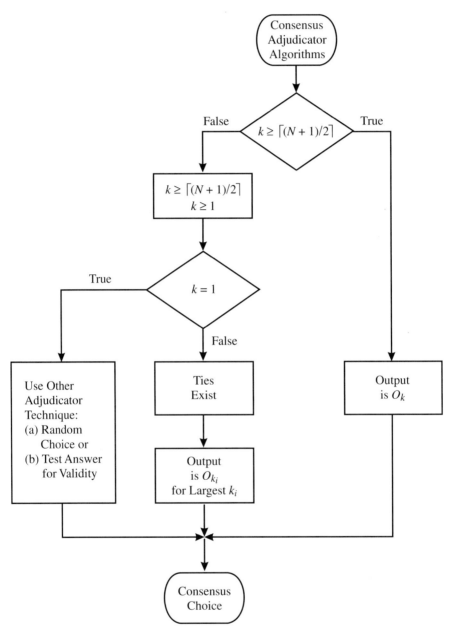

**Figure 4.15** Flow chart based on the consensus voting algorithm in McAllister and Vouk [1996, p. 578].

in pairs; if these pairs do not check, they are switched out of the circuit. The most practical application is where $n = 2$ and $N = 4$. For discussion purposes, we call the elements digital circuits $A$, $B$, $C$, and $D$. Circuit $A$ is compared with circuit $B$; circuit $C$ is compared with circuit $D$. The output of the $AB$ pair is then compared with the output of the $CD$ pair—an activity that I refer to as *pairwise comparison*. The software analog I call $N$ *self-checking programming*. The reader should reflect that this is essentially the same logic used in the Stratus system fault detection described in Section 3.11.

Assuming that all the comparitors are perfect, the pairwise comparison described in the preceding paragraph for $N = 4$ will succeed if (a), all four elements succeed ($ABCD$); (b), if three elements succeed ($\overline{A}BCD + A\overline{B}CD + AB\overline{C}D + ABC\overline{D}$); and (c), if two elements fail but in opposite pairs ($\overline{AB}CD + AB\overline{CD}$). In the case of (a), all elements succeed and no failures are present; in (b), on the other hand, the one failure means that one pair of elements disconnects itself but that the remaining pair continues to operate successfully. There are six ways for two failures to occur, but only the two ways given in (c) mean that a single pair fails because one failure in each pair represents a system failure. If each of the four elements is identical with a probability of success of $p$, the probability of success can be obtained as follows from the binomial distribution:

$$R(\text{pairwise}:4) = B(4:4) + B(3:4) + (2/6)B(2:4) \tag{4.62a}$$

Substituting terms from Eq. (4.60) into Eq. (4.62a) yields

$$\begin{aligned}R(\text{pairwise}:4) &= (p^4) + (4p^3 - 4p^4) + (1/3)(6p^2 - 12p^3 + 6p^4) \\ &= p^2(2 - p^2)\end{aligned} \tag{4.62b}$$

Equation (4.62b) is compared with other systems in Table 4.12, where we see that the pairwise voting is slightly worse than it is for TMR.

There are various other combinations of advanced voting techniques de-

**TABLE 4.12 Comparison of Reliabilities for Various Voting Systems**

| Single Element | Pairwise-out-of-Voting | Two-out-of-Four | TMR Voting |
|---|---|---|---|
| $p$ | $p^2(2 - p^2)$ | $p^2(3p^2 - 8p + 6)$ | $p^2(3 - 2p)$ |
| 1 | 1 | 1 | 1 |
| 0.8 | 0.8704 | 0.9728 | 0.896 |
| 0.6 | 0.590 | 0.8208 | 0.648 |
| 0.4 | 0.2944 | 0.5248 | 0.352 |
| 0.2 | 0.0784 | 0.1808 | 0.104 |
| 0 | 0 | 0 | 0 |

scribed by McAllister and Vouk [1996], who also compute and compare the reliability of many of these systems by assuming independent as well as dependent failures.

### 4.11.6 Adaptive Voting

Another technique for voting makes use of the fact that some circuit failure modes are intermittent or transient. In such a case, one does not wish to lock out (i.e., ignore) a circuit when it is behaving well (but when it is malfunctioning, it should be ignored). The technique of adaptive voting can be used to automatically switch between these situations [Pierce, 1961; Shooman, 1990, p. 324; Siewiorek, 1992, pp. 174, 178–182].

An ordinary majority voter may be visualized as a device that takes the average of the outputs and gives a one output if the average is $> 0.5$ and a zero output if the average is $\leq 0.5$. (In the case of software outputs, a range of values not limited to the range 0–1 will occur, and one can deal with various point estimates such as the average, the arithmetic mean of the min and max values, or, as McAllister and Vouk suggest, the median.) An adaptive voter may be viewed as a weighted sum where each outpt $x_i$ is weighted by a coefficient. The coefficient $a_i$ could be adjusted to equal the probability that the output $x_i$ was correct. Thus the test quantity of the adaptive voter (with an even number of elements) would be given by

$$\frac{a_1 x_1 + a_2 x_2 + \cdots + a_{2n+1} x_{2n+1}}{a_1 + a_2 + \cdots + a_{2n+1}} \qquad (4.63)$$

The coefficients $a_i$ can be adjusted dynamically by taking statistics on the agreement between each $x_i$ and the voter output over time. Another technique is to periodically insert test inputs and compare each output $x_i$ with the known (i.e., precomputed) correct output. If some $x_i$ is frequently in error, it should be disconnected. The adaptive voter adjusts $a_i$ to be a very small number, which is in essence the same thing. The reliability of the adaptive-voter scheme is superior to the ordinary voter; however, there are design issues that must be resolved to realize an adaptive voter in practice.

The reader will appreciate that there are many choices for an adjudicator algorithm that yield an associated set of architectures. However, cost, volume, weight, and simplicity considerations generally limit the choices to a few of the simpler configurations. For example, when majority voting is used, it is generally limited to TMR or, in the case of the Space Shuttle example, 4-level voting with lockout. The most complex arrangement the author can remember is a 5-level majority logic system used to control the *Apollo* Mission's main *Saturn* engine. For the Space Shuttle and Carousel navigation system examples, the astronauts/pilots had access to other information, such as previous problems with individual equipment and ground-based measurements or observations. Thus the accessibility of individual outputs and possible tests allow human operators to organize a wide variety of behaviors. Presently, commercial airliners are switching from inertial navigation systems to navigation using the satellite-based Global Positioning System (GPS). Handheld GPS receivers have dropped in price to the $100–$200 range, so one can imagine every airline pilot keeping one in his or her flight bag as a backup. A similar trend occurred in the 1780s when pocket chronometers dropped in price to less than £65. Ship captains of the East India Company as well as those of the Royal Navy (who paid out of their own pockets) eagerly bought these accurate watches to calculate longitude while at sea [Sobel, 1995, p. 162].

## REFERENCES

Arsenault, J. E., and J. A. Roberts. *Reliability and Maintainability of Electronic Systems.* Computer Science Press, Rockville, MD, 1980.

Avizienis, A., H. Kopetz, and J.-C. Laprie (eds.). *Dependable Computing and Fault-Tolerant Systems.* Springer-Verlag, New York, 1987.

Battaglini, G., and B. Ciciani. Realistic Yield-Evaluation of Fault-Tolerant Programmable-Logic Arrays. *IEEE Transactions on Reliability* (September 1998): 212–224.

Bell, C. G., and A. Newel-Pierce. *Computer Structures: Readings and Examples.* McGraw-Hill, New York, 1971.

Calabro, S. R. *Reliability Principles and Practices.* McGraw-Hill, New York, 1962.

Cardan, J. *The Book of my Life* (trans. J. Stoner). Dover, New York, 1963.

Cardano, G. *Ars Magna.* 1545.

Grisamone, N. T. Calculation of Circuit Reliability by Use of von Neuman Redundancy Logic Analysis. *IEEE Proceedings of the Fourth Annual Conference on Electronic Reliability,* October 1993. IEEE, New York, NY.

Hall, H. S., and S. R. Knight. *Higher Algebra.* 1887. Reprint, Macmillan, New York, 1957.

Iyanaga, S., and Y. Kawanda (eds.). *Encyclopedic Dictionary of Mathematics.* MIT Press, Cambridge, MA, 1980.

Knox-Seith, J. K. A Redundancy Technique for Improving the Reliability of Digital Systems. Stanford Electronics Laboratory Technical Report No. 4816-1. Stanford University, Stanford, CA, December 1963.

McAllister, D. F., and M. A. Vouk. "Fault-Tolerant Software Reliability Engineering." In *Handbook of Software Reliability Engineering,* M. R. Lyu (ed.). McGraw-Hill, New York, 1996, ch. 14, pp. 567–614.

Miller, E. Reliability Aspects of the Variable Instruction Computer. *IEEE Transactions on Electronic Computing* 16, 5 (October 1967): 596.

Moore, E. F., and C. E. Shannon. Reliable Circuits Using Less Reliable Relays. *Journal of the Franklin Institute* 2 (October 1956).

Pham, H. (ed.). *Fault-Tolerant Software Systems: Techniques and Applications.* IEEE Computer Society Press, New York, 1992.

Pierce, W. H. Improving the Reliability of Digital Systems by Redundancy and Adaptation. Stanford Electronics Laboratory Technical Report No. 1552-3, Stanford, CA: Stanford University, July 17, 1961.

Pierce, W. H. *Failure-Tolerant Computer Design*. Academic Press, Rockville, NY, 1965.

Randell, B. *The Origins of Digital Computers—Selected Papers*. Springer-Verlag, New York, 1975.

Shannon, C. E., and J. McCarthy (eds.). Probabilistic Logics and the Synthesis of Reliable Organisms from Unreliable Components. In *Automata Studies*, by J. von Neuman. Princeton University Press, Princeton, NJ, 1956.

Shiva, S. G. *Introduction to Logic Design*. Scott Foresman and Company, Glenview, IL, 1988.

Shooman, M. L. *Probabilistic Reliability: An Engineering Approach*. McGraw-Hill, New York, 1968.

Shooman, M. L. *Probabilistic Reliability: An Engineering Approach*, 2d ed. Krieger, Melbourne, FL, 1990.

Siewiorek, D. P., and R. S. Swarz. *The Theory and Practice of Reliable System Design*. The Digital Press, Bedford, MA, 1982.

Siewiorek, D. P., and R. S. Swarz. *Reliable Computer Systems Design and Evaluation*, 2d ed. The Digital Press, Bedford, MA, 1992.

Siewiorek, D. P., and R. S. Swarz. *Reliable Computer Systems Design and Evaluation*, 3d ed. A. K. Peters, www.akpeters.com, 1998.

Sobel, D. *Longitude*. Walker and Company, New York, 1995.

Toy, W. N. Dual Versus Triplication Reliability Estimates. *AT&T Technical Journal* (November/December 1987): 15–20.

Traverse, P. AIRBUS and ATR System Architecture and Specification. In *Software Diversity in Computerized Control Systems*, U. Voges (ed.), vol. 2 of *Dependable Computing and Fault-Tolerant Systems*, A. Avizienis (ed.). Springer-Verlag, New York, 1987, pp. 95–104.

Vouk, M. A., D. F. McAllister, and K. C. Tai. Identification of Correlated Failures of Fault-Tolerant Software Systems. *Proceedings of COMSAC '85*, 1985, pp. 437–444.

Vouk, M. A., A. Pradkar, D. F. McAllister, and K. C. Tai. Modeling Execution Times of Multistage $N$-Version Fault-Tolerant Software. *Proceedings of COMSAC '90*, 1990, pp. 505–511. (Also printed in Pham [1992], pp. 55–61.)

Vouk, M. A. et al. An Empirical Evaluation of Consensus Voting and Consensus Recovery Block Reliability in the Presence of Failure Correlation. *Journal of Computer and Software Engineering* 1, 4 (1993): 367–388.

Wakerly, J. F. *Digital Design Principles and Practice*, 2d ed. Prentice-Hall, Englewood Cliffs, NJ, 1994.

Wakerly, J. F. *Digital Design Principles 2.1*. Student CD-ROM package. Prentice-Hall, Englewood Cliffs, NJ, 2001.

## PROBLEMS

**4.1.** Derive the equation analogous to Eq. (4.9) for a four-element majority voting scheme.

## PROBLEMS 197

**4.2.** Derive the equation analogous to Eq. (4.9) for a five-element majority voting scheme.

**4.3.** Verify the reliability functions sketched in Fig. 4.2.

**4.4.** Compute the reliability of a 3-level majority voting system for the case where the failure rate is constant, $\lambda = 10^{-4}$ failures per hour, and $t = 1,000$ hours. Compare this with the reliability of a single system.

**4.5.** Repeat problem 4.4 for a 5-level majority voting system.

**4.6.** Compare the results of problem 4.4 with a single system: two elements in parallel, two elements in standby.

**4.7.** Compare the results of problem 4.5 with a single system: two elements in parallel, two elements in standby.

**4.8.** What should the reliability of the voter be if it increases the probability of failure of the system of problem 4.4 by 10%?

**4.9.** Compute the reliability at $t = 1,000$ hours of a system composed of a series connection of module 1 and module 2, each with a constant failure rate of $\lambda_1 = 0.5 \times 10^{-4}$ failures per hour. If we design a 3-level majority voting system that votes on the outputs of module 2, we have the same system as in problem 4.4. However, if we vote at the outputs of modules 1 and 2, we have an improved system. Compute the reliability of this system and compare it with problem 4.4.

**4.10.** Expand the reliability functions in series in the high-reliability region for the TMR 3–2–1 system and the TMR 3–2 system for the three systems of Fig. 4.3. [Include more terms than in Eqs. (4.14)–(4.16).]

**4.11.** Compute the MTTF for the expansions of problem 4.10, compare these with the exact MTTF for these systems, and comment.

**4.12.** Verify that an expansion of Eqs. (4.3a, b) leads to seven terms in addition to the term one, and that this leads to Eqs. (4.5a, b) and (4.6a, b).

**4.13.** The approximations used in plotting Fig. 4.3 are less accurate for the larger values of $\lambda t$. Recompute the values using the exact expressions and comment on the accuracy of the approximations.

**4.14.** Inspection of Fig. 4.4 shows that $N$-modular redundancy is of no advantage over a single unit at $t = 0$ (they both have a reliability of 1) and at $\lambda t = 0.69$ (they both have a reliability of 0.5). The maximum advantage of $N$-modular redundancy is realized somewhere in between them. Compute the ratio of the $N$-modular redundancy given by Eq. (4.17) divided by the reliability of a single system that equals $p$. Maximize (i.e., differentiate this ratio with respect to $p$ and set equal to 0) to solve for the value of $p$ that gives the biggest improvement in reliability. Since $p = e^{-\lambda t}$, what is the value of $\lambda t$ that corresponds to the optimum value of $p$?

**4.15.** Repeat problem 4.14 for the case of component redundancy and majority voting as shown in Fig. 4.5 by using the reliability equation given in Eq. (4.18).

**4.16.** Verify Grisamone's results given in Table 4.1.

**4.17.** Develop a reliability expression for the system of Fig. 4.8 assuming that (1): All circuits $A_i$, $B_i$, $C_i$, and the voters $V_i$ are independent circuits or independent integrated circuit chips.

**4.18.** Develop a reliability expression for the system of Fig. 4.8 assuming that (2): All circuits $A_i$, $B_i$, and $C_i$ are independent circuits or independent integrated circuit chips and the voters $V_i$, $V_i'$, and $V_i''$ are all on the same chip.

**4.19.** Develop a reliability expression for the system of Fig. 4.8 assuming that (3): All voters $V_i$, $V_i'$, and $V_i''$ are independent circuits or independent integrated circuit chips and circuits $A_i$, $B_i$, and $C_i$ are all on the same chip.

**4.20.** Develop a reliability expression for the system of Fig. 4.8 assuming that (4): All circuits $A_i$, $B_i$, and $C_i$ and all voters $V_i$, $V_i'$, and $V_i''$ are all on the same chip.

**4.21.** Section 4.5.3 discusses the difference between various failure models. Compare the reliability of a 1-bit TMR system under the following failure model assumptions:

(a) The failures are always s-a-1.

(b) The failures are always s-a-0.

(c) The circuits fail so that they always give the complement of the correct output.

(d) The circuits fail at a transient rate $\lambda_t$ and produce the complement of the correct output.

**4.22.** Repeat problem 4.21, but instead of calculating the reliability, calculate the probability that any one transmission is in error.

**4.23.** The circuit of Fig. 4.10 for a 32-bit word leads to a 512-gate circuit as described in this chapter. Using the information in Fig. B7, calculate the reliability of the voter and warning circuit. Using Eq. (4.19) and assuming that the voter reliability decreases the system reliability to 90% of what would be achieved with a perfect voter, calculate $p_c$. Again using Fig. B7, calculate the equivalent gate complexity of the digital circuit in the TMR scheme.

**4.24.** Repeat problem 4.10 for an $r$-level voting system.

**4.25.** Drive a set of Markov equations for the model given in Fig. 4.11 and show that the solution of each equation leads to Eqs. (4.25a–c).

**4.26.** Formulate a four-state model related to Fig. 4.11, as discussed in the text, where the component states *two failures* and *three failures* are not merged but are distinct. Solve the model for the four-state probabilities and show that the first two states are identical with Eqs. (4.25a, b) and that the sum of the third and fourth states equals Eq. (4.25c).

**4.27.** Compare the effects of repair on TMR reliability by plotting Eq. (4.27e), including the third term, with Eq. (4.27d). Both equations are to be plotted versus time for the cases where $\mu = 10\lambda$, $\mu = 25\lambda$, and $\mu = 100\lambda$.

**4.28.** Over what time range will the graphs in the previous problem be valid? (Hint: When will the next terms in the series become significant?)

**4.29.** The logic function for a voter was simplified in Eq. (4.23) and Table 4.5. Suppose that all four minterms given in Table 4.5 were included without simplification, which provides some redundancy. Compare the reliability of the unminimized voter with the minimized voter (cf. Shooman [1990, p. 324]).

**4.30.** Make a model for coupler reliability and for a TMR voter. Compare the reliability of two elements in parallel with that for a TMR.

**4.31.** Repeat problem 4.30 when both systems include repair.

**4.32.** Compare the MTTF of the systems in Table 3.4 with TMR and 5MR voter systems.

**4.33.** Repeat problem 4.32 for Table 3.5.

**4.34.** Compute the initial reliability for the systems of Tables 3.4 and 3.5 and compare with TMR and 5MR voter systems.

**4.35.** Sketch and compare the initial reliabilities of TMR and 5MR Eqs. (4.27d) and (4.39b). Both equations are to be plotted versus time for the cases where $\mu = 0$, $\mu = 10\lambda$, $\mu = 25\lambda$, and $\mu = 100\lambda$. Note that for $\mu = 100\lambda$ and for points where the reliability has decreased to 0.99 or 0.95, the series approximations may need additional terms.

**4.36.** Check the values in Table 4.6.

**4.37.** Check the series expansions and the values in Table 4.7.

**4.38.** Plot the initial reliability of the four systems in Table 4.7. Calculate the next term in the series expansion and evaluate the time at which it represents a 10% correction in the unreliability. Draw a vertical bar on the curve at this point. Repeat for each of the systems yielding a comparison of the reliabilities and a range of validity of the series expressions.

**4.39.** Compare the voter circuit and reliability of (a) a TMR system, (b) a 5MR system, and (c) five parallel elements with a coupler. Assume the voters and the coupler are imperfect. Compute and plot the reliability.

**4.40.** What time interval will be needed before the repair terms in the comparison made in problem 4.39 become significant?

**4.41.** It is assumed that a standby element cannot fail when it is in standby. However, this is not always true for many reasons; for example, batteries discharge in standby, corrosion can occur, and insulation can break down, all of which may significantly change the comparison. How large can the standby failure rate be and still be ignored?

**4.42.** The reliability of the coupling device in a standby or parallel system is more complex than the voter reliability in a TMR circuit. These effects on availability may be significant. How large can the coupling failure rate be and still be ignored?

**4.43.** Repair in any of these systems is predicted by knowing when a system has failed. In the case of TMR, we gave a simple logic circuit that would detect which element has failed. What is the equivalent detection circuit in the case of a parallel or standby system and what are the effects?

**4.44.** Check the values in Table 4.9.

**4.45.** Check the values in Table 4.10.

**4.46.** Add another line to Table 4.10 for 5-level modular redundancy.

**4.47.** Check the computations given in Tables 4.11 and 4.12.

**4.48.** Determine the range of $p$ for which the various systems in Table 4.11 are superior to a single element.

**4.49.** Determine the range of $p$ for which the various systems in Table 4.12 are superior to a single element.

**4.50.** Explain how a system based on the adaptive voting algorithm of Eq. (4.63) will operate if 50% of all failures are transient and clear in a short period of time.

**4.51.** Explain how a system based on the adaptive voting algorithm of Eq. (4.63) will operate if it is basically a TMR system and 50% of all element one failures are transient and 25% of all elements two and three failures are transient.

**4.52.** Repeat and verify the availability computations in the last paragraph of Section 4.9.2.

**4.53.** Compute the auto availability of a two-car family in which both the husband and wife need a car every day. Repeat the computation if a single car will serve the family in a pinch while the other car gets repaired. (See the brief discussion of auto reliability in Section 3.10.1 for failure and repair rates.)

**4.54.** At the end of Section 4.9.2 before the final numerical example, three factors not included in the model were listed. Discuss how you would model these effects for a more complex Markov model.

**4.55.** Can you suggest any approximate procedures to determine if any of the effects in problem 4.54 are significant?

**4.56.** Repeat problem 4.39 for the system availability. Make approximations where necessary.

**4.57.** Repeat problem 4.30 for system availability.

**4.58.** Repeat the derivation of Eq. (4.26c).

**4.59.** Repeat the derivation of Eq. (4.37).

**4.60.** Check the values given in Table 4.9.

**4.61.** Derive Eq. (4.59).

# 5

# SOFTWARE RELIABILITY AND RECOVERY TECHNIQUES

## 5.1 INTRODUCTION

The general approach in this book is to treat reliability as a system problem and to decompose the system into a hierarchy of related subsystems or components. The reliability of the entire system is related to the reliability of the components by some sort of structure function in which the components may fail independently or in a dependent manner. The discussion that follows will make it abundantly clear that software is a major "component" of the system reliability,[1] $R$. The reason that a separate chapter is devoted to software reliability is that the probabilistic models used for software differ from those used for hardware; moreover, hardware and software (and human) reliability can be combined only at a very high system level. (Section 5.8.5 discusses a macrosoftware reliability model that allows hardware and software to be combined at a lower level.) Specifically, if the hardware, software, and human failures are *independent* (often, this is not the case), one can express the system reliability, $R_{SY}$, as the product of the hardware reliability, $R_H$, the software reliability, $R_S$, and the human operator reliability, $R_O$. Thus, if independence holds, one can model the reliability of the various factors separately and combine them: $R_{SY} = R_H \times R_S \times R_O$ [Shooman, 1983, pp. 351–353].

This chapter will develop models that can be used for the software reliability. These models are built upon the principles of continuous random variables

---

[1]Another important "component" of system reliability is human reliability if an operator is involved in any control, monitoring, input, or similar task. A discussion of human reliability models is beyond the scope of this book; the reader is referred to Dougherty and Fragola [1988].

developed in Appendix A, Sections A6 and A7, and Appendix B, Section B3; the reader may wish to review these concepts while reading this chapter.

Clearly every system that involves a digital computer also includes a significant amount of software used to control system operation. It is hard to think of a modern business system, such as that used for information, transportation, communication, or government, that is not heavily computer-dependent. The microelectronics revolution has produced microprocessors and memory chips that are so cheap and powerful that they can be included in many commercial products. For example, a 1999 luxury car model contained 20–40 microprocessors (depending on which options were installed), and several models used local area networks to channel the data between sensors, microprocessors, displays, and target devices [*New York Times*, August 27, 1998]. Consumer products such as telephones, washing machines, and microwave ovens use a huge number of embedded microcomponents. In 1997, 100 million microprocessors were sold, but this was eclipsed by the sale of 4.6 billion embedded microcomponents. Associated with each microprocessor or microcomponent is memory, a set of instructions, and a set of programs [Pollack, 1999].

### 5.1.1 Definition of Software Reliability

One can define *software engineering* as the body of engineering and management technologies used to develop *quality, cost-effective, schedule-meeting* software. *Software reliability* measurement and estimation is one such technology that can be defined as the measurement and prediction of the *probability* that the software will perform its *intended function (according to specifications) without error* for a *given period of time*. Oftentimes, the design, programming, and testing techniques that contribute to high software reliability are included; however, we consider these techniques as part of the design process for the development of *reliable software*. Software reliability complements reliable software; both, in fact, are important topics within the discipline of software engineering. *Software recovery* is a set of fail-safe design techniques for ensuring that if some serious error should crash the program, the computer will automatically recover to reinitialize and restart its program. The software succeeds during software recovery if no crucial data is lost, or if an operational calamity occurs, but the recovery transforms a total failure into a benign or at most a troubling, nonfatal "hiccup."

### 5.1.2 Probabilistic Nature of Software Reliability

On first consideration, it seems that the outcome of a computer program is a deterministic rather than a probabilistic event. Thus one might say that the output of a computer program is *not* a random result. In defining the concept of a random variable, Cramer [Chapter 13, 1991] talks about spinning a coin as an experiment and the outcome (heads or tails) as the event. If we can control all aspects of the spinning and repeat it each time, the result will always be the same; however, such control needs to be so precise that it is practically impossible to repeat the experiment in an identical manner. Thus the event (heads or tails) is a random variable. The remainder of this section develops a similar argument for software reliability where the random element in the software is the changing set of inputs.

Our discussion of the probabilistic nature of software begins with an example. Suppose that we write a computer program to solve the roots $r_1$ and $r_2$ of a quadratic equation, $Ax^2 + Bx + C = 0$. If we enter the values 1, 5, and 6 for $A$, $B$, and $C$, respectively, the roots will be $r_1 = -2$ and $r_2 = -3$. A single test of the software with these inputs confirms the expected results. Exact repetition of this experiment with the *same values* of $A$, $B$, and $C$ will always yield the same results, $r_1 = -2$ and $r_2 = -3$, unless there is a hardware failure or an operating system problem. Thus, in the case of this computer program, we have defined a deterministic experiment. No matter how many times we repeat the computation with the same values of $A$, $B$, and $C$, we obtain the same result (assuming we exclude outside influences such as power failures, hardware problems, or operating system crashes unrelated to the present program). Of course, the real problem here is that after the first computation of $r_1 = -2$ and $r_2 = -3$ we do no useful work to repeat the same identical computation. To do useful work, we must vary the values of $A$, $B$, and $C$ and compute the roots for other input values. Thus the probabilistic nature of the experiment, that is, the correctness of the values obtained from the program for $r_1$ and $r_2$, is dependent on the input values $A$, $B$, and $C$ in addition to the correctness of the computer program for this particular set of inputs.

The reader can readily appreciate that when we vary the values of $A$, $B$, and $C$ over the range of possible values, either during test or operation, we would soon see if the software developer achieved an error-free program. For example, was the developer wise enough to treat the problem of imaginary roots? Did the developer use the quadratic formula to solve for the roots? How, then, was the case of $A = 0$ treated where there is only one root and the quadratic formula "blows up" (i.e., leads to an exponential overflow error)? Clearly, we should test for all these values during development to ensure that there are no residual errors in the program, regardless of the input value. This leads to the concept of exhaustive testing, which is always infeasible in a practical problem. Suppose in the quadratic equation example that the values of $A$, $B$, and $C$ were restricted to integers between $+1,000$ and $-1,000$. Thus there would be 2,000 values of $A$ and a like number of values of $B$ and $C$. The possible input space for $A$, $B$, and $C$ would therefore be $(2,000)^3 = 8$ billion values.[2] Suppose that

---

[2] In a real-time system, each set of input values enters when the computer is in a different "initial state," and all the initial states must also be considered. Suppose that a program is designed to sum the values of the inputs for a given period of time, print the sum, and reset. If there is a high partial sum, and a set of inputs occurs with large values, overflow may be encountered. If the partial sum were smaller, this same set of inputs would therefore cause no problems. Thus, in the general case, one must consider the input space to include all the various combinations of inputs and *states* of the system.

we solve for each value of roots, substitute in the original equation to check, and only print out a result if the roots when substituted do not yield a zero of the equation. If we could process 1,000 values per minute, the exhaustive test would require 8 million minutes, which is 5,556 days or 15.2 years. This is hardly a feasible procedure: any such computation for a practical problem involves a much larger test space and a more difficult checking procedure that is impossible in any practical sense. In the quadratic equation example, there was a ready means of checking the answers by substitution into the equation; however, if the purpose of the program is to calculate satellite orbits, and if 1 million combinations of input parameters are possible, then a person(s) or computer must independently obtain the 1 million right answers and check them all! Thus the probabilistic nature of software reliability is based on the varying values of the input, the huge number of input cases, the initial system states, and the impossibility of exhaustive testing.

The basis for software reliability is quite different than the most common causes of hardware reliability. Software development is quite different from hardware development, and the source of software errors (random discovery of latent design and coding defects) differs from the source of most hardware errors (equipment failures). Of course, some complex hardware does have latent design and assembly defects, but the *dominant mode* of hardware failures is equipment failures. Mechanical hardware can jam, break, and become worn-out, and electrical hardware can burn out, leaving a short or open circuit or some other mode of failure. Many who criticize probabilistic modeling of software complain that instructions do not wear out. Although this is a true statement, the random discovery of latent software defects is indeed just as damaging as equipment failures, even though it constitutes a different mode of failure.

The development of models for software reliability in this chapter begins with a study of the software development process in Section 5.3 and continues with the formulation of probabilistic models in Section 5.4.

## 5.2 THE MAGNITUDE OF THE PROBLEM

Modeling, predicting, and measuring software reliability is an important quantitative approach to achieving high-quality software and growth in reliability as a project progresses. It is an important management and engineering design metric; most software errors are at least troublesome—some are very serious—so the major flaws, once detected, must be removed by localization, redesign, and retest.

The seriousness and cost of fixing some software problems can be appreciated if we examine the Year 2000 Problem (Y2K). The largely overrated fears occurred because during the early days of the computer revolution in the 1960s and 1970s, computer memory was so expensive that programmers used many tricks and shortcuts to save a little here and there to make their programs operate with smaller memory sizes. In 1965, the cost of magnetic-core computer memory was expensive at about $1 per word and used a significant operating current. (Presently, microelectronic memory sells for perhaps $1 per megabyte and draws only a small amount of current; assuming a 16-bit word, this cost has therefore been reduced by a factor of about 500,000!) To save memory, programmers reserved only 2 digits to represent the last 2 digits of the year. They did not anticipate that any of their programs would survive for more than 5–10 years; moreover, they did not contemplate the problem that for the year 2000, the digits "00" could instead represent the year 1900 in the software. The simplest solution was to replace the 2-digit year field with a 4-digit one. The problem was the vast amount of time required not only to *search for the numerous instances* in which the year was used as input or output data or used in intermediate calculations in *existing software*, but also to *test* that the changes have been successful and have not introduced any *new* errors. This problem was further exacerbated because many of these older software programs were poorly documented, and in many cases they were translated from one version to another or from one language to another so they could be used in modern computers without the need to be rewritten. Although only minor problems occurred at the start of the new century, hundreds of millions of dollars had been expended to make a few changes that would only have been trivial if the software programs had been originally designed to prevent the Y2K problem.

Sometimes, however, efforts to avert Y2K software problems created problems themselves. One such case was that of the 7-Eleven convenience store chain. On January 1, 2001, the point-of-sale system used in the 7-Eleven stores read the year "2001" as "1901," which caused it to reject credit cards if they were used for automatic purchases (manual credit card purchases, in addition to cash and check purchases, were not affected). The problem was attributed to the system's software, even though it had been designed for the 5,200-store chain to be Y2K-compliant, had been subjected to 10,000 tests, and worked fine during 2000. (The chain spent 8.8 million dollars—0.1% of annual sales—for Y2K preparation from 1999 to 2000.) Fortunately, the bug was fixed within 1 day [The Associated Press, January 4, 2001].

Another case was that of Norway's national railway system. On the morning of December 31, 2000, none of the new 16 airport-express trains and 13 high-speed signature trains would start. Although the computer software had been checked thoroughly before the start of 2000, it still failed to recognize the correct date. The software was reset to read December 1, 2000, to give the German maker of the new trains 30 days to correct the problem. None of the older trains were affected by the problem [*New York Times*, January 3, 2001].

Before we leave the obvious aspects of the Y2K problem, we should consider how deeply entrenched some of these problems were in *legacy software*: old programs that are used in their original form or rejuvenated for extended use. Analysts have found that some of the old IBM 9020 computers used in outmoded components of air traffic control systems contain an algorithm in their microcode for switching between the two redundant cooling pumps each month to even the wear. (For a discussion of cooling pumps in typical IBM computers, see Siewiorek [1992, pp. 493, 504].) Nobody seemed to know how this calendar-sensitive algorithm would behave in the year 2000! The engineers and programmers who wrote the microcode for the 9020s had retired before 2000, and the obvious answer—replace the 9020s with modern computers—proceeded slowly because of the cost. Although no major problems occurred, the scare did bring to the attention of many managers the potential problems associated with the use of legacy software.

Software development is a lengthy, complex process, and before the focus of this chapter shifts to model building, the development process must be studied.

## 5.3 SOFTWARE DEVELOPMENT LIFE CYCLE

Our goal is to make a probabilistic model for software, and the first step in any modeling is to understand the process [Boehm, 2000; Brooks, 1995; Pfleerer, 1998; Schach, 1999; and Shooman, 1983]. A good approach to the study of the software development process is to define and discuss the various phases of the software development life cycle. A common partitioning of these phases is shown Table 5.1. The life cycle phases given in this table apply directly to the technique of program design known as structured procedural programming (SPP). In general, it also applies with some modification to the newer approach known as object-oriented programming (OOP). The details of OOP, including the popular design diagrams used for OOP that are called the universal modeling language (UMLs), are beyond the scope of this chapter; the reader is referred to the following references for more information: [Booch, 1999; Fowler, 1999; Pfleerer, 1998; Pooley, 1999; Pressman, 1997; and Schach, 1999]. The remainder of this section focuses on the SPP design technique.

### 5.3.1 Beginning and End

The beginning and end of the software development life cycle are the start of the project and the discard of the software. The start of a project is generally driven by some event; for example, the head of the Federal Aviation Administration (FAA) or of some congressional committee decides that the United States needs a new air traffic control system, or the director of marketing in a company proposes to a management committee that to keep the company's competitive edge, it must develop a new database system. Sometimes, a project starts with a written *needs document*, which could be an internal memorandum, a long-range plan, or a study of needed improvements in a particular field. The necessity is sometimes a business expansion or evolution; for example, a company buys a new subsidiary business and finds that its old payroll program will not support the new conglomeration, requiring an updated payroll program. The needs document generally specifies *why* new software is

**TABLE 5.1 Project Phases for the Software Development Life Cycle**

| Phase | Description |
|---|---|
| Start of project | Initial decision or motivation for the project, including overall system parameters. |
| Needs | A study and statement of the need for the software and what it should accomplish. |
| Requirements | Algorithms or functions that must be performed, including functional parameters. |
| Specifications | Details of how the tasks and functions are to be performed. |
| Design of prototype | Construction of a prototype, including coding and testing. |
| Prototype: System test | Evaluation by both the developer and the customer of how well the prototype design meets the requirements. |
| Revision of specifications | Prototype system tests and other information may reveal needed changes. |
| Final design | Design changes in the prototype software in response to discovered deviations from the original specifications or the revised specifications, and changes to improve performance and reliability. |
| Code final design | The final implementation of the design. |
| Unit test | Each major unit (module) of the code is individually tested. |
| Integration test | Each module is successively inserted into the pretested control structure, and the composite is tested. |
| System test | Once all (or most) of the units have been integrated, the system operation is tested. |
| Acceptance test | The customer designs and witnesses a test of the system to see if it meets the requirements. |
| Field deployment | The software is placed into operational use. |
| Field maintenance | Errors found during operation must be fixed. |
| Redesign of the system | A new contract is negotiated after a number of years of operation to include changes and additional features. The aforementioned phases are repeated. |
| Software discard | Eventually, the software is no longer updated or corrected but discarded, perhaps to be replaced by new software. |

needed. Generally, old software is discarded once new, improved software is available. However, if one branch of an organization decides to buy new software and another branch wishes to continue with its present version, it may be difficult to define the end of the software's usage. Oftentimes, the discarding takes place many years beyond what was originally envisioned when the software was developed or purchased. (In many ways, this is why there was a Y2K problem: too few people ever thought that their software would last to the year 2000.)

### 5.3.2 Requirements

The project formally begins with the drafting of a *requirements document* for the system in response to the needs document or equivalent document. Initially, the requirements constitute high-level system requirements encompassing both the hardware and software. In a large project, as the requirements document "matures," it is expanded into separate hardware and software requirements; the requirements will specify *what* needs to be done. For an air traffic control system (ATC), the requirements would deal with the ATC centers that they must serve, the present and expected future volume of traffic, the mix of aircraft, the types of radar and displays used, and the interfaces to other ATC centers and the aircraft. Present travel patterns, expected growth, and expected changes in aircraft, airport, and airline operational characteristics would also be reflected in the requirements.

### 5.3.3 Specifications

The project specifications start with the requirements and the details of *how* the software is to be designed to satisfy these requirements. Continuing with our air traffic control system example, there would be a *hardware specifications document* dealing with (a) what type of radar is used; (b) the kinds of displays and display computers that are used; (c) the distributed computers or microprocessors and memory systems; (d) the communications equipment; (e) the power supplies; and (f) any networks that are needed for the project. The *software specifications document* will delineate (a) what tracking algorithm to use; (b) how the display information for the aircraft will be handled; (c) how the system will calculate any potential collisions; (d) how the information will be displayed; and (e) how the air traffic controller will interact with both the system and the pilots. Also, the exact nature of any required records of a technical, managerial, or legal nature will be specified in detail, including how they will be computed and archived. Particular projects often use names different from requirements and specifications (e.g., system requirements versus software specifications and high-level versus detailed specifications), but their content is essentially the same. A combined hardware–software specification might be used on a small project.

It is always a difficult task to define when requirements give way to specifications, and in the practical world, some specifications are mixed in the requirements document and some sections of the specifications document actually seem like requirements. In any event, it is important that the *why*, the *what*, and the *how* of the project be spelled out in a set of documents. The *completeness* of the set of documents is more important than exactly how the various ideas are *partitioned* between requirements and specifications.

Several researchers have outlined or developed experimental systems that use a formal language to write the specifications. Doing so has introduced a formalism and precision that is often lacking in specifications. Furthermore, since the formal specification language would have a grammar, one could build an automated specification checker. With some additional work, one could also develop a simulator that would in some way synthetically execute the specifications. Doing so would be very helpful in many ways for uncovering missing specifications, incomplete specifications, and conflicting specifications. Moreover, in a very simple way, it would serve as a preliminary execution of the software. Unfortunately, however, such projects are only in the experimental or prototype stages [Wing, 1990].

### 5.3.4 Prototypes

Most innovative projects now begin with a prototype or rapid prototype phase. The purpose of the prototype is multifaceted: developers have an opportunity to try out their design ideas, the difficult parts of the project become rapidly apparent, and there is an early (imperfect) working model that can be shown to the customer to help identify errors of omission and commission in the requirements and specification documents. In constructing the prototype, an initial control structure (the main program coordinating all the parts) is written and tested along with the interfaces to the various components (subroutines and modules). The various components are further decomposed into smaller subcomponents until the *module* level is reached, at which time programming or coding at the module level begins. The nature of a module is described in the paragraphs that follow.

A module is a block of code that performs a well-described function or procedure. The length of a module is a frequently debated issue. Initially, its length was defined as perhaps 50–200 source lines of code (SLOC). The SLOC length of a module is not absolute; it is based on the coder's "intellectual span of control." Since a program listing contains about 50 lines, this means that a module would be 1–4 pages long. The reasoning behind this is that it would be difficult to read, analyze, and trace the control structures of a program that extend beyond a few pages and keep all the logic of the program in mind; hence the term *intellectual span of control*. The concept of a module, module interface, and rough bounds on module size are more directly applicable to an SPP approach than to that of an OOP; however, as with very large and complex modules, very large and complex objects are undesirable.

Sometimes, the prototype progresses rapidly since old code from related projects can be used for the subroutines and modules, or a "first draft" of the software can be written even if some of the more complex features are left out. If the old code actually survives to the final version of the program, we speak of such code as *reused* code or *legacy* code, and if such reuse is significant, the development life cycle will be shortened somewhat and the cost will be reduced. Of course, the prototype code must be tested, and oftentimes when a prototype is shown to the customer, the customer understands that some features are not what he or she wanted. It is important to ascertain this as early as possible in the project so that revisions can be made in the specifications that will impact the final design. If these changes are delayed until late in the project, they can involve major changes in the code as well as significant redesign and extensive retesting of the software, for which large cost overruns and delays may be incurred. In some projects, the contracting is divided into two phases: delivery and evaluation of the prototype, followed by revisions in the requirements and specifications and a second contract for the delivered version of the software. Some managers complain that designing a prototype that is to be replaced by a final design is doing a job twice. Indeed it is; however, it is the best way to develop a large, complex project. (See Chapter 11, "Plan to Throw One Away," of Brooks [1995].) The cost of the prototype is not so large if one considers that much of the prototype code (especially the control structure) can be modified and reused for the final design and that the prototype test cases can be reused in testing the final design. It is likely that the same manager who objects to the use of prototype software would heartily endorse the use of a prototype board (breadboard), a mechanical model, or a computer simulation to "work out the bugs" of a hardware design without realizing that the software prototype is the software analog of these well-tried hardware development techniques.

Finally, we should remark that not all projects need a prototype phase. Consider the design of a fourth payroll system for a customer. Assume that the development organization specializes in payroll software and had developed the last three payroll systems for the customer. It is unlikely that a prototype would be required by either the customer or the developer. More likely, the developer would have some experts with considerable experience study the present system, study the new requirements, and ask many probing questions of the knowledgeable personnel at the customer's site, after which they could write the specifications for the final software. However, this payroll example is not the usual case; in most cases, prototype software is generally valuable and should be considered.

### 5.3.5 Design

Design really begins with the needs, requirements, and specifications documents. Also, the design of a prototype system is a very important part of the design process. For discussion purposes, however, we will refer to the final design stage as program design. In the case of SPP, there are two basic design approaches: top–down and bottom–up. The top–down process begins with the complete system at level 0; then, it decomposes this into a number of subsystems at level 1. This process continues to levels 2 and 3, then down to level $n$ where individual modules are encountered and coded as described in the following section. Such a decomposition can be modeled by a hierarchy diagram (H-diagram) such as that shown in Fig. 5.1(a). The diagram, which resembles an inverted tree, may be modeled as a mathematical graph where each "box" in the diagram represents a *node* in the graph and each line connecting the boxes represents a *branch* in the graph. A node at level $k$ (the predecessor) has several successor nodes at level

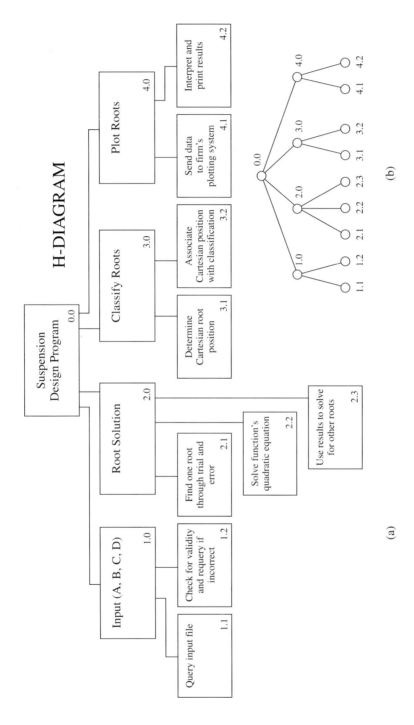

**Figure 5.1** (a), An H-diagram depicting the high-level architecture of a program to be used in designing the suspension system of a high-speed train, assuming that the dynamics can be approximately modeled by a third-order system (characteristic polynomial is a cubic); and (b), a graph corresponding to (a).

($k + 1$) (sometimes, the terms *ancestor* and *descendant* or *parent* and *child* are used). The graph has no loops (cycles), all nodes are connected (you can traverse a sequence of branches from any node to any other node), and the graph is undirected (one can traverse all branches in either direction). Such a graph is called a *tree* (free tree) and is shown in Fig. 5.1(b). For more details on trees, see Cormen [p. 91ff.].

The example of the H-diagram given in Fig. 5.1 is for the top-level architecture of a program to be used in the hypothetical design of the suspension system for a high-speed train. It is assumed that the dynamics of the suspension system can be approximated by a third-order differential equation and that the stability of the suspension can be studied by plotting the variation in the roots of the associated third-order characteristic polynomial ($Ax^3 + Bx^2 + Cx + D = 0$), which is a function of the various coefficients $A$, $B$, $C$, and $D$. It is also assumed that the company already has a plotting program (4.1) that is to be reused. The block (4.2) is to determine whether the roots have any positive real parts, since this indicates instability. In a different design, one could move the function 4.2 to 2.4. Thus the H-diagram can be used to discuss differences in high-level design architecture of a program. Of course, as one decomposes a problem, modules may appear at different levels in the structure, so the H-diagram need not be as symmetrical as that shown in Fig. 5.1.

One feature of the top–down decomposition process is that the decision of how to design lower-level elements is delayed until that level is reached in the design decomposition and the final decision is delayed until coding of the respective modules begins. This hiding process, called *information hiding*, is beneficial, as it allows the designer to progress with his or her design while more information is gathered and design alternatives are explored before a commitment is made to a specific approach. If at each level $k$ the project is decomposed into very many subproblems, then that level becomes cluttered with many concepts, at which point the tree becomes very wide. (The number of successor nodes in a tree is called the degree of the predecessor node.) If the decomposition only involves two or three subproblems (degree 2 or 3), the tree becomes very deep before all the modules are reached, which is again cumbersome. A suitable value to pick for each decomposition is 5–9 subprograms (each node should have degree 5–9). This is based on the work of the experimental psychologist Miller [1956], who found that the classic human senses (sight, smell, hearing, taste, and touch) could discriminate 5–9 logarithmic levels. (See also Shooman [1983, pp. 194, 195].) Using the 5–9 decomposition rule provides some bounds to the structure of the design process for an SPP.

Assume that the program size is $N$ source lines of code (SLOC) in length. If the graph is symmetrical and all the modules appear at the lowest level $k$, as shown in Fig. 5.1(a), and there are $r = 5$–$9$ successors at each node, then:

1. All the levels above $k$ represent program interfaces.
2. At level 0, there are between $5^0 = 1$ and $9^0 = 1$ interfaces. At level 1, the

top level node has between $5^1 = 5$ and $9^1 = 9$ interfaces. Also at level 2 are between $5^2 = 25$ and $9^2 = 81$ interfaces. Thus, for $k$ levels starting with level 0, the sum of the geometric progression $r^0 + r^1 + r^2 + \cdots + r^k$ is given by the equations that follow. (See Hall and Knight [1957, p. 39] or a similar handbook for more details.)

$$\text{Sum} = (r^k - 1)/(r - 1) \quad (5.1a)$$

and for $r = 5$ to 9, we have

$$(5^k - 1)/4 \leq \text{number of interfaces} \leq (9^k - 1)/8 \quad (5.1b)$$

3. The number of modules at the lowest level is given by

$$5^k \leq \text{number of modules} \leq 9^k \quad (5.1c)$$

4. If each module is of size $M$, the number of lines of code is

$$M \times 5^k \leq \text{number of SLOC} \leq M \times 9^k \quad (5.1d)$$

Since modules generally vary in size, Eq. (5.1d) is still approximately correct if $M$ is replaced by the average value $\overline{M}$.

We can better appreciate the use of Eqs. (5.1a–d) if we explore the following example. Suppose that a module consists of 100 lines of code, in which case $M = 100$, and it is estimated that a program design will take about 10,000 SLOC. Using Eq. (5.1c, d), we know that the number of modules must be about 100 and that the number of levels are bounded by $5^k = 100$ and $9^k = 100$. Taking logarithms and solving the resulting equations yields $2.09 \leq k \leq 2.86$. Thus, starting with the top-level 0, we will have about 2 or 3 successor levels. Similarly, we can bound the number of interfaces by Eq. (5.1b), and substitution of $k = 3$ yields the number of interfaces between 31 and 91. Of course, these computations are for a symmetric graph; however, they give us a rough idea of the size of the H-diagram design and the number of modules and interfaces that must be designed and tested.

### 5.3.6 Coding

Sometimes, a beginning undergraduate student feels that coding is the most important part of developing software. Actually, it is only one of the sixteen phases given in Table 5.1. Previous studies [Shooman, 1983, Table 5.1] have shown that coding constitutes perhaps 20% of the total development effort. The preceding phases of design—"start of project" through the "final design"—entail about 40% of the development effort; the remaining phases, starting with the unit (module) test, are another 40%. Thus coding is an important part of the development process, but it does not represent a large fraction of the cost of developing software. This is probably the first lesson that the software engineering field teaches the beginning student.

The phases of software development that follow coding are various types of testing. The design is an SPP, and the coding is assumed to follow the structured programming approach where the minimal basic control structures are as follows: IF THEN ELSE and DO WHILE. In addition, most languages also provide DO UNTIL, DO CASE, BREAK, and PROCEDURE CALL AND RETURN structures that are often called extended control structures. Prior to the 1970s, the older, dangerous, and much-abused control structure GO TO LABEL was often used indiscriminately and in a poorly thought-out manner. One major thrust of structured programming was to outlaw the GO TO and improve program structure. At the present, unless a programmer must correct, modify, or adapt a very old (legacy) code, he or she should never or very seldom encounter a GO TO. In a few specialized cases, however, an occasional well-thought-out, carefully justified GO TO is warranted [Shooman, 1983].

Almost all modern languages support structured programming. Thus the choice of a language is based on other considerations, such as how familiar the programmers are with the language, whether there is legacy code available, how well the operating system supports the language, whether the code modules are to be written so that they may be reused in the future, and so forth. Typical choices are C, Ada, and Visual Basic. In the case of OOP, the most common languages at the present are C++ and Ada.

### 5.3.7 Testing

Testing is a complex process, and the exact nature of it depends on the design philosophy and the phase of the project. If the design has progressed under a top–down structured approach, it will be much like that outlined in Table 5.1. If the modern OOP techniques are employed, there may be more testing of interfaces, objects, and other structures within the OOP philosophy. If proof of program correctness is employed, there will be many additional layers added to the design process involving the writing of proofs to ensure that the design will satisfy a mathematical representation of the program logic. These additional phases of design may replace some of the testing phases.

Assuming the top–down structured approach, the first step in testing the code is to perform unit (module) testing. In general, the first module to be written should be the main control structure of the program that contains the highest interface levels. This main program structure is coded and tested first. Since no additional code is generally present, sometimes "dummy" modules, called *test stubs*, are used to test the interfaces. If legacy code modules are available for use, clearly they can serve to test the interfaces. If a prototype is to be constructed first, it is possible that the main control structure will be designed well enough to be reused largely intact in the final version.

Each functional unit of code is subjected to a test, called *unit* or *module testing*, to determine whether it works correctly by itself. For example, suppose that company $X$ pays an employee a base weekly salary determined by the employee's number of years of service, number of previous incentive awards, and number of hours worked in a week. The basic pay module in the payroll program of the company would have as inputs the date of hire, the current date, the number of hours worked in the previous week, and historical data on the number of previous service awards, various deductions for withholding tax, health insurance, and so on. The unit testing of this module would involve formulating a number of hypothetical (or real) work records for a week plus a number of hypothetical (or real) employees. The base pay would be computed with pencil, paper, and calculator for these test cases. The data would serve as inputs to the module, and the results (outputs) would be compared with the precomputed results. Any discrepancies would be diagnosed, the internal cause of the error (fault) would be located, and the code would be redesigned and rewritten to correct the error. The tests would be repeated to verify that the error had been eliminated. If the first code unit to be tested is the program control structure, it would define the software interfaces to other modules. In addition, it would allow the next phase of software testing—the integration test—to proceed as soon as a number of units had been coded and tested. During the integration test, one or more units of code would be added to the control structure (and any previous units that had been integrated), and functional tests would be performed along a path through the program involving the new unit(s) being tested. Generally, only one unit would be integrated at a time to make localizing any errors easier, since they generally come from within the new module of code; however, it is still possible for the error to be associated with the other modules that had already completed the integration test. The integration test would continue until all or most of the units have been integrated into the maturing software system. Generally, module and many integration test cases are constructed by examining the code. Such tests are often called *white box* or *clear box* tests (the reason for these names will soon be explained).

The system test follows the integration test. During the system test, a scenario is written encompassing an entire operational task that the software must perform. For example, in the case of air traffic control software, one might write a scenario that replicates aircraft arrivals and departures at Chicago's O'Hare Airport during a slow period—say, between 11 and 12 P.M. This would involve radar signals as inputs, the main computer and software for the system, and one or more display processors. In some cases, the radar would not be present, but simulated signals would be fed to the computer. (Anyone who has seen the physical size of a large, modern radar can well appreciate why the radar is not physically present, unless the system test is performed at an air traffic control center, which is unlikely.) The display system is a "desk-size" console likely to be present during the system test. As the system test progresses, the software gradually approaches the time of release when it can be placed into operation. Because most system tests are written based on the requirements and specifications, they do not depend on the nature of the code; they are as if the code were hidden from view in an opaque or black box. Hence such functional tests are often called *black box* tests.

On large projects (and sometimes on smaller ones), the last phase of testing is acceptance testing. This is generally written into the contract by the customer. If the software is being written "in house," an acceptance test would be performed if the company software development procedures call for it. A typical acceptance test would contain a number of operational scenarios performed by the software on the intended hardware, where the location would be chosen from (a) the developer's site, (b) the customer's site, or (c) the site at which the system is to be deployed. In the case of air traffic control (ATC), the ATC center contains the present on-line system $n$ and the previous system, $n-1$, as a backup. If we call the new system $n+1$, it would be installed alongside $n$ and $n-1$ and operate on the same data as the on-line system. Comparing the outputs of system $n+1$ with system $n$ for a number of months would constitute a very good acceptance test. Generally, the criterion for acceptance is that the software must operate on real or simulated system data for a specified number of hours or be subjected to a certain number of test inputs. If the acceptance test is passed, the software is accepted and the developer is paid; however, if the test is failed, the developer resumes the testing and correcting of software errors (including those found during the acceptance test), and a new acceptance test date is scheduled.

Sometimes, "third party" testing is used, in which the customer hires an outside organization to make up and administer integration, system, or acceptance tests. The theory is that the developer is too close to his or her own work and cannot test and evaluate it in an unbiased manner. The third party test group is sometimes an independent organization within the developer's company. Of course, one wonders how independent such an in-house group can be if it and the developers both work for the same boss.

The term *regression testing* is often used, describing the need to retest the software with the previous test cases after each new error is corrected. In theory, one must repeat all the tests; however, a selected subset is generally used in the retest. Each project requires a test plan to be written early in the development cycle in parallel with or immediately following the completion of specifications. The test plan documents the tests to be performed, organizes the test cases by phase, and contains the expected outputs for the test cases. Generally, testing costs and schedules are also included.

When a commercial software company is developing a product for sale to the general business and home community, the later phases of testing are often somewhat different, for which the terms *alpha testing* and *beta testing* are often used. Alpha testing means that a test group within the company evaluates the software before release, whereas beta testing means that a number of "selected customers" with whom the developer works are given early releases of the software to help test and debug it. Some people feel that beta testing is just a way of reducing the cost of software development and that it is not a thorough way of testing the software, whereas others feel that the company still does adequate testing and that this is just a way of getting a lot of extra field testing done in a short period of time at little additional cost.

During early field deployment, additional errors are found, since the actual

**218**  SOFTWARE RELIABILITY AND RECOVERY TECHNIQUES

operating environment has features or inputs that cannot be simulated. Generally, the developer is responsible for fixing the errors during early field deployment. This responsibility is an incentive for the developer to do a thorough job of testing before the software is released because fixing errors after it is released could cost 25–100 times as much as that during the unit test. Because of the high cost of such testing, the contract often includes a warranty period (of perhaps 1–2 years or longer) during which the developer agrees to fix any errors for a fee.

If the software is successful, after a period of years the developer and others will probably be asked to provide a proposal and estimate the cost of including additional features in the software. The winner of the competition receives a new contract for the added features. If during initial development the developer can determine something about possible future additions, the design can include the means of easily implementing these features in the future, a process for which the term "putting hooks" into the software is often used. Eventually, once no further added features are feasible or if the customer's needs change significantly, the software is discarded.

### 5.3.8  Diagrams Depicting the Development Process

The preceding discussion assumed that the various phases of software development proceed in a sequential fashion. Such a sequence is often called *waterfall* development because of the appearance of the symbolic model as shown in Fig. 5.2. This figure does not include a prototype phase; if this is added to the development cycle, the diagram shown in Fig. 5.3 ensues. In actual practice, portions of the system are sometimes developed and tested before the remaining portions. The term *software build* is used to describe this process; thus one speaks of build 4 being completed and integrated into the existing system composed of builds 1–3. A diagram describing this build process, called the incremental model of software development, is given in Fig. 5.4. Other related models of software development are given in Schach [1999].

Now that the general features of the development process have been described, we are ready to introduce software reliability models related to the software development process.

## 5.4  RELIABILITY THEORY

### 5.4.1  Introduction

In Section 5.1, software reliability was defined as the probability that the software will perform its intended function, that is, the *probability of success*, which is also known as the reliability. Since we will be using the principles of reliability developed in Appendix B, Section B3, we summarize the development of reliability theory that is used as a basis for our software reliability models.

RELIABILITY THEORY   219

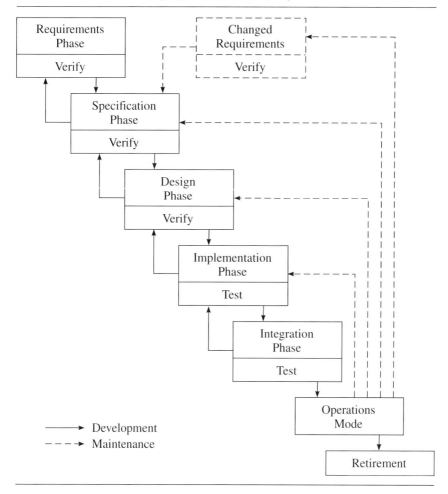

**Figure 5.2**  Diagram of the waterfall model of software development.

### 5.4.2  Reliability as a Probability of Success

The reliability of a system (hardware, software, human, or a combination thereof) is the probability of success, $P_s$, which is unity minus the probability of failure, $P_f$. If we assume that $t$ is the time of operation, that the operation starts at $t = 0$, and that the time to failure is given by $t_f$, we can then express the reliability as

$$R(t) = P_s = P(t_f \geq t) = 1 - P_f = 1 - P(0 \leq t_f \leq t) \quad (5.2)$$

## SOFTWARE LIFE-CYCLE DEVELOPMENT MODELS (RAPID PROTOTYPE MODEL)

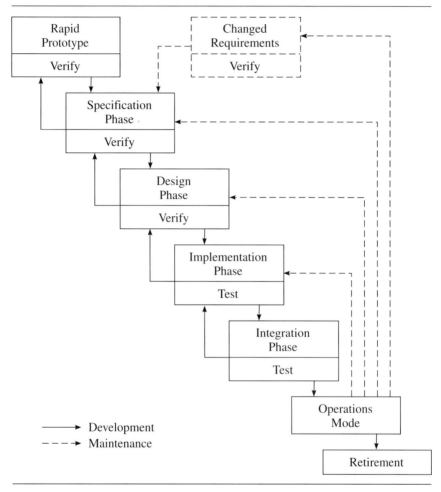

**Figure 5.3** Diagram of the rapid prototype model of software development.

The notation, $P(0 \leq t_f \leq t)$, in Eq. (5.2) stands for the probability that the time to failure is less than or equal to $t$. Of course, time is always a positive value, so the time to failure is always equal to or greater than 0. Reliability can also be expressed in terms of the cumulative *probability distribution function* for the *random variable time to failure*, $F(t)$, and the *probability density function*, $f(t)$ (see Appendix A, Section A6). The density function is the derivative of the distribution function, $f(t) = dF(t)/dt$, and the distribution function is the

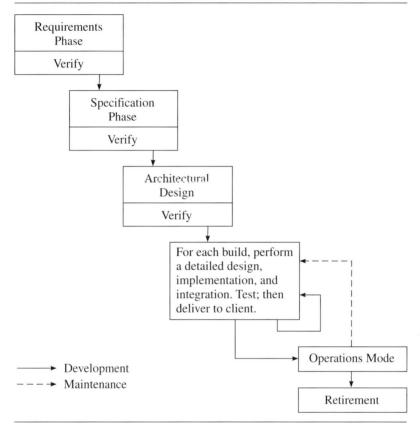

**Figure 5.4** Diagram of the incremental model of software development.

integral of the density function, $F(t) = 1 - \int f(t)\,dt$. Since by definition $F(t) = P(0 \le t_f \le t)$, Eq. (5.2) becomes

$$R(t) = 1 - F(t) = 1 - \int f(t)\,dt \qquad (5.3)$$

Thus reliability can be easily calculated if the probability density function for the time to failure is known. Equation (5.3) states the simple relationships among $R(t)$, $F(t)$, and $f(t)$; given any one of the functions, the other two are easy to calculate.

### 5.4.3 Failure-Rate (Hazard) Function

Equation (5.3) expresses reliability in terms of the traditional *mathematical probability functions*, $F(t)$, and $f(t)$; however, *reliability engineers* have found these functions to be generally ill-suited for study if we want intuition, failure data interpretation, and mathematics to agree. Intuition suggests that we study another function—a conditional probability function called the failure rate (hazard), $z(t)$. The following analysis develops an expression for the reliability in terms of $z(t)$ and relates $z(t)$ to $f(t)$ and $F(t)$.

The probability density function can be interpreted from the following relationship:

$$P(t < t_f < t + dt) = P(\text{failure in interval } t \text{ to } t + dt) = f(t)\, dt \tag{5.4}$$

One can relate the probability functions to failure data analysis if we begin with $N$ items placed on the life test at time $t$. The number of items surviving the life test up to time $t$ is denoted by $n(t)$. At any point in time, the probability of failure in interval $dt$ is given by (number of failures)/$N$. (To be mathematically correct, we should say that this is only true in the limit as $dt \to 0$.) Similarly, the reliability can be expressed as $R(t) = n(t)/N$. The number of failures in interval $dt$ is given by $[n(t) - n(t + dt)]$, and substitution in Eq. (5.4) yields

$$\frac{n(t) - n(t + dt)}{N} = f(t)\, dt \tag{5.5}$$

However, we can also write Eq. (5.4) as

$$f(t)\, dt = P(\text{no failure in interval 0 to } t)$$
$$\times P(\text{failure in interval } dt\,|\,\text{no failure in interval 0 to } t) \tag{5.6a}$$

The last expression in Eq. (5.6a) is a conditional failure probability, and the symbol | is interpreted as "given that." Thus $P$(failure in interval $dt$ | no failure in interval 0 to $t$) is the probability of failure in 0 to $t$ given that there was no failure up to $t$, that is, the item is working at time $t$. By definition, $P$(failure in interval $dt$ | no failure in interval 0 to $t$) is called the *hazard* function, $z(t)$; its more popular name is the *failure-rate* function. Since the probability of no failure is just the reliability function, Eq. (5.6a) can be written as

$$f(t)\, dt = R(t) \times z(t)\, dt \tag{5.6b}$$

This equation relates $f(t)$, $R(t)$, and $z(t)$; however, we will develop a more convenient relationship shortly.

Substitution of Eq. (5.6b) into Eq. (5.5) along with the relationship $R(t) = n(t)/N$ yields

$$\frac{n(t) - n(t+dt)}{N} = R(t)z(t)\,dt = \frac{n(t)}{N} z(t)\,dt \qquad (5.7)$$

Solving Eqs. (5.5) and (5.7) for $f(t)$ and $z(t)$, we obtain

$$f(t) = \frac{n(t) - n(t+dt)}{N\,dt} \qquad (5.8)$$

$$z(t) = \frac{n(t) - n(t+dt)}{n(t)\,dt} \qquad (5.9)$$

Comparing Eqs. (5.8) and (5.9), we see that $f(t)$ reflects the rate of failure based on the original number $N$ placed on test, whereas $z(t)$ gives the *instantaneous* rate of failure based on the number of survivors at the beginning of the interval.

We can develop an equation for $R(t)$ in terms of $z(t)$ from Eq. (5.6b):

$$z(t) = \frac{f(t)}{R(t)} \qquad (5.10)$$

and from Eq. (5.3), differentiation of both sides yields

$$\frac{dR(t)}{dt} = -f(t) \qquad (5.11)$$

Substituting Eq. (5.11) into (5.10) and solving for $z(t)$ yields

$$z(t) = -\frac{dR(t)}{dt} \bigg/ R(t) \qquad (5.12)$$

This differential equation can be solved by integrating both sides, yielding

$$\ln\{R(t)\} = -\int z(t)\,dt \qquad (5.13a)$$

Eliminating the natural logarithmic function in this equation by exponentiating both sides yields

$$R(t) = e^{-\int z(t)\,dt} \qquad (5.13b)$$

which is the form of the reliability function that is used in the following model development.

If one substitutes limits for the integral, a dummy variable, $x$, is required inside the integral, and a constant of integration must be added, yielding

$$R(t) = e^{-\int_0^t z(x)\,dx + A} = Be^{-\int_0^t z(x)\,dx} \tag{5.13c}$$

As is normally the case in the solution of differential equations, the constant $B = e^{-A}$ is evaluated from the initial conditions. At $t = 0$, the item is good and $R(t = 0) = 1$. The integral from 0 to 0 is 0; thus $B = 1$ and Eq. (5.13c) becomes

$$R(t) = e^{-\int_0^t z(x)\,dx} \tag{5.13d}$$

### 5.4.4 Mean Time To Failure

Sometimes, the *complete information* on failure behavior, $z(t)$ or $f(t)$, is not needed, and the reliability can be represented by the mean time to failure (MTTF) rather than the more detailed reliability function. A point estimate (MTTF) is given instead of the complete time function, $R(t)$. A rough analogy is to rank the strength of a hitter in baseball in terms of his or her batting average, rather than the complete statistics of how many times at bat, how many first-base hits, how many second-base hits, and so on.

The mean value of a probability function is given by the expected value, $E(t)$, of the random variable, which is given by the integral of the product of the random variable (time to failure) and its density function, which has the following form:

$$\text{MTTF} = E(t) = \int_0^\infty t f(t)\,dt \tag{5.14}$$

Some mathematical manipulation of Eq. (5.14) involving integration by parts [Shooman, 1990] yields a simpler expression:

$$\text{MTTF} = E(t) = \int_0^\infty R(t)\,dt \tag{5.15}$$

Sometimes, the mean time to failure is called mean time between failure (MTBF), and although there is a minor difference in their definitions, we will use the terms interchangeably.

### 5.4.5 Constant-Failure Rate

In general, a choice of the failure-rate function defines the reliability model. Such a choice should be made based on past studies that include failure-rate data or reasonable engineering assumptions. In several practical cases, the failure rate is constant in time, $z(t) = \lambda$, and the mathematics becomes quite simple. Substitution into Eqs. (5.13d) and (5.15) yields

$$R(t) = e^{-\int_0^t \lambda\, dx} = e^{-\lambda t} \tag{5.16}$$

$$\text{MTTF} = E(t) = \int_0^\infty e^{-\lambda t}\, dt = \frac{1}{\lambda} \tag{5.17}$$

The result is particularly simple: the reliability function is a decreasing exponential function where the exponent is the negative of the failure rate $\lambda$. A smaller failure rate means a slower exponential decay. Similarly, the MTTF is just the reciprocal of the failure rate, and a small failure rate means a large MTTF.

As an example, suppose that past life tests have shown that an item fails at a constant-failure rate. If 100 items are tested for 1,000 hours and 4 of these fail, then $\lambda = 4/(100 \times 1{,}000) = 4 \times 10^{-5}$. Substitution into Eq. (5.17) yields MTTF = 25,000 hours. Suppose we want the reliability for 5,000 hours; in that case, substitution into Eq. (5.16) yields $R(5{,}000) = e^{-(4/100{,}000)\times 5{,}000} = e^{-0.2} = 0.82$. Thus, if the failure rate were constant at $4 \times 10^{-5}$, the MTTF is 25,000 hours, and the reliability (probability of no failures) for 5,000 hours is 0.82.

More complex failure rates yield more complex results. If the failure rate increases with time, as is often the case in mechanical components that eventually "wear out," the hazard function could be modeled by $z(t) = kt$. The reliability and MTTF then become the equations that follow [Shooman, 1990].

$$R(t) = e^{-\int_0^t kx\, dx} = e^{-kt^2/2} \tag{5.18}$$

$$\text{MTTF} = E(t) = \int_0^\infty e^{-kt^2/2}\, dt = \sqrt{\frac{\pi}{2k}} \tag{5.19}$$

Other choices of hazard functions would give other results.

The reliability mathematics of this section applies to hardware failure and human errors, and also to software errors if we can characterize the software errors by a failure-rate function. The next section discusses how one can formulate a failure-rate function for software based on a software error model.

## 5.5 SOFTWARE ERROR MODELS

### 5.5.1 Introduction

Many reliability models discussed in the remainder of this chapter are related to the number of residual errors in the software; therefore, this section discusses software error models. Generally, one speaks of *faults* in the code that cause *errors* in the software operation; it is these errors that lead to system *failure*. Software engineers differentiate between a fault, a software error, and a software-caused system failure only when necessary, and the slang expression "software bug" is commonly used in normal conversation to describe a software problem.[3]

Software errors occur at many stages in the software life cycle. Errors may occur in the *requirements-and-specifications* phase. For example, the specifications might state that the time inputs to the system from a precise cesium atomic clock are in hours, minutes, and seconds when actually the clock output is in hours and decimal fractions of an hour. Such an erroneous specification might be found early in the development cycle, especially if a hardware designer familiar with the cesium clock is included in the specification review. It is also possible that such an error will not be found until a system test, when the clock output is connected to the system. Errors in requirements and specifications are identified as separate entities; however, they will be *added to the code faults* in this chapter. If the range safety officer has to destroy a satellite booster because it is veering off course, it matters little to him or her whether the problem lies in the specifiations or whether it is a coding error.

Errors occur in the *program logic*. For example, the THEN and ELSE clauses in an IF THEN ELSE statement may be interchanged, creating an error, or a loop is erroneously executed $n-1$ times rather than the correct value, which is $n$ times. When a program is coded, *syntax errors* are always present and are caught by the compiler. Such syntax errors are too frequent, embarrassing, and universal to be considered errors.

Actually, *design errors* should be recorded once the program management reviews and endorses a preliminary design expressed by a set of design representations (H-diagrams, control graphs, and maybe other graphical or abbreviated high-level control-structure code outlines called *pseudocodes*) in addition to requirements and specifications. Often, a formal record of such changes is not kept. Furthermore, errors found by code reading and testing at the middle (unit) code level (called *module errors*) are often not carefully kept. A change in the preliminary design and the occurrence of module test errors should *both* be carefully recorded.

Oftentimes, the standard practice is not to start counting software errors,

---

[3]The origin of the word "bug" is very interesting. In the early days of computers, many of the machines were constructed of vacuum tubes and relays, used punched cards for input, and used machine language or assembly language. Grace Hopper, one of the pioneers who developed the language COBOL and who spent most of her career in the U.S. Navy (rising to the rank of admiral), is generally credited with the expression. One hot day in the summer of 1945 at Harvard, she was working on the Mark II computer (successor to the pioneering Mark I) when the machine stopped. Because there was no air conditioning, the windows were opened, which permitted the entry of a large moth that (subsequent investigation revealed) became stuck between the contacts of one of the relays, thereby preventing the machine from functioning. Hopper and the team removed the moth with tweezers; later, it was mounted in a logbook with tape (it is now displayed in the Naval Museum at the Naval Surface Weapons Center in Dahlgren, Virginia). The expression "bug in the system" soon became popular, as did the term "debugging" to denote the fixing of program errors. It is probable that "bug" was used before this incident during World War II to describe system or hardware problems, but this incident is clearly the origin of the term "software bug" [Billings, 1989, p. 58].

regardless of their cause, until the software comes under *configuration control*, generally at the start of integration testing. Configuration control occurs when a technical manager or management board is put in charge of the official version of the software and records any changes to the software. Such a change (error fix) is submitted in writing to the configuration control manager by the programmer who corrected the error and retested the code of the *module* with the design change. The configuration control manager retests the present version of the *software system* with the inserted change; if he or she agrees that it corrects the error and does not seem to cause any problems, the error is added to the official log of found and corrected errors. The code change is added to the official version of the program at the next compilation and release of a new, official version of the software. It is desirable to start recording errors earlier in the program than in the configuration control stage, but better late than never! The origin of configuration control was probably a reaction to the early days of program *patching*, as explained in the following paragraph.

In the early days of programming, when the compilation of code for a large program was a slow, laborious procedure, and configuration control was not strongly enforced, programmers inserted their own changes into the compiled version of the program. These additions were generally done by inserting a machine language GO TO in the code immediately before the beginning of the bad section, transferring program flow to an unused memory block. The correct code in machine language was inserted into this block, and a GO TO at the end of this correction block returned the program flow to an address in the compiled code immediately after the old, erroneous code. Thus the error was bypassed; such insertions were known as *patches*. Oftentimes, each programmer had his or her own collection of patches, and when a new compilation of software was begun, these confusing, sometimes overlapping and chaotic sets of patches had to be analyzed, recoded in higher-level language, and officially inserted in the code. No doubt configuration control was instituted to do away with this terrible practice.

### 5.5.2 An Error-Removal Model

A software error-removal model can be formulated at the beginning of an integration test (system test). The variable $\tau$ is used to represent the number of months of development time, and one arbitrarily calls the start of configuration control $\tau = 0$. At $\tau = 0$, we assume that the software contains $E_T$ total errors. As testing progresses, $E_c(\tau)$ errors are corrected, and the remaining number of errors, $E_r(\tau)$, is given by

$$E_r(\tau) = E_T - E_c(\tau) \qquad (5.20)$$

If some corrections made to discovered errors are imperfect, or if new errors are caused by the corrections, we call this *error generation*. Equation (5.20) is based on the assumption that there is no error generation—a situation that is

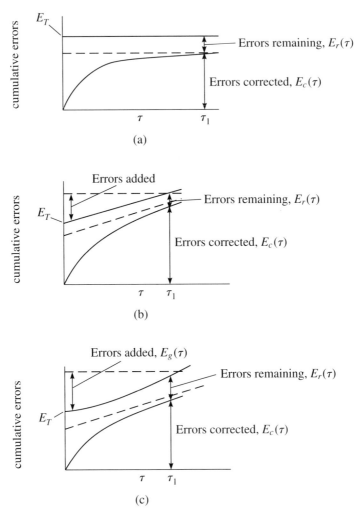

**Figure 5.5** Cumulative errors debugged versus months of debugging. (a) Approaching equilibrium, horizontal asymptote, no generation of new errors; (b) approaching equilibrium, generation rate of new errors equal to error-removal rate; and (c) diverging process, generation rate of new errors exceeding error-removal rate.

illustrated in Fig. 5.5(a). Note that in the figure a line drawn through any time $\tau$ parallel to the y-axis is divided into two line segments by the error-removal curve. The segment below the curve represents the errors that have been corrected, whereas the segment above the curve extending to $E_T$ represents the remaining number of errors, and these line segments correspond to the terms in Eq. (5.20). Suppose the software is released at time $\tau_1$, in which case the figure shows that not all the errors have been removed, and there is still a small residual number remaining. If all the coding errors could be removed, there clearly would be no code-related reasons for software failures (however, there would still be requirements-and-specifications errors). By the time integration testing is reached, we assume that the number of requirements-and-specifications errors is very small and that the number of code errors gradually decreases as the test process finds more errors to be subsequently corrected.

### 5.5.3 Error-Generation Models

In Fig. 5.5(b), we assume that there is some error generation and that the error discovery and correction process must be more effective or must take longer to leave the software with the same number of residual errors at release as in Fig. 5.5(a). Figure 5.5(c) depicts an extraordinary situation in which the error removal and correction initially exceeds the error generation; however, generation does eventually exceed correction, and the residual number of errors increases. In this case, the most obvious choices are to release at time $\tau_1$ and suffer poor reliability from the number of residual errors, or else radically change the test and correction process so that the situation of Fig. 5.5(a) or (b) ensues and then continue testing. One could also return to an earlier saved release of the software where error generation was modest, change the test and correction process, and, starting with this baseline, return to testing. The last and most unpleasant choice is to discard the software and start again. (Quantitative error-generation models are given in Shooman [1983, pp. 340–350].)

### 5.5.4 Error-Removal Models

Various models can be proposed for the error-correction function, $E_c(\tau)$, given in Eq. (5.20). The direct approach is to use the raw data. Error-removal data collected over a period of several months can be plotted. Then, an empirical curve can be fitted to the data, which can be extrapolated to forecast the future error-removal behavior. A better procedure is to propose a model based on past observations of error-removal curves and use the actual data to determine the model parameters. This blends the past information on the general shape of error-removal curves with the early data on the present project, and it also makes the forecasting less vulnerable to a few atypical data values at the start of the program (the *statistical noise*). Generally, the procedure takes a smaller number of observations, and a useful model emerges early in the development cycle—soon after $\tau = 0$. Of course, the estimate of the model parameters will have an associated statistical variance that will be larger at the beginning, when only a few data values are available, and smaller later in the project after more data is collected. The parameter variance will of course affect the range of the forecasts. If the project in question is somewhat like the previous projects, the chosen model will in effect filter out some of the statistical noise and yield better forecasts. However, what if for some reason the project is quite different from the previous ones? The "inertia" of the model will temporarily mask these differences. Also, suppose that in the middle of testing some of the test personnel or strategies are changed and the error-removal curve is significantly changed (for better or for worse). Again, the model inertia will temporarily mask these changes. Thus it is important to plot the actual data and examine it while one is using the model and making forecasts. There are many statistical tests to help the observer determine if differences represent statistical noise or different behavior; however, plotting, inspection, and thinking are all the initial basic steps.

One must keep in mind that with modern computer facilities, complex modeling and statistical parameter estimation techniques are easily accomplished; the difficult part is collecting enough data for accurate, stable estimates of model parameters and for interpretation of the results. Thus the focus of this chapter is on understanding and interpretation, not on complexity. In many cases, the error removal is too scant or inaccurate to support a sophisticated model over a simple one, and the complex model shrouds our understanding. Consider this example: Suppose we wish to estimate the math skills of 1,000 first-year high-school students by giving them a standardized test. It is too expensive to test all the students. If we decide to test 10 students, it is unlikely that the most sophisticated techniques for selecting the sample or processing the data will give us more than a wide range of estimates. Similarly, if we find the funds to test 250 students, then any elementary statistical techniques should give us good results. Sophisticated statistical techniques may help us make a better estimate if we are able to test, say, 50 students; however, the simpler techniques should still be computed first, since they will be understood by a wider range of readers.

***Constant Error-Removal Rate.*** Our development starts with the simplest models. Assuming that the error-detection rate is constant leads to a single-parameter error-removal model. In actuality, even if the removal rate were constant, it would fluctuate from week to week or month to month because of statistical noise, but there are ample statistical techniques to deal with this. Another factor that must be considered is the delay of a few days or, occasionally, a few weeks between the discovery of errors and their correction. For simplicity, we will assume (as most models do) that such delays do not cause problems.

If one assumes a constant error-correction (removal) rate of $\rho_0$ errors/month [Shooman, 1972, 1983], Eq. (5.20) becomes

$$E_r(\tau) = E_T - \rho_0 \tau \tag{5.21}$$

We can also derive Eq. (5.21) in a more basic fashion by letting the error-removal rate be given by the derivative of the number of errors remaining. Thus, differentiation of Eq. (5.20) yields

$$\text{error-correction rate} = \frac{dE_r(\tau)}{d\tau} = -\frac{dE_c(\tau)}{d\tau} \tag{5.22a}$$

Since we assume that the error-correction rate is constant, Eq. (5.22a) becomes

$$\text{error-correction rate} = \frac{dE_r(\tau)}{d\tau} = -\frac{dE_c(\tau)}{d\tau} = -\rho_0 \quad (5.22b)$$

Integration of Eq. (5.22b) yields

$$E_r(\tau) = C - \rho_0 \tau \quad (5.22c)$$

The constant $C$ is evaluated from the initial condition at $\tau = 0$, $E_r(\tau) = E_T = C$, and Eq. (5.22c) becomes

$$E_r(\tau) = E_T - \rho_0 \tau \quad (5.22d)$$

which is, of course, identical to Eq. (5.21). The cumulative number of errors corrected is given by the second term in the equation, $E_c(\tau) = \rho_0 \tau$.

Although there is some data to support a constant error-removal rate [Shooman and Bolsky, 1975], most practitioners observe that the error-removal rate decreases with development time, $\tau$.

Note that in the foregoing discussion we always assumed that the same effort is applied to testing and debugging over the interval in question. Either the same number of programmers is working on the given phase of development, the same number of worker hours is being expended, or the same number and difficulty level of tests is being employed. Of course, this will vary from day to day; we are really talking about the average over a week or a month. What would really destroy such an assumption is if two people worked on testing during the first two weeks in a month and six tested during the last two weeks of the month. One could always deal with such a situation by substituting for $\tau$ the number of worker hours, $WH$; $\rho_0$ would then become the number of errors removed per worker hour. One would think that $WH$ is always available from the business records for the project. However, this is sometimes distorted by the "big project phenomenon," which means that sometimes the manager of big project Z is told by his boss that there will be four programmers not working on the project who will charge their salaries to project Z for the next two weeks because they have no project support and Z is the only project that has sufficient resources to cover their salaries. In analyzing data, one should always be alert to the fact that such anomalies can occur, although the record of $WH$ is generally reliable.

As an example of how a constant error-removal rate can be used, consider a 10,000-line program that enters the integration test phase. For discussion purposes, assume we are omniscient and know that there are 130 errors. Suppose that the error removal proceeds at the rate of 15 per month and that the error-removal curve will be as shown in Fig. 5.6. Suppose that the schedule calls for release of the software after 8 months. There will be 130 − 120 = 10 errors left after 8 months of testing and debugging, but of course this information

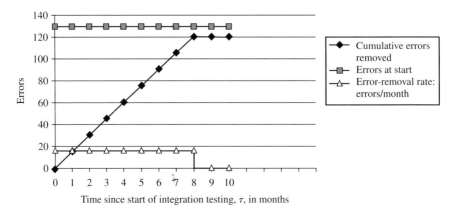

**Figure 5.6** Illustration of a constant error-removal rate.

is unknown to the test team and managers. The error-removal rate in Fig. 5.6 remains constant up to 8 months, then drops to 0 when testing and debugging is stopped. (Actually, there will be another phase of error correction when the software is released to the field and the field errors are corrected; however, this is ignored here.) The number of errors remaining is represented by the vertical line between the cumulative errors removed and the number of errors at the start.

How significant are the 10 residual errors? It depends on how often they occur during operation and how they affect the program operation. A complete discussion of these matters will have to wait until we develop the software reliability models in subsequent sections. One observation that makes us a little uneasy about this constant error-removal model is that the cumulative error-removal curve given in Fig. 5.6 is linearly increasing and does not give us an indication that most of the residual errors have been removed. In fact, if one tested for about an additional two-thirds of a month, another 10 errors would be found and removed, and all the errors would be gone. Philosophically, removal of *all* errors is hard to believe; practical experience shows that this is rare, if at all possible. Thus we must look for a more realistic error-removal model.

***Linearly Decreasing Error-Removal Rate.*** Most practitioners have observed that the error-removal rate decreases with development time, $\tau$. Thus the next error-removal model we introduce is one that decreases with development time, and the simplest choice for a decreasing model is a linear decrease. If we assume that the error-removal rate decreases linearly as a function of time, $\tau$ [Musa, 1975, 1987], then instead of Eq. (5.22a) we have

$$\frac{dE_r(\tau)}{d\tau} = -(K_1 - K_2\tau) \tag{5.23a}$$

which represents a linearly decreasing error-removal rate. At some time $\tau_0$, the linearly decreasing failure rate should go to 0, and substitution into Eq. (5.23a) yields $K_2 = K_1/\tau_0$. Substitution into Eq. (5.23a) yields

$$\frac{dE_r(\tau)}{d\tau} = -K_1\left(1 - \frac{\tau}{\tau_0}\right) = -K\left(1 - \frac{\tau}{\tau_0}\right) \quad (5.23b)$$

which clearly shows the linear decrease. For convenience, the subscript on $K$ was dropped since it was no longer needed. Integration of Eq. (5.23b) yields

$$E_r(\tau) = C - K\tau\left(1 - \frac{\tau}{2\tau_0}\right) \quad (5.23c)$$

The constant $C$ is evaluated from the initial condition at $\tau = 0$, $E_r(\tau) = E_T = C$, and Eq. (5.23c) becomes

$$E_r(\tau) = E_T - K\tau\left(1 - \frac{\tau}{2\tau_0}\right) \quad (5.23d)$$

Inspection of Eq. (5.23b) shows that $K$ is determined by the initial error-removal rate at $\tau = 0$.

We now repeat the example introduced above to illustrate a linearly decreasing error-removal rate. Since we wish the removal of 120 errors after 8 months to compare with the previous example, we set $E_T = 130$, and at $\tau = 8$, $E_r(\tau = 8)$ is equal to 10. Solving for $K$, we obtain a value of 30, and the equations for the error-correction rate and number of remaining errors become

$$\frac{dE_r(\tau)}{d\tau} = -30\left(1 - \frac{\tau}{8}\right) \quad (5.24a)$$

$$E_r(\tau) = 130 - 30\tau\left(1 - \frac{\tau}{16}\right) \quad (5.24b)$$

The error-removal curve will be as shown in Fig. 5.7 and decreases to 0 at 8 months. Suppose that the schedule calls for release of the software after 8 months. There will be $130 - 120 = 10$ errors left after 8 months of testing and debugging, but of course this information is unknown to the test team and managers. The error-removal rate in Fig. 5.7 drops to 0 when testing and debugging is stopped. The number of errors remaining is represented by the vertical line between the cumulative errors removed and the number of errors at the start. These results give an error-removal curve that seems to become asymptotic as we approach 8 months of testing and debugging. Of course, the

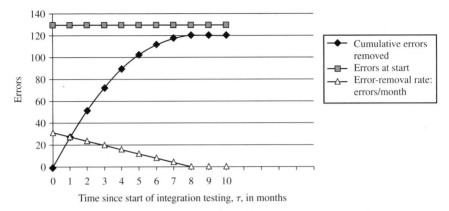

**Figure 5.7** Illustration of a linearly decreasing error-removal rate.

decrease to 0 errors removed in 8 months was chosen to match the previous constant error-removal example. In practice, however, the numerical values of parameters $K$ and $\tau_0$ would be chosen to match experimental data taken during the early part of the testing. The linear decrease of the error rate still seems somewhat artificial, and a final model with an exponentially decreasing error rate will now be developed.

***Exponentially Decreasing Error-Removal Rate.*** The notion of an exponentially decreasing error rate is attractive since it predicts a harder time in finding errors as the program is perfected. Programmers often say they observe such behavior as a program nears release. In fact, one can derive such an exponential curve based on simple assumptions. Assume that the number of errors corrected, $E_c(\tau)$, is exactly equal to the number of errors detected, $E_d(\tau)$, and that the rate of error detection is proportional to the number of remaining errors [Shooman, 1983, pp. 332–335].

$$\frac{dE_d(\tau)}{d\tau} = \alpha E_r(\tau) \tag{5.25a}$$

Substituting for $E_r(\tau)$, from Eq. (5.20) and letting $E_d(\tau) = E_c(\tau)$ yields

$$\frac{dE_c(\tau)}{d\tau} = \alpha[E_T - E_c(\tau)] \tag{5.25b}$$

Rearranging the differential equation given in Eq. (5.25b) yields

$$\frac{dE_c(\tau)}{d\tau} + \alpha E_c(\tau) = \alpha E_T \tag{5.25c}$$

To solve this differential equation, we obtain the homogeneous solution by setting the right-hand side equal to 0 and substituting the trial solution $E_c(\tau) = Ae^{a\tau}$ into Eq. (5.25c). The only solution is when $a = \alpha$. Since the right-hand side of the equation is a constant, the homogeneous solution is a constant. Adding the homogeneous and particular solutions yields

$$E_c(\tau) = Ae^{-\alpha\tau} + B \qquad (5.25d)$$

We can determine the constants $A$ and $B$ from initial conditions or by substitution back into Eq. (5.25c). Substituting the initial condition into Eq. (5.25d) when $\tau = 0$, $E_c = 0$ yields $A + B = 0$ or $A = -B$. Similarly, when $\tau \to \infty$, $E_c \to E_T$, and substitution yields $B = E_T$. Thus Eq. (5.25d) becomes

$$E_c(\tau) = E_T(1 - e^{-\alpha\tau}) \qquad (5.25e)$$

Substitution of Eq. (5.25e) into Eq. (5.20) yields

$$E_r(\tau) = E_T e^{-\alpha\tau} \qquad (5.25f)$$

We continue with the example introduced above to illustrate a linearly decreasing error-removal rate starting with $E_T = 130$ at $\tau = 0$. To match the previous results, we assume that $E_r(\tau = 8)$ is equal to 10, and substitution into Eq. (5.25f) gives $10 = 130e^{-8\alpha}$. Solving for $\alpha$ by taking natural logarithms of both sides yields the value $\alpha = 0.3206$. Substitution of these values leads to the following equations:

$$\frac{dE_r(\tau)}{d\tau} = -\alpha E_T e^{-\alpha\tau} = -41.68 e^{-0.3206\tau} \qquad (5.26a)$$

$$E_r(\tau) = 130 e^{-0.3206\tau} \qquad (5.26b)$$

The error-removal curve is shown in Fig. 5.8. The rate starts at 41.68 at $\tau = 0$ and decreases to 3.21 at $\tau = 8$. Theoretically, the error-removal rate continues to decrease exponentially and only reaches 0 at infinity. We assume, however, that testing stops after $\tau = 8$ and the removal rate falls to 0. The error-removal curve climbs a little more steeply than that shown in Fig. 5.7, but they both reach 120 errors removed after 8 months and stay constant thereafter.

***Other Error-Removal-Rate Models.*** Clearly, one could continue to evolve many other error-removal-rate models, and even though the ones discussed in this section should suffice for most purposes, we should mention a few other approaches in closing. All of these models assume a constant number of worker hours expended throughout the integration test and error-removal phase. On many projects, however, the process starts with a few testers, builds to a peak, and then uses fewer personnel as the release of the software nears. In such a case, an S-shaped error-removal curve ensues. Initially, the shape is

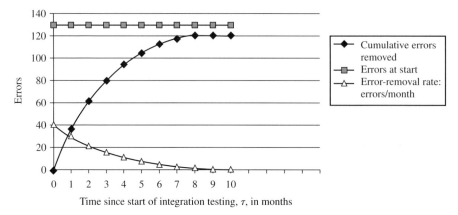

**Figure 5.8** Illustration of an exponentially decreasing error-removal rate.

concave upward until the main force is at work, at which time it is approximately linear; then, toward the end of the curve, it becomes concave downward. One way to model such a curve is to use piecewise methods. Continuing with our error-removal example, suppose that the error-removal rate starts at 2 per month at $\tau = 0$ and increases to 5.4 and 14.77 after 1 and 2 months, respectively. Between 2 and 6 months it stays constant at 15 per month; in months 7 and 8, it drops to 5.52 and 2 per month. The resultant curve is given in Fig. 5.9. Since fewer people are used during the first 2 and last 2 months, fewer errors are removed (about 90 for the numerical values used for the purpose of illustration). Clearly, to match the other error-removal models, a larger number of personnel would be needed in months 3–6.

The next section relates the reliability of the software to the error-removal-rate models that were introduced in this section.

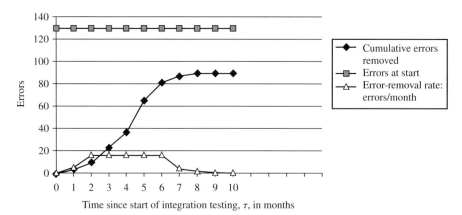

**Figure 5.9** Illustration of an S-shaped error-removal rate.

## 5.6 RELIABILITY MODELS

### 5.6.1 Introduction

In the preceding sections, we established the mathematical basis of the reliability function and related it to the failure-rate function. Also, a number of error-removal models were developed. Both of these efforts were preludes to formulating a software reliability model. Before we become absorbed in the details of reliability model development, we should review the purpose of software reliability models.

Software reliability models are used to answer two main questions during product development: When should we stop testing? and Will the product function well and be considered reliable? Both are technical management questions; the former can be restated as follows: When are there few enough errors so that the software can be released to the field (or at least to the last stage of testing)? To continue testing is costly, but to release a product with too many errors is more costly. The errors must be fixed in the field at high cost, and the product develops a reputation for unreliability that will hurt its acceptance. The software reliability models to be developed quantify the number of errors remaining and especially provide a prediction of the field reliability, helping technical and business management reach a decision regarding when to release the product. The contract or marketing plan contains a release date, and penalties may be assessed by a contract for late delivery. However, we wish to avoid the dilemma of the on-time release of a product that is too "buggy" and thus defective.

The other job of software reliability models is to give a prediction of field reliability as early as possible. Two many software products are released and, although they operate, errors occur too frequently; in retrospect, the projects become failures because people do not trust the results or tire of dealing with frequent system crashes. Most software products now have competitors, so consequently an unreliable product loses out or must be fixed up after release at great cost. Many software systems are developed for a single user for a special purpose, for example, air traffic control, IRS tax programs, social services' record systems, and control systems for radiation-treatment devices. Failures of such systems can have dire consequences and huge impact. Thus, given requirements and a quality goal, the types of reliability models we seek are those that are easy to understand and use and also give reasonable results. The relative accuracy of two models in which one predicts one crash per week and another predicts two crashes per week may seem vastly different in a mathematical sense. However, suppose a good product should have less than one crash a month or, preferably, a few crashes per year. In this case, both models tell the same story—the software is not nearly good enough! Furthermore, suppose that these predictions are made early in the testing when only a little failure data is available and the variance produces a range of estimates that vary by more than two to one. The real challenge is to get practitioners to collect data, use simple models, and make predictions to guide the program. One can always apply more sophisticated models to the same data set once the basic ideas are understood. The biggest mistake is to avoid making a reliability estimate because (a) it does not work, (b) it is too costly, and (c) we do not have the data. None of these reasons is correct or valid, and this fact represents poor management. The next biggest mistake is to make a model, obtain poor reliability predictions, and ignore them because they are too depressing.

### 5.6.2 Reliability Model for Constant Error-Removal Rate

The basic simplicity and some of the drawbacks of the simple constant error-removal model were discussed in the previous section on error-removal models. Even with these limitations, this is the simplest place to start for us to develop most of the features of software reliability models based on this model before we progress to more complex ones [Shooman, 1972].

The major assumption needed to relate an error-removal model to a software reliability model is how the failure rate is related to the remaining number of errors. For the remainder of this chapter, we assume that the failure rate is proportional to the remaining number of errors:

$$z(\tau) = k E_r(\tau) \qquad (5.27)$$

The bases of this assumption are as follows:

1. It seems reasonable to assume that more residual errors in the software result in higher software failure rates.
2. Musa [1987] has experimental data supporting this assumption.
3. If the rate of error discovery is a random process dependent on input and initial conditions, then the discovery rate is proportional to the number of residual errors.

If one combines Eq. (5.27) with one of the software error-removal models of the previous section, then a software reliability model is defined. Substitution of the failure rate into Eqs. (5.13d) and (5.15) yields a reliability model $R(t)$ and an expression for the MTTFs.

As an example, we begin with the constant error-removal model, Eq. (5.22d),

$$E_r(\tau) = E_T - \rho_0 \tau \qquad (5.28a)$$

Using the assumption of Eq. (5.27), one obtains

$$z(\tau) = k E_r(\tau) = k(E_T - \rho_0 \tau) \qquad (5.29)$$

and the reliability and MTTF expressions become

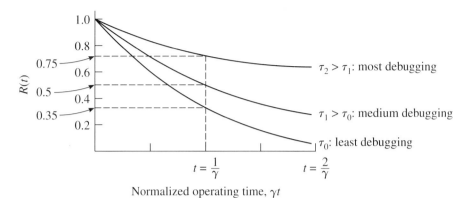

**Figure 5.10** Variation of reliability function $R(t)$ with operating time $t$ for fixed values of debugging time $\tau$. Note the time axis, $t$, is normalized.

$$R(t) = e^{-\int k(E_t - \rho_0 \tau) dt} = e^{-k(E_T - \rho_0 \tau)t} \tag{5.30a}$$

$$\text{MTTF} = \frac{1}{k(E_T - \rho_0 \tau)} \tag{5.30b}$$

The two preceding equations mathematically define the constant error-removal rate software reliability model; however, there is still much to be said in an engineering sense about how we apply this model. We must have a procedure for estimating the model parameters, $E_T$, $k$, and $\rho_0$, and we must interpret the results. For discussion purposes, we will reverse the order: we assume that the parameters are known and discuss the reliability and MTTF functions first. Since the parameters are assumed to be known, the exponent in Eq. (5.30a) is just a function of $\tau$; for convenience, we can define $k(E_T - \rho_0 \tau) = \gamma(\tau)$. Thus, as $\tau$ increases, $\gamma$ decreases. Equation (5.30a) therefore becomes

$$R(t) = e^{-\gamma t} \tag{5.31}$$

Equation (5.31) is plotted in Fig. 5.10 in terms of the normalized time scale $\gamma t$.

Let us assume that the project receives a minimum amount of testing and debugging during $\tau_0$ months. There would still be quite a few errors left, and the reliability would be mediocre. In fact, Fig. 5.10 shows (see vertical dotted line) that when $t = 1/\gamma$, the reliability is 0.35, meaning that there is a 65% chance that a failure occurs in the interval $0 \leq t \leq 1/\gamma$ and a 35% chance that no errors occurs in this interval. This is rather poor and would not be satisfactory in any normal project. If predicted early in the integration test process, changes would be made. One can envision more vigorous testing that would

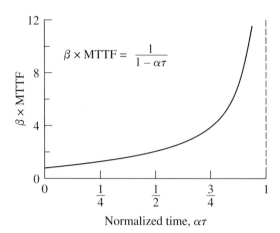

**Figure 5.11** Plot of MTTF versus debugging time $\tau$, given by Eq. (5.32). Note the time axis, $\tau$, and the MTTF axis are both normalized.

increase the parameter $\rho_0$ and remove errors faster or, as we will discuss now, just test longer. Assume that the integration test period is lengthened to $\tau_1 > \tau_0$ months. More errors will be removed, $\gamma$ will be smaller, and the exponential curve will decrease more slowly as shown by the middle curve in the figure. There would be a 50% chance that a failure occurs in the interval $0 \leq t \leq 1/\gamma$ and a 50% chance that no error occurs in this interval—better, but still not good enough. Suppose the test is lengthened further to $\tau_2 > \tau_1$ months, yielding a success probability of 75%. This might be satisfactory in some projects but would still not be good enough for really high reliability projects, so one should explore major changes. A different error-removal model would yield a different reliability function, predicting either higher or lower reliability, but the overall interpretation of the curves would be substantially the same. The important point is that one would be able to predict (as early as possible in testing) an operational reliability and compare this with the project specifications or observed reliabilities for existing software that serves a similar function.

Similar results, but from a slightly different viewpoint, are obtained by studying the MTTF function. Normalization will again be used to simplify the plotting of the MTTF function. Note how $\alpha$ and $\beta$ are defined in Eq. (5.32) and that $\tau = 1$ represents the point where all the errors have been removed and the MTTF approaches infinity. Note that the MTTF function initially increases almost linearly and slowly as shown in Fig. 5.11. Later, when the number of errors remaining is small, the function increases rapidly. The behavior of the MTTF function is the same as the function $1/x$, as $x \to 0$. The importance of this effect is that the majority of the improvement comes at the end of the testing cycle; thus, without a model, a manager may say that based on data before the "knee" of the curve, there is only slow progress in improving the MTTF, so why not release the software and fix additional bugs in the field?

Given this model, one can see that with a little more effort, rapid progress is expected once the knee of the curve is passed, and a little more testing should yield substantial improvement. The fact that the MTTF approaches infinity as the number of errors approaches 0 is somewhat disturbing, but this will be remedied when other error-removal models are introduced.

$$\text{MTTF} = \frac{1}{k(E_T - \rho_0\tau)} = \frac{1}{kE_T(1 - \rho_0\tau/E_T)} = \frac{1}{\beta(1 - \alpha\tau)} \qquad (5.32)$$

One can better appreciate this model if we use the numerical data from the example plotted in Fig. 5.6. The parameters $E_T$ and $\rho_0$ given in the example are 130 and 15, but the parameter $k$ must still be determined. Suppose that $k = 0.000132$, in which case Eq. (5.30a) becomes

$$R(t) = e^{-0.000132(130 - 15\tau)t} \qquad (5.33)$$

At $\tau = 8$, the equation becomes

$$R(t) = e^{-0.00132t} \qquad (5.34a)$$

The preceding is plotted as the middle curve in Fig. 5.12. Suppose that the software operates for 300 hours; then the reliability function predicts that there is a 67% chance of no software failures in the interval $0 \leq t \leq 300$. If we assume that these software reliability estimates are being made early in the testing process (say, after 2 months), one could predict the effects—good and bad—of debugging for more or less than $\tau = 8$ months. (Again, we ask the reader to be patient about where all these values for $E_T$, $\rho_0$, and $k$ are coming from. They would be derived from data collected on the program during the first 2 months of testing. The discussion of the parameter estimation process has purposely been separated from the interpretation of the models to avoid confusion.)

Frequently, management wants the technical staff to consider shortening the test period, since doing so would save project-development money and help keep the project on time. We can use the software reliability model to illustrate the effect (often disastrous) of such a change. If testing and debugging are shortened to only 6 months, Eq. (5.33) would become

$$R(t) = e^{-0.00528t} \qquad (5.34b)$$

Equation (5.34b) is plotted as the lower curve in Fig. 5.12. At 300 hours, there is only a 20.5% chance of no errors, which is clearly unacceptable. One might also show management the beneficial effects of slightly longer testing and debugging time. If we debugged for 8.5 months, then Eq. (5.34) would become

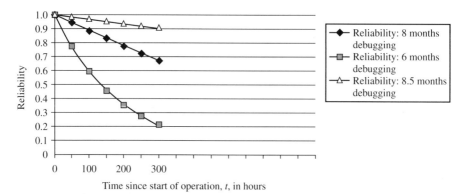

**Figure 5.12** Reliability functions for constant error-removal rate and 6, 8, and 8.5 months of debugging. See Eqs. (5.34a–c).

$$R(t) = e^{-0.00033t} \qquad (5.34c)$$

Equation (5.34c) is plotted as the upper curve in Fig. 5.12, and the reliability at 300 hours is 90.6%—a very significant improvement. Thus the technical people on the project should lobby for a slightly longer integration test period.

The overall interpretation of Fig. 5.12 leads to sensible conclusions; however, the constant error-removal model breaks down when $\tau$ is allowed to approach 8.67 months of testing. We see that Eq. (5.33) predicts that all the errors have been removed and that the reliability becomes unity. This effect becomes even clearer when we examine the MTTF function, and it is a good reason to progress shortly to the reliability models related to both the linearly decreasing and exponentially decreasing error-removal models.

The MTTF function is given by Eq. (5.32), and substituting the numerical values $E_T = 130$, $\rho_0 = 15$, and $k = 0.000132$ (corresponding to 8 months of debugging) yields

$$\text{MTTF} = \frac{1}{k(E_T - \rho_0 \tau)} = \frac{1}{0.000132(130 - 15\tau)} = \frac{7575.75}{(130 - 15\tau)} \qquad (5.35)$$

The MTTF function given in Eq. (5.35) is plotted in Fig. 5.13 and listed in Table 5.2. The dramatic differences in the MTTF predicted by this model as the number of remaining errors rapidly approaches 0 seem difficult to believe and represent another reason to question constant error-removal-rate models.

### 5.6.3 Reliability Model for a Linearly Decreasing Error-Removal Rate

We now develop a reliability model for the linearly decreasing error-removal rate as we did with the constant error-removal-rate model. The linearly decreas-

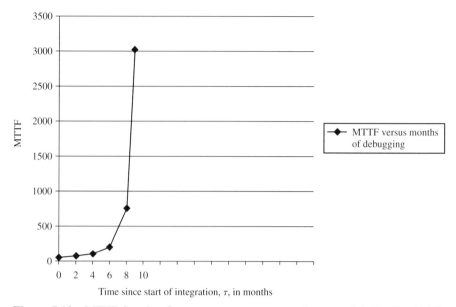

**Figure 5.13** MTTF function for a constant error-removal-rate model. See Eq. (5.35).

ing error-removal-rate model is given by Eq. (5.23d). Continuing with the example in use, we let $E_T = 130$, $K = 30$, and $\tau_0 = 8$, which led to Eq. (5.24b), and substitution yields the failure-rate function Eq. (5.29):

$$z(\tau) = kE_r(\tau) = kE_r(\tau) = k[130 - 30\tau(1 - \tau/16)] \qquad (5.36)$$

and also yields the reliability function:

**TABLE 5.2  MTTF for Constant Error-Removal Model**

| | |
|---|---|
| Total months of debugging | 8 |
| Formula for MTTF | $\dfrac{7{,}575.76}{130 - 15\tau}$ |
| Elapsed months of debugging, $\tau$: | MTTF |
| 0 | 58.28 |
| 2 | 75.76 |
| 4 | 108.23 |
| 6 | 189.39 |
| 8 | 757.58 |
| 8.5 | 3,030.30 |

**244** SOFTWARE RELIABILITY AND RECOVERY TECHNIQUES

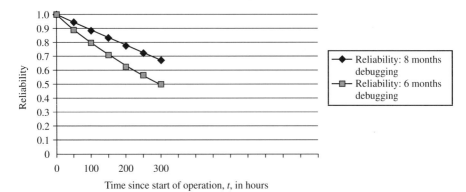

**Figure 5.14** Reliability functions for the linearly decreasing error-removal-rate model and 6 and 8 months of debugging. See Eqs. (5.37c, d).

$$R(t) = e^{-\int_0^t z(x)\,dx} = e^{-k[130 - 30\tau(1 - \tau/16)]t} \quad (5.37\text{a})$$

If we use the same value for $k$ as in the constant error-removal-rate reliability model, $k = 0.000132$, then Eq. (5.37a) becomes

$$R(t) = e^{-0.000132[130 - 30\tau(1 - \tau/16)]t} \quad (5.37\text{b})$$

If we debug for 8 months, substitution of $\tau = 8$ into Eq. (5.37b) yields

$$R(t) = e^{-0.00132t} \quad (5.37\text{c})$$

Similarly, if $\tau = 6$, substitution into Eq. (5.37b) yields

$$R(t) = e^{-0.00231t} \quad (5.37\text{d})$$

Note that since we have chosen a linearly decreasing error model that goes to 0 at $\tau = 8$ months, there is no additional error removal between 8 and 8.5 months. (Again, this may seem a little strange, but this effect will disappear when we consider the exponentially decreasing error-rate model in the next section.) The reliability functions given in Eqs. (5.37c, d) are plotted in Fig. 5.14. Note that the reliability curve for 8 months of debugging is identical to the curve for the constant error-removal model given in Fig. 5.12. This occurs because we have purposely chosen the linearly decreasing error model to have the same area (cumulative errors removed) over 8 months as the constant error-removal-rate model (the area of the triangle is the same as the area of the rectangle). In the case of 6 months of debugging, the reliability function associated with the linearly decreasing error-removal model is better than that of the constant error-removal model. This is because the linearly decreasing model starts

## RELIABILITY MODELS

**TABLE 5.3 MTTF for Linearly Decreasing Error-Removal Model**

| | |
|---|---|
| Total months of debugging | 8 |
| Formula for MTTF | $\dfrac{7{,}575.76}{[130 - 30\tau(1 - \tau/16)]}$ |
| Elapsed months of debugging, $\tau$: | MTTF |
| 0 | 58.28 |
| 2 | 97.75 |
| 4 | 189.39 |
| 6 | 432.9 |
| 8 | 757.58 |

out at a higher removal rate and decreases; thus, over 6 months of debugging we take advantage of the higher error-removal rates at the beginning, whereas over 8 months the lower error-removal rates at the end balance the larger error-removal rates at the beginning. We will now develop the MTTF function for the linear error-removal case.

The MTTF function is derived by substitution of Eq. (5.37a) into Eq. (5.15). Note that the integration in Eq. (5.15) is done with respect to $t$ and the function $z$ in Eq. (5.36), which multiplies $t$ in the exponent of Eq. (5.37a) is a function of $\tau$ (*not* $t$), so it is a constant in the integration used to determine MTTF. The result is

$$\text{MTTF} = \frac{1}{k[130 - 30\tau(1 - \tau/16)]} \tag{5.38a}$$

We substitute the value chosen for $k$, $k = 0.000132$, and $\tau = 8$ into Eq. (5.38a), yielding

$$\text{MTTF} = \frac{7575.76}{[130 - 30\tau(1 - \tau/16)]} \tag{5.38b}$$

The results of Eq. (5.38b) are given in Table 5.3 and Fig. 5.15. By comparing Figs. 5.13 and 5.15 or, better, Tables 5.2 and 5.3, one observes that because of the way in which the constants were picked, the MTTF curves for the linearly decreasing error-removal and the constant error-removal models agree when $\tau = 0$ and 8. For intermediate values of $\tau = 2, 4, 6$, and so on, the MTTF for the linearly decreasing error-removal model is higher because of the initially higher error-removal rate. Since the linearly decreasing error-removal model was chosen to go to 0 at $\tau = 8$, the values of MTTF for $\tau > 8$ really stay at 757.58. The model presented in the next section will remedy this counterintuitive result.

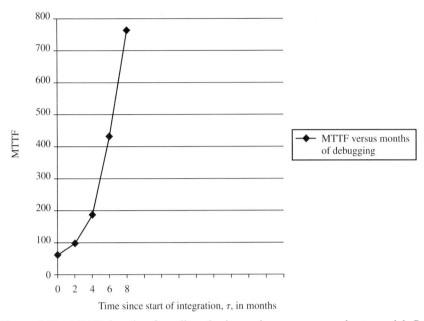

**Figure 5.15** MTTF function for a linearly decreasing error-removal-rate model. See Eq. (5.38b).

### 5.6.4 Reliability Model for an Exponentially Decreasing Error-Removal Rate

An exponentially decreasing error-removal-rate model was introduced in Section 5.5.4, and the general shape of this function removed some of the anomalies of the constant and the linearly decreasing models. Also, it was shown in Eqs. (5.25a–e) that this exponential model was the result of assuming that error detection was proportional to the number of errors present. In addition, many practitioners as well as theoretical modelers have observed that the error-removal rate decreases at a declining rate as testing increases (i.e., as $\tau$ increases), which fits in with the hypothesis—one that is not too difficult to conceive—that early errors removed in a computer program are uncovered by tests. Later errors are more subtle and more "deeply embedded," requiring more time and effort to formulate tests to uncover them. An exponential error-removal model has been proposed to represent these phenomena.

Using the same techniques as those of the preceding sections, we will now develop a reliability model based on the exponentially decreasing error-removal model. The number of remaining errors is given in Eq. (5.25f):

$$E_r(\tau) = E_T e^{-\alpha\tau} \qquad (5.39a)$$

$$z(\tau) = k E_T e^{-\alpha\tau} \qquad (5.39b)$$

and substitution into Eq. (5.13d) yields the reliability function.

$$R(t) = e^{-\int kE_T e^{-\alpha\tau}\,dt} = e^{-(kE_T e^{-\alpha\tau})t} \tag{5.40}$$

The preceding equation seems a little peculiar since it is an exponential function raised to a power that in turn is an exponential function. However, it is really not that complicated, and this is where the mathematical assumptions that seem to be reasonable lead. To better understand the result, we will continue with the running example that was introduced previously.

To make our comparison between models, we have chosen constants that cause the error-removal function to begin with 130 errors at $\tau = 0$ and decrease to 10 errors at $\tau = 8$ months. Thus Eq. (5.39a) becomes

$$E_r(\tau = 8) = 10 = 130 e^{-\alpha 8} \tag{5.41a}$$

Solving this equation for $\alpha$ yields $\alpha = 0.3206$. If we require the reliability function to yield a reliability of 0.673 at 300 hours of operation after $\tau = 8$ months of debugging, substitution into Eq. (5.40) yields an equation allowing us to solve for $k$.

$$R(300) = 0.673 = e^{-(k 130 e^{-0.3206 \times 8})300} \tag{5.41b}$$

The value of $k = 0.000132$ is the same as that determined previously for the other models. Thus Eq. (5.40) becomes

$$R(t) = e^{-(0.01716 e^{-0.3206\tau})t} \tag{5.42a}$$

The reliability function for $\tau = 8$ months is

$$R(t) = e^{-(0.00132)t} \qquad (\tau = 8) \tag{5.42b}$$

Similarly, for $\tau = 6$ and 8.5 months, substitution into Eq. (5.42a) yields the reliability functions:

$$R(t) = e^{-(0.002507)t} \qquad (\tau = 6) \tag{5.42c}$$

$$R(t) = e^{-(0.001125)t} \qquad (\tau = 8.5) \tag{5.42d}$$

Equations (5.42b–d) are plotted in Fig. 5.16. The reliability function for 8 months of debugging is, of course, identical to the previous two models because of the way we have chosen the parameters. The reliability function for $\tau = 6$ months of debugging yields a reliability of 0.47 at 300 hours of operation, which is considerably better than the 0.21 reliability in the constant error-removal-rate model. This occurs because the exponentially decreasing

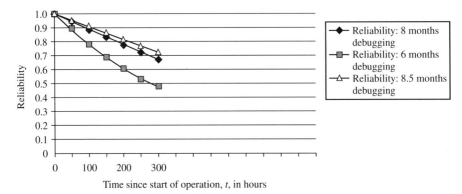

**Figure 5.16** Reliability functions for exponentially decreasing error-removal rate and 6, 8, and 8.5 months of debugging. See Eqs. (5.42b–d).

error-removal model eliminates more errors early and fewer errors later than the constant error-removal model; thus the loss of debugging between $6 < \tau < 8$ months is less damaging. This is the same reason why for $\tau = 8.5$ months of debugging the constant error-removal-rate model does better $[R(t = 300) = 0.91]$ than $[R(t = 300) = 0.71]$ for the exponential model. If we compare the exponential model with the linearly decreasing one, we find identical results at $\tau = 8$ months and very similar results at $\tau = 6$ months, where the linearly decreasing model yields $[R(t = 300) = 0.50]$ and the exponential model yields $[R(t = 300) = 0.47]$. This is reasonable since the initial portion of an exponential function is approximately linear. As was discussed previously, the linearly decreasing model is assumed to make no debugging progress after $\tau = 8$ months; thus no comparisons at $\tau = 8.5$ months are relevant.

The MTTF function for the exponentially decreasing model is computed by substituting Eq. (5.40) into Eq. (5.15) or more simply by observing that it is the reciprocal of the exponent given in Eq. (5.40):

$$\text{MTTF} = \frac{1}{kE_T e^{-\alpha\tau}} \tag{5.43a}$$

Substitution of the parameters $k = 0.000132$, $E_T = 130$, and $\alpha = 0.3206$ into Eq. (5.43a) yields

$$\text{MTTF} = \frac{58.28}{e^{-0.3206\tau}} = 58.28 e^{0.3206\tau} \tag{5.43b}$$

The MTTF curve given in Eq. (5.43b) is compared with those of Figs. 5.13 and 5.15 in Fig. 5.17. Note that it is easier to compare the behavior of the three models introduced so far by inspecting the MTTF functions, than by comparing the reliability functions. For the purpose of comparison, we have constrained

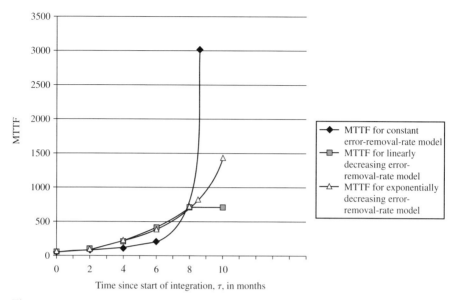

**Figure 5.17** MTTF function for constant, linearly decreasing, and exponentially decreasing error-removal-rate models.

all the reliability functions to have the same reliability at $t = 300$ hours (0.67); of course, all the reliability curves start at unity at $t = 0$. Thus, the only comparison we can make is how fast the reliability curves decay between $t = 0$ and $t = 300$ hours. Comparison of the MTTF curves yields a bit more information since the curves are plotted versus $\tau$, which is the *resource variable*. All three curves in Fig. 5.17 start at 58 hours and increase to 758 hours after 8 months of testing and debugging; however, the difference in the concave upward curvature between $\tau = 2$ and 8 months is quite apparent. The linearly decreasing and exponentially decreasing curves are about the same because at $\tau = 6$ months, the linear curve achieves an MTTF of 433 hours and the exponential curve is 399 hours, whereas the constant model only reaches 139 errors. Thus, if we had data for the first 2 months of debugging and wished to predict the progress as we approached the release time $\tau = 8$ months, any of the three models would yield approximately the same results. In applying the models, one would plot the actual error-removal rate and choose a model that best matches the actual data (experience would lead us to guess that this would be the exponential model). The real differences among the models are obvious in the region between $\tau = 8$ and 10 months. The constant error-removal model climbs to $\infty$ when the debugging time approaches 8.66 months, which is anomalous. The linearly decreasing model ceases to make progress after 8 months, which is again counterintuitive. Only the exponentially decreasing model continues to display progress after 8 months at a reasonable rate. Clearly, other more advanced reliability models can be (and have been) developed.

However, the purpose of this development is to introduce simple models that can easily be applied and interpreted, and a worthwhile working model appears to be the exponentially decreasing error-removal-rate model. The next section deals with the very important issue of how we estimate the constants of the model.

## 5.7 ESTIMATING THE MODEL CONSTANTS

### 5.7.1 Introduction

The previous sections assumed values for the various model constants; for example, $k$, $E_T$, and $\alpha$ in Eq. (5.40). In this section, we discuss the way to estimate values for these constants based on current project data (measurements) or past data. One can view this parameter estimation procedure as curve fitting to experimental data or as statistical parameter estimation. Essentially, this is the same idea from a slightly different viewpoint and using different methods; however, the end result is the same: to determine parameters of the model based on early measurements of the project (or past data) that allow prediction of the future of the project. Before we begin our discussion of parameter estimation, it is useful to consider other phases of the project.

In the previous section, we focused on the integration test phase. Software reliability models, however, can be applied to other phases of the project. Reliability predictions are most useful when they are made in the very early stages of the project, but during these phases so little detailed information is known that any predictions have a wide range of uncertainty (nevertheless, they are still useful guides). Toward the end of the project, during early field deployment, a rash of software crashes indicates that more expensive (at this late date) debugging must be done. The use of a software reliability model can predict *quantitatively* how much more work must be done. If conditions are going well during deployment, the model can quantify how well, which is especially important if the contract contains a cost incentive. The same models already discussed can be used during the deployment phase. To apply software reliability to the earlier module (unit) test phase, another type of reliability model must be employed (this is discussed in Section 5.8 on other models). Perhaps the most challenging and potentially most useful phase for software reliability modeling is during the contracting and early design phases. Because no code has been written and none can be tested, any estimates that can be made depend on past project data. In fact, we will treat reliability model constant estimation based on past data as a general technique and call it *handbook estimation*.

### 5.7.2 Handbook Estimation

The simplest use of past data in reliability estimation may be illustrated as follows. Suppose your company specializes in writing payroll programs for large organizations, and in the last 10 years you have written 78 systems of various sizes and complexities. In the last 5 years, reliability data has been kept and analyzed for 27 different systems. The data has been compiled along with explanations and analyses in a report that is called the company's *Reliability Handbook*. The most significant events recorded in this handbook are system crashes that occur between one and four times per year for the 27 different projects. In addition, data is recorded on minor errors that occur more frequently. A new client, *company X*, wants to have its antiquated, inadequate payroll program updated, and this new project is being called *system $\beta$*. Company X wants a quote for the development of system $\beta$, and the reliability of the system is to be included in the quote along with performance details, development of system $\beta$, and the reliability of the system is to be included in the quote along with performance details, and development schedule, the price, and so on. A study of the handbook reveals that the less complex systems have an MTTF of one-half to one year. System $\beta$ looks like a project of simple to medium complexity. It seems that the company could safely say that the MTTF for the system should be about one-half year but might vary from one-quarter to one year. This is a very comfortable situation, but suppose that the only recorded reliability data is on two systems. One data set represents in-house data; the other is a copy of a reliability report written by a conscientious customer during the first two years of operation who shared the report with you. Such data is better than nothing, but it is too weak to draw very detailed conclusions. The best action to take is to search for other data sources for system $\beta$ and make it a company decision to improve your future position by beginning the collection of data on all new projects as well as those currently under development, and query past customers to see if they have any data to be shared. You could even propose that the "business data processing professional organization" to which you belong sponsors a reliability data collection process to be run by an industry committee. This committee could start the process by collecting papers reporting on relevant systems that have appeared in the literature. An anonymous questionnaire could be circulated to various knowledgeable people, encouraging them to contribute data with sufficient technical details to make listing these projects in a composite handbook useful, but not enough information so that the company or project can be identified. Clearly, the largest software development companies have such handbooks and the smaller companies do not. The subject of hardware reliability started in the late 1940s with the collection of component and some system reliability data spearheaded by Department of Defense funds. Unfortunately, no similar efforts have been sponsored to date in the software reliability field by Department of Defense funds or professional organizations. For a modest initial collection of such data, see Shooman [1983, p. 368, Table 5.10] and Musa [1987, p. 116, Table 5.2].

From the data that does exist, we are able to compute a rough estimate for the parameter $E_T$ first introduced in Eq. (5.20) and present in all the models developed to this point. It seems unreasonable to report the same value for $E_T$

for both large and small programs; thus Shooman and Musa both normalize the value by dividing by the total number of source instructions $I_T$. For the data from Shooman, we exclude the values for the end-of-integration testing, acceptance testing, and simulation testing. This results in a mean value for $E_T/I_T$ of $5.14 \times 10^{-3}$ and a standard deviation of $4.23 \times 10^{-3}$ for seven data points. Similarly, we make the same computation for the data in Table 5.2 of Musa [1987] for the 25 system test values and obtain a mean value for $E_T/I_T$ of $7.85 \times 10^{-3}$ and a standard deviation of $5.27 \times 10^{-3}$. These values are in rough agreement, considering the diverse data sources and the imperfection in defining what constitutes not only an error but the phases of development as well. Thus we can state that based on these two data sets we would expect a mean value of about $5-9 \times 10^{-3}$ for $E_T/I_T$ and a range from $\mu - \sigma$ (lowest for Shooman data) of about $1 \times 10^{-3}$ to $\mu + \sigma$ (highest for Musa data) of about $13 \times 10^{-3}$. Of course, to obtain the value of $E_T$ for any of the models, we would multiply these values by the value of $I_T$ for the project in question.

What about handbook data for the initial estimation of any of the other model parameters? Unfortunately, little such data exists in collected form. For typical values, see Shooman [1983, p. 368, Table 5.10] and Musa [1987].

### 5.7.3 Moment Estimates

The best way to proceed with parameter estimation for a reliability model is to plot the error-removal rate versus $\tau$ on a simple graph with whatever intervals are used in recording the data (generally, daily or weekly). One could employ various statistical means to test which model best fits the data: a constant, a linear, an exponential, or another model, but inspection of the graph is generally sufficient to make such a determination.

***Constant Error-Removal-Rate Data.*** Suppose that the error-removal data looks approximately constant and that the time axis is divided into regular or irregular intervals, $\Delta\tau_i$, corresponding to the data, and that in each interval there are $E_c(\Delta\tau_i)$ corrected errors. Thus the data for the error-correction rate is a sequence of values $E_c(\Delta\tau_i)/\Delta\tau_i$. The simplest way to estimate the value of $\rho_0$ is to take the mean value of the error-correction rates:

$$\rho_0 = \frac{1}{i} \sum_i \frac{E_c(\Delta\tau_i)}{\Delta\tau_i} \qquad (5.44)$$

Thus, by examining Eqs. (5.30a, b), we see that there are two additional parameters to estimate: $k$ and $E_T$.

The estimate given in Eq. (5.44) utilizes the mean value that is the first moment and belongs to a general class of statistical estimates called *moment estimates*. The general idea of applying moment estimation to the evaluation of parameters for probability distributions (models) is to first compute a number of moments of the probability distribution equal to the number of parameters to be estimated. The moments are then computed from the numerical data; the first moment formula is equated to the first moment of the data, the second moment formula is equated to the second moment of the data, and so on until enough equations are formulated to solve for the parameters. Since we wish to estimate $k$ and $E_T$ in Eqs. (5.30a, b), two moment equations are needed. Rather than compute the first and second moments, we use a slight variation in the method and compute the first moment at two different values of $\tau_i$, $\tau_1$, and $\tau_2$. Since the random variable is time to failure, the first moment (mean) is given by Eq. (5.30b). To compute the mean of the data, we require a set of test data from which we can calculate mean time to failure. The best data would of course be operational data, but since the software is being integrated, it would be difficult to place it into operation. The next best data is *simulated* operational data, generally obtained by testing the software in a simulated operational mode by using specially prepared software. Such software is generally written for use at the end of the test cycle when comprehensive system tests are performed. It is best that such software be developed early in the test cycle so that it is available for "reliability testing" during integration. Such simulation testing is time-consuming, it can be employed during off hours (e.g., second and third shift) so that it does not interrupt the normal development schedule. (Musa [1987] has written extensively on the use of ordinary integration test results when simulation testing is not available. This subject will be discussed later.) Simulation testing is based on a number of scenarios representing different types of operation and results in $n$ total runs, with $r$ failures and $n - r$ successes. The $n - r$ successful runs represent $T_1, T_2, \ldots, T_{n-r}$ hours of successful operation and the $r$ unsuccessful runs represent $t_1, t_2, \ldots, t_r$ hours of successful operation before the failures occur. Thus the testing produces $H$ total hours of successful operation.

$$H = \sum_{i=1}^{n-r} T_i + \sum_{i=1}^{r} t_i \tag{5.45}$$

Assuming that the failure rate is constant over the test interval (no debugging occurs while we are testing), the failure rate is given by $z = \lambda$:

$$\lambda = \frac{r}{H} \tag{5.46a}$$

and since the MTTF is the reciprocal,

$$\text{MTTF} = \frac{1}{\lambda} = \frac{H}{r} \tag{5.46b}$$

Thus, applying the moment method reduces to matching Eqs. (5.30b) and (5.46b) at times $\tau_a$ and $\tau_b$ in the development cycle, yielding

$$\text{MTTF}_a = \frac{H_a}{r_a} = \frac{1}{k(E_T - \rho_0 \tau_a)} \qquad (5.47a)$$

$$\text{MTTF}_b = \frac{H_b}{r_b} = \frac{1}{k(E_T - \rho_0 \tau_b)} \qquad (5.47b)$$

Because $\rho_0$ is already known, the two preceding equations can be solved for the parameters $k$ and $E_T$, and our model is complete. [One could have skipped the evaluation of $\rho_0$ using Eq. (5.44) and generated a third MTTF equation similar to Eqs. (5.47a, b) at a third development time $\tau_3$. The three equations could then have been solved for the three parameters. The author feels that fitting as many parameters as possible from the error-removal data followed by using the test data to estimate the remaining data is a superior procedure.] If we apply this model as integration continues, a sequence of test data will be accumulated and the question arises: Which two sets of test data will be used in Eqs. (5.47a, b)—the last two or the first and the last? This issue is settled if we use least-squares or maximum-likelihood methods of estimation (which will soon be discussed) since they both use all available sets of test data. In any event, the use of the moment estimates described in this section is always a good starting point in building a model, even if more advanced methods will be used later. The reader must realize that the significant costs and waiting periods for applying such models are associated with the test results. The analysis takes at most one-half of a day, and if calculation programs are used, even less time than that. Thus it is suggested that several models be calculated and compared as the project progresses whenever new test data is available.

***Linearly Decreasing Error-Removal-Rate Data.*** Suppose that inspection of the error-removal data reveals that the error-removal rate decreases in an approximately linear manner. Examination of Eq. (5.23b) shows that there are two parameters in the error-removal-rate model: $K$ and $\tau_0$. In addition, there is the parameter $E_T$ and, from Eq. (5.27), the additional parameter $k$. We have several choices regarding the evaluation of these four constants. One can use the error-removal-rate curve to evaluate two of these parameters, $K$ and $\tau_0$, and use the test data to evaluate $k$ and $E_T$ as was done in the previous section in Eqs. (5.47a, b).

The simplest procedure is to evaluate $K$ and $\tau_0$ using the error-removal rates during the first two test intervals. The error-removal rate is found by differentiating [cf. Eqs. (5.23d) and (5.24a)].

$$\frac{dE_r(\tau)}{d\tau} = K\left(1 - \frac{\tau}{2\tau_0}\right) \qquad (5.48a)$$

If we adopt the same notation as used in Eq. (5.44), the error-removal rate becomes $E_c(\Delta\tau_i)/\Delta\tau_i$. If we match Eq. (5.48a) at the midpoints of the first two intervals, $\tau_a/2$ and $\tau_a + \tau_b/2$, the following two equations result:

$$\frac{E_c(\Delta\tau_a)}{\Delta\tau_a} = K\left(1 - \frac{\tau_a}{4\tau_0}\right) \tag{5.48b}$$

$$\frac{E_c(\Delta\tau_b)}{\Delta\tau_b} = K\left(1 - \frac{\tau_a + \tau_b/2}{2\tau_0}\right) \tag{5.48c}$$

and they can be solved for $K$ and $\tau_0$. This leaves the two parameters $k$ and $E_T$, which can be evaluated from test data in much the same way as Eqs. (5.47a, b). The two equations are

$$\text{MTTF}_a = \frac{H_a}{r_a} = \frac{1}{k\left[E_T - K\left(1 - \frac{\tau_a}{2\tau_0}\right)\right]} \tag{5.49a}$$

$$\text{MTTF}_b = \frac{H_b}{r_b} = \frac{1}{k\left[E_T - K\left(1 - \frac{\tau_b}{2\tau_0}\right)\right]} \tag{5.49b}$$

***Exponentially Decreasing Error-Removal-Rate Data.*** Suppose that inspection of the error-removal data reveals that the error-removal rate decreases in an approximately exponential manner. One good way of testing this assumption is to plot the error-removal-rate data on a log–log graph by computer or on graph paper. An exponential curve rectifies on log–log axes. (There are more sophisticated statistical tests to check how well a set of data fits an exponential curve. See Shooman [1983, p. 28, problem 1.3] or Hoel [1971].) If Eq. (5.40) is examined, we see that there are three parameters to estimate $k$, $E_T$, and $\alpha$. As before, we can estimate some of these parameters from the error-removal-rate data and some from simulation test data. One can probably investigate which parameters should be estimated from one set of data and which from the other sets should be estimated via theoretical arguments; however, the practical approach is to use the better data to estimate as many parameters as possible. Error-removal data is universally collected whenever the software comes under configuration control, but simulation test data requires more effort and expense. Error-removal data is therefore more plentiful, allowing the estimation of as many model parameters as possible. Examination of Eq. (5.25e) reveals that $E_T$ and $\alpha$ can be estimated from the error data. Estimation equations for $E_T$ and $\alpha$ begin with Eq. (5.25e). Taking the natural logarithm of both sides of the equation yields

$$\ln\{E_r(\tau)\} = \ln\{E_T\} - \alpha\tau \tag{5.50a}$$

If we have two sets of error-removal data at $\tau_a$ and $\tau_b$, Eq. (5.50a) can be used to solve for the two parameters. Substitution yields

$$\ln\{E_r(\tau_a)\} = \ln\{E_T\} - \alpha\tau_a \tag{5.50b}$$
$$\ln\{E_r(\tau_b)\} = \ln\{E_T\} - \alpha\tau_b \tag{5.50c}$$

Subtracting the second equation from the first and solving for $\alpha$ yields

$$\alpha = \frac{\ln\{E_c(\tau_a)\} - \ln\{E_c(\tau_b)\}}{\tau_b - \tau_a} \tag{5.51}$$

Knowing the value of $\alpha$, one could substitute into either Eq. (5.50b) or (5.50c) to solve for $E_T$. However, there is a simple way to use information from both equations (which should be a better estimate) by adding the two equations and solving for $E_T$.

$$\ln\{E_T\} = \frac{\ln\{E_c(\tau_a)\} + \ln\{E_c(\tau_b)\} + \alpha(\tau_a + \tau_b)}{2} \tag{5.52}$$

Once we know $E_T$ and $\alpha$, one set of integration test data can be used to determine $k$. From Eq. (5.43a), we proceed in the same manner as Eq. (5.47a); however, only one test time is needed.

$$\text{MTTF}_a = \frac{H_a}{r_a} = \frac{1}{kE_T e^{-\alpha\tau_a}} \tag{5.53}$$

### 5.7.4 Least-Squares Estimates

The moment estimates of the preceding sections have a number of good attributes:

1. They require the least amount of data.
2. They are computationally simple.
3. They serve as a good starting point for more complex estimates.

The computational simplicity is not too significant in this era of cheap, fast computers. Nevertheless, it is still a good idea to use a calculator, pencil, and paper to get a feeling for data values before a more complex, less transparent, more accurate computer algorithm is used.

The main drawback of moment estimates is the lack of clear direction for how to proceed when several data sets are available. The simplest procedure in such a case is to use least-squares estimation. A complete development of least-squares estimation appears in Shooman [1990] and is applied to software reliability modeling in Shooman [1983, pp. 372–374]. However, computer mathematics packages such as Mathematica, Mathcad, Macsyma, and Maple all have least-squares programs that are simple to use; any increased complexity is buried within the program, and computational time is not significant with modern computers. We will briefly discuss the use of least-squares estimation for the case of an exponentially decreasing error-removal rate.

Examination of Eq. (5.50a) shows that on log–log paper, the equation becomes a straight line. It is recommended that the data be initially plotted and a straight line be fitted by inspection through the data. When $\tau = 0$, the y-axis intercept, $E_c(\tau = 0)$ is equal to $E_T$, and the slope of the straight line is $-\alpha$. Once these initial estimates have been determined, one can use a least-squares program to find the mean values of the parameters and their variances.

In a similar manner, one can determine the value of $k$ by substitution in Eq. (5.53) for one set of simulation data. Assuming that we have several sets of simulation data at $\tau_j = a, b, \ldots$, we can write the equation as

$$\ln\{\text{MTTF}_j\} = \frac{H_j}{r_j} = -[\ln\{k\} + \ln\{E_T\} - \alpha\tau_j] \tag{5.54}$$

The preceding equation is used as the basis of a least-squares estimation to determine the mean value and variance of $k$. Again, it is useful to plot Eq. (5.54) and fit a straight line to the data as a precursor to program estimation.

### 5.7.5 Maximum-Likelihood Estimates

In England in the 1930s, Fisher developed the elegant theory called maximum-likelihood estimation (MLE) for estimating the values of parameters of probability distributions from data [Shooman, 1983, pp. 537–540; Shooman, 1990, pp. 80–96]. We can explain some of the ideas underlying MLE in a simple fashion. If $R(t)$ is the reliability function, then $f(t)$ is the associated density function for the time to failure, and the parameters are $\theta_1, \theta_2$, and so forth, and we have $f(\theta_1, \theta_2, \ldots, \theta_i, t)$. The data are the several values of time to failure $t_1, t_2, \ldots, t_i$, and the task is to estimate the best values for $\theta_1, \theta_2, \ldots, \theta_i$ from the data. Suppose there are two parameters, $\theta_1$ and $\theta_2$, and three values of time data: $t_1 = 50$, $t_2 = 200$, and $t_3 = 490$. If we know the values of $\theta_1$ and $\theta_2$, then the probability of obtaining the test values is related to the joint likelihood function (assuming independence), $L(\theta_1, \theta_2) = f(\theta_1, \theta_2, 50) \cdot f(\theta_1, \theta_2, 200) \cdot f(\theta_1, \theta_2, 490)$. Fisher's brilliant procedure was to compute values of $\theta_1$ and $\theta_2$, which maximized $L$. To find the maximum of $L$, one computes the partial derivatives of $L$ with respect to $\theta_1$ and $\theta_2$ and sets these values to zero. The resultant equations are solved for the MLE values of $\theta_1$ and $\theta_2$. If there are more than two parameters, more partial derivative equations are needed. The application of MLE to software reliability models is discussed in Shooman [1983, pp. 370–372, 544–548].

The advantages of MLE estimates are as follows:

1. They automatically handle multiple data sets.
2. They provide variance estimates.

3. They have some sophisticated statistical evaluation properties.

Note that least-squares estimation also possesses the first two properties. Some of the disadvantages of MLE estimates are as follows:

1. They are more complex and more difficult to understand than moment or least-squares estimates.
2. MLE estimates involve the solution of a set of complex equations that often requires numerical solution. (Moment or least-squares estimates can be used as starting values to expedite the numerical solution.)

The way of overcoming the first problem in the preceding list is to start with moment or least-squares estimates to develop insight, whereas the second problem requires development of a computer estimation program, which takes some development effort. Fortunately, however, such programs are available; among them are SMERFS [Farr, 1991; Lyu, 1996, pp. 733–735]; SoRel [Lyu, 1996, pp. 737–739]; CASRE [Lyu, 1996, pp. 739–745]; and others [Strark, Appendix A in Lyu, 1996, pp. 729–745].

## 5.8 OTHER SOFTWARE RELIABILITY MODELS

### 5.8.1 Introduction

Since the first software reliability models were introduced [Jelinski and Moranda, 1972; Shooman, 1972], there have been many software reliability models developed. The ones introduced in the preceding section are simple to understand and apply. In fact, depending on how one counts, the 4 models (constant, linearly decreasing, exponentially decreasing, and S-shaped) along with the 3 parameter estimation methods (moment, least-squares, and MLE) actually form a group of 12 models. Some of the other models developed in the literature are said to have better "mathematical properties" than these simple models. However, the real test of a model is how well it performs, that is, if data is taken between months 1 and 2 of an 8-month project, how well does it predict at the end of month 2 the growth in MTTF or the decreasing failure rate between months 3 and 8. Also, how does the prediction improve after data for months 3 and 4 is added, and so forth.

### 5.8.2 Recommended Software Reliability Models

Software reliability models are not used as universally in software development as they should be. Some reasons that project managers give for this are the following:

1. It costs too much to do such modeling and I can't afford it within my project budget.

2. There are so many software reliability models to use that I don't know which is best; therefore, I choose not to use any.
3. We are using the most advanced software development strategies and tools and produce high-quality software; thus we don't need reliability measurements.
4. Even if a model told me that the reliability will be poor, I would just test some more and remove more errors.
5. If I release a product with too many errors, I can always fix those that get discovered during early field deployment.

Almost all of these responses are invalid. Regarding response (1), it does not cost that much to employ software reliability models. During integration testing, error collection is universally done, and the analysis is relatively inexpensive. The only real cost is the scheduling of the simulation/system test early in integration testing, and since this can be done during off hours (second and third shift), it is not that expensive and does not delay development. (Why do managers always state that there is not enough money to do the job right, yet always find lots of money to fix residual errors that should have been eliminated much earlier in the development process?) Response (3) has been the universal cry of software development managers since the dawn of software, and we know how often this leads to grief. Responses (4) and (5) are true and have some merit; however, the cost of fixing a lot of errors at these late stages is prohibitive, and the delivery schedule and early reputation of a product are imperiled by such an approach. This leaves us with response (2), which is true and for which some of the models are mathematically sophisticated. This is one of the reasons why the preceding section's treatment of software reliability models focused on the simplest models and methods of parameter estimation in the hope that the reader would follow the development and absorb the principles.

As a direct rebuttal to response (2), a group of experienced reliability modelers (including this author) began work in the early 1990s to produce a document called *Recommended Practice for Software Reliability* (a software reliability standard) [AIAA/ANSI, 1993]. This standard recommends four software reliability models: the Schneidewind model, the generalized exponential model [Shooman, April 1990], the Musa/Okumoto model, and the Littlewood/Verrall model. A brief study of the models shows that the generalized exponential model is identical with the three models discussed previously in this chapter. The basic development described in the previous section corresponds to the earliest software reliability models [Jelinski and Moranda, 1972; Shooman, 1972], and the constant error-removal-rate model [Shooman, 1972]. The linearly decreasing error-removal-rate model is essentially Musa's basic model [1975], and the exponentially decreasing error-removal-rate model is Musa's logarithmic model [1987]. Comprehensive parameter estimation equations appear in the AIAA/ANSI standard [1993] and in Shooman [1990]. The reader is referred to these references for further details.

### 5.8.3 Use of Development Test Data

Several authors, notably Musa, have observed that it would be easiest to use development test data where the tests are performed and the system operates for $T$ hours rather than simulating real operation where the software runs for $t$ hours of operation. We assume that development tests stress the system more "rapidly" than simulated testing—that $T = Ct$ and that $C > 1$. In practice, Musa found that values of 10–15 are typical for $C$. If we introduce the parameter $C$ into the exponentially decreasing error-rate model (Musa's logarithmic model), we have an additional parameter to estimate. Parameters $E_T$ and $\alpha$ can be estimated from the error-removal data; $k$ and $C$, from the development test data. This author feels that the use of simulation data not requiring the introduction of $C$ is superior; however, the use of development data and the necessary introduction of the fourth parameter $C$ is certainly convenient. If such a method is to be used, a handbook with data listing previous values of $C$ and judicious choices from the previous results would be necessary for accurate prediction.

### 5.8.4 Software Reliability Models for Other Development Stages

The software reliability models introduced so far are immediately applicable to integration testing or early field deployment stages. (Later field deployment, too, is applicable, but by then it is often too late to improve a bad product; a good product is apparent to everybody and needs little further debugging.) The earlier one can employ software reliability, the more useful the models are in predicting the future. However, during unit (module testing), other models are required [Shooman, 1983, 1990].

Software reliability estimation is of great use in the specification and early design phases as a means of estimating how good the product can be made. Such estimates depend on the availability of field data on other similar past projects. Previous project data would be tabulated in a "handbook" of previous projects, and such data can be used to obtain initial values of parameters for the various models by matching the present project with similar historical projects. Such handbook data does exist within the databases of large software development organizations, but this data is considered proprietary and is only available to workers within the company. The existence of a "software reliability handbook" in the public domain would require the support of a professional or government organization to serve as a sponsor.

Assuming that we are working within a company where such data is available early in the project (perhaps even during the proposal phase), early estimates can be made based on the use of historical data to estimate the model parameters. Accuracy of the parameters depends on the closeness of the match between handbook projects and the current one in question. If a few projects are acceptable matches, one can estimate the parameter range.

If one is fortunate enough to possess previous data and, later, to obtain system test data, one is faced with the decision regarding when the previous project data is to be discarded and when the system test data can be used to estimate model parameters. The initial impulse is to discard neither data set but to average them. Indeed, the statistical approach would be to use Bayesian estimation procedures (see Mood and Graybill [1963, p. 187]), which may be viewed as an elaborate statistical-weighting scheme. A more direct approach is to use a linear-weighting scheme. Assume that the historical project data leads to a reliability estimate for the software given by $R_0(t)$, and the reliability estimate from system test data is given by $R_1(t)$. The composite estimate is given by

$$R(t) = a_0 R_0(t) + a_1 R_1(t) \tag{5.55}$$

It is not difficult to establish that $a_0 + a_1$ should be set equal to unity. Before test data is available, $a_0$ will be equal to unity and $a_1$ will be 0; as test data becomes available, $a_0$ will approach 0 and $a_1$ will approach unity. The weighting procedure is derived by minimizing the variance of $R(t)$, assuming that the variance of $R_0(t)$ is given by $\sigma_0^2$ and that of $R_1(t)$ by $\sigma_1^2$. The end result is a set of weighting formulas given by the equations that follow. (For details, see Shooman [1971].)

$$a_0 = \frac{\dfrac{1}{\sigma_0^2}}{\dfrac{1}{\sigma_0^2} + \dfrac{1}{\sigma_1^2}} \tag{5.56a}$$

$$a_1 = \frac{\dfrac{1}{\sigma_1^2}}{\dfrac{1}{\sigma_0^2} + \dfrac{1}{\sigma_1^2}} \tag{5.56b}$$

The reader who has studied electric-circuit theory can remember the form of these equations by observing that they are analogous to how resistors combine in parallel. To employ these equations, the analyst must estimate a value of $\sigma_0^2$ based on the variability of the previous project data and use the value of $\sigma_1^2$ given by applying the least-squares (or another) method to the system test data.

The problems at the end of this chapter provide further exploration of other models, the parameter estimation, the numerical differences among the methods, and the effect on the reliability and MTTF functions. For further details on software reliability models, the reader is referred to AIAA/ANSI standard [1993], Musa [1987], and Lyu [1996].

### 5.8.5 Macro Software Reliability Models

Most of the software reliability models in the literature are black box models. There is one clear box model that relates the software reliability to some features of the program structure [Shooman, 1983, pp. 377–384; Shooman, 1991]. This model decomposes the software into major execution paths of the control structure. The software failure rate is developed in terms of the frequency of path execution, the probability of error along a path, and the traversal time for the path. For more details, see Shooman [1983, 1991].

## 5.9 SOFTWARE REDUNDANCY

### 5.9.1 Introduction

Chapters 3 and 4 discussed in detail the various ways one can employ redundancy to enhance the reliability of the hardware. After a little thought, we raise the question: Can we employ software redundancy? The answer is yes; however, there are several issues that must be explored. A good way to introduce these considerations is to assume that one has a TMR system composed of three identical digital computers and a voter. The preceding chapter detailed the hardware reliability for such a system, but what about the software? If each computer contains a copy of the same program, then when one computer experiences a software error, the other two should as well. Thus the three copies of the software provide no redundancy. The system model would be a hardware TMR system in series with the software reliability, and the system reliability, $R_{sys}$, would be given by the product of the hardware voting system, $R_{TMR}$, and the software reliability, $R_{software}$, assuming independence between the hardware and software errors. We should actually speak of two types of software errors. The first type is the *most common* one due to a scenario with a set of inputs that uncovers a latent fault in the software. Clearly, all copies of the same software will have that same fault and should process the scenario identically; thus there is no software redundancy. However, *some* software errors are due to the interaction of the inputs, the state of the hardware, and any residual faults. By the state of the hardware we mean the storage values in registers (maybe other storage devices) at the time the scenario is begun. Since these storage values are dependent on when the computer is powered up and cleared as well as the past data processed, the states of the three processors may differ. There may be a small amount of redundancy due to these effects, but we will ignore state-dependent errors.

Based on the foregoing discussion, the only way one can provide software reliability is to write different independent versions of the software. The cost is higher, of course, and there is always the chance that even independent programming groups will incorporate the same (common mode) software errors, degrading the amount of redundancy provided. A complete discussion appears in Shooman [1990, pp. 582–587]. A summary of the relevant analysis appears in the following paragraphs, as well as an example of how modular hardware and software redundancy is employed in the Space Shuttle orbital flight control system.

### 5.9.2 *N*-Version Programming

The official term for separately developed but functionally identical versions of software is *N-version software*. We provide only a brief summary of these techniques here; the reader is referred to the following references for details: Lala [1985, pp. 103–107]; Pradhan [1986, pp. 664–667]; and Siewiorek [1982, pp. 119–121, 169–175]. The term *N*-version programming was probably coined by Chen and Avizienis [1978] to liken the use of redundant software to *N*-modular redundancy in hardware. To employ this technique, one writes two or more independent versions of the program and uses them in a voting-type arrangement. The heart of the matter is to discuss what we mean by independent software. Suppose we have three processors in a TMR arrangement, all running the same program. We assume that hardware and software failures are independent except for natural or manmade disasters that can affect all three computers (earthquake, fire, power failure, sabotage, etc.). In the case of software error, we would expect all three processors to err in the same manner and the voter to dutifully pass on the same erroneous output without detection of an error. (As was discussed previously, the only possible differences lie in the rare case in which the processors have different states.) To design independent programs to achieve software reliability, we need independent development groups (probably in different companies), different design approaches, and perhaps even different languages. A simplistic example would be the writing of a program to find the roots of a quadratic equation, $f(x)$, which has only real roots. The obvious approach would be to use the quadratic formula. A different design would be to use the theorem from the theory of equations, which states that if $f(a) > 0$ and if $f(b) < 0$, then at least one root lies between $a$ and $b$. One could bisect the interval $(a, b)$, check the sign of $f([a + b]/2)$, and choose a new, smaller interval. Once iteration determines the first root, polynomial division can be used to determine the second root. We could ensure further diversity of the two approaches by coding one in C++ and the other in Ada. There are some difficulties in ensuring independent versions and in synchronizing different versions, as well as possible problems in comparing the outputs of different versions.

It has been suggested that the following procedures be followed to ensure that we develop independent versions:

1. Each programmer works from the same requirements.
2. Each programmer or programming group works independently of the others, and communication between groups is not permitted except by passing messages (which can be edited or blocked) through the contracting organization.

3. Each version of the software is subjected to the same comprehensive acceptance tests.

Dependence among errors in various versions can occur for a variety of reasons, such as the following:

1. Identical misinterpretation of the requirements.
2. Identical, incorrect treatment of boundary problems.
3. Identical (or equivalent), incorrect designs for difficult portions of the problem.

The technique of *N*-version programming has been used or proposed for a variety of situations, such as the following:

1. For Space Shuttle flight control software (discussed in Section 5.9.3).
2. For the slat-and-flap control system of A310 Airbus Industry aircraft.
3. For point switching, signal control, and traffic control in the Göteborg area of the Swedish State Railway.
4. For nuclear reactor control systems (proposed by several authors).

If the software versions are *independent*, we can use the same mathematical models as were introduced in Chapter 4. Consider the triple-modular redundant (TMR) system as an example. If we assume that there are three independent versions of the software and that the voting is perfect, then the reliability of the TMR system is given by

$$R = p_i^2(3 - 2p_i) \tag{5.57}$$

where $p_i$ is the identical reliability of each of the three versions of the software. We assume that all of the software faults are independent and affect only one of the three versions.

Now, we consider a simple model of dependence. If we assume that there are two different ways in which common-mode dependencies exist, that is, requirements and program, then we can make the model given in Fig. 5.18. The reliability expression for this model is given by Shooman [1988].

$$R = p_{cmr}p_{cms}[p_i^2(3 - 2p_i)] \tag{5.58}$$

This expression is the same mathematical formula as that of a TMR system with an imperfect voter (i.e., the common-mode errors play an analogous role to voter failures).

The results of the above analysis will be more meaningful if we evaluate the effects of common-mode failures for a set of data. Although common

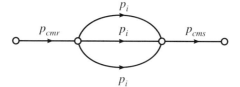

where

$p_i$ = 1 − Probability of an independent-mode-software fault
$p_{cmr}$ = 1 − Probability of a common-mode-requirements error
$p_{cms}$ = 1 − Probability of a common-mode-software fault

**Figure 5.18** Reliability model of a triple-modular program including common-mode failures.

mode data is hard to obtain, Chen and Avizienis [1978] and Pradhan [1986, p. 665] report some practical data for 12 different sets of 3 independent programs written for solving a differential equation for temperature over a two-dimensional region. From these results, we deduce that the individual program reliabilities were $p_i = 0.851$, and substitution into Eq. (5.58) yields $R = 0.94$ for the TMR system. Thus the unreliability of the single program, $(1 − 0.851) = 0.149$, has been reduced to $(1 − 0.94) = 0.06$; the decrease in unreliability $(0.149/0.06)$ is a factor of 2.48 (the details of the computation are in Shooman [1990, pp. 583–587]). This data did not include any common-mode failure information; however, the second example to be discussed does include this information.

Some data gathered by Knight and Leveson [1986] discussed 27 different versions of a program, all of which were subjected to 200 acceptance tests. Upon acceptance, the program was subjected to one million test runs (see also McAllister and Vouk [1996]).

Five of the programs tested without error, and the number of errors in the others ranged up to 9,656 for program number 22, which had a demonstrated $p_i = (1 − 9,656/1,000,000) = 0.990344$. If there were no common-mode errors, substitution of this value for $p_i$ into Eq. (5.57) yields $R = 0.99972$. The improvement in unreliability, $1 − R$, is $0.009656/0.00028$, or a factor of 34.5.

The number of common occurrences was also recorded for each error, allowing one to estimate the common-mode probability. By treating all the common mode situations as if they affected all the programs (a worst-case assumption), we have as the estimate of common mode (sum of the number of multiple failure occurrences)/(number of tests) = $1,255/1,000,000 = 0.001255$. The probability of common-mode error is given by $p_{cmr}p_{cms} = 1 − 0.001255 = 0.998745$. Substitution into Eq. (5.58) yields $R = 0.99846$. The improvement in $1 − R$ would now be from $0.009656$ to $0.00154$, and the improvement factor is 6.27—still substantial, but a significant decrease from the 34.5 that was achieved without common-mode failures. (The details are given in Shooman

[1990, pp. 582–587].) Another case is computed in which the initial value of $p_i = (1 - 1,368/1,000,000) = 0.998632$ is much higher. In this case, TMR produces a reliability of 0.99999433 for an improvement in unreliability by a factor of 241. However, the same estimate of common-mode failures reduces this factor to only 1.1! Clearly, such a small improvement factor would not be worth the effort, and either the common-mode failures must be reduced or other methods of improving the software reliability should be pursued. Although this data varies from program to program, it does show the importance of common-mode failures. When one wishes to employ redundant software, clearly one must exercise all possible cautions to minimize common-mode failures. Also, it is suggested that modeling be done at the outset of the project using the best estimates of independent and common-mode failure probabilities and that this continue throughout the project based on the test results.

### 5.9.3 Space Shuttle Example

One of the best known examples of hardware and software reliability is the Space Shuttle Orbiter flight control system. Once in orbit, the flight control system must maintain the vehicle's altitude (rotations about 3 axes fixed in inertial space). Typically, one would use such rotations to lock onto a view of the earth below, travel along a line of sight to an object that the Space Shuttle is approaching, and so forth. The Space Shuttle uses a combination of various large and small gas jets oriented about the 3 axes to produce the necessary rotations. Orbit-change maneuvers, including the crucial reentry phase, are also carried out by the flight control system using somewhat larger orbit-maneuvering system (OMS) engines. There is much hardware redundancy in terms of sensors, various groupings of the small gas jets, and even the use of a combination of small gas jets for sustained firing should the OMS engines fail. In this section, we focus on the computer hardware and software in this system, which is shown in Fig. 5.19.

There are five identical computers in the system, denoted as Hardware A, B, C, D, and E, and two different software systems, denoted by Software A and B. Computers A–D are connected in a voting arrangement with lockout switches at the inputs to the voter as shown. Each of these computers uses the complete software system—Software A. The four computers and associated software comprise the primary avionics software system (PASS), which is a two-out-of-four system. If a failure in one computer occurs and is confirmed by subsequent analysis and by disagreement with the other three computers as well as by other tests and telemetered data to Ground Control, this computer is then disconnected by the crew from the arrangement, and the remaining system becomes a TMR system. Thus this system will sustain two failures and still be functional rather than tolerating only a single failure, as is the case with an ordinary TMR system. Because of all the monitoring and test programs available in space and on the ground, it is likely that even after two failures, if a third malfunction occurred, it would still be possible to determine and switch

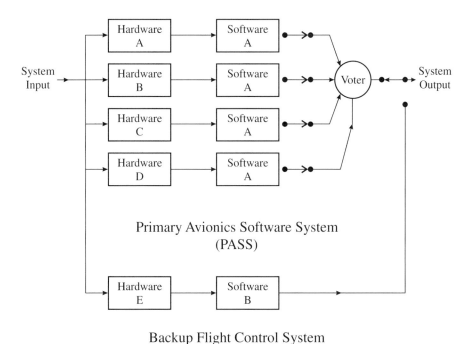

**Figure 5.19** Hardware and software redundancy in the Space Shuttle's avionics control system.

to the one remaining good computer. Thus the PASS has a very high level of hardware redundancy, although it is vulnerable to common-mode software failures in Software A. To guard against this, a backup flight control system (BFS) is included with a fifth computer and independent Software B. Clearly, Hardware E also supplies additional computer redundancy. In addition to the components described, there are many replicated sensors, actuators, controls, data buses, and power supplies.

The computer self-test features detect 96% of the faults that could occur. Some of the built-in test and self-test features include the following:

- Bus time-out tests: If the computer does not perform a periodic operation on the bus, and the timer has expired, the computer is labeled as failed.
- Comparisons: Check sum is computed, and the computer is labeled as failed if there are two successive miscompares.
- Watchdog timers: Processors set a timer, and if the timer completes its count before it is reset, the computer is labeled as failed and is locked out.

To provide as much independence as possible, the two versions of the software were developed by different organizations. The programs were both written in the HAL/S language developed by Intermetrics. The primary system was written by IBM Federal Systems Division, and the backup software was written by Rockwell and Draper Labs. Both Software A and Software B perform all the critical functions, such as ascent to orbit, descent from orbit, and reentry, but Software A also includes various noncritical functions, such as data logging, that are not included in the backup software.

In addition to the redundant features of Software A and B, great emphasis has been applied to the life-cycle management of the Space Shuttle software. Although the software for each mission is unique, many of its components are reused from previous missions. Thus, if an error is found in the software for flight number 76, all previous mission software (all of which is stored) containing the same code is repaired and retested. Also, the reason why such an error occurred is analyzed, and any possibilities for similar mechanisms to cause errors in the rest of the code for this mission and previous missions are investigated. This great care, along with other features, resulted in the Space Shuttle software team being one of the first organizations to earn the highest rating of "level 5" when it was examined by the Software Engineering Institute of Carnegie Mellon University and judged with respect to the capability maturity model (CMM) levels. The reduction in error rate for the first 11 flights indicates the progress made and is shown in Fig. 5.20. An early reliability study of ground-based Space Shuttle software appears in Shooman [1984]; the model predicted the observed software error rate on flight number 1.

The more advanced voting techniques discussed in Section 4.11 also apply to *N*-version software. For a comprehensive discussion of voting techniques, see McAllister and Vouk [1996].

## 5.10 ROLLBACK AND RECOVERY

### 5.10.1 Introduction

The term *recovery technique* includes a class of approaches that attempts to detect a software error and, in various ways, retry the computation. Suppose, for example, that the track of an aircraft on the display in an air traffic control system becomes corrupted. If the previous points on the path and the current input data are stored, then the computation of the corrupted points can be retried based on the stored values of the current input data. Assuming that no critical situation is in progress (e.g., a potential air collision), the slight delay in recomputing and filling in these points causes no harm. At the very worst, these few points may be lost, but the software replaces them by a projected flight path based on the past path data, and soon new actual points are available. This is also a highly acceptable solution. The worst outcomes that must be strenuously avoided are from those cases in which the errors terminate the track or cause the entire display to crash. Some designers would call such recovery techniques *rollback* because the com-

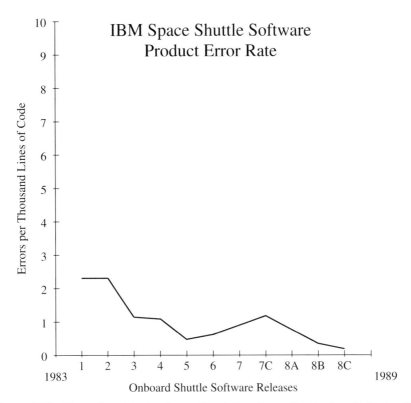

**Figure 5.20** Errors found in the Space Shuttle's software for the first 11 flights. The IBM Federal Systems Division (now United Space Alliance), wrote and maintained the onboard Space Shuttle control software, twice receiving the George M. Low Trophy, NASA's excellence award for quality and productivity. This graph was part of the displays at various trade shows celebrating the awards. See Keller [1991] and Schneidewind [1992] for more details.

putation backs up to the last set of previous valid data and attempts to reestablish computations in the problem interval and resume computations from there on. Another example that fits into this category is the familiar case in which one uses a personal computer with a word processing program. Suppose one issues a print command and discovers that the printer is turned off or the printer cable is disconnected. Most (but not all) modern software will give an error message and return control to the user, whereas some older programs lock the keyboard and will not recover once the cable is connected or the printer is turned on. The only recourse is to reboot the computer or to power down and then up again. Sometimes, though, the last lines of code since the last manual or autosave operation are lost in either process.

All of these techniques attempt to detect a software error and, in various ways, retry the computation. The basic assumption is that the problem is not a hard error but a transient error. A transient software error is one due to a software fault that results only in a system error for particular system states. Thus, if we repeat the computation again and the system state has changed, there is a good probability that the error will not be repeated on the second trial.

Recovery techniques are generally classified as forward or backward error-recovery techniques. The general philosophy of *forward error recovery* is to continue operation while knowing that there is an error in computation and correct for this error a little later. Techniques such as this work only in certain circumstances; for example, in the case of a tracking algorithm for an air traffic control system. In the case of *backward error recovery*, we wish to restart or roll back the computation process to some point before the occurrence of the error and restart the computation. In this section, we discuss four types of backward error recovery:

1. Reboot/restart techniques
2. Journaling techniques
3. Retry techniques
4. Checkpoint techniques

For a more complete discussion of the topics introduced in this section, see Sieworek [1982] and Section 3.10.

### 5.10.2 Rebooting

The simplest—but weakest—recovery technique from the implementation standpoint is to reboot or restart the system. The process of rebooting is well known to users of PCs who, without thinking too much about it, employ it one or more times a week to recover from errors. Actually, this raises a philosophical point: Is it better to have software that is well debugged and has very few errors that occur infrequently, or is having software with more residual errors that can be cleared by frequent rebooting also acceptable? The author remembers having a conversation with Ed Yourdon about an old computer when he was preparing a paper on reliability measurements [Yourdon, 1972]. Yourdon stated that a lot of computer crashes during operation were not recorded for the Burroughs B5500 computer (popular during the mid-1960s) because it was easy to reboot; the operator merely pushed the HALT button to stop the system and pushed the LOAD button to load a fresh version of the operating system. Furthermore, Yourdon stated, "The restart procedure requires two to five minutes. This can be contrasted with most IBM System/360s, where a restart usually required fifteen to thirty minutes." As a means of comparison, the author collected some data on reboot times that appears in Table 5.4.

It would seem that a restarting time of under one minute is now considered acceptable for a PC. It is more difficult to quantify the amount of information that is lost when a crash occurs and a reboot is required. We consider three typical applications: (a) word processing, (b) reading and writing e-mail, and

**TABLE 5.4  Typical Computer Reboot Times**

| Computer | Operating System | Reboot Time |
| --- | --- | --- |
| IBM System/360[a] | "OS-360" | 15–30 min |
| Burroughs 5500[a] | "Burroughs OS" | 2–5 min |
| Digital PC 360/20 | Windows 3.1 | 41 sec |
| IBM Compatible Pentium '90 | Windows '95 | 54 sec |
| IBM Notebook Celeron 300 | Windows '98 + Office | 80 sec |

[a] From Yourdon [1972].

(c) a Web search. We assume that word processing is being done on a PC and that applications (b) and (c) are being conducted from home via modem connections and a high-speed line to a server at work (a more demanding situation than connection from a PC to a server via a local area network where all three facilities are in a work environment). As stated before, the loss during word processing due to a "lockup and reboot" depends on the text lost since the last manual or autosave operation. In addition, there is the lost time to reload the word processing software. These losses become significant when the crash frequency becomes greater than, say, one or two per month. Choosing small intervals between autosaves, keeping backup documents, and frequently printing out drafts of new additions to a long document are really necessities. A friend of the author's who was president of a company that wrote and published technical documents for clients had a disastrous fire that destroyed all of his computer hardware, paper databases, and computer databases. Fortunately, he had about 70% of the material stored on tape and disks *in another location that was unaffected*, and it took almost a year to restore his business to full operation. The process of reading and writing e-mail is even more involved. A crash often severs the communication connection between the PC and the server, which must then be reestablished. Also, the e-mail program must be reentered. If a write operation was in progress, many e-mail programs do not save the text already entered. A Web search that locks up may require only the reissuing of the search, or it may require reacquisition of the server providing the connection. Different programs provide a wide variety of behaviors in response to such crashes. Not only is time lost, but any products that were being read, saved, or printed during the crash are lost as well.

### 5.10.3  Recovery Techniques

A reboot operation is similar to recovery. However, reboot generally involves the action of a human operator who observes that something is wrong with the system and attempts to correct the problem. If this attempt is unsuccessful, the operator issues a manual reboot command. The term *recovery* generally means that the system itself senses operational problems and issues a reboot command. In some cases, the software problem is more severe and a simple reboot is insufficient. Recovery may involve the reloading of some or all of the operating system. If this is necessary on a PC, the BIOS stored in ROM provides a basic means of communication to enable such a reloading. The most serious problems could necessitate a lower-level fix of the disk that stores the operating system. If we wish such a process to be autonomous, a special software program must be included that performs these operations in response to an "initiate recovery command." Some of the clearest examples of such recovery techniques are associated with robotic space-research vehicles.

Consider a robotic deep-space mission that loses control and begins to spin or tumble in space. The solar cells lose generating capacity, and the antennae no longer point toward Earth. The system must be designed from the start to recover from such a situation, as battery power provides a limited amount of time for such recovery to take place. Once the spacecraft is stabilized, the solar cells must be realigned with the Sun and the antennae must be realigned with Earth. This is generally provided by a small, highly secure kernel in the operating system that takes over in such a situation. In addition to hardware redundancy for all critical equipment, the software is generally subjected to a proof-of-correctness and an unusually high level of testing to ensure that it will perform its intended task. Many of NASA's spacecraft have recovered from such situations, but some have not. The main point of this discussion is that reboot or recovery for all these examples must be contained in the requirements and planned for during the entire design, not added later in the process as almost an afterthought.

### 5.10.4 Journaling Techniques

Journaling techniques are slightly more complex and somewhat better than reboot or restart techniques. Such techniques are also somewhat quicker to employ than reboot or restart techniques since only a subset of the inputs must be saved. To employ these techniques requires that

1. a copy of the original database, disk, and filename be stored,
2. all transactions (inputs) that affect the data must be stored during execution, and
3. the process be backed up to the beginning and the computation be retried.

Clearly, items (2) and (3) require a lot of storage; in practice, journaling can only be executed for a given time period, after which the inputs and the process must be erased and a new journaling time period created. The choice of the time period between journaling refreshes is an important design parameter. Storage of inputs and processes is continuous during operation regardless of the time period. The commands to refresh the journaling process should not absorb too much of the operating time budget for the system. The main trade-off will be between the amount of storage and the amount of processing time for computational retry, which increases with the length of the journaling period versus the impact of system overhead for journaling, which decreases as the interval between journaling refresh increases. It is possible that the storage requirements dominate and the optimum solution is to refresh when storage is filled up.

These techniques of journaling are illustrated by an example. The Xerox Alto personal computer used an editor called Bravo. Journaling was used to recover if a computer crash occurred during an editing session. Most modern PC-based word processing systems use a different technique to avoid loss of data during a session. A timer is set, and every few minutes the data in the input buffer (representing new input data since the last manual or automatic save operation) is stored. The addition of journaling to the periodic storage process would ensure no data loss. (Perhaps the keystrokes that occurred immediately preceding a crash would be lost, but this at most would constitute the last word or the last command.)

### 5.10.5 Retry Techniques

Retry techniques are quicker than those discussed previously, but they are more complex since more redundant process-state information must be stored. Retry is begun immediately after the error is detected. In the case of transient errors, one waits for the transient to die out and then initiates retry, whereas in the case of hard errors, the approach is to reconfigure the system. In either case, the operation affected by the error is then retried, which requires a complete knowledge of the system state (kept in storage) before the operation was attempted. If the interrupted operation or the error has irrevocably modified some data, the retry fails. Several examples of retry operation are as follows:

1. Disk controllers generally use disk-read reentry to minimize the number of disk-read errors. Consider the case of an MS-DOS personal computer system executing a disk-read command when an error is encountered. The disk-read operation is terminated, and the operator is asked whether he or she wishes to retry or abort. If the retry command is issued and the transient error has cleared, recovery is successful. However, if there is a hard error (e.g., a damaged floppy), retry will not clear the problem, and other processes must be employed.
2. The Univac 1100/60 computer provided retry for macroinstructions after a failure.
3. The IBM System/360 provided extensive retry capabilities, performing retries for both CPU and I/O operations.

Sometimes, the cause of errors is more complex and the retry may not work. Consider the following example that puzzled and plagued the author for a few months. A personal computer with a bad hard-disk sector worked fine with all programs except with a particular word processor. During ordinary save operations, the operating system must have avoided the bad sector in storing disk files. However, the word processor automatically saved the workspace every few minutes. Small text segments in the workspace were fine, but medium-sized text segments were sometimes subjected to disk-read errors during the autosave operation but not during a normal (manually issued) save command. In response to the error message "abort or retry," a simple retry response generally worked the first time or, at worst, required an abort followed by a save command. With large text segments in the workspace, real trouble occurred: When a disk-read error was encountered during automatic saving, one or more paragraphs of text *from previous word processing sessions* that were stored in the buffer were often randomly inserted into the present workspace, thereby corrupting the document. This is a graphic example of a retry failure. The author was about to attempt to lock out the bad disk sectors so they would not be used; however, the problem disappeared with the arrival of the second release of the word processor. Most likely, the new software used a slightly different buffer autosave mechanism.

### 5.10.6 Checkpointing

One advantage of checkpoint techniques is that they can generally be implemented using only software, as contrasted with retry techniques that may require additional dedicated hardware in addition to the necessary software routines. Also in the case of retry, the entire time history of the system state during the relevant period is saved, whereas in checkpointing the time history of the system state is saved only at specific points (checkpoints); thus less storage is required. A major disadvantage of checkpointing is the amount and difficulty of the programming that is required to employ checkpoints. The steps in the checkpointing process are as follows:

1. After the error is detected, recovery is initiated as soon as transient errors die out or, in the case of hard errors, the system is reconfigured.
2. The system is rolled back to the most recent checkpoint, and the system state is set to the stored checkpoint state and the process is restarted. If the operation is successfully restored, the process continues, and only some time and any new input data during the recovery process are lost. If operation is not restored, rollback to an earlier checkpoint can be attempted.
3. If the interrupted operation or the error has irrevocably modified some data, the checkpoint technique fails.

One better-developed example of checkpointing is within the Guardian operating system used for the Tandem computer system. The system consists of a *primary* process that does all the work and a *backup* process that operates on the same inputs and is ready to take over if the primary process fails. At critical points, the primary process sends checkpoint messages to the backup process.

For further details on the Guardian operating system, the reader is referred to Siewiorek [1992, pp. 635–648]. Also, see the discussion in Section 3.10.

Some comments are necessary with respect to the way customers generally use Tandem computer systems and the Guardian operating system:

1. The initial interest in the Tandem computer system was probably due to the marketing value of the term "NonStop architecture" that was used to describe the system. Although proprietary studies probably exist, the author does not know of any reliability or availability studies in the open literature that compared the Tandem architecture with a competitive system such as a Digital Equipment VAX Cluster or an IBM system configured for high reliability. Thus it is not clear how these systems compared to the competition, although most users are happy.
2. Once the system was studied by potential customers, one of the most important selling points was its modular structure. If the capacity of an existing Tandem system was soon to be exceeded, the user could simply buy additional Tandem machines, connect them in parallel, and easily integrate the expanded capacity with the existing system, which sometimes could be accomplished without shutting down system operation. This was a clear advantage over competitors, so it was built into the basic design.
3. The use of the Guardian operating system's checkpointing features could easily be turned on or off in configuring the system. Many users turned this feature off because it slowed down the system somewhat, but more importantly because to use it required some complex system programming to be added to the application programs. Newer Tandem systems have made such programming easier to use, as discussed in Section 3.10.1.

### 5.10.7 Distributed Storage and Processing

Many modern computer systems have a client–server architecture—typically, PCs or workstations are the clients, and the server is a more powerful processor with large disk storage attached. The clients and server are generally connected by local area networks (LANs). In fact, processing and data storage both tend to be decentralized, and several servers with their sets of clients are often connected by another network. In such systems, there is considerable theoretical and practical interest in devising algorithms to synchronize the various servers and to prevent two or more users from colliding when they attempt to access data from the same file. Even more important is the prevention of system lockup when one user is writing to a device and another user tries to read the device. For more information, the reader is referred to Bhargava [1987] and to the literature.

# REFERENCES

AIAA/ANSI R-013-1992. Recommended Practice Software Reliability. The American Institute of Aeronautics and Astronautics, The Aerospace Center, Washington, DC, ISBN 1-56347-024-1, February 23, 1993.

The Associated Press. "Y2K Bug Bites 7-Eleven Late." *Newsday*, January 4, 2001, p. A49.

Basili, V., and D. Weiss. A Methodology for Collecting Valid Software Engineering Data. *IEEE Transactions on Software Engineering* 10, 6 (1984): 42–52.

Bernays, A. "Carrying On About Carry-Ons." *New York Times*, January 25, 1998, p. 33 of Travel Section.

Beiser, B. *Software Testing Techniques*, 2d ed. Van Nostrand Reinhold, New York, 1990.

Bhargava, B. K. *Concurrency Control and Reliability in Distributed Systems*. Van Nostrand Reinhold, New York, 1987.

Billings, C. W. *Grace Hopper Naval Admiral and Computer Pioneer*. Enslow Publishers, Hillside, NJ, 1989.

Boehm, B. *Software Engineering Economics*. Prentice-Hall, Englewood Cliffs, NJ, 1981.

Boehm, B. et al. Avoiding the Software Model-Crash Spiderweb. New York: *IEEE Computer Magazine* (November 2000): 120–122.

Booch, G. et al. The Unified Modeling Language User Guide. Addison-Wesley, Reading, MA, 1999.

Brilliant, S. S., J. C. Knight, and N. G. Leveson. The Consistent Comparison Problem in $N$-Version Software. *ACM SIGSOFT Software Engineering Notes* 12, 1 (January 1987): 29–34.

Brooks, F. P. *The Mythical Man-Month Essays on Software Engineering*. Addison-Wesley, Reading, MA, 1995.

Butler, R. W., and G. B. Finelli. The Infeasibility of Experimental Quantification of Life-Critical Real-Time Software Reliability. *IEEE Transactions on Software Reliability Engineering* 19 (January 1993): 3–12.

Chen, L., and A. Avizienis. $N$-Version Programming: A Fault-Tolerance Approach to Reliability of Software Operation. *Digest of Eighth International Fault-Tolerant Computing Symposium*, Toulouse, France, 1978. IEEE Computer Society, New York, pp. 3–9.

Chillarege, R., and D. P. Siewiorek. Experimental Evaluation of Computer Systems Reliability. *IEEE Transactions on Reliability* 39, 4 (October 1990).

Cormen, T. H. et al. *Introduction to Algorithms*. McGraw-Hill, New York, 1992.

Cramer, H. *Mathematical Methods of Statistics*. Princeton University Press, Princeton, NJ, 1991.

Dougherty, E. M. Jr., and J. R. Fragola. *Human Reliability Analysis*. Wiley, New York, 1988.

Everett, W. W., and J. D. Musa. A Software-Reliability Engineering Practice. New York, *IEEE Computer Magazine* 26, 3 (1993): 77–79.

Fowler, M., and K. Scott. *UML Distilled Second Edition*. Addison-Wesley, Reading, MA, 1999.

Fragola, J. R., and M. L. Shooman. Significance of Zero Failures and Its Effect on Risk Decision Making. *Proceedings International Conference on Probabilistic Safety Assessment and Management*, New York, NY, September 13–18, 1998, pp. 2145–2150.

Garman, J. R. The "Bug" Heard 'Round The World. ACM SIGSOFT Software Engineering Notes (October 1981): 3–10.

Hall, H. S., and S. R. Knight. *Higher Algebra*, 1887. Reprint, Macmillan, New York, 1957.

Hamlet, D., and R. Taylor. Partition Testing does not Inspire Confidence. *IEEE Transactions on Software Engineering* 16, 12 (1990): 1402–1411.

Hatton, L. *Software Faults and Failure*. Addison-Wesley, Reading, MA, 2000.

Hecht, H. Fault-Tolerant Software. *IEEE Transactions on Reliability* 28 (August 1979): 227–232.

Hiller, S., and G. J. Lieberman. *Operations Research*. Holden-Day, San Francisco, 1974.

Hoel, P. G. *Introduction to Mathematical Statistics*. Wiley, New York, 1971.

Howden, W. E. Functional Testing. *IEEE Transactions on Software Engineering* 6, 2 (March 1980): 162–169.

*IEEE Computer Magazine*, Special Issue on Managing OO Development. (September 1996.)

Jacobson, I. et al. Making the Reuse Business Work. *IEEE Computer Magazine*, New York (October 1997): 36–42.

Jacobson, I. *The Road to the Unified Software Development Process*. Cambridge University Press, New York, 2000.

Jelinski, Z., and P. Moranda. "Software Reliability Research." In *Statistical Computer Performance Evaluation*, W. Freiberger (ed.). Academic Press, New York, 1972, pp. 465–484.

Kahn, E. H. et al. Object-Oriented Programming for Structured Procedural Programmers. *IEEE Computer Magazine*, New York (October 1995): 48–57.

Kanon, K., M. Kaaniche, and J.-C. Laprie. Experiences in Software Reliability: From Data Collection to Quantitative Evaluation. *Proceedings of the Fourth International Symposium on Software Reliability Engineering (ISSRE '93)*, 1993. IEEE, New York, NY, pp. 234–246.

Keller, T. W. et al. Practical Applications of Software Reliability Models. *Proceedings International Symposium on Software Reliability Engineering*, IEEE Computer Society Press, Los Alamitos, CA, 1991, pp. 76–78.

Knight, J. C., and N. G. Leveson. An Experimental Evaluation of Independence in Multiversion Programming. *IEEE Transactions on Software Engineering* 12, 1 (January 1986): 96–109.

Lala, P. K. *Fault Tolerant and Fault Testable Hardware Design*. Prentice-Hall, Englewood Cliffs, NJ, 1985.

Lala, P. K. *Self-Checking and Fault-Tolerant Digital Design*. Academic Press, division of Elsevier Science, New York, 2000.

Leach, R. J. *Introduction to Software Engineering.* CRC Press, Boca Raton, FL, 2000.

Littlewood, B. *Software Reliability: Achievement and Assessment.* Blackwell, Oxford, U.K., 1987.

Lyu, M. R. *Software Fault Tolerance.* Wiley, Chichester, U.K., 1995.

Lyu, M. R. (ed.). *Handbook of Software Reliability Engineering.* McGraw-Hill, New York, 1996.

McAllister, D. F., and M. A. Voulk. "Fault-Tolerant Software Reliability Engineering." In *Handbook of Software Reliability Engineering,* M. R. Lyu (ed.). McGraw-Hill, New York, 1996, ch. 14, p. 567–609.

Miller, G. A. The Magical Number Seven, Plus or Minus Two: Some Limits on our Capacity for Processing Information. *The Psychological Review* 63 (March 1956): 81–97.

Mood, A. M., and F. A. Graybill. *Introduction to the Theory of Statistics,* 2d ed. McGraw-Hill, New York, 1963.

Musa, J. A Theory of Software Reliability and its Application. *IEEE Transactions on Software Engineering* 1, 3 (September 1975): 312–327.

Musa, J., A. Iannino, and K. Okumoto. *Software Reliability: Measurement, Prediction, Application.* McGraw-Hill, New York, 1987.

Musa, J. Sensitivity of Field Failure Intensity to Operational Profile Errors. *Proceedings of the 5th International Symposium on Software Reliability Engineering,* Monterey, CA, 1994. IEEE, New York, NY, pp. 334–337.

*New York Times,* "Circuits Section." August 27, 1998, p. G1.

*New York Times,* "The Y2K Issue Shows Up, a Year Late." January 3, 2001, p. A3.

Pfleerer, S. L. *Software Engineering Theory and Practice.* Prentice-Hall, Upper Saddle River, NJ, 1998, pp. 31–33, 181, 195–198, 207.

Pooley, R., and P. Stevens. *Using UML: Software Engineering with Objects and Components.* Addison-Wesley, Reading, MA, 1999.

Pollack, A. "Chips are Hidden in Washing Machines, Microwaves. . . ." New York Times, *Media and Technology Section,* January 4, 1999, p. C17.

Pradhan, D. K. *Fault-Tolerant Computing Theory and Techniques,* vols. I and II. Prentice-Hall, Englewood Cliffs, NJ, 1986.

Pradhan, D. K. *Fault-Tolerant Computing Theory and Techniques,* vol. I, 2d ed. Prentice-Hall, Englewood Cliffs, NJ, 1993.

Pressman, R. H. *Software Engineering: A Practitioner's Approach,* 4th ed. McGraw-Hill, New York, 1997, pp. 348–363.

Schach, S. R. *Classical and Object-Oriented Software Engineering with UML and C++,* 4th ed. McGraw-Hill, New York, 1999.

Schach, S. R. *Classical and Object-Oriented Software Engineering with UML and Java.* McGraw-Hill, New York, 1999.

Schneidewind, N. F., and T. W. Keller. Application of Reliability Models to the Space Shuttle. *IEEE Software* (July 1992): 28–33.

Shooman, M. L., and M. Messinger. Use of Classical Statistics, Bayesian Statistics, and Life Models in Reliability Assessment. Consulting Report, U.S. Army Research Office, June 1971.

Shooman, M. L. Probabilistic Models for Software Reliability Prediction. In *Statistical Computer Performance Evaluation*, W. Freiberger (ed.). Academic Press, New York, 1972, pp. 485–502.

Shooman, M. L., and M. Bolsky. Types, Distribution, and Test and Correction Times for Programming Errors. *Proceedings 1975 International Conference on Reliable Software*. IEEE, New York, NY, Catalog No. 75CHO940-7CSR, p. 347.

Shooman, M. L. *Software Engineering: Design, Reliability, and Management*. McGraw-Hill, New York, 1983, ch. 5.

Shooman, M. L., and G. Richeson. Reliability of Shuttle Mission Control Center Software. *Proceedings Annual Reliability and Maintainability Symposium*, 1984. IEEE, New York, NY, pp. 125–135.

Shooman, M. L. Validating Software Reliability Models. *Proceedings of the IFAC Workshop on Reliability, Availability, and Maintainability of Industrial Process Control Sysems*. Pergamon Press, division of Elsevier Science, New York, 1988.

Shooman, M. L. A Class of Exponential Software Reliability Models. *Workshop on Software Reliability*. IEEE Computer Society Technical Committee on Software Reliability Engineering, Washington, DC, April 13, 1990.

Shooman, M. L., *Probabilistic Reliability: An Engineering Approach*, 2d ed. Krieger, Melbourne, FL, 1990, Appendix H.

Shooman, M. L. A Micro Software Reliability Model for Prediction and Test Apportionment. *Proceedings International Symposium on Software Reliability Engineering*, 1991. IEEE, New York, NY, p. 52–59.

Shooman, M. L. Software Reliability Models for Use During Proposal and Early Design Stages. *Proceedings ISSRE '99, Symposium on Software Reliability Engineering*. IEEE Computer Society Press, New York, 1999.

*Spectrum*, Special Issue on Engineering Software. IEEE Computer Society Press, New York (April 1999).

Siewiorek, D. P., and R. S. Swarz. *The Theory and Practice of Reliable System Design*. The Digital Press, Bedford, MA, 1982.

Siewiorek, D. P., and R. S. Swarz. *Reliable Computer Systems Design and Evaluation*, 2d ed. The Digital Press, Bedford, MA, 1992.

Siewiorek, D. P. and R. S. Swarz. *Reliable Computer Systems Design and Evaluation*, 3d ed. A. K. Peters, www.akpeters.com, 1998.

Stark, G. E. Dependability Evaluation of Integrated Hardware/Software Systems. *IEEE Transactions on Reliability* (October 1987).

Stark, G. E. et al. Using Metrics for Management Decision-Making. *IEEE Computer Magazine*, New York (September 1994).

Stark, G. E. et al. An Examination of the Effects of Requirements Changes on Software Maintenance Releases. *Software Maintenance: Research and Practice*, vol. 15, August 1999.

Stork, D. G. Using Open Data Collection for Intelligent Software. *IEEE Computer Magazine*, New York (October 2000): 104–106.

Tai, A. T., J. F. Meyer, and A. Avizienis. *Software Performability, From Concepts to Applications*. Kluwer Academic Publishers, Hingham, MA, 1995.

Wing, J. A Specifier's Introduction to Formal Methods. New York: *IEEE Computer Magazine* 23, 9 (September 1990): 8–24.

Yanini, E. *New York Times*, Business Section, December 7, 1997, p. 13.

Yourdon, E. Reliability Measurements for Third Generation Computer Systems. *Proceedings Annual Reliability and Maintainability Symposium*, 1972. IEEE, New York, NY, pp. 174–183.

## PROBLEMS

**5.1.** Consider a software project with which you are familiar (past, in-progress, or planned). Write a few sentences or a paragraph describing the phases given in Table 5.1 for this project. Make sure you start by describing the project in succinct form.

**5.2.** Draw an H-diagram similar to that shown in Fig. 5.1 for the software of problem 5.1.

**5.3.** How well does the diagram of problem 5.2 agree with Eqs. (5.1 a–d)? Explain.

**5.4.** Write a short version of a test plan for the project of problem 5.1. Include the number and types of tests for the various phases. (Note: A complete test plan will include test data and expected answers.)

**5.5.** Would (or did) the development follow the approach of Figs. 5.2, 5.3, or 5.4? Explain.

**5.6.** We wish to develop software for a server on the Internet that keeps a database of locations for new cars that an auto manufacturer is tracking. Assume that as soon as a car is assembled, a reusable electronic box is installed in the vehicle that remains there until the car is delivered to a purchaser. The box contains a global positioning system (GPS) receiver that determines accurate location coordinates from the GPS satellites and a transponder that transmits a serial number and these coordinates via another satellite to the server. The server receives these transponder signals and stores them in a file. The server has a geographical database so that it can tell from the coordinates if each car is (a) in the manufacturer's storage lot, (b) in transit, or (c) in a dealer's showroom or lot. The database is accessed by an Internet-capable cellular phone or any computer with Internet access [Stork, 2000, p. 18].

(a) How would you design the server software for this system? (Figs. 5.2, 5.3, or 5.4?)

(b) Draw an H-diagram for the software.

**5.7.** Repeat problem 5.3 for the software in problem 5.6.

**5.8.** Repeat problem 5.4 for the software in problem 5.6.

**5.9.** Repeat problem 5.5 for the software in problem 5.6.

**5.10.** A component with a constant-failure rate of $4 \times 10^{-5}$ is discussed in Section 5.4.5.
  (a) Plot the failure rate as a function of time.
  (b) Plot the density function as a function of time.
  (c) Plot the cumulative distribution function as a function of time.
  (d) Plot the reliability as a function of time.

**5.11.** It is estimated that about 100 errors will be removed from a program during the integration test phase, which is scheduled for 12 months duration.
  (a) Plot the error-removal curve assuming that the errors will follow a constant-removal rate.
  (b) Plot the error-removal curve assuming that the errors will follow a linearly decreasing removal rate.
  (c) Plot the error-removal curve assuming that the errors will follow an exponentially decreasing removal rate.

**5.12.** Assume that a reliability model is to be fitted to problem 5.11. The number of errors remaining in the program at the beginning of integration testing is estimated to be 120. From experience with similar programs, analysts believe that the program will start integration testing with an MTTF of 150 hours.
  (a) Assuming a constant error-removal rate during integration, formulate a software reliability model.
  (b) Plot the reliability function versus time at the beginning of integration testing—after 4, 8, and 12 months of debugging.
  (c) Plot the MTTF as a function of the integration test time, $\tau$.

**5.13.** Repeat problem 5.12 for a linearly decreasing error-removal rate.

**5.14.** Repeat problem 5.12 for an exponentially decreasing error-removal rate.

**5.15.** Compare the reliability functions derived in problems 5.12, 5.13, and 5.14 by plotting them on the same time axis for $\tau = 0$, $\tau = 4$, $\tau = 8$, and $\tau = 12$ months.

**5.16.** Compare the MTTF functions derived in problems 5.12, 5.13, and 5.14 by plotting them on the same time axis versus $\tau$.

**5.17.** After 1 month of integration testing of a program, the MTTF = 10 hours, and 15 errors have been removed. After 2 months, the MTTF = 15 hours, and 25 total errors have been removed.
  (a) Assuming a constant error-removal rate, fit a model to this data set. Estimate the parameters by using moment-estimation techniques [Eqs. (5.47a, b)].
  (b) Sketch MTTF versus development time $\tau$.

**282** SOFTWARE RELIABILITY AND RECOVERY TECHNIQUES

      **(c)** How much integration test time will be required to achieve a 100-hour MTTF? How many errors will have been removed by this time and how many will remain?

**5.18.** Repeat problem 5.17 assuming a linearly decreasing error-rate model and using Eqs. (5.49a, b).

**5.19.** Repeat problem 5.17 assuming an exponentially decreasing error-rate model and using Eqs. (5.51) and (5.52).

**5.20.** After 1 month of integration testing, 20 errors have been removed, the MTTF of the software is measured by testing it with the use of simulated operational data, and the MTTF = 10 hours. After 2 months, the MTTF = 20 hours, and 50 total errors have been removed.

    **(a)** Assuming a constant error-removal rate, fit a model to this data set. Estimate the parameters by using moment-estimation techniques [Eqs. (5.47a, b)].

    **(b)** Sketch the MTTF versus development time $\tau$.

    **(c)** How much integration test time will be required to achieve a 60-hour MTTF? How many errors will have been removed by this time and how many will remain?

    **(d)** If we release the software when it achieves a 60-hour MTTF, sketch the reliability function versus time.

    **(e)** How long can the software operate, if released as in part (d) above, before the reliability drops to 0.90?

**5.21.** Repeat problem 5.20 assuming a linearly decreasing error-rate model and using Eqs. (5.49a, b).

**5.22.** Repeat problem 5.20 assuming an exponentially decreasing error-rate model and using Eqs. (5.51) and (5.52).

**5.23.** Assume that the company developing the software discussed in problem 5.17 has historical data for similar systems that show an average MTTF of 50 hours with a variance $\sigma^2$ of 30 hours. The variance of the reliability modeling is assumed to be 20 hours. Using Eqs. (5.55) and (5.56a, b), compute the reliability function.

**5.24.** Assume that the model of Fig. 5.18 holds for three independent versions of reliable software. The probability of error for 10,000 hours of operation of each version is 0.01. Compute the reliability of the TMR configuration assuming that there are no common-mode failures. Recompute the reliability of the TMR configuration if 1% of the errors are due to common-mode requirement errors and 1% are due to common-mode software faults.

# 6
# NETWORKED SYSTEMS RELIABILITY

## 6.1 INTRODUCTION

Many physical problems (e.g., computer networks, piping systems, and power grids) can be modeled by a network. In the context of this chapter, the word *network* means a physical problem that can be modeled as a mathematical graph composed of nodes and links (directed or undirected) where the branches have associated physical parameters such as flow per minute, bandwidth, or megawatts. In many such systems, the physical problem has sources and sinks or inputs and outputs, and the proper operation is based on connection between inputs and outputs. Systems such as computer or communication networks have many nodes representing the users or resources that desire to communicate and also have several links providing a number of interconnected pathways. These many interconnections make for high reliability and considerable complexity. Because many users are connected to such a network, a failure affects many people; thus the reliability goals must be set at a high level.

This chapter focuses on computer networks. It begins by discussing the several techniques that allow one to analyze the reliability of a given network, after which the more difficult problem of optimum network design is introduced. The chapter concludes with a brief introduction to one of the most difficult cases to analyze—where links can be disabled because of two factors: (a) link congestion (a situation in which flow demand exceeds flow capacity and a link is blocked or an excessive queue builds up at a node), and (b) failures from broken links.

A new approach to reliability in interconnected networks is called survivability analysis [Jia and Wing, 2001]. The concept is based on the design of a network so it is robust in the face of abnormal events—the system must survive and not crash. Recent research in this area is listed on Jeannette M. Wing's Web site [Wing, 2001].

The mathematical techniques used in this chapter are properties of mathematical graphs, tie sets, and cut sets. A summary of the relevant concepts is given in Section B2.7, and there is a brief discussion of some aspects of graph theory in Section 5.3.5; other concepts will be developed in the body of the chapter. The reader should be familiar with these concepts before continuing with this chapter. For more details on graph theory, the reader is referred to Shooman [1983, Appendix C]. There are of course other approaches to network reliability; for these, the reader is referred to the following references: Frank [1971], Van Slyke [1972, 1975], and Colbourn [1987, 1993, 1995]. It should be mentioned that the cut-set and tie-set methods used in this chapter apply to reliability analyses in general and are employed throughout reliability engineering; they are essentially a theoretical generalization of the block diagram methods discussed in Section B2. Another major approach is the use of fault trees, introduced in Section B5 and covered in detail in Dugan [1996].

In the development of network reliability and availability we will repeat for clarity some of the concepts that are developed in other chapters of this book, and we ask for the reader's patience.

## 6.2 GRAPH MODELS

We focus our analytical techniques on the reliability of a communication network, although such techniques also hold for other network models. Suppose that the network is composed of computers and communication links. We represent the system by a mathematical graph composed of nodes representing the computers and edges representing the communications links. The terms used to describe graphs are not unique; oftentimes, notations used in the mathematical theory of graphs and those common in the application fields are interchangeable. Thus a mathematics textbook may talk of vertices and arcs; an electrical-engineering book, of nodes and branches; and a communications book, of sites and interconnections or links. In general, these terms are synonymous and used interchangeably.

In the most general model, both the nodes and the links can fail, but here we will deal with a simplified model in which only the links can fail and the nodes are considered perfect. In some situations, communication can go only in one direction between a node pair; the link is represented by a directed edge (an arrowhead is added to the edge), and one or more directed edges in a graph result in a directed graph (digraph). If communication can occur in both directions between two nodes, the edge is nondirected, and a graph without any directed nodes is an ordinary graph (i.e., nondirected, not a digraph). We will consider both directed and nondirected graphs. (Sometimes, it is useful to view

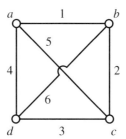

**Figure 6.1** A four-node graph representing a computer or communication network.

a nondirected graph as a special case of a directed graph in which each link is represented by two identical parallel links, with opposite link directions.)

When we deal with nondirected graphs composed of $E$ edges and $N$ nodes, the notation $G(N, E)$ will be used. A particular node will be denoted as $n_i$ and a particular edge denoted as $e_j$. We can also identify an edge by naming the nodes that it connects; thus, if edge $j$ is between nodes $s$ and $t$, we may write $e_j = (n_s, n_t) = e(s, t)$. One also can say that edge $j$ is *incident on nodes s and t*. As an example, consider the graph of Fig. 6.1, where $G(N = 4, E = 6)$. The nodes $n_1$, $n_2$, $n_3$, and $n_4$ are $a$, $b$, $c$, and $d$. Edge 1 is denoted by $e_1 = e(n_1, n_2) = (a, b)$, edge 2 by $e_2 = e(n_2, n_3) = (b, c)$, and so forth. The example of a network graph shown in Fig. 6.1 has four nodes ($a$, $b$, $c$, $d$) and six edges (1, 2, 3, 4, 5, 6). The edges are undirected (directed edges have arrowheads to show the direction), and since in this particular example all possible edges between the four nodes are shown, it is called a complete graph. The total number of edges in a graph with $n$ nodes is the number of combinations of $n$ things taken two at a time $= n!/[(2!)(n - 2)!]$. In the example of Fig. 6.1, the total number of edges in $4!/[(2!)(4 - 2)!] = 6$.

In formulating the network model, we will assume that each link is either good or bad and that there are no intermediate states. Also, independence of link failures is assumed, and no repair or replacement of failed links is considered. In general, the links have a high reliability, and because of all the multiple (redundant) paths, the network has a very high reliability. This large number of parallel paths makes for high complexity; the efficient calculation of network reliability is a major problem in the analysis, design, or synthesis of a computer communication network.

## 6.3 DEFINITION OF NETWORK RELIABILITY

In general, the definition of reliability is *the probability that the system operates successfully for a given period of time under environmental conditions* (see Appendix B). We assume that the systems being modeled operate continuously and that the time in question is the clock time since the last failure or restart of the system. The environmental conditions include not only temperature, atmosphere, and weather, but also system load or traffic. The term *successful operation* can have many interpretations. The two primary ones are related to how many of the $n$ nodes can communicate with each other. We assume that as time increases, a number of the $m$ links fail. If we focus on communication between a pair of nodes where $s$ is the source node and $t$ is the target node, then successful operation is defined as the presence of one or more operating paths between $s$ and $t$. This is called the two-terminal problem, and the probability of successful communication between $s$ and $t$ is called two-terminal reliability. If successful operation is defined as all nodes being able to communicate, we have the all-terminal problem, for which it can be stated that node $s$ must be able to communicate with all the other $n-1$ nodes, since communication between any one node $s$ and all others nodes, $t_1, t_2, \ldots, t_{n-1}$, is equivalent to communication between all nodes. The probability of successful communication between node $s$ and nodes $t_1, t_2, \ldots, t_{n-1}$ is called the all-terminal reliability.

In more formal terms, we can state that the all-terminal reliability is the probability that node $n_i$ can communicate with node $n_j$ for all pairs $n_i n_j$ (where $i \neq j$). We wish to show that this is equivalent to the proposition that node $s$ can communicate with all other nodes $t_1 = n_2, t_2 = n_3, \ldots, t_{n-1} = n_n$. Choose any other node $n_x$ (where $x \neq 1$). By assumption, $n_x$ can communicate with $s$ because $s$ can communicate with all nodes and communication is in both directions. However, once $n_x$ reaches $s$, it can then reach all other nodes because $s$ is connected to all nodes. Thus all-terminal connectivity for $x = 1$ results in all-terminal connectivity for $x \neq 1$, and the proposition is proved.

In general, reliability, $R$, is the probability of successful operation. In the case of networks, we are interested in all-terminal reliability, $R_{\text{all}}$:

$$R_{\text{all}} = P(\text{that all } n \text{ nodes are connected}) \tag{6.1}$$

or the two-terminal reliability:

$$R_{st} = P(\text{that nodes } s \text{ and } t \text{ are connected}) \tag{6.2}$$

Similarly, $k$-terminal reliability is the probability that a subset of $k$ nodes $2 \leq k \leq n$) are connected. Thus we must specify what type of reliability we are discussing when we begin a problem.

We stated previously that repairs were not included in the analysis of network reliability. This is not strictly true; for simplicity, no repair was assumed. In actuality, when a node-switching computer or a telephone communications line goes down, each is promptly repaired. The metric used to describe a repairable system is availability, which is defined as the probabilty that *at any instant of time t*, the system is up and available. Remember that in the case of reliability, there were no failures in the *interval* 0 to $t$. The notation is $A(t)$, and availability and reliability are related as follows by the union of events:

$$A(t) = P(\text{no failure in interval 0 to } t + 1 \text{ failure and}$$
$$1 \text{ repair in interval 0 to } t + 2 \text{ failures and}$$
$$2 \text{ repairs in interval 0 to } t + \cdots) \qquad (6.3)$$

The events in Eq. (6.3) are all mutually exclusive; thus Eq. (6.3) can be expanded as a sum of probabilities:

$$A(t) = P(\text{no failure in interval 0 to } t)$$
$$+ P(1 \text{ failure and 1 repair in interval 0 to } t)$$
$$+ P(2 \text{ failures and 2 repairs in interval 0 to } t) + \cdots \qquad (6.4)$$

Clearly,

- The first term in Eq. (6.4) is the reliability, $R(t)$
- $A(t) = R(t) = 1$ at $t = 0$
- For $t > 0$, $A(t) > R(t)$
- $R(t) \to 0$ as $t \to \infty$
- It is shown in Appendix B that $A(t) \to A_{ss}$ as $t \to \infty$ and, as long as repair is present, $A_{ss} > 0$

Availability is generally derived using Markov probability models (see Appendix B and Shooman [1990]). The result of availability derivations for a single element with various failure and repair probability distributions can become quite complex. In general, the derivations are simplified by assuming exponential probability distributions for the failure and repair times (equivalent to constant-failure rate, $\lambda$, and constant-repair rate, $\mu$). Sometimes, the mean time to failure (MTTF) and the mean time to repair (MTTR) are used to describe the repair process and availability. In many cases, the terms mean time between failure (MTBF) and mean time between repair (MTBR) are used instead of MTTF and MTTR. For constant-failure and -repair rates, the mean times become MTBF = $1/\lambda$ and MTBR = $1/\mu$. The solution for $A(t)$ has an exponentially decaying transient term and a constant steady-state term. After a few failure repair cycles, the transient term dies out and the availability can be represented by the simpler steady-state term. For the case of constant-failure and -repair rates for a single item, the steady-state availability is given by the equation that follows (see Appendix B).

$$A_{ss} = \mu/(\lambda + \mu) = \text{MTBF}/(\text{MTBF} + \text{MTBR}) \qquad (6.5)$$

Since the MTBF $\gg$ MTBR in any well-designed system, $A_{ss}$ is close to unity. Also, alternate definitions for MTTF and MTTR lead to slightly different but equivalent forms for Eq. (6.5) (see Kershenbaum [1993].)

Another derivation of availability can be done in terms of system uptime, $U(t)$, and system downtime, $D(t)$, resulting in the following different formula for availability:

$$A_{ss} = U(t)/[U(t) + D(t)] \tag{6.6}$$

The formulation given in Eq. (6.6) is more convenient than that of Eq. (6.5) if we wish to estimate $A_{ss}$ based on collected field data. In the case of a computer network, the availability computations can become quite complex if the repairs of the various elements are coupled, in which case a single repairman might be responsible for maintaining, say, two nodes and five lines. If several failures occur in a short period of time, a queue of failed items waiting for repairs might build up and the downtime is lengthened, and the term "repairman-coupled" is used. In the ideal case, if we assume that each element in the system has its own dedicated repairman, we can guarantee that the elements are *decoupled* and that the *steady-state availabilities can be substituted into probability expressions in the same way as reliabilities are*. In a practical case, we do not have individual repairmen, but if the repair rate is much larger than the failure rate of the several components for which the repairman supports, then approximate decoupling is a good assumption. Thus, in most network reliability analyses there will be no distinction made between reliability and availability; the two terms are used interchangeably in the network field in a loose sense. Thus a reliability analyst would make a combinatorial model of a network and insert reliability values for the components to calculate system reliability. Because decoupling holds, he or she would substitute component availabilities in the same model and calculate the system availability; however, a network analyst would perform the same availability computation and refer to it colloquially as "system reliability." For a complete discussion of availability, see Shooman [1990].

## 6.4 TWO-TERMINAL RELIABILITY

The evaluation of network reliability is a difficult problem, but there are several approaches. For any practical problem of significant size, one must use a computational program. Thus all the techniques we discuss that use a "pencil-paper-and-calculator" analysis are preludes to understanding how to write algorithms and programs for network reliability computation. Also, it is always valuable to have an analytical solution of simpler problems for use to test reliability computation programs until the user becomes comfortable with such a program. Since two-terminal reliability is a bit simpler than all-terminal reliability, we will discuss it first and treat all-terminal reliability in the following section.

### 6.4.1 State-Space Enumeration

Conceptually, the simplest means of evaluating the two-terminal reliability of a network is to enumerate all possible combinations where each of the $e$ edges can be good or bad, resulting in $2^e$ combinations. Each of these combinations of good and bad edges can be treated as an event $E_i$. These events are all mutually

TWO-TERMINAL RELIABILITY 289

exclusive (disjoint), and the reliability expression is simply the probability of the union of the subset of these events that contain a path between $s$ and $t$.

$$R_{st} = P(E_1 + E_2 + E_3 \cdots) \tag{6.7}$$

Since each of these events is mutually exclusive, the probability of the union becomes the sum of the individual event probabilities.

$$R_{st} = P(E_1) + P(E_2) + P(E_3) + \cdots \tag{6.8}$$

[Note that in Eq. (6.7) the symbol + stands for union ($\cup$), whereas in Eq. (6.8), the + represents addition. Also throughout this chapter, the intersection of $x$ and $y$ ($x \cap y$) is denoted by $x \cdot y$, or just $xy$.]

As an example, consider the graph of a complete four-node communication network that is shown in Fig. 6.1. We are interested in the two-terminal reliability for node pair $a$ and $b$; thus $s = a$ and $t = b$. Since there are six edges, there are $2^6 = 64$ events associated with this graph, all of which are presented in Table 6.1. The following definitions are used in constructing Table 6.1:

$E_i$ = the event $i$
$j$ = the success of edge $j$
$j'$ = the failure of edge $j$

The term *good* means that there is at least one path from $a$ to $b$ for the given combination of good and failed edges. The term *bad*, on the other hand, means that there are no paths from $a$ to $b$ for the given combination of good and failed edges. The result—good or bad—is determined by inspection of the graph.

Note that in constructing Table 6.1, the following observations prove helpful: Any combination where edge 1 is good represents a connection, and at least three edges must fail (edge 1 plus two others) for any event to be bad.

Substitution of the good events from Table 6.1 into Eq. (6.8) yields the two-terminal reliability from $a$ to $b$:

$$\begin{aligned} R_{ab} = &[P(E_1)] + [P(E_2) + \cdots + P(E_7)] + [P(E_8) + P(E_9) + \cdots + P(E_{22})] \\ &+ [P(E_{23}) + P(E_{24}) + \cdots + P(E_{34}) + P(E_{37}) + \cdots + P(E_{42})] \\ &+ [P(E_{43}) + P(E_{44}) + \cdots + P(E_{47}) + P(E_{50}) + P(E_{56})] + [P(E_{58})] \end{aligned} \tag{6.9}$$

The first bracket in Eq. (6.9) has one term where all the edges must be good, and if all edges are identical and independent, and they have a probability of success of $p$, then the probability of event $E_1$ is $p^6$. Similarly, for the second bracket, there are six events of probability $qp^5$ where the probability of failure $q = 1 - p$, etc. Substitution in Eq. (6.9) yields a polynomial in $p$ and $q$:

$$R_{ab} = p^6 + 6qp^5 + 15q^2p^4 + 18q^3p^3 + 7q^4p^2 + q^5p \tag{6.10}$$

**TABLE 6.1  The Event-Space for the Graph of Fig. 6.1** ($s = a, t = b$)

| | | |
|---|---|---|
| No failures: | $\binom{6}{0} = \dfrac{6!}{0!6!} = 1$ | |

| | |
|---|---|
| $E_1 = 123456$ | Good |

| | | |
|---|---|---|
| One failure: | $\binom{6}{1} = \dfrac{6!}{1!5!} = 6$ | |

| | |
|---|---|
| $E_2 = 1'23456$ | Good |
| $E_3 = 12'3456$ | Good |
| $E_4 = 123'456$ | Good |
| $E_5 = 1234'56$ | Good |
| $E_6 = 12345'6$ | Good |
| $E_7 = 123456'$ | Good |

| | | |
|---|---|---|
| Two failures: | $\binom{6}{2} = \dfrac{6!}{2!4!} = 15$ | |

| | |
|---|---|
| $E_8 = 1'2'3456$ | Good |
| $E_9 = 1'23'456$ | Good |
| $E_{10} = 1'234'56$ | Good |
| $E_{11} = 1'2345'6$ | Good |
| $E_{12} = 1'23456'$ | Good |
| $E_{13} = 12'3'456$ | Good |
| $E_{14} = 12'34'56$ | Good |
| $E_{15} = 12'345'6$ | Good |
| $E_{16} = 12'3456'$ | Good |
| $E_{17} = 123'4'56$ | Good |
| $E_{18} = 123'45'6$ | Good |
| $E_{19} = 123'456'$ | Good |
| $E_{20} = 1234'5'6$ | Good |
| $E_{21} = 1234'56'$ | Good |
| $E_{22} = 12345'6'$ | Good |

Continued . . .

| | | |
|---|---|---|
| Three failures: | $\binom{6}{3} = \dfrac{6!}{3!3!} = 20$ | |

| | |
|---|---|
| $E_{23} = 1234'5'6'$ | Good |
| $E_{24} = 123'45'6'$ | Good |
| $E_{25} = 123'4'56'$ | Good |
| $E_{26} = 123'4'5'6$ | Good |
| $E_{27} = 12'345'6'$ | Good |
| $E_{28} = 12'34'56'$ | Good |
| $E_{29} = 12'34'5'6$ | Good |
| $E_{30} = 12'3'456'$ | Good |
| $E_{31} = 12'3'45'6$ | Good |
| $E_{32} = 12'3'4'56$ | Good |

**TABLE 6.1** (Continued)

| | |
|---|---|
| $E_{33} = 1'2345'6'$ | Good |
| $E_{34} = 1'234'56'$ | Good |
| $E_{35} = 1'234'5'6$ | Bad |
| $E_{36} = 1'2'3456'$ | Bad |
| $E_{37} = 1'2'345'6$ | Good |
| $E_{38} = 1'2'34'56$ | Good |
| $E_{39} = 1'23'456'$ | Good |
| $E_{40} = 1'23'45'6$ | Good |
| $E_{41} = 1'23'4'56$ | Good |
| $E_{42} = 1'2'3'456$ | Good |

Four failures: $\binom{6}{4} = \dfrac{6!}{4!2!} = 15$

| | |
|---|---|
| $E_{43} = 123'4'5'6'$ | Good |
| $E_{44} = 12'34'5'6'$ | Good |
| $E_{45} = 12'3'45'6'$ | Good |
| $E_{46} = 12'3'4'56'$ | Good |
| $E_{47} = 12'3'4'5'6$ | Good |
| $E_{48} = 1'234'5'6'$ | Bad |
| $E_{49} = 1'23'45'6'$ | Bad |
| $E_{50} = 1'23'4'56'$ | Good |
| $E_{51} = 1'23'4'5'6$ | Bad |
| $E_{52} = 1'2'345'6'$ | Bad |
| $E_{53} = 1'2'34'56'$ | Bad |
| $E_{54} = 1'2'34'5'6$ | Bad |
| $E_{55} = 1'2'3'456'$ | Bad |
| $E_{56} = 1'2'3'45'6$ | Good |
| $E_{57} = 1'2'3'4'56$ | Bad |

Continued . . .

Five failures: $\binom{6}{5} = \dfrac{6!}{5!1!} = 6$

| | |
|---|---|
| $E_{58} = 12'3'4'5'6'$ | Good |
| $E_{59} = 1'23'4'5'6'$ | Bad |
| $E_{60} = 1'2'34'5'6'$ | Bad |
| $E_{61} = 1'2'3'45'6'$ | Bad |
| $E_{62} = 1'2'3'4'56'$ | Bad |
| $E_{63} = 1'2'3'4'5'6$ | Bad |

Six failures: $\binom{6}{6} = \dfrac{6!}{6!0!} = 1$

| | |
|---|---|
| $E_{64} = 1'2'3'4'5'6'$ | Bad |

Substitutions such as those in Eq. (6.10) are prone to algebraic mistakes; as a necessary (but not sufficient) check, we evaluate the polynomial for $p = 1$ and $q = 0$, which should yield a reliability of unity. Similarly, evaluating the polynomial for $p = 0$ and $q = 1$ should yield a reliability of 0. (Any network has a reliability of unity regardless of its topology if all edges are perfect; it has a reliability of 0 if all its edges have failed.)

Numerical evaluation of the polynomial for $p = 0.9$ and $q = 0.1$ yields

$$R_{ab} = 0.9^6 + 6(0.1)(0.9)^5 + 15(0.1)^2(0.9)^4 + 18(0.1)^3(0.9)^3$$
$$+ 7(0.1)^4(0.9)^2 + (0.1)^5(0.9) \tag{6.11a}$$
$$R_{ab} = 0.5314 + 0.35427 + 0.0984 + 0.0131 + 5.67 \times 10^{-4} + 9 \times 10^{-6} \tag{6.11b}$$
$$R_{ab} = 0.997848 \tag{6.11c}$$

Usually, event-space-reliability calculations require much effort and time even though the procedure is clear. The number of events builds up exponentially as $2^e$. For $e = 10$, we have 1,024 terms, and if we double the $e$, there are over a million terms. However, we seek easier methods.

### 6.4.2 Cut-Set and Tie-Set Methods

One can reduce the amount of work in a network reliability analysis below the $2^e$ complexity required for the event-space method if one focuses on the minimal cut sets and minimal tie sets of the graph (see Appendix B and Shooman [1990, Section 3.6.5]). The tie sets are the groups of edges that form a path between $s$ and $t$. The term *minimal* implies that no node or edge is traversed more than once, but another way of defining this is that minimal tie sets have no subsets of edges that are a tie set. If there are $i$ tie sets between $s$ and $t$, then the reliability expression is given by the expansion of

$$R_{st} = P(T_1 + T_2 + \cdots + T_i) \tag{6.12}$$

Similarly, one can focus on the minimal cut sets of a graph. A cut set is a group of edges that break *all* paths between $s$ and $t$ when they are removed from the graph. If a cut set is minimal, no subset is also a cut set. The reliability expression in terms of the $j$ cut sets is given by the expansion of

$$R_{st} = 1 - P(C_1 + C_2 + \cdots + C_j) \tag{6.13}$$

We now apply the above theory to the example given in Fig. 6.1. The minimal cut sets and tie sets are found by inspection for $s = a$ and $t = b$ and are given in Table 6.2.

Since there are fewer cut sets, it is easier to use Eq. (6.13) rather than Eq. (6.12); however, there is no general rule for when $j < i$ or vice versa.

TWO-TERMINAL RELIABILITY 293

**TABLE 6.2 Minimal Tie Sets and Cut Sets for the Example of Fig. 6.1 ($s = a, t = b$)**

| Tie Sets | Cut Sets |
|---|---|
| $T_1 = 1$ | $C_1 = 1'4'5'$ |
| $T_2 = 52$ | $C_2 = 1'6'2'$ |
| $T_3 = 46$ | $C_3 = 1'5'6'3'$ |
| $T_4 = 234$ | $C_4 = 1'2'3'4'$ |
| $T_5 = 536$ | — |

$$R_{ab} = 1 - P(C_1 + C_2 + C_3 + C_4) \quad (6.14a)$$
$$R_{ab} = 1 - P(1'4'5' + 1'6'2' + 1'5'3'6' + 1'2'3'4') \quad (6.14b)$$
$$\begin{aligned}R_{ab} = 1 &- [P(1'4'5') + P(1'6'2') + P(1'5'3'6') + P(1'2'3'4')] \\ &+ [P(1'2'4'5'6') + P(1'3'4'5'6') + P(1'2'3'4'5') \\ &+ P(1'2'3'5'6') + P(1'2'3'4'6') + P(1'2'3'4'5'6')] \\ &- [P(1'2'3'4'5'6') + P(1'2'3'4'5'6') + P(1'2'3'4'5'6') \\ &+ P(1'2'3'4'5'6')] + [P(1'2'3'4'5'6')] \quad (6.14c)\end{aligned}$$

The expansion of the probability of a union of events that occurs in Eq. (6.14) is often called the inclusion–exclusion formula. [See Eq. (A11).]

Note that in the expansions in Eqs. (6.12) or (6.13), ample use is made of the theorems $x \cdot x = x$ and $x + x = x$ (see Appendix A). For example, the second bracket in Eq. (6.14c) has as its second term $P(c_1 c_3) = P([1'4'5'][1'5'6'3']) = P(1'3'4'5'6')$, since $1' \cdot 1' = 1'$ and $5' \cdot 5' = 5'$. The reader should note that this point is often overlooked (see Appendix D, Section D3), and it may or may not make a numerical difference.

If all the edges have equal probabilities of failure $= q$ and are independent, Eq. (6.14c) becomes

$$R_{ab} = 1 - [2q^3 + 2q^4] + [5q^5 + q^6] - [4q^6] + [q^6]$$
$$R_{ab} = 1 - 2q^3 - 2q^4 + 5q^5 - 2q^6 \quad (6.15)$$

The necessary checks, $R_{ab} = 1$ for $q = 0$ and $R_{ab} = 0$ for $q = 1$, are valid.

For $q = 0.1$, Eq. (6.15) yields

$$R_{ab} = 1 - 2 \times 0.1^3 - 2 \times 0.1^4 + 5 \times 0.1^5 - 2 \times 0.1^6 = 0.997848 \quad (6.16)$$

Of course, the result of Eq. (6.16) is identical to Eq. (6.11c). If we substitute tie sets into Eq. (6.12), we would get a different though equivalent expression.

The expansion of Eq. (6.13) has a complexity of $2^j$ and is more complex than Eq. (6.12) if there are more cut sets than tie sets. At this point, it would seem that we should analyze the network and see how many tie sets and cut sets exist between $s$ and $t$, and assuming that $i$ and $j$ are manageable numbers (as is the case in the example to follow), then either Eq. (6.12) or Eq. (6.13) is feasible. In a very large problem (assume $i < j < e$), even $2^i$ is too large to deal with, and the approximations of Section 6.4.3 are required. Of course, large problems will utilize a network reliability computation program, but an approximation can be used to check the program results or to speed up the computation in a truly large problem [Colbourn, 1987, 1993; Shooman, 1990].

The complexity of the cut-set and tie-set methods depends on two factors: the order of complexity involved in finding the tie sets (or cut sets) and the order of complexity for the inclusion–exclusion expansion. The algorithms for finding the number of cut sets are of polynomial complexity; one discussed in Shier [1991, p. 63] is of complexity order $O(n+e+ie)$. In the case of cut sets, the finding algorithms are also of polynomial complexity, and Shier [1991, p. 69] discusses one that is of order $O([n+e]j)$. Observe that the notation $O(f)$ is called the order of $f$ or "big $O$ of $f$." For example, if $f = 5x^3 + 10x^2 + 12$, the order of $f$ would be the dominating term in $f$ as $x$ becomes large, which is $5x^3$. Since the constant 5 is a multiplier independent of the size of $x$, it is ignored, so $O(5x^3 + 10x^2 + 12) = x^3$ (see Rosen [1999, p. 105]).

In both cases, the dominating complexity is that of expansion for the inclusion–exclusion algorithm for Eqs. (6.12) and (6.13), where the orders of complexity are exponential, $O(2^i)$ or $O(2^j)$ [Colbourn, 1987, 1993]. This is the reason why approximate methods are discussed in the next two sections. In addition, some of these algorithms are explored in the problems at the end of this chapter.

If we examine Eqs. (6.12) and (6.13), we see that the complexity of these expressions is a function of the cut sets or tie sets, the number of edges in the cut sets or tie sets, and the number of "brackets" that must be expanded (the number of terms in the union of cut sets or tie sets—i.e., in the inclusion–exclusion formula). We can approximate the cut-set or tie-set expression by dropping some of the less-significant brackets of the expansion, by dropping some of the less-significant cut sets or tie sets, or by both.

### 6.4.3 Truncation Approximations

The inclusion–exclusion expansions of Eqs. (6.12) and (6.13) sometimes yield a sequence of probabilities that decrease in size so that many of the higher-order terms in the sequence can be neglected, resulting in a simpler approximate formula. These terms are products of probabilities, so if these probabilities are small, the higher-order product terms can be neglected. In the case of tie-set probabilities, this is when the probabilities of success are small—the so-called low-reliability region, which is *not* the region of practical interest. Cut-set analysis is preferred since this is when the probabilities of failure are small—the so-called high-reliability region, which is *really* the region of practical interest. Thus cut-set approximations are the ones most frequently used

## TWO-TERMINAL RELIABILITY 295

in practice. If only the first bracket in Eq. (6.14c) is retained in addition to the unity term, one obtains the same expression that would have ensued *had the cuts been disjoint* (but they are not). Thus we will call the retention of only the first two terms the *disjoint approximation*.

In Shooman [1990, Section 3.6.5], it is shown that a disjoint cut-set approximation is a lower bound. For the example of Fig. 6.1, we obtain Eq. (6.17) for the disjoint approximation, and assuming $q = 0.1$:

$$R_{ab} \geq 1 - [2q^3 + 2q^4] = 1 - 0.002 - 0.0002 = 0.9978 \qquad (6.17)$$

which is quite close to the exact value given in Eq. (6.16). If we include the next bracket in Eq. (6.14c), we get a closer approximation at the expense of computing $[j + \binom{j}{2}] = [j(j-1)/2]$ terms.

$$R_{ab} \leq 1 - [2q^3 + 2q^4] + [5q^5 + q^6]$$
$$= 0.9978 + 5 \times 0.1^5 + 0.1^6 = 0.997851 \qquad (6.18)$$

Equation (6.18) is not only an approximation but an upper bound. In fact, as more terms are included in the inclusion–exclusion formula, we obtain a set of alternating bounds (see Shooman [1990, Section 3.6.5]). Note that Eq. (6.17) is a sharp lower bound and that Eq. (6.18) is ever sharper, but both equations effectively bracket the exact result. Clearly, the sharpness of these bounds increases as $q_i = 1 - p_i$ decreases for the $i$ edges of the graph.

$$0.997800 \leq R_{ab} \leq 0.997851 \qquad (6.19)$$

We can approximate $R_{ab}$ by the midpoint of the two bounds.

$$R_{ab} = \frac{0.997800 + 0.997851}{2} = 0.9978255 \qquad (6.20)$$

The accuracy of the preceding approximation can be evaluated by examining the deviation in the computed probability of failure $F_{ab} = 1 - R_{ab}$. In the region of high reliability, all the values of $R_{ab}$ are very close to unity, and differences are misleadingly small. Thus, as our error criterion, we will use

$$\% \text{ error} = \frac{|F_{ab}(\text{estimate}) - F_{ab}(\text{exact})|}{F_{ab}(\text{exact})} \times 100 \qquad (6.21)$$

Of course, the numerator of Eq. (6.21) would be the same if we took the differences in the reliabilities. Evaluation of Eq. (6.21) for the results given in Eqs. (6.16) and (6.20) yields

$$\% \text{ error} = \frac{|0.0021745 - 0.002152|}{0.002152} \times 100\% = 1.05 \qquad (6.22)$$

Clearly, this approximation is good for this example and will be good in most cases. Of course, in approximate evaluations of a large network, we do not know the exact reliability, but we can still approximate Eq. (6.21) by using the difference between the two-term and three-term approximations. For the numerator and the average of the denominator:

$$\% \text{ error} \approx \frac{|0.002200 - 0.002149|}{0.0021745} \times 100\% = 2.35 \quad (6.23)$$

A moment's reflection leads to the conclusion that the highest-order approximation will always be the closest and should be used in the denominator of an error bound. The numerator, on the other hand, should be the difference between the two highest-order terms. Thus, for our example,

$$\% \text{ error} \approx \frac{|0.002200 - 0.002149|}{0.0021749} \times 100\% = 2.37 \quad (6.24)$$

Therefore, a practical approach in designing a computer analysis program is to ask the analyst to input the accuracy he or she desires, then to compute a succession of bounds involving more and more terms in the expansion of Eq. (6.13) at each stage. An equation similar to Eq. (6.24) would be used for the last two terms in the computation to determine when to stop computing alternating bounds. The process truncates when the error approximation yields an estimated value that has a smaller error bound that that of the required error. We should take note that the complexity of the "one-at-a-time" approximation is of order $j$ (number of cut sets) and that of the "two-at-a-time" approximation is of order $j^2$. Thus, even if the error approximation indicates that more terms are needed, the complexity will only be of order $j^3$ or perhaps $j^4$. The inclusion–exclusion complexity is therefore reduced from order $2^j$ to a polynomial in $j$ (perhaps $j^2$ or $j^3$).

### 6.4.4 Subset Approximations

In the last section, we discussed approximation by truncating the inclusion–exclusion expression. Now we discuss approximation by exclusion of low-probability cut sets or tie sets. Clearly, the occurrence probability of the lower-order (fewer edges) cut sets is higher than the higher-order (more edges) ones. Thus, we can approximate Eq. (6.14a) dropping $C_3$ and $C_4$ fourth-order cut sets and retaining the third-order cut set to yield an upper bound (since we have dropped cut sets, we are making an optimistic approximation).

$$R_{ab} \leq 1 - P(C_1 + C_2) = 1 - P(C_1) - P(C_2) + P(C_1)P(C_2)$$
$$= 1 - P(1'4'5') - P(1'6'2') + P(1'2'4'5'6') \quad (6.25a)$$

For $q = 0.1$,

$$R_{ab} \leq 1 - 2 \times 0.1^3 + 0.1^5 = 0.99801 \tag{6.25b}$$

We can use similar reasoning to develop a lower bound (we drop tie sets, thereby making a pessimistic approximation) by dropping all but the "one-hop" tie sets ($T_1$) and the "two-hop" tie sets ($T_2, T_3$)—compare Eq. (6.12) and Table 6.2.

$$R_{ab} \geq P(T_1 + T_2 + T_3) = P(1 + 25 + 46) = P(1) + P(25) + P(46)$$
$$- [P(125) + P(146) + P(2456)] + [P(12456)] \tag{6.26a}$$

For $p = 0.9$,

$$R_{ab} \geq p + 2p^2 - 2p^3 - p^4 + p^5 = 0.9 + 2 \times 0.9^2 - 2 \times 0.9^3 - 0.9^4 + 0.9^5$$
$$= 0.99639 \tag{6.26b}$$

Given Eq. (6.25b) and (6.26b), we can bound $R_{ab}$ by

$$0.99639 \leq R_{ab} \leq 0.99801 \tag{6.27}$$

and approximate $R_{ab}$ by the midpoint of the two bounds:

$$R_{ab} = \frac{0.99639 + 0.99801}{2} = 0.9971955 \tag{6.28}$$

The error bound for this approximation is computed in the same manner as Eq. (6.23).

$$\% \text{ error} \approx \frac{|0.00361 - 0.00199|}{0.0028045} \times 100\% = 57.8 \tag{6.29}$$

The percentage error is larger than in the case of the truncation approximations, but it remains small enough for the approximation to be valid. The complexity is still exponential—of order $2^x$; however, $x$ is now a small integer and $2^x$ is of modest size. Furthermore, the tie-set and cut-set algorithms take less time since we now do not need to find all cut sets and tie sets—only those of order $\leq x$. Of course, one can always combine both approximation methods by dropping out higher-order cut sets and then also truncating the expansion. For more details on network reliability approximations, see Murray [1992, 1993].

### 6.4.5 Graph Transformations

Anyone who has studied electric-circuit theory in a physics or engineering class knows that complex networks of resistors can be reduced to an equivalent single resistance through various combinations of series, parallel, and $Y$–$\Delta$ transformations. Such knowledge has obviously stimulated the development

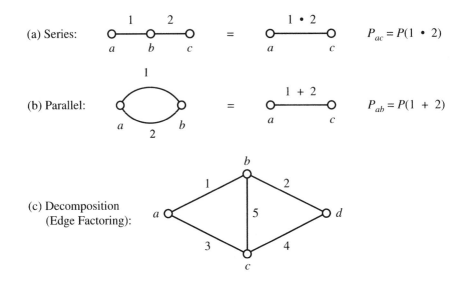

Expand about 5: $R_{ad} = [P(5)P(G_1) + P(5')P(G_2)]$

**Figure 6.2** Illustration of series, parallel, and decomposition transformations for two-terminal pair networks.

of equivalent network transformations: some that are remarkably similar, and some that are quite different, especially in the case of all-terminal reliability (to be discussed later). We must remember that these are not flow transformations but *probability* transformations.

This method of calculating network reliability is based on transforming the network into a simpler network (or set of networks) by successively applying transformations. Such transformations are simpler for two-terminal reliability than for all-terminal reliability. For example, for two-terminal reliability, we can use the transformations given in Fig. 6.2. In this figure, the series transformation indicates that we replace two branches in series with a single branch that is denoted by the intersection of the two original branches (1 · 2). In the parallel transformation, we replace the two parallel branches with a single-series branch that is denoted by the union of the two parallel branches (1 + 2). The edge-factoring case is more complex; the obvious branch to factor about is edge 5, which complicates the graph. Edge 5 is considered good and has a probability of 1 (shorted), and the graph decomposes to $G_1$. If edge 5 is bad, however, it is assumed that no transmission can occur and that it has a probability of 0 (open circuit), and the graph decomposes to $G_2$. Note that both $G_1$ and $G_2$ can now be evaluated by using combinations of series and parallel transformations. These three transformations—series, parallel, and decomposition—are all that is needed to perform the reliability analysis for many networks.

Now we discuss a more difficult network configuration. In the first transformation in Fig. 6.2(a) series, we readily observe that *both* (intersection) edges 1 and 2 must be up for a connection between *a* and *c* to occur. However, this transformation only works if there is *no third edge connected to node b*; if a third edge exists, a more elaborate transformation is needed (which will be discussed in Section 6.6 on all-terminal reliability). Similarly, in the case of the parallel transformation, nodes *a* and *b* are connected in *either* (union) 1 or 2 is up.

Assume that any failures of edge 1 and edge 2 are independent and the probabilities of success for edges 1 and 2 are $p_1$ and $p_2$ (probabilities of failure are $q_1 = 1 - p_1$, $q_2 = 1 - p_2$). Then for the series subnetwork of Fig. 6.2(a), $p_{ac} = p_1 p_2$, and for the parallel subnetwork in Fig. 6.2(b), $p_{ab} = p_1 + p_2 - p_1 p_2 = 1 - q_1 q_2$.

The case of decomposition (called the keystone component method in system reliability [Shooman, 1990] or the edge-factoring method in network reliability) is a little more subtle; it is used to eliminate an edge $x$ from a graph. Since all edges must either be up or down, we reduce the original network to two other networks $G_1$ (given that edge $x$ is up) and $G_2$ (given that edge $x$ is down). In general, one uses series and parallel transformations first, resorting to edge-factoring only when no more series or parallel transformations can be made. In the subnetwork of Fig. 6.2(c), we see that neither series nor parallel transformation is immediately possible because of edge 5, for which reason decomposition should be used.

The mathematical basis of the decomposition transformation lies in the laws of conditional probability and Bayesian probability [Mendenhall, 1990, pp. 64–65]. These laws lead to the following probability equation for terminal pair *st* and edge *x*.

$P$(there is a path between $s$ and $t$)
$= P(x \text{ is good}) \times P(\text{there is a path between } s \text{ and } t | x \text{ is good})$
$+ P(x \text{ is bad}) \times P(\text{there is a path between } s \text{ and } t | x \text{ is bad})$  (6.30)

The preceding equation can be rewritten in a more compact notation as follows:

$$P_{st} = P(x)P(G_1) + P(x')P(G_2) \tag{6.31}$$

The term $P(G_1)$ is the probability of a connection between $s$ and $t$ for the modified network where $x$ is good, that is, the terminals at either end of edge $x$ are connected to the graph [see Fig. 6.2(c)]. Similarly, the term $P(G_2)$ is the probability that there is a connection between $s$ and $t$ for the modified network $G_2$ where $x$ is bad, that is, the edge $x$ is removed from the graph [again, see Fig. 6.2(c)]. Thus Eq. (6.31) becomes for $st = ab$:

$$P_{st} = p_5(1 - q_1 q_3)(1 - q_2 q_4) + q_5(p_1 p_2 + p_3 p_4 - p_1 p_2 p_3 p_4) \tag{6.32}$$

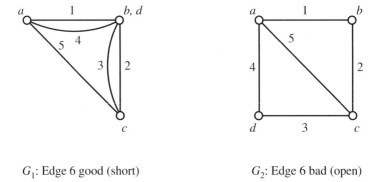

$G_1$: Edge 6 good (short)          $G_2$: Edge 6 bad (open)

**Figure 6.3** Decomposition subnetworks for the graph of Fig. 6.1 expanded about edge 6.

Of course, in most examples, networks $G_1$ and $G_2$ are a bit more complex, and sometimes transformations are recursively computed. More examples of transformations appear in the problems at the end of this chapter; for a complete discussion of transformations, see Satyanarayana [1985] and A. M. Shooman [1992].

We can illustrate the use of the three transformations of Fig. 6.2 on the network given in Fig. 6.1, where we begin by decomposing about edge 6.

$$R_{ab} = P(6) \cdot P[G_1] + P(6') \cdot P[G_2] \tag{6.33}$$

The networks $G_1$ and $G_2$ are shown in Fig. 6.3. Note that for edge 6 good (up), nodes $b$ and $d$ merge in $G_1$, whereas for edge 6 bad (down), edge 6 is simply removed from the network.

We now calculate $P(G_1)$ and $P(G_2)$ for a connection between nodes $a$ and $b$ with the aid of the series and parallel transformations of Fig. 6.2:

$$\begin{aligned} P(G_1) &= P(1 + 4 + 52 + 53) = [P(1) + P(4) + P(52) + P(53)] \\ &\quad - [P(14) + P(152) + P(153) + P(452) + P(453) + P(523)] \\ &\quad + [P(4534) + P(1523) + P(1453) + P(1452)] - [P(14523)] \end{aligned} \tag{6.34}$$

$$\begin{aligned} P(G_2) &= P(1 + 25 + 243) = [P(1) + P(25) + P(243)] \\ &\quad - [P(125) + P(1243) + P(2543)] + [P(12543)] \end{aligned} \tag{6.35}$$

Assuming that all edges are identical and independent with probabilities of success and failure of $p$ and $q$, substitution into Eqs. (6.33), (6.34), and (6.35) yields

$$R_{ab} = p[2p + p^2 - 5p^3 + 4p^4 - p^5] + q[p + p^2 - 2p^4 + p^5] \tag{6.36}$$

Substitution of $p = 0.9$ and $q = 0.1$ into Eq. (6.36) yields

$$R_{ab} = 0.9[0.99891] + 0.1[0.98829] = 0.997848 \qquad (6.37)$$

Of course, this result agrees with the previous computation given in Eq. (6.16).

## 6.5 NODE PAIR RESILIENCE

All-terminal reliability, in which all the node pairs can communicate, is discussed in the next section. Also, $k$-terminal reliability will be treated as a specified subset ($2 \leq k \leq$ all-terminal pairs) of all-terminal reliability. In this section, another metric, essentially one between two-terminal and all-terminal, is discussed.

Van Slyke and Frank [1972] proposed a measure they called resilience for the expected number of node pairs that can communicate (i.e., they are connected by one or more tie sets). Let $s$ and $t$ represent a node pair. The number of node pairs in a network with $N$ nodes is the number of combinations of $N$ choose 2, that is, the number of combinations of 2 out of $N$.

$$\text{Number of node pairs} = \binom{N}{2} = \frac{N!}{2!(N-2)!} = \frac{N(N-1)}{2} \qquad (6.38)$$

Our notation for the set of $s, t$ node pairs contained in an $N$ node network is $\{s, t\} \subset N$, and the expected number of node pairs that can communicate is denoted as resilience, res($G$):

$$\text{res}(G) = \sum_{\{s,t\} \subset N} R_{st} \qquad (6.39)$$

We can illustrate a resilience calculation by applying Eq. (6.39) to the network of Fig. 6.1. We begin by observing that if $p = 0.9$ for each edge, symmetry simplifies the computation. The node pairs divide into two categories: the edge pairs ($ab$, $ad$, $bc$, and $cd$) and the diagonal pairs ($ac$ and $bd$). The edge-pair reliabilities were already computed in Eqs. (6.36) and (6.37). For the diagonals, we can use the decomposition given in Fig. 6.3 (where $s = a$ and $t = c$) and compute $R_{ac}$ as shown in the following equations:

$$P(G_1) = P[5 + (1+4)(2+3)]$$
$$= P(5) + P(1+4)P(2+3) - P(5)P(1+4)P(2+3)$$
$$= [p + (1-p)(2p-p^2)^2] \qquad (6.40a)$$
$$P(6)P(G_1) = p[p + (1-p)(2p-p^2)^2] = 0.898209 \qquad (6.40b)$$
$$P(G_2) = P[5 + 12 + 43] = P(5) + P(12) + P(43)$$
$$- [P(512) + P(543) + P(1243)]$$
$$+ [P(51243)] = p + 2p^2 - 2p^3 - p^4 + p^5 \qquad (6.41a)$$
$$P(6')P(G_2) = q[p + 2p^2 - 2p^3 - p^4 + p^5] = 0.099639 \qquad (6.41b)$$
$$R_{ac} = 0.98209 + 0.099639 = 0.997848 \qquad (6.42)$$

Substitution of Eqs. (6.37) and (6.42) into (6.39) yields

$$\text{res}(G) = 4 \times 0.997848 + 2 \times 0.997848 = 5.987 \qquad (6.43)$$

Note for this particular example that because all edge reliabilities are equal and because it is a complete graph, symmetry would have predicted that $R_{st}$ values were the same for node pairs $ab$, $ad$, $bc$, and $cd$, and similarly for node pairs $ac$ and $bd$. Clearly, for a very reliable network, the resilience will be close to the maximum $N(N-1)/2$, which for this example is 6. In fact, it may be useful to normalize the resilience by dividing it by $N(N-1)/2$ to yield a "normalized" resilience metric. In our example, $\text{res}(G)/6 = 0.997848$. In general, if we divide Eq. (6.39) by Eq. (6.38), we obtain the average reliability for all the two-terminal pairs in the network. Although this requires considerable computation, the metric may be useful when the $p_i$ are unequal.

## 6.6 ALL-TERMINAL RELIABILITY

The all-terminal reliability problem is somewhat more difficult than the two-terminal reliability problem. Essentially, we must modify the two-terminal problem to account for all-terminal pairs. Each of the methods of Section 6.4 is discussed in this section for the case of all-terminal reliability.

### 6.6.1 Event-Space Enumeration

We may proceed as we did in Section 6.4.1 except that now we examine all the good events for two-terminal reliability and strike out (i.e., classify as bad) those that do not connect *all* the terminal pairs. By applying these restrictions to Table 6.1, we obtain Table 6.3. From this table, we can formulate an all-terminal reliability expression similar to the two-terminal case.

## ALL-TERMINAL RELIABILITY

**TABLE 6.3  Modification of Table 6.1 for the All-Terminal Reliability Problem**

| Event | Connection ab | Connection ac | Connection ad | Term |
|---|---|---|---|---|
| $E_1$ | √ | √ | √ | $p^6$ |
| $E_2, E_3, \ldots, E_7$ | √ | √ | √ | $qp^5$ |
| $E_8, E_9, \ldots, E_{22}$ | √ | √ | √ | $q^2p^4$ |
| $E_{23}, E_{24}, E_{26}, E_{27}$ $E_{28}, E_{29}, E_{30}$ $E_{32}, E_{33}, E_{34}, E_{37}$ $E_{38}, E_{39}, E_{40}, E_{41}$ $E_{42}$ | √ | √ | √ | $q^3p^3$ |
| Other 26 fail for at least 1 terminal pair | — | — | — | $q^3p^3, q^4p^2,$ $q^3p^1, q^6$ |

$$R_{\text{all-terminal}} = R_{\text{all}} = \sum_{\substack{i=1 \\ i \neq 25, 31, 35, 36}}^{42} P(E_i) \quad (6.44)$$

Note that events 25, 31, 35, and 36 represent the only failure events with three edge failures. These four cases involve isolation of each of the four vertices. All other failure events involve four or more failures.

Substituting the terms from Table 6.3 into Eq. (6.44) yields

$$R_{\text{all}} = p^6 + 6qp^5 + 15q^2p^4 + 16q^3p^3 = 0.9^6 + 6 \times 0.1 \times 0.9^5 + 15$$
$$\times\ 0.1^2 \times 0.9^4 + 16 \times 0.1^3 \times 0.9^9 = 0.531441$$
$$+ 0.354294 + 0.098415 + 0.011664 \quad (6.45a)$$
$$R_{\text{all}} = 0.995814 \quad (6.45b)$$

Of course, the all-terminal reliability is lower than the two-terminal reliability computed previously.

### 6.6.2  Cut-Set and Tie-Set Methods

One can also compute all-terminal reliability using cut- and tie-set methods either via exact computations or via the various approximations. The computations become laborious even for the modest-size problem of Fig. 6.1. Thus we will set up the *exact* calculations and discuss the solution rather than carry out the computations. Exact calculations for a practical network would be performed via a network modeling program; therefore, the purpose of this section is to establish the understanding of how computations are performed and also to serve as a background for the *approximate* methods that follow.

**TABLE 6.4 Cut Sets and Tie Sets for All-Terminal Reliability Computations**

| Pair $ab$ | Pair $ad$ | Pair $ac$ |
|---|---|---|
| \multicolumn{3}{c}{Tie Sets} | | |
| $T_1 = 1$ | $T_6 = 5$ | $T_{11} = 4$ |
| $T_2 = 25$ | $T_7 = 12$ | $T_{12} = 53$ |
| $T_3 = 46$ | $T_8 = 43$ | $T_{13} = 16$ |
| $T_4 = 234$ | $T_9 = 163$ | $T_{14} = 123$ |
| $T_5 = 356$ | $T_{10} = 462$ | $T_{15} = 526$ |
| \multicolumn{3}{c}{Cut Sets} | | |
| $C_1 = 1'5'4'$ | $C_5 = 5'4'1'$ | $C_9 = 4'6'3'$ |
| $C_2 = 1'2'6'$ | $C_6 = 5'3'2'$ | $C_{10} = 4'5'1'$ |
| $C_3 = 1'3'5'6'$ | $C_7 = 5'6'4'2'$ | $C_{11} = 4'1'2'3'$ |
| $C_4 = 1'2'3'4'$ | $C_8 = 5'6'1'3'$ | $C_{12} = 4'2'5'6'$ |

We begin by finding the tie sets and cut sets for the terminal pairs $ab$, $ac$, and $ad$ (see Fig. 6.1 and Table 6.4). Note that if node $a$ is connected to all other nodes, there is a connection between all nodes.

In terms of tie sets, we can write

$$P_{\text{all}} = P([\text{path } ab] \cdot [\text{path } ad] \cdot [\text{path } ac]) \tag{6.46}$$

$$P_{\text{all}} = P([T_1 + T_2 + \cdots + T_5] \cdot [T_6 + T_7 + \cdots + T_{10}]$$
$$\cdot [T_{11} + T_{12} + \cdots + T_{15}]) \tag{6.47}$$

The expansion of Eq. (6.47) involves 125 intersections followed by complex calculations involving expansion of the union of the resulting events (inclusion–exclusion); clearly, hand computations are starting to become intractable. A similar set of equations can be written in terms of cut sets. In this case, interrupting path $ab$, $ad$, or $ac$ is sufficient to generate all-terminal cut sets.

$$P_{\text{all}} = 1 - P([\text{no path } ab] + [\text{no path } ad] + [\text{no path } ac]) \tag{6.48}$$

$$P_{\text{all}} = 1 - P([C_1 + C_2 + C_3 + C_4] + [C_5 + C_6 + C_7 + C_8]$$
$$+ [C_9 + C_{10} + C_{11} + C_{12}]) \tag{6.49}$$

The expansion of Eq. (6.49) involves the expansion of the union for 12 events (there are $2^{12}$ terms; see Section A4.2) and the disjoint or reduced approximation or computer solution are the only practical approaches.

### 6.6.3 Cut-Set and Tie-Set Approximations

The difficulty in expanding Eqs. (6.47) and (6.48) makes approximations almost imperative in any pencil-paper-and-calculator analysis. We can begin by simplifying Eq. (6.49) by observing that cut sets $C_1 = C_5 = C_{10}$, $C_4 = C_{11}$, $C_7 = C_{12}$, and $C_3 = C_8$; then, $C_5$, $C_{10}$, $C_{11}$, $C_{12}$, and $C_8$ can be dropped, thereby reducing Eq. (6.49) to seven cut sets. Since all edges are assumed to have equal reliabilities, $p = 1 - q$, and the disjoint approximation for Eq. (6.49) yields

$$P_{\text{all}} \geq 1 - P(C_1 + C_2 + C_3 + C_4 + C_6 + C_7 + C_9) \quad (6.50a)$$

and substituting $q = 0.1$ yields

$$P_{\text{all}} \geq 1 - 4q^3 - 3q^4 = 1 - 4(0.1)^3 - 3(0.1)^4 = 0.9957 \quad (6.50b)$$

To obtain an upper bound, we add the 21 terms in the second bracket in the expansion of Eq. (6.49) to yield

$$P_{\text{all}} \leq 0.9957 + 17q^5 + 4q^6 = 0.99600 \quad (6.51)$$

If we average Eqs. (6.50b) and (6.51), we obtain $P_{\text{all}} \approx 0.995759$, which is $(0.000174 \times 100/0.0004) = 4.21\%$ in error [cf. Eq. (6.24)]. In this case, the approximation yields excellent results.

### 6.6.4 Graph Transformations

In the case of all-terminal reliability, the transformation schemes must be defined in a more careful manner than was done for the two-terminal case in Fig. 6.2. The problem arises in the case in which a series transformation is to be performed. As noted in part (a) of Fig. 6.2, the series transformation *eliminates* node $b$, causing no trouble in the two-terminal reliability computation where node $b$ is not an initial or terminal vertex (for $R_{st}$ where neither $s$ nor $t$ is $b$). This is the crux of the matter, since we must still include node $b$ in the all-terminal computation. Of course, eliminating node $b$ does not invalidate the transmission between nodes $a$ and $c$. If we continue to use Eq. (6.46) to define all-terminal reliability, the transformations given in Table 6.2 are correct; however, we must evaluate all the events in the brackets of Eq. (6.46) and their intersections. Essentially, this reduces the transformation procedure to an equivalent tree with one node with incidence $N - 1$ (called a *root* or *central* node) and the remainder of incidence 1 (called *pendant* nodes). The tree is then evaluated.

The more common procedure for all-terminal transformation is to reduce the network to two nodes, $s$ and $t$, where the reliability of the equivalent $s$–$t$ edge is the network all-terminal reliability. [A. M. Shooman, 1992]. A simple example (Fig. 6.4) clarifies the differences in these two approaches.

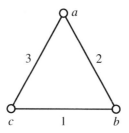

$P(1) = P(2) = P(3) = p = 0.9$

**Figure 6.4** An example illustrating all-terminal reliability transformations.

We begin our discussion of Fig. 6.4 by using event-space methods to calculate the all-terminal reliability. The events are $E_1 = 123$, $E_2 = 1'23$, $E_3 = 12'3$, $E_4 = 123'$, $E_5 = 1'2'3$, $E_6 = 1'23'$, $E_7 = 12'3'$, and $E_8 = 1'2'3'$. By inspection, we see that the good events (which connect $a$ to $b$ and $a$ to $c$) are, namely, $E_1$, $E_2$, $E_3$, and $E_4$. Note that any event with two or more edge failures isolates the vertex connected to these two edges and is a cut set.

$$R_{\text{all}} = P(E_1 + E_2 + E_3 + E_4) = P(E_1) + P(E_2) + P(E_3) + (E_4)$$
$$= P(123) + P(1'23) + P(12'3) + P(123')$$
$$= p^3 + 3qp^2 = 0.9^3 + 3 \times 0.1(0.9)^2 = 0.972 \quad (6.52)$$

To perform all-terminal reliability transformations in the conventional manner, we choose two nodes, $s$ and $t$, and reduce the network to an equivalent $st$ edge. We can reduce any network using a combination of the three transformations shown in Fig. 6.5. Note that the series transformation has a denominator term $[1 - p(1')p(2')]$, which is the probability that the node that disappears (node $b$) is still connected. The other transformations are the same as the two-terminal case. Also, once the transformation process is over, the resulting probability $p_{st}$ must be multiplied by the connection probability of all nodes that have disappeared via the series transformation (for a proof of these procedures, see A. M. Shooman [1992]). We will illustrate the process by solving the network given in Fig. 6.4.

We begin to transform Fig. 6.4 by choosing the $st$ nodes to be $c$ and $b$; thus we wish to use the series transformation to eliminate node $a$. The transformation yields an edge that is in parallel with edge 2 and has a probability of

$$P_{cab} = \frac{P(3)P(2)}{1 - P(3')P(2')} = \frac{0.9 \times 0.9}{1 - 0.1 \times 0.1} = 0.8181818 \quad (6.53)$$

Combining the two branches ($cab$ and $cb$) in parallel yields

## ALL-TERMINAL RELIABILITY

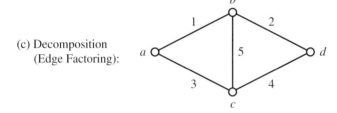

(a) Series: $P_{ac} = p_1 p_2 / (1 - q_1 q_2)$

(b) Parallel: $P_{ab} = P(1 + 2)$

(c) Decomposition (Edge Factoring):

Expand about 5: $R_{ad} = [P(5)P(G_1) + P(5')P(G_2)]$

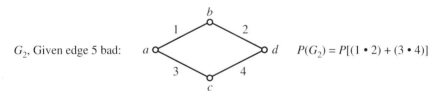

$G_1$, Given edge 5 good: $P(G_1) = P[(1 + 3)] \times P[(2 + 4)]$

$G_2$, Given edge 5 bad: $P(G_2) = P[(1 \cdot 2) + (3 \cdot 4)]$

**Figure 6.5** All-terminal graph-reduction techniques.

$$P_{st} = P(cb + cab) = p(cb) + p(cab) - p(cb)p(cab)$$
$$= 0.9 + 0.8181818 - 0.9 \times 0.8181818 = 0.981811 \quad (6.54)$$

The $P_{st}$ value must be multiplied by the probability that $a$ is connected $P(2 + 3) = P(2) + P(3) - P(23) = 0.9 + 0.9 - 0.9 \times 0.9 = 0.99$.

$$R_{\text{all}} = 0.99 \times 0.9818118 = 0.971993682 \quad (6.55)$$

**308** NETWORKED SYSTEMS RELIABILITY

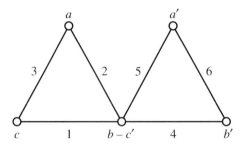

**Figure 6.6** A second all-terminal reliability example.

Of course, Eq. (6.55) yields the same result as Eq. (6.52). As a second example, suppose that you have the same three nodes as $acb$ of Fig. 6.4 followed by a second identical triangle labeled $c'a'b'$, and that nodes $b$ and $c'$ are the same node (see Fig. 6.6). The reliability is given by the square of the value in Eq. (6.45), that is,

$$R_{\text{all}} = 0.971993682^2 = 0.9447717178 \tag{6.56}$$

If we use transformations, we have for $P_{st} = P_{cb'} = 0.9818118^2 = 0.9639544$. This must be multiplied by 0.99 to correct for node $a$ that has disappeared and 0.99 for missing node $a'$, yielding $0.99^2 \times 0.9639544 = 0.9447717$. A comprehensive reliability computation program for two-terminal and all-terminal reliability using the three transformations in Figs. 6.2 or 6.5, as well as more advanced transformations, is given in A. M. Shooman [1992].

### 6.6.5 k-Terminal Reliability

Up until now we have discussed two-terminal and all-terminal reliability. One can define a more general concept of $k$-terminal reliability, where $k$ terminals must be connected. When $k = 2$, we have two-terminal reliability; when $k = n$, we have all-terminal reliability. Thus $k$-terminal reliability can be viewed as a more general concept. See A. M. Shooman [1991] for a detailed development of this concept.

### 6.6.6 Computer Solutions

The foregoing sections introduced the concepts of network reliability computations. Clearly, the example of Fig. 6.1 is much simpler than most practical networks of interest, and, in general, a computer solution is required. Current research focuses on efficient computer algorithms for network reliability computation. For instance, Murray [1992] has developed a computer program that finds network cut sets and tie sets using different algorithms and various reduced cut-set and tie-set methods to compute reliability. A. M. Shooman

[1991, 1992] has developed a computer program for network reliability that is based on the transformations of Section 6.5 and is modified for the *k*-terminal problem, in addition to other more advanced transformations. His model includes the possibility of node failure.

## 6.7 DESIGN APPROACHES

The previous sections treated the problem of network analysis. This section, however, treats the problem of design. We assume that there is a group of $N$ cities or sites represented by nodes that are to be connected by $E$ edges to form the network and that the various links have different costs and reliabilities. Our job is to develop design procedures that yield a good network—that is, at low cost with high reliability. (The reader is referred to the books by Colbourn [1987] and Shier [1991] for a comprehensive introduction to the literature in this field.)

The problem stated in the previous paragraph is actually called the *topological design problem* and is an abstraction of the *complete design problem*. The complete design problem includes many other considerations, such as the following:

1. The capacity in bits per second of each line is an important consideration, and connections between nodes can fail if the number of messages per minute is too large for the capacity of the line, if congestion ensues, and if a queue of waiting messages forms that causes unacceptable delays in transmission.
2. If messages do not go through because of interrupted transmission paths or excessive delays, information is fed back to various nodes. An algorithm, called the *routing algorithm*, is stored at one or more nodes, and alternate routes are generally invoked to send such messages via alternate paths.
3. When edge or node failures occur, messages are rerouted, which may cause additional network congestion.
4. Edges between nodes are based on communication lines (twisted-copper and fiber-optic lines as well as coaxial cables, satellite links, etc.) and represent a discrete (rather than continuous) choice of capacities.
5. Sometimes, the entire network design problem is divided into a backbone network design (discussed previously) and a terminal concentrator design (for the connections within a building).
6. Political considerations sometimes govern node placement. For example, suppose that we wish to connect the cities New York and Baltimore and that engineering considerations point to a route via Atlantic City. For political considerations, however, it is likely that the route would instead go through Philadelphia.

For more information on this subject, the reader is referred to Kirshenbaum [1993]. The remainder of this chapter considers the topological design of a backbone network.

### 6.7.1 Introduction

The general problem of network design is quite complex. If we start with $n$ nodes to connect, we have the problem of determining which set of edges (arcs) is optimum. Optimality encompasses parameters such as best reliability, lowest cost, shortest delay, maximum bandwidth, and most flexibility for future expansion. The problem is too complex for direct optimization of any practical network, so subsets and/or approximate optimizations are used to reduce the general problem to a tractable one.

At the present, most network design focuses on minimizing cost with constraints on time delay and/or throughput. Semiquantitative reliability constraints are often included, such as that contained in the statement "the network should be at least two-connected," that is, there should be no fewer than two paths between each node pair. This may not produce the best design when reliability is of great importance, and furthermore, a two-connected criterion may not yield a minimum cost design. This section approaches network design as a problem in maximization of network reliability within a specified cost budget.

High-capacity fiber-optic networks make it possible to satisfy network throughput constraints with significantly fewer lines than required with older technology. Thus such designs have less inherent redundancy and generally lower reliability. (A detailed study requires a comparison of the link reliabilities of conventional media versus fiber optics.) In such cases, reliability therefore must be the focus of the design process from the outset to ensure that adequate reliability goals are met at a reasonable cost.

Assume that network design is composed of three phases: (a) a backbone design, (b) a local access design, and (c) a local area network within the building (however, here we focus our attention on backbone design). Assume also that backbone design can be broken into two phases: $A_1$ and $A_2$. Phase $A_1$ is the choice of an initial connected configuration, whereas phase $A_2$ is augmentation with additional arcs to improve performance under constraint(s). Each new arc in general increases the reliability, cost, and throughput and may decrease the delay.

### 6.7.2 Design of a Backbone Network Spanning-Tree Phase

We begin our discussion of design by focusing on phase $A_1$. A *connected graph* is one where all nodes are connected by branches. A connected graph with the smallest number of arcs is a *spanning tree*. In general, a *complete network* (all edges possible) with $n$ nodes will have $n^{(n-2)}$ spanning trees, each with $e_s = (n-1)$ edges (arcs). For example, for the four-node network of Fig. 6.1, there are $4^{(4-2)} = 16$ spanning trees with $(4-1) = 3$ arcs; these are shown in Fig. 6.7.

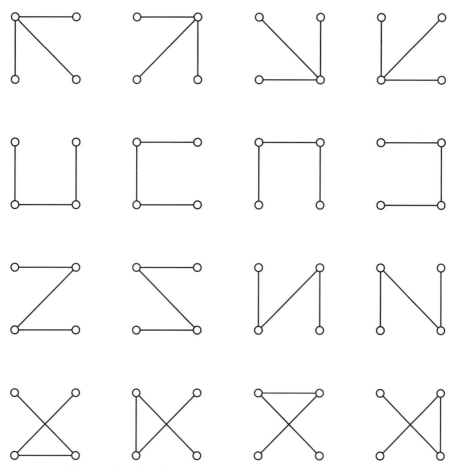

**Figure 6.7** The 16 spanning trees for the network of Fig. 6.1.

The all-terminal reliability of a spanning tree is easy to compute since removal of any one edge disconnects at least one terminal pair. Thus, each edge is one of the $(n-1)$ cut sets of the spanning tree, and all the $(n-1)$ edges are the single tie set. All edges of the spanning tree must work for all the terminals to be connected, and the all-terminal reliability of a spanning tree with $n$ nodes and $(n-1)$ independent branches with probabilities $p_i$ is the probability that all branches are good.

$$R_{\text{all}} = \prod_{i=1}^{n-1} p_i \tag{6.57}$$

If all the branch reliabilities are independent and identical, $p_i = p$, and Eq. (6.57) becomes

**312** NETWORKED SYSTEMS RELIABILITY

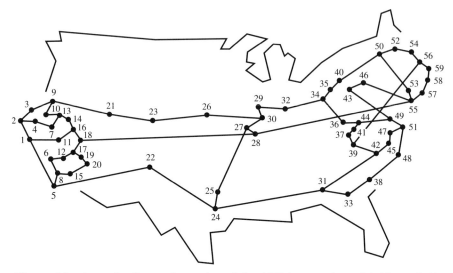

**Figure 6.8** A graph of an early version of the ARPA network model (circa 1979).

$$R_{\text{all}} = p^{(n-1)} \qquad (6.58)$$

Thus, for identical branches, all the spanning trees have the same reliability. If the branch reliabilities, $p_i$, differ, then we can use the spanning tree with the highest reliability for the first-stage design of phase $A_1$. In the case of the spanning trees of Fig. 6.7, we can compute each of the reliabilities using Eq. (6.57) and select the most reliable one as the starting point. If there are a large number of nodes, then an exhaustive search for the most reliable tree is no longer feasible. For example, a graph of an early version of the Military Advanced Research Planning Agency (ARPA) network shown in Fig. 6.8 has 59 nodes. If we were to start over and ask what network we would obtain by using the design techniques under discussion, we would need to consider $59^{57}$ = $8.68 \times 10^{100}$ designs. Fortunately, Kruskal's and Prim's algorithms can be used for finding the spanning tree of a graph with the minimum weights. The use of these algorithms is discussed in the next section.

For most small and medium networks, the graph model shown in Fig. 6.1 is adequate; for large networks, however, the use of a computer is mandated, and for storing network topology in a computer program, matrix (array) techniques are most commonly used. Two types of matrices are commonly employed: adjacency matrices and incidence matrices [Dierker, 1986; Rosen, 1999]. In a network with $n$ nodes and $e$ edges, an adjacency matrix is an $n \times n$ matrix where the rows and columns are labeled with the node numbers. The entries are either a zero, indicating no connection between the nodes, or a one, indicating an arc between the nodes. An adjacency matrix for the graph of Fig.

## (a) Adjacency Matrix

|  | | a | b | c | a' | b' |
|---|---|---|---|---|---|---|
| | | | | Nodes | | |
| Nodes | a | 0 | 1 | 1 | 0 | 0 |
| | b | 1 | 0 | 1 | 1 | 1 |
| | c | 1 | 1 | 0 | 0 | 0 |
| | a' | 0 | 1 | 0 | 0 | 1 |
| | b' | 0 | 1 | 0 | 1 | 0 |

## (b) Incidence Matrix

|  | | 1 | 2 | 3 | 4 | 5 | 6 |
|---|---|---|---|---|---|---|---|
| | | | | Edges | | | |
| Nodes | a | 0 | 1 | 1 | 0 | 0 | 0 |
| | b | 1 | 1 | 0 | 1 | 1 | 0 |
| | c | 1 | 0 | 1 | 0 | 0 | 0 |
| | a' | 0 | 0 | 0 | 0 | 1 | 1 |
| | b' | 0 | 0 | 0 | 1 | 0 | 1 |

**Figure 6.9** Adjacency and incidence matrices for the example of Fig. 6.6.

6.6 is shown in Fig. 6.9(a). Note that the main diagonal is composed of all zeroes since self-loops are generally not present in communications applications, but they may be common in other applications of graphs. Also, the adjacency matrix is applicable to simple graphs that do not have multiple edges between nodes. (An adjacency matrix can be adopted to represent a graph with multiple edges between nodes if each entry represents a *list* of all the connecting edges. Also, if the adjacency matrix can be made nonsymmetrical and if entries of +1 and −1 for branches leaving and entering nodes can be introduced, then the adjacency matrix can represent a directed graph.) The sum of all the entries in a row of an adjacency matrix is the degree of the node associated with the row. This degree is the number of edges that are incident on the node.

One can also represent a graph by an incidence matrix that has $n$ rows and $e$ columns. A zero in any location indicates that the edge is not incident on the associated node, whereas a one indicates that the edge is incident on the node. (Multiple edges and self-loops can be represented by adding columns for these additional edges, and directed edges can be represented by +1 and −1 entries.) An incidence matrix for the graph of Fig. 6.6 is shown in Fig. 6.9(b).

**314** NETWORKED SYSTEMS RELIABILITY

### 6.7.3 Use of Prim's and Kruskal's Algorithms

A weighted graph is one in which one or more parameters are associated with each edge of the graph. In a network, the associated parameters are commonly cost, reliability, length, capacity, and delay. A common problem is to find a minimum spanning tree—a spanning tree with the smallest possible sum of the weights of the edges. Either Prim's or Kruskal's algorithms can be used to find a minimum spanning tree [Cormen, 1992; Dierker, 1986; Kershenbaum, 1993; and Rosen, 1999]. Both are "greedy" algorithms—an optimum choice is made at each step of the procedure.

In Section 6.7.2, we discussed the use of a spanning tree as the first phase of design. An obvious choice is the use of a minimum-cost spanning tree for phase $A_1$, but in most cases an exhaustive search of the possible spanning trees is impractical and either Kruskal's or Prim's algorithm can be used. Similarly, if the highest-reliability spanning tree is desired for phase $A_1$, these same algorithms can be used. Examination of Eq. (6.57) shows that the highest reliability is obtained if we select the set of $p_i$s that has a maximum sum, since this will also result in a maximum product. [This is true because the log of Eq. (6.57) is the sum of the logs of the individual probabilties, and the maximum of the log of a function is the same as the maximum of the function.] Prim's and Kruskal's algorithms can also be used to find a maximum spanning tree—a spanning tree with the largest possible sum of the weights of the edges. Thus it will find the highest-reliability spanning tree. Another approach is to find the spanning tree with the minimum probabilities of failure, which also maximizes the reliability.

A simple statement of Prim's algorithm is to select an initial edge of minimum (maximum) weight, by search or by first ordering all edges by weight, and to use this edge to begin the tree. Subsequent edges are selected as minimum (maximum) weight edges from the set of edges that is connected to a node of the tree but does not form a circuit (loop). The process is continued until $(n - 1)$ edges have been selected. For a description of Kruskal's algorithm (a similar procedure), and the choice between Kruskal's and Prim's algorithms, the reader is directed to the references [Cormen, 1992; Dierker, 1986; Kershenbaum, 1993; and Rosen, 1999]. The use of Prim's algorithm in design is illustrated by the following problem.

We wish to design a network that connects six cities represented by the graph nodes of Fig. 6.10(a). The edge reliabilities and edge costs are given in Fig. 6.10(b) and (c), which are essentially weighted incidence matrices in which the entries to the left of the diagonal are deleted for clarity because they are symmetrical about the main diagonal.

We begin our design by finding the lowest-cost spanning tree using Prim's algorithm. To start, we order the edge costs as shown in the first column of Table 6.5. The algorithm proceeds as follows until 5 edges forming a tree are selected from the 15 possible edges to form the *lowest-cost* spanning tree: select 1–2; select 1–4; select 2–3; we cannot select 2–4 since it forms a loop

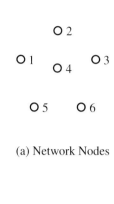

(a) Network Nodes

|   | 1 | 2 | 3 | 4 | 5 | 6 |
|---|---|---|---|---|---|---|
| 1 |   | 0.94 | 0.91 | 0.96 | 0.93 | 0.92 |
| 2 |   |   | 0.94 | 0.97 | 0.91 | 0.92 |
| 3 |   |   |   | 0.94 | 0.90 | 0.94 |
| 4 |   |   |   |   | 0.93 | 0.96 |
| 5 |   |   |   |   |   | 0.91 |

(b) Edge Reliabilities

|   | 1 | 2 | 3 | 4 | 5 | 6 |
|---|---|---|---|---|---|---|
| 1 |   | 10 | 25 | 10 | 20 | 30 |
| 2 |   |   | 10 | 10 | 25 | 20 |
| 3 |   |   |   | 20 | 40 | 10 |
| 4 |   |   |   |   | 20 | 10 |
| 5 |   |   |   |   |   | 30 |

(c) Edge Costs

**Figure 6.10** A network design example.

(delete selection); select 3–6; we cannot select 4–6 since it forms a loop (delete selection); and, finally, select 1–5 to complete the spanning tree. The sequence of selections is shown in the second and third columns of Table 6.5. Note that the remaining 8 edges 2–6 through 3–5 are not considered. One could have chosen edge 4–5 instead of 1–5 for the last step and achieved a different tree with the same cost. The total cost of this minimum-cost tree is 10 + 10 + 10 + 10 + 20 = 60 units. The reliability of this network can be easily calculated as the product of the edge reliabilities: 0.94 × 0.96 × 0.94 × 0.94 × 0.93 = 0.71415. The resulting network is shown in Fig. 6.11(a).

Now, we repeat the design procedure by calculating a maximum-reliability

**316** NETWORKED SYSTEMS RELIABILITY

**TABLE 6.5 Prim's Algorithm for Minimum Cost**[a]

| Edge | Cost | Selected Step No. | Deleted Step No. |
|---|---|---|---|
| 1–2 | 10 | 1 | — |
| 1–4 | 10 | 2 | — |
| 2–3 | 10 | 3 | — |
| 2–4 | 10 | — | 4 |
| 3–6 | 10 | 5 | — |
| 4–6 | 10 | — | 6 |
| 1–5 | 20 | 7 | — |
| 2–6 | 20 | | |
| 3–4 | 20 | | |
| 4–5 | 20 | | |
| 1–3 | 25 | | |
| 2–5 | 25 | | |
| 1–6 | 30 | | |
| 5–6 | 30 | | |
| 3–5 | 40 | | |

[a] See Fig. 6.10.

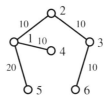

Cost = 60, Reliability = 0.7415

(a) Minimum Cost Spanning Tree

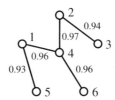

Cost = 60, Reliability = 0.7814

(b) Maximum Reliability Spanning Tree

**Figure 6.11** Two spanning tree designs for the example of Fig. 6.10.

**TABLE 6.6 Prim's Algorithm for Maximum Reliability**[a]

| Edge | Reliability | Selected Step No. | Deleted Step No. |
|---|---|---|---|
| 2–4 | 0.97 | 1 | — |
| 1–4 | 0.96 | 2 | — |
| 4–6 | 0.96 | 3 | — |
| 1–2 | 0.94 | — | 4 |
| 2–3 | 0.94 | 5 | — |
| 3–4 | 0.94 | — | 6 |
| 3–6 | 0.94 | — | 7 |
| 1–5 | 0.93 | 8 | — |
| 4–5 | 0.93 | | |
| 1–6 | 0.92 | | |
| 2–6 | 0.92 | | |
| 1–3 | 0.91 | | |
| 2–5 | 0.91 | | |
| 5–6 | 0.91 | | |
| 3–5 | 0.90 | | |

[a] See Fig. 6.10.

tree using Prim's algorithm. The result is shown in Table 6.6 and the resultant tree is in Fig. 6.11(b). The reliability optimization has produced a superior design with the same cost but a higher reliability. In general, the two procedures will produce designs having lower costs and lower reliabilities along with higher reliabilities and higher costs. In such cases, engineering trade-offs must be performed for design selection. Perhaps a different solution with less-than-optimum cost and reliability would be the best solution. Since the design procedure for the spanning tree phase is not too difficult, a group of spanning trees can be computed and carried forward to the enhancement phase, and a design trade-off can be performed on the final designs.

Since both cost and reliability matter in the optimization, and also since there are many ties, we can return to Table 6.5 and re-sort the edges by cost and wherever ties occur by reliability. The result is Table 6.7, which in this case yields the same design as Table 6.6.

The decision to use a spanning tree for the first stage of design is primarily based on the existence of a simple algorithm (Prim's or Kruskal's) to obtain phase $A_1$ of the design. Other procedures, however, could be used. For example, one could begin with a Hamiltonian circuit (tour), a network containing $N$ edges and one circuit that passes through each node only once. A Hamiltonian circuit has one more edge than a spanning tree. (The reader is referred to the problems at the end of this chapter for a consideration of Hamiltonian tours for phase $A_1$ of the design.) Hamiltonian tours do not exist for all networks [Dierker, 1986; Rosen, 1999], but if we consider that all edges are potentially possible, that is, a complete graph, Hamiltonian tours will exist [Frank, 1971].

**TABLE 6.7 Prim's Algorithm for Minimum Cost Edges First Sorted by Cost and Then by Reliability**[a]

| Edge | Cost | Reliability | Selected Step No. | Deleted Step No. |
|---|---|---|---|---|
| 2–4 | 10 | 0.97 | 1 | — |
| 1–4 | 10 | 0.96 | 2 | — |
| 4–6 | 10 | 0.96 | 3 | — |
| 1–2 | 10 | 0.94 | — | 4 |
| 2–3 | 10 | 0.94 | 5 | — |
| 3–6 | 10 | 0.94 | — | 6 |
| 3–4 | 20 | 0.94 | — | 7 |
| 1–5 | 20 | 0.93 | 8 | — |
| 4–5 | 20 | 0.93 | | |
| 2–6 | 20 | 0.92 | | |
| 1–3 | 25 | 0.91 | | |
| 2–5 | 25 | 0.91 | | |
| 1–6 | 30 | 0.92 | | |
| 5–6 | 30 | 0.91 | | |
| 3–5 | 40 | 0.90 | | |

[a]See Fig. 6.10.

### 6.7.4 Design of a Backbone Network: The Enhancement Phase

The first phase of design, $A_1$, will probably produce a connected network with a smaller all-terminal reliability than required. To improve the reliability, new branches are added. Unfortunately, the effect on the network all-terminal reliability is now a function of not only the reliability of the added branch but also of its location in the network. Thus we must evaluate the network reliability for each of the proposed choices and pick the most cost-effective solution. The use of network reliability calculation programs greatly aids such a trial-and-error procedure [Murray, 1992; A. M. Shooman, 1992]. There are also various network design programs that incorporate reliability and other calculations to arrive at a network design [Kershenbaum, 1993].

In general, one designs a network with only a fraction of all the edges that would be available in a complete graph, which is given by

$$e_c = \binom{n}{2} = \frac{n(n-1)}{2} \tag{6.59}$$

If phase $A_1$ of the design is a spanning tree with $n - 1$ edges, there are $e_r$ remaining edges given by

$$e_r = e_c - e_t = \frac{n(n-1)}{2} - (n-1) = \frac{(n-1)(n-2)}{2} \tag{6.60}$$

For the example given in Fig. 6.10, we have 6 nodes, 15 possible edges in the complete graph, 5 edges in the spanning tree, and 10 possible additional arcs that can be assigned during the enhancement phase. We shall experiment with a few examples of enhancement and leave further discussion for the examples at the end of this chapter. Three attractive choices for enhancement are the additional cost (= 10 edges not used in Table 6.7), edge 1–2, and edge 3–6. A simplifying technique will be used to evaluate the reliability of the enhanced network. We let $R_{A_1}$ be the reliability of the network created by phase $A_1$ and let $X$ represent the success of the added branch ($X'$ is the event failure of the added branch). Using the decomposition (edge-factoring) theorem given in Fig. 6.5(c), we can write the reliability of the enhanced network $R_{A_2}$ as

$$R_{A_2} = P(X)P(\text{Network}|X) + P(X')P(\text{Network}|X') \quad (6.61)$$

The term $P(\text{Network}|X')$ is the reliability of the network with added edge $X$ failed (open-circuited) $= R_{A_1}$, that is, what it was before we added the enhancement edge. If $P(X)$ is denoted by $p_x$, then $P(X') = 1 - p_x$. Lastly, $P(\text{Network}|X)$ is the reliability of the network with $X$ good, that is, with both nodes for edge $X$ merged (one could say that edge $X$ was "shorted," which simplifies the computation). Thus Eq. (6.61) becomes

$$R_{A_2} = p_x P(\text{Network}|X \text{ shorted}) + (1 - p_x) R_{A_1} \quad (6.62)$$

Evaluation of the $X$ shorted term for the addition of edge 1–2 or 3–6 is given in Fig. 6.12. Note that in Fig. 6.12(a), there are parallel branches between (1 = 2) and 4, which are reduced in parallel. Similarly, in Fig. 6.12(b), edges 4–2 and (2 = 6) and 3 are reduced in series and then in parallel with 4 − (6 = 3). Note that in the computations given in Fig. 6.12(b), the series transformation and the overall computation must be "conditioned" by the terms shown in { } [see Fig. 6.5(a) and Eqs. (6.53)–(6.55)]. Substitution into Eq. (6.62) yields

$$R_{A_2} = (0.94)[0.838224922] + (0.6)[0.7814]$$
$$= 0.8348 \quad \text{(for addition of edge 1–2)} \quad (6.63)$$
$$R_{A_2} = (0.94)[0.8881] + (0.6)[0.7814]$$
$$= 0.8817 \quad \text{(for addition of edge 3–6)} \quad (6.64)$$

The addition of edge 3–6 seems to be more cost-effective than if edge 1–2 were added. In general, the objective of the various design methods is to choose branches that optimize the various parameters of the computer network. The reader should consult the references for further details.

### 6.7.5 Other Design Approaches

Although network design is a complex, difficult problem, substantial work has been done in this field over the years. A number of algorithms in addition to

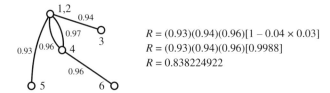

$R = (0.93)(0.94)(0.96)[1 - 0.04 \times 0.03]$
$R = (0.93)(0.94)(0.96)[0.9988]$
$R = 0.838224922$

(a) Addition of edges 1–2

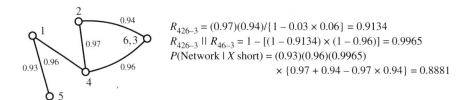

$R_{426-3} = (0.97)(0.94)/\{1 - 0.03 \times 0.06\} = 0.9134$
$R_{426-3} \parallel R_{46-3} = 1 - [(1 - 0.9134) \times (1 - 0.96)] = 0.9965$
$P(\text{Network} \mid X \text{ short}) = (0.93)(0.96)(0.9965)$
$\times \{0.97 + 0.94 - 0.97 \times 0.94\} = 0.8881$

(b) Addition of edges 3–6

**Figure 6.12** Evaluation of the $X$ shorted term.

Prim's and Kruskal's are used for design approaches (some of these algorithms are listed in Table 6.8). The reader may consult the following references for further details: Ahuja [1982]; Colbourn [1987, 1993, 1995]; Corman [1992]; Frank [1971]; Kershenbaum [1993]; and Tenenbaum [1981, 1996].

The various design approaches presently in use are described in Ahuja [1982] and Kershenbaum [1993]. A topological design algorithm usually consists of two parts: the first initializes the topology and the second iteratively improves it (we have used the terms *connected toplogy* and *augmentation to improve performance* to refer to the first and second parts). Design algorithms begin by obtaining a feasible network topology. Augmentation adds further edges to the design to improve such various design parameters as network delay, cost, reliability, and capacity. The optimizing step is repeated until one of the constraints is exceeded. A number of algorithms and heuristics can be applied. Most procedures start with a minimum spanning tree, but because the spanning tree algorithms do not account for link-capacity constraints, they are often called *unconstrained solutions*. The algorithms then modify the starting topology until it satisfies the constraints.

The algorithms described by Ahuja's book [1982] include Kruskal's, Chandy–Russell's, and Esau–Williams's, in addition to Kershenbaum–Chou's generalized heuristic. Kershenbaum's book [1993] discusses these same algorithms, but, in addition, it describes Sharma's algorithm, commenting that this algorithm is widely used in practice. The basic procedure for the

**TABLE 6.8  Various Network Design Algorithms**

### Tree-Traversal Algorithms

(1) Breath first search (BFS)
(2) Depth first search (DFS)
(3) Connected-components algorithm
(4) Minimum spanning tree:
    (a) "greedy" algorithm
    (b) Kruskal's algorithm
    (c) Prim's algorithm

### Shortest-Path Algorithms

(1) Dijkstra's algorithm
(2) Bellman's algorithm
(3) Floyd's algorithm
(4) Incremental-shortest-path algorithms

### Single-Commodity-Network Flows

(1) Ford–Fulkerson algorithm
(2) Minimum-cost flows

Esau–Williams's algorithm is to use a "greedy" type of algorithm to construct a minimum spanning tree. One problem that arises is that nodes may connect to the center of one component (group) of nodes and leave nodes that are stranded far from a center (a costly connection). The Esau–Williams's algorithm aids in eliminating this problem by implementing a trade-off between the connection to a center and the interconnection between two components.

Kershenbaum [1993] discusses other algorithms (Gerla's, Frank's, Chou's, Eckl's, and Maruyama's) that he calls *branch-exchange* algorithms. The design starts with a feasible topology and then locally modifies it by adding or dropping links; alternatively, however, the design may start with a complete graph and identify links to drop. One can decide which links to drop by finding the flow in each link, computing the cost-to-flow ratio, and removing the link with the largest ratio. If an improvement is obtained, the exchange is accepted; if not, another is tried. Kershenbaum speaks of evaluation in terms of cost and/or delay but not reliability. Thus the explicit addition of reliability to the design procedure results in different solutions. In all cases, these will emphasize reliability, but some cases will produce an improved solution.

## REFERENCES

Ahuja, V. *Design and Analysis of Computer Communication Networks.* McGraw-Hill, New York, 1982.

Bateman, K., and E. Cortes. Availability Modeling of FDDI Networks. *Proceedings*

*Annual Reliability and Maintainability Symposium*, 1989. IEEE, New York, NY, pp. 389–395.

Boesch, F. T. Synthesis of Reliable Networks—A Survey. *IEEE Transactions on Reliability* 35 (1986): 240–246.

Brecht, T. B., and C. J. Colbourn. Improving Reliability Bounds in Computer Networks. *Networks* 16 (1986): 369–380.

Cardwell, R. H., C. Monma, and T. Wu. Computer-Aided Design Procedures for Survivable Fiber-Optic Networks. *IEEE Journal on Selected Areas in Communications* 7, 8 (October 1989).

Chandy, K. M., and R. A. Russell. The Design of Multipoint Linkages in a Teleprocessing Tree Network. *IEEE Transactions on Computers* 21, 10 (October 1972).

Clark, B. N., E. M. Neufeld, and C. J. Colbourn. Maximizing the Mean Number of Communicating Vertex Pairs in Series-Parallel Networks. *IEEE Transactions on Reliability* 35 (1986): 247–251.

Cohn, O. et al. Algorithms for Backbone Network Design. *Proceedings ORSA/TIMS*, Las Vegas, NV, May 1990.

Colbourn, C. J. *The Combinatorics of Network Reliablity*. Oxford University Press, New York, 1987.

Colbourn, C. J. Analysis and Synthesis Problems for Network Resilience. *Mathematical Computer Modeling* 17, 11 (1993): 43–48, Pergamon Press, division of Elsevier Science, New York.

Colbourn, C. J. et al. *Network Reliability a Computational Environment*. CRC Press, Boca Raton, FL, 1995.

Cormen, T. H. et al. *Introduction to Algorithms*. McGraw-Hill, New York, 1990.

Dierker, P. F., and W. L. Voxman. *Discrete Mathematics*. Harcourt Brace Jovanovich, Austin, TX, 1986.

Dugan, J. B. "Software System Analysis Using Fault Trees." In *Handbook of Software Reliability Engineering*, M. R. Lyu (ed.). McGraw-Hill, New York, 1996, ch. 15.

Frank, H., and I. T. Frisch. *Communication, Transmission, and Transportation Networks*. Addison-Wesley, Reading, MA, 1971.

Hou, W., and I. G. Okogbaa. Reliability Analysis for Integrated Networks with Unreliable Nodes and Software Failures in the Time Domain. *Proceedings Annual Reliability and Maintainability Symposium*, 2000. IEEE, New York, NY, pp. 113–117.

Jia, S., and J. M. Wing. Survivability Analysis of Networked Systems. From the Carnegie Mellon Web site: www.cs.cmu.edu/~wing/ (2001).

Kershenbaum, A. *Telecommunications Network Design Algorithms*. McGraw-Hill, New York, 1993.

Kershenbaum, A., and W. Chou. A Unified Algorithm for Designing Multidrop Teleprocessing Networks. *IEEE Transactions on Communications* 22, 11 (November 1974).

Kershenbaum, A., and R. Van Slyke. Computing Minimum Spanning Trees Efficiently. *Proceedings of the 1972 ACM Conference*, Boston, 1972.

Kershenbaum, A., and R. Van Slyke. Recursive Analysis of Network Reliability. *Networks* 3, 1 (1973).

Lawler, E. *Combinatorial Optimization: Network and Materials*. Holt, Rinehart, and Winston, Austin, TX, 1976.

Liu, H.-L., and M. L. Shooman. Simulation of Computer Network Reliability with Congestion. *Proceedings Annual Reliability and Maintainability Symposium*, 1999. IEEE, New York, NY, pp. 208–213.

Mendenhall, W. et al. Mathematical Statistics with Applications, 4th ed. PWS-Kent, Boston, 1990.

Messinger, M., and M. L. Shooman. Techniques for Optimum Spares Allocation: A Tutorial Review (Dynamic Programming). *IEEE Transactions on Reliability* 19, 4 (November 1970).

Monma, C. L., and D. F. Shallcross. Methods for Designing Communications Networks with Certain Two-Connected Survivability Constraints. *Operations Research* 37 (1989): 531–541.

Murray, K. "Path Set and Cut Set Approximations and Bounds for Network Reliability Analysis." Ph.D. thesis, Polytechnic University, 1992.

Murray, K., A. Kershenbaum, and M. L. Shooman. "A Comparison of Two Different Path Set and Cut Set Approximations for Network Reliability Analysis. *Proceedings Annual Reliability and Maintainability Symposium*, 1993. IEEE, New York, NY.

Naor, D. et al. A Fast Algorithm for Optimally Increasing Edge-Connectivity. *Proceedings IEEE Symposium on Foundations of Computer Science*, St. Louis, MO, 1990. IEEE, New York, NY.

Ponta, A. et al. Reliability Analysis of Torino Sud District Heating System [Network]. *Proceedings Annual Reliability and Maintainability Symposium*, 1999. IEEE, New York, NY, pp. 336–342.

Provan, J. S. The Complexity of Reliability Computations in Planar and Acyclic Graphs. *SIAM J. Computing* 15 (1986): 36–61.

Provan, J. S., and M. O. Ball. Computing Network Reliability in Time Polynomial in the Number of Cuts. *Operations Research* 32 (1984): 516–526.

Rosen, K. H. *Discrete Mathematics and its Applications*. WCB–McGraw-Hill, New York, 1999.

Rosenthal, A. Computing the Reliability of Complex Networks. *SIAM J. of Applied Mathematics* 32, 2 (March 1977).

Satyanarayana, A. A Unified Formula for Analysis of Some Network Reliability Problems. *IEEE Transactions on Reliability* 31 (1982): 23–32.

Satyanarayana, A., and R. K. Wood. A Linear Time Algorithm for Computing $k$-Terminal Reliability on Series-Parallel Networks. *SIAM J. Computing* 14 (1985): 818–832.

Schwartz, M. *Computer Communication Network Design and Analysis*. Prentice-Hall, Englewood Cliffs, NJ, 1977.

Schweber, W. *Electronic Communication Systems*. Prentice-Hall, Upper Saddle River, NJ, 1999.

Sharma, R. *Network Topology Optimization*. Van Nostrand Reinhold, New York, 1991.

Shier, D. R. *Network Reliability and Algebraic Structures*. Oxford University Press, New York, 1991.

Shooman, A. M. "Exact Graph-Reduction Algorithms for Network Reliability Analysis." Ph.D. thesis, Polytechnic University, 1992.

Shooman, A. M., and A. Kershenbaum. Exact Graph-Reduction Algorithms for Network Reliability Analysis. *Proceedings IEEE/GLOBECOM Conference*, December 1991. IEEE, New York, NY.

Shooman, A. M., and A. Kershenbaum. Methods for Communication-Network Reliability Analysis: Probabilistic Graph Reduction. *Proceedings Annual Reliability and Maintainability Symposium*, 1992. IEEE, New York, NY, pp. 441–448.

Shooman, M. L. *Probabilistic Reliability: An Engineering Approach*. McGraw-Hill, New York, 1968.

Shooman, M. L. *Probabilistic Reliability: An Engineering Approach*, 2d ed. Krieger, Melbourne, FL, 1990.

Shooman, M. L., and E. Cortes. Reliability and Availability Modeling of Coupled Communication Networks—A Simplified Approach. *Proceedings Annual Reliability and Maintainability Symposium*, Orlando, FL, 1991. IEEE, New York, NY.

Tenenbaum, A. S. *Computer Networks*. Prentice-Hall, Englewood Cliffs, NJ, 1981.

Tenenbaum, A. S. *Computer Networks*, 3d ed. Prentice-Hall, Englewood Cliffs, NJ, 1996.

Van Slyke, R., and H. Frank. Network Reliability Analysis: Part I. *Networks* 1, 2 (1972).

Van Slyke, R., H. Frank, and A. Kershenbaum. Network Reliability Analysis: Part II. *SIAM* (1975).

Wing, J. M. Carnegie Mellon Web site: www.cs.cmu/~wing/ (2001).

## PROBLEMS

**6.1.** Consider a computer network connecting the cities of Boston, Hartford, New York, Philadelphia, Pittsburgh, Baltimore, and Washington. (See Fig. P1.)

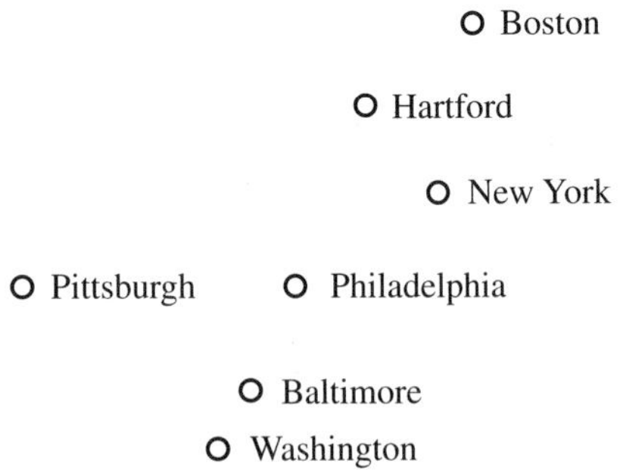

O  Boston

O  Hartford

O  New York

O  Pittsburgh     O  Philadelphia

O  Baltimore

O  Washington

**Figure P1**

(a) What is the minimum number of lines to connect all the cities?

(b) What is the best tree to pick?

**6.2.**

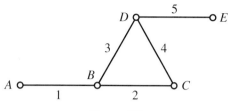

**Figure P2**

You are to evaluate the reliability of the network shown in Fig. P2. Edge reliabilities are all 0.9 and independent.
(a) Find the two-terminal reliability, $R_{AC}$:
   (1) using state-space techniques,
   (2) using tie-set techniques,
   (3) using transformations.
(b) Find the all-terminal reliability, $R_{all}$:
   (1) using state-space techniques,
   (2) using cut-set approximations,
   (3) using transformations.

**6.3.** The four-node network shown in Fig. P3 is assumed to be a complete graph (all possible edges are present). Assume that all edge reliabilities are 0.9.

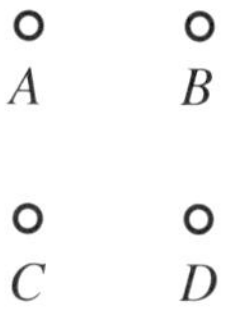

○    ○
A    B

○    ○
C    D

**Figure P3**

(a) By enumeration, find all the trees you can define for this network.
(b) By enumeration, find all the Hamiltonian tours you can find for this network.
(c) Compute the all-terminal reliabilities for the tree networks.
(d) Compute the all-terminal reliabilities for the Hamiltonian networks.
(e) Compute the two-terminal reliabilities $R_{AB}$ for the trees.

(f) Compute the two-terminal reliabilities $R_{AB}$ for the Hamiltonian tours.

**6.4.** Assume that the edge reliabilities for the network given in problem 6.3 are $R(AB) = 0.95$; $R(AC) = 0.92$; $R(AD) = 0.90$; $R(BC) = 0.90$; $R(BD) = 0.80$; and $R(CD) = 0.80$.
  (a) Repeat problem 6.3(c) for the preceding edge reliabilities.
  (b) Repeat problem 6.3(d) for the preceding edge reliabilities.
  (c) Repeat problem 6.3(e) for the preceding edge reliabilities.
  (d) Repeat problem 6.3(f) for the preceding edge reliabilities.
  (e) If the edge costs are $C(AB) = C(AC) = 10$; $C(AD) = C(BC) = 8$; and $C(BD) = C(CD) = 7$, which of the networks is most cost-effective?

**6.5.** Prove that the number of edges in a complete, simple $N$ node graph is given by $N(N-1)/2$. (Hint: Use induction or another approach.)

**6.6.**

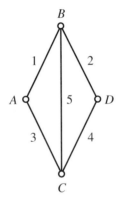

**Figure P4**

For the network shown in Fig. P4, find
  (a) the adjacency matrix,
  (b) the incidence matrix.

**6.7.** A simple algorithm for finding the one-link and two-link (one-hop and two-hop) paths between any two nodes is based on the properties of the incidence matrix.
  (a) If a one-hop path (one-link tie set) exists between any two nodes, $n_1$ and $n_2$, there will be ones in the $n_1$ and $n_2$ rows for the column corresponding to the one-link tie set.
  (b) If a two-hop tie set exists between any two nodes, it can be found

by taking the sum of the matrix entries for the two columns. The numeral 1 appears in the column for the connected nodes and 2 for the intermediate node.

(c) Write a "pseudocode" algorithm for this approach to finding one- and two-hop tie sets.

(d) Does this generalize to $n$ hops? Explain the difficulties.

**6.8.** Can the algorithm of problem 6.7 be adapted to a directed graph? (See Shooman [1990, p. 609].)

**6.9.** What is the order of the algorithm of problem 6.7?

**6.10.** Write a simple program in any language to implement the algorithm of problem 6.9 for up to six nodes.

**6.11.** Compute the two-terminal reliability $R_{ad}$ for the network of problem 6.6 by using the program of problem 6.10.

(a) Assume that all links have a success probability of 0.9.

(b) Assume that links 1, 2, and 3 have a success probability of 0.95 and that links 4 and 5 have a success probability of 0.85.

**6.12.** Check the reliability of problem 6.11 by using any analytical technique.

**6.13.** This homework problem refers to the network described in Fig. 6.10 in the text. All links are potentially possible, and the matrices define the link costs and reliabilities. Assume the network in question is composed of links 1–2; 2–3; 3–6; 6–5; 5–1; 1–4; and 4–6. Compute the two-terminal reliabilities $R_{26}$ and $R_{23}$ by using the following methods:

(a) state-space,

(b) tie-set,

(c) cut-set,

(d) keystone-component,

(e) edge-transformation.

**6.14.** Assume that you are doing a system design for a reliable, low-cost network with the same geometry, potential arc costs, and reliabilities as those given in problem 6.13. Compare the different designs obtained by plotting reliability versus cost, where we assume that the maximum cost budget is 100. Use the following approaches to network design:

(a) Start with a minimum-cost tree and add new minimum-cost edges.

(b) Start with a minimum-cost tree and add new highest-reliability edges.

(c) Start with a highest-reliability tree and add new highest-reliability edges.

(d) Start with a highest-reliability Hamiltonian tour and add new highest-reliability edges.

(e) Start with a minimum-cost Hamiltonian tour and add new highest-reliability edges.

**6.15.** For a Hamiltonian tour in which the $n$ branch probabilities are independent and equal, show that the all-terminal reliability is given by the binomial distribution:

$$R_{\text{all}} = P(\text{no branch failure + one branch failure})$$
$$= p^n + np^{(n-1)}(1-p)$$

**6.16.** Repeat problem 6.15 for a Hamiltonian tour in which the branch probabilities are all different. Instead of the binomial distribution, you will have to write out a series of terms that yields a formula different from that of problem 6.15, but one that also reduces to the same formula for equal probabilities.

**6.17.** Clearly, a Hamiltonian network of $n$ identical elements has a higher reliability than that of a spanning tree for $n$ nodes. Show that the improvement ratio is $[p + n(1-p)]$.

**6.18.** In this problem, we make a rough analysis to explore how the capacity of communication lines affects a communication network.

(a) Consider the nodes of problem 6.1 and connect them in a Hamiltonian tour network in Table P1: Boston–Hartford–New York–Philadelphia–Pittsburgh–Baltimore–Washington–Boston. Assume that the Hartford–Baltimore–Pittsburgh traffic is small (2 units each way) compared to the Boston–New York–Philadelphia–Washington traffic (10 units each way). Fill in Table P1 with total traffic units, assuming that messages are always sent (routed) via the shortest possible way. Assume full duplex (simultaneous communication in both directions).

(b) Assume a break on the Philadelphia–Pittsburgh line. Some messages must now be rerouted over different, longer paths because of this break. For example, the Washington–Boston line must handle all of the traffic between Philadelphia, New York, Boston, and Washington. Recompute the traffic table (Table P2).

**TABLE P1 Traffic Table Before Break**

| | Boston | Hartford | New York | Philadelphia | Pittsburgh | Baltimore | Washington |
|---|---|---|---|---|---|---|---|
| Boston | | | | | | | |
| Hartford | | | | | | | |
| New York | | | | | | | |
| Philadelphia | | | | | | | |
| Pittsburgh | | | | | | | |
| Baltimore | | | | | | | |
| Washington | | | | | | | |

**TABLE P2  Traffic Table After Break**

| | Boston | Hartford | New York | Philadelphia | Pittsburgh | Baltimore | Washington |
|---|---|---|---|---|---|---|---|
| Boston | | | | | | | |
| Hartford | | | | | | | |
| New York | | | | | | | |
| Philadelphia | | | | | | | |
| Pittsburgh | | | | | | | |
| Baltimore | | | | | | | |
| Washington | | | | | | | |

# 7
# RELIABILITY OPTIMIZATION

## 7.1 INTRODUCTION

The preceding chapters of this book discussed a wide range of different techniques for enhancing system or device fault tolerance. In some applications, only one of these techniques is practical, and there is little choice among the methods. However, in a fair number of applications, two or more techniques are feasible, and the question arises regarding which technique is the most cost-effective. To address this problem, if one is given two alternatives, one can always use one technique for design A and use the other technique for design B. One can then analyze both designs A and B to study the trade-offs. In the case of a standby or repairable system, if redundancy is employed at a component level, there are many choices based on the number of spares and which component will be spared. At the top level, many systems appear as a series string of elements, and the question arises of how we are to distribute the redundancy in a cost-effective manner among the series string. Specifically, we assume that the number of redundant elements that can be added is limited by cost, weight, volume, or some similar constraint. The object is to determine the set of redundant components that still meets the constraint and raises the reliability by the largest amount. Some authors refer to this as redundancy optimization [Barlow, 1965]. Two practical works—Fragola [1973] and Mancino [1986]—are given in the references that illustrate the design of a system with a high degree of parallel components. The reader should consult these papers after studying the material in this chapter.

In some ways, this chapter can be considered an extension of the material in Chapter 4. However, in this chapter we discuss the *optimization approach*, where rather than having the redundancy apply to a single element, it is distributed over the entire system in such a way that it optimizes reliability. The optimization approach has been studied in the past, but is infrequently used in practice for many reasons, such as (a) the system designer does not understand the older techniques and the resulting mathematical formulation; (b) the solution takes too long; (c) the parameters are not well known; and (d) constraints change rapidly and invalidate the previous solution. We propose a technique that is clear, simple to explain, and results in the rapid calculation of a family of good suboptimal solutions along with the optimal solution. The designer is then free to choose among this family of solutions, and if the design features or parameters change, the calculations can be repeated with modest effort.

We now postulate that the design of fault-tolerant systems can be divided into three classes. In the first class, only one design approach (e.g., parallel, standby, voting) is possible, or intuition and experience points only to a single approach. Thus it is simple to decide on the level of redundancy required to meet the design goal or the level allowed by the constraint. To simplify our discussion, we will refer to cost, but we must keep in mind that all the techniques to be discussed can be adapted to any other single constraint or, in many cases, multiple constraints. Typical multiple constraints are cost, reliability, volume, and weight. Sometimes, the optimum solution will not satisfy the reliability goal; then, either the cost constraint must be increased or the reliability goal must be lowered. In the second class, if there are two or three alternative designs, we would merely repeat the optimization for each class as discussed previously and choose the best result. The second class is one in which there are many alternatives within the design approach because we can apply redundancy at the subsystem level to many subsystems. The third class, where a mixed strategy is being considered, also has many combinations. To deal with the complexity of the third-class designs, we will use computer computations and an optimization approach to guide us in choosing the best alternative or set of alternatives.

## 7.2 OPTIMUM VERSUS GOOD SOLUTIONS

Because of practical considerations, an approximate optimization yielding a good system is favored over an exact one yielding the best solution. The parameters of the solution, as well as the failure rates, weight, volume, and cost, are generally only known approximately at the beginning of a design; moreover, in some cases, we only know the function that the component must perform, not how that function will be implemented. Thus the range of possible parameters is often very broad, and to look for an exact optimization when the parameters are known only over a broad range may be an elegant mathematical formulation but is not a practical engineering solution. In fact, sometimes choosing the exact optimum can involve considerable risk if the solution is very sensitive to small changes in parameters.

To illustrate, let us assume that there are two design parameters, $x$ and $y$, and the resulting reliability is $z$. We can visualize the solution as a surface in $x$, $y$, $z$ space, where the reliability is plotted along the vertical $z$-axis as the two design parameters vary in the horizontal $xy$ plane. Thus our solution is a surface lying above the $xy$ plane and the height ($z$) of the surface is our reliability that ranges between 0 and unity. Suppose our surface has two maxima: one where the surface is a tall, thin spire with the reliability $zs = 0.98$ at the peak, which occurs at $xs$, $ys$, and the other where the surface is a broad one and where the reliability reaches $zb = 0.96$ at a small peak located at $xb$, $yb$ in the center of a broad plateau having a height of 0.94. Clearly, if we choose the spire as our design and if parameters $x$ or $y$ are a little different than $xs$, $ys$, the reliability may be much lower—below 0.96 and even below 0.94—because of the steep slopes on the flanks of the spire. Thus the maximum of 0.96 is probably a better design and has less risk, since even if the parameters differ somewhat from $xb$, $yb$, we still have the broad plateau where the reliability is 0.94. Most of the exact optimization techniques would choose the spire and not even reveal the broad peak and plateau as other possibilities, especially if the points $xs$, $ys$ and $xb$, $yb$ were well-separated. Thus it is important to find a means of calculating the sensitivity of the solution to parameter variations or calculating a range of good solutions close to the optimum.

There has been much emphasis in the theoretical literature on how to find an exact optimization. The brute force approach is to enumerate all possible combinations and calculate the resulting reliability; however, except for small problems, this approach requires long or intractable computations. An alternate approach uses dynamic programming to reduce the number of possible combinations that must be evaluated by breaking the main optimization into a sequence of carefully formulated suboptimizations [Bierman, 1969; Hiller, 1974; Messinger, 1970]. The approach that this chapter recommends is the use of a two-step procedure. We assume that the problem in question is a large system. Generally, at the top level of a large system, the problem can be modeled as a series connection of a number of subsystems. The process of apportionment (see Lloyd [1977, Appendix 9A]) is used to allocate the system reliability (or availability) goal among the various subsystems and is the first step of the procedure. This process should reduce a large problem into a number of smaller subproblems, the optimization of which we can approach by using a bounded enumeration procedure. One can greatly reduce the size of the solution space by establishing a sequence of bounds; the resulting subsystem optimization is well within the power of a modern PC, and solution times are reasonable. Of course, the first step in the process—that of apportionment—is generally a good one, but it is not necessarily an optimum one. It does, however, fit in well with the philosophy alluded to in the previous section that a broad, easy-to-achieve, easy-to-understand suboptimum is preferred in a practical case. As described later in this chapter, allocation tends to divert more resources to the "weakest link in the chain."

There are other important practical arguments for simplified semioptimum techniques instead of exact mathematical optimization. In practice, optimizing a design is a difficult problem for many reasons. Designers, often harried by schedule and costs, look for a feasible solution to meet the performance parameters; thus reliability may be treated as an afterthought. This approach seldom leads to a design with optimum reliability—much less a good suboptimal design. The opposite extreme is the classic optimization approach, in which a mathematical model of the system is formulated along with constraints on cost, volume, weight, and so forth, where all the allowable combinations of redundant parallel and standby components are permitted and where the underlying integer programming problem is solved. The latter approach is seldom taken for the previously stated reasons: (a) the system designer does not understand the mathematical formulation or the solution process; (b) the solution takes too long; (c) the parameters are not well known; and (d) the constraints rapidly change and invalidate the previous solution. Therefore, clear, simple, and rapid calculation of a family of good suboptimal solutions is a sensible approach. The study of this family should reveal which solutions, if any, are very sensitive to changes in the model parameters. Furthermore, the computations are simple enough that they can be repeated should significant changes occur during the design process. Establishing such a range of solutions is an ideal way to ensure that reliability receives adequate consideration among the various conflicting constraints and system objectives during the trade-off process—the preferred approach to choosing a good, well-balanced design.

## 7.3 A MATHEMATICAL STATEMENT OF THE OPTIMIZATION PROBLEM

One can easily define the classic optimization approach as a mathematical model of the system that is formulated along with constraints on cost, volume, weight, and so forth, in which all the allowable combinations of redundant parallel and standby components are permitted and the underlying integer programming problem must be solved.

We begin with a series model for the system with $k$ components where $x_1$ is the event success of element one, $\bar{x}_1$ is the event failure of element one, and $P(x_1) = 1 - P(\bar{x}_1)$ is the probability of success of element one, which is the reliability, $r_1$ (see Fig. 7.1). Clearly, the components in the foregoing mathematical model can be subsystems if we wish.

The system reliability is given by the probability of the event in which all the components succeed (the intersection of their successes):

$$R_s = P(x_1 \cap x_2 \cap \cdots \cap x_k) \qquad (7.1\text{a})$$

If we assume that all the elements are independent, Eq. (7.1a) becomes

## A MATHEMATICAL STATEMENT OF THE OPTIMIZATION PROBLEM     335

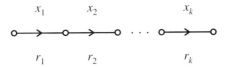

**Figure 7.1**  A series system of $k$ components.

$$R_s = \prod_{i=1}^{k} R_i \qquad (7.1\text{b})$$

We will let the single constraint on our design be the cost for illustrative purposes, and the total cost, $c$, is given by the sum of the individual component costs, $c_i$:

$$c = \sum_{i=1}^{k} c_i \qquad (7.2)$$

We assume that the system reliability given by Eq. (7.1b) is below the system specifications or goals, $R_g$, and that the designer must improve the reliability of the system to meet these specifications. (In the highly unusual case where the initial design exceeds the reliability specifications, the initial design can be used with a built-in safety factor, or else the designer can consider using cheaper shorter-lifetime parts to save money; the latter is sometimes a risky procedure.) We further assume that the maximum allowable system cost, $c_0$, is in general sufficiently greater than $c$ so that the funds can be expended (e.g., redundant components added) to meet the reliability goal. If the goal cannot be reached, the best solution is the one with the highest reliability within the allowable cost constraint.

In the case where more than one solution exceeds the reliability goal within the cost constraint, it is useful to display a number of "good" solutions. Since we wish the mathematical optimization to serve a practical engineering design process, we should be aware that the designer may choose to just meet the reliability goal with one of the suboptimal solutions and save some money. Alternatively, there may be secondary factors that favor a good suboptimal solution (e.g., the sensitivity and risk factors discussed in the preceding section).

There are three conventional approaches to improving the reliability of the system posed in the preceding paragraph:

1. Improve the reliability of the basic elements, $r_i$, by allocating some or all of the cost budget, $c_0$, to fund redesign for higher reliability.
2. Place components in parallel with the subsystems that operate contin-

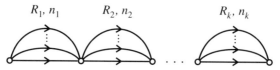

**Figure 7.2** The choice of redundant components to optimize the reliability of the series system of Fig. 7.1.

uously (see Fig. 7.2). This is ordinary parallel redundancy (hot redundancy).

3. Place components in parallel (standby) with the $k$ subsystems and switch them in when an on-line failure is detected (cold redundancy).

There are also strategies that combine these three approaches. Such combined approaches, as well as reliability improvement by redesign, are discussed later in this chapter and also in the problems. Most of the chapter focuses on the second and third approaches of the preceding list—hot and cold redundancy.

## 7.4 PARALLEL AND STANDBY REDUNDANCY

### 7.4.1 Parallel Redundancy

Assuming that we employ parallel redundancy (ordinary redundancy, hot redundancy) to optimize the system reliability, $R_s$, we employ $n_k$ elements in parallel to raise the reliability of each subsystem that we denote by $R_k$ (see Fig. 7.2).

The reliability of a parallel system of $n_k$ independent components is most easily formulated in terms of the probability of failure $(1 - r_i)^{n_i}$. For the structure of Fig. 7.2 where all failures are independent, Eq. (7.1b) becomes

$$R_s = \prod_{i=1}^{k} (1 - [1 - r_i]^{n_i}) \qquad (7.3)$$

and Eq. (7.2) becomes

$$c = \sum_{i=1}^{k} n_i c_i \qquad (7.4)$$

We can develop a similar formulation for standby redundancy.

### 7.4.2 Standby Redundancy

In the case of standby systems, it is well known that the probability of failure is governed by the Poisson distribution (see Section A5.4).

$$P(x;\mu) = \frac{\mu^x e^{-\mu}}{x!} \tag{7.5}$$

where

$x$ = the number of failures
$\mu$ = the expected number of failures

A standby subsystem succeeds if there are fewer failures than the number of available components, $x_k < n_k$; thus, for a system that is to be improved by standby redundancy, Eqs. (7.3) and (7.4) becomes

$$R_s = \prod_{i=1}^{k} \sum_{x_k=0}^{x_k=n_k-1} P(x_k;\mu_k) \tag{7.6}$$

and, of course, the system cost is still computed from Eq. (7.4).

## 7.5 HIERARCHICAL DECOMPOSITION

This section examines the way a designer deals with a complex problem and attempts to extract the engineering principles that should be employed. This leads to a number of viewpoints, from which some simple approaches emerge. The objective is to develop an approach that allows the designer to decompose a complex system into a manageable architecture.

### 7.5.1 Decomposition

Systems engineering generally deals with large, complex structures that, when taken as a whole (in the gestalt), are often beyond the "intellectual span of control." Thus the first principle in approaching such a design is to decompose the problem into a hierarchy of subproblems. This initial decomposition stops when the complexity of the resulting components is reduced to a level that puts it within the "intellectual span of control" of one manager or senior designer. This approach is generally called *divide and conquer* and is presented for use on complex problems in books on algorithms [Aho, 1974, p. 60; Cormen, 1992, p. 12]. The term probably comes from the ancient political maxim *divide et impera* ("divide and rule") cited by Machiavelli [Bartlett, 1968, p. 150b], or possibly early principles of military strategy.

### 7.5.2 Graph Model

Although the decomposition of a large system is generally guided by experience and intuition, there are some guidelines that can be used to guide the process. We begin by examining the structure of the decomposition. One can

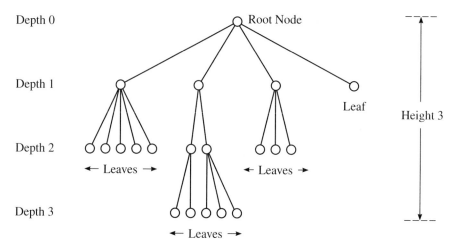

**Figure 7.3** A tree model of a hierarchical decomposition illustrating some graph nomenclature.

describe a hierarchical block diagram of a system in more precise terms if we view it as a mathematical *graph* [Cormen, 1992, pp. 93–94]. We replace each box in the block diagram by a *vertex (node)* and leaving the connecting lines that form the *edges (branches)* of the graph. Since information can flow in both directions, this is an *undirected* graph; if information can flow in only one direction, however, the graph is a *directed graph*, and an arrowhead is drawn on the edge to indicate the direction. A *path* in the graph is a continuous sequence of vertices from the *start vertex* to the *end vertex*. If the end vertex is the same as the start vertex, then this (closed) path is called a *cycle (loop)*. A graph without cycles where all the nodes are connected is called a *tree* (the graph corresponding to a hierarchical block diagram is a tree). The top vertex of a tree is called the *root (root node)*. In general, a node in the tree that corresponds to a component with subcomponents is called a *parent* of the subcomponents, which are called *children*. The root node is considered to be at *depth 0 (level 0)*; its children are at *depth 1 (level 1)*. In general, if a parent node is at level $n$, then its children are at level $n + 1$. The largest depth of any vertex is called the *depth of the tree*. The number of children that a parent has is the *out-degree*, and the number of parents connected to a child is the *in-degree*. A node that has no children is the end node (terminal node) of a path from the root node and is called a *leaf node (external node)*. Nonleaf nodes are called *internal nodes*. An example illustrating some of this nomenclature is given in Fig. 7.3.

### 7.5.3 Decomposition and Span of Control

If we wish our decomposition to be modeled by a tree, then the in-degree must always be one to prevent cycles or inputs to a stage entering from more than

# HIERARCHICAL DECOMPOSITION

one stage. Sometimes, however, it is necessary to have more than one input to a node, in which case one must worry about synchronization and coupling between the various nodes. Thus, if node $x$ has inputs from nodes $p$ and $q$, then any change in either $p$ or $q$ will affect node $x$. Imposing this restriction on our hierarchical decomposition leads to simplicity in the interfacing of the various system elements.

We now discuss the appropriate size of the out-degree. If we wish to decompose the system, then the minimum size of the out-degree at each node must be two, although this will result in a tree of great height. Of course, if any node has a great number of children (a large out-degree), we begin to strain the intellectual span of control. The experimental psychologist Miller [1956] studied a large number of experiments related to sensory perception and concluded that humans can process about 5–9 levels of "complexity." (A discussion of how Miller's numbers relate to the number of mental discriminations that one can make appears in Shooman [1983, pp. 194, 195].) If we specify the out-degree to be seven for each node and all the leaves (terminal nodes) to be at level (depth) $h$, then the number of leaves at level $h$ ($NL_h$) is given by

$$NL_h = 7^h \tag{7.7}$$

In practice, each leaf is the lowest level of replaceable unit, which is generally called a line replaceable unit (LRU). In the case of software, we would probably call the analog of an LRU a module or an object. The total number of nodes, $N$, in the graph can be computed if we assume that all the leaves appear at level $h$.

$$N = NL_0 + NL_1 + NL_2 + \cdots + NL_h \tag{7.8a}$$

If each parent node has seven children, Eq. (7.8a) becomes

$$N = 1 + 7 + 7^2 + \cdots + 7^h \tag{7.8b}$$

Using the formula for the sum of the terms in a geometric progression,

$$N = a(r^n - 1)/(r - 1) \tag{7.9a}$$

where

$r = $ the common ratio (in our case, 7)
$n = $ the number of terms (in our case, $h + 1$)
$a = $ the first term (in our case, 1)

Substitution in Eq. (7.9a) yields

$$N = (7^{h+1} - 1)/6 \tag{7.9b}$$

If $h = 2$, we have $N = (7^3 - 1)/6 = 57$. We can check this by substitution in Eq. (7.8b), yielding $1 + 7 + 49 = 57$.

### 7.5.4 Interface and Computation Structures

Another way of viewing a decomposition structure is to think in terms of two classes of structures, interfaces, and computational elements—a breakdown that applies to either hardware or software. In the case of hardware, the computational elements are LRUs; for software, they are modules or classes. In the case of hardware, the interfaces are analog or digital signals (electrical, light, sound) passed from one element (depth, level) to another; the joining of mechanical surfaces, hydraulics or pneumatic fluids; or similar physical phenomena. In the case of software, the interfaces are generally messages, variables, or parameters passed between procedures or objects. Both hardware and software have errors (failure rates, reliability) associated with either the computational elements or the interfaces. If we again assume that leaves appear only at the lowest level of the tree, the number of computational elements is given by the last term in Eq. (7.8a), $NL_h$. In counting interfaces, there is the interface out of an element at level $i$ and the interface to the corresponding element at level $i + 1$. In electrical terms, we might call this the output impedance and the corresponding input impedance. In the case of software, we would probably be talking about the passing of parameters and their scope between a procedure call and the procedure that is called, or else the passing of messages between classes and objects. For both hardware and software, we count the interface (information-out–information-in) pair as a single interface. Thus all modules except level 0 have a single associated interface pair. There is no structural interface at level 0; however, let us consider the *system specifications* as a single interface at level 0. Thus, we can use Eqs. (7.8) and (7.9) to count the number of interfaces, which is equivalent to the number of elements. Continuing the foregoing example where $h = 2$, we have $7^2 = 49$ computational elements and $(7^3 - 1)/6 = 57$ interfaces. Of course, in a practical example, not all the leaves will appear at depth (level) $h$, since some of the paths will terminate before level $h$; thus the preceding computations and formulas can only be considered upper bounds on an actual (less-idealized) problem.

One can use these formulas for many interfaces and computational units to conjecture models for complexity, errors, reliability, and cost.

### 7.5.5 System and Subsystem Reliabilities

The structure of the system at level 1 in the graph model of the hierarchical decomposition is a group of subsystems equal in number to the out-degree of the root node. Based on Miller's work, we have decided to let the out-degree be 7 (or 5 to 9). As an example, let us consider an overview of an air traffic control (ATC) system for an airport [Gilbert, 1973, p. 39, Fig. 61]. Level 0 in our decomposition is the "air traffic control system." At level 1, we have the major subsystems that are given in Table 7.1.

An expert designer of a new ATC system might view things a little differently (in fact, two expert designers working for different companies might

### HIERARCHICAL DECOMPOSITION 341

**TABLE 7.1  A Typical Air Traffic Control System at Level 1**

- Tracking radar and associated computer.
- Air traffic control (ATC) displays and display computers.
- Voice communications with pilot.
- Transponders on the aircraft (devices that broadcast a digital identification code and position information).
- Communications from other ATC centers.
- Weather information.
- The main computer.

each come up with a slightly different model even at level 1), but the list in Table 7.1 is sufficient for our discussions. We let $X_1$ represent the success of the tracking radar, $X_2$ represent the success of the controller displays, and so on up to $X_7$, which represents the success of the main computer. We can now express the reliability of the system in terms of events $X_1$–$X_7$. At this high a level, the system will only succeed if all the subsystems succeed; thus the system reliability, $R_s$, can be expressed as

$$R_s = P(X_1 \cap X_2 \cap \cdots \cap X_7) \tag{7.10}$$

If the seven aforementioned subsystems are statistically independent, then Eq. (7.10) becomes

$$R_s = P(X_1)P(X_2)\cdots P(X_7) \tag{7.11}$$

In all likelihood, the independent assumption at this high level is valid; it is unlikely that one could postulate mechanisms whereby the failure of the tracking radar would cause failure of the controller displays. The common mode failure mechanisms that would lead to dependence (such as a common power system or a hurricane) are quite unlikely. System designers would be aware that a common power system is a vulnerable point and therefore would not design the system with this feature. In all likelihood, the systems will have independent computer systems. Similarly, it is unlikely that a hurricane would damage both the tracking radar and the controller displays; the radar should be designed for storm resistance, and the controller displays should be housed in a stormproof building; moreover, the occurrence of a hurricane should be much less frequent than that of other possible forms of failure modes. Thus it is a reasonable engineering assumption that statistical independence exists, and Eq. (7.11) is a valid simplification of Eq. (7.10).

Because of the nature of the probabilities, that is, they are bounded by 0 and 1, and also because of the product nature of Eq. (7.11), we can bound each of the terms. There is an infinite number of values of $P(X_1), P(X_2), \ldots, P(X_7)$ that satisfies Eq. (7.11); however, the smallest value of $P(X_1)$ occurs when

$P(X_2), \ldots, P(X_7)$ assume their largest values—that is, unity. We can repeat this solution for each of the subsystems to yield a set of minimum values.

$$P(X_1) \geq R_s \qquad (7.12a)$$

$$P(X_2) \geq R_s \qquad (7.12b)$$

and so on up to

$$P(X_7) \geq R_s \qquad (7.12c)$$

These minimum bounds are true in general for any subsystem if the system structure is series; thus we can write

$$P(X_i) \geq R_s \qquad (7.13)$$

The equality only holds in Eqs. (7.12) and (7.13) if all the other subsystems have a reliability equal to unity (i.e., they never fail); thus, in the real world, the equality conditions can be deleted. These minimum bounds will play a large role in the optimization technique developed later in this chapter.

## 7.6 APPORTIONMENT

As was discussed in the previous section, one of the first tasks in approaching the design of a large, complex system is to decompose it. Another early task is to establish reliability allocations or budgets for the various subsystems that emerge during the decomposition, a process often referred to as *apportionment* or *allocation*. At this point, we must discuss the difference between a mathematician's and an engineer's approach to optimization. The mathematician would ask for a precise system model down to the LRU level, the failure rate, and cost, weight, volume, etc., of each LRU; then, the mathematician invokes an optimization procedure to achieve the exact optimization. The engineer, on the other hand, knows that this is too complex to calculate and understand in most cases and therefore seeks an alternate approach. Furthermore, the engineer knows that many of the details of lower-level LRUs will not be known until much later and that estimates of their failure rates at that point would be rather vague, so he or she adopts a much simpler design approach: beginning a top–down process to apportion the reliability goal among the major subsystems at depth 1.

Apportionment has historically been recognized as an important reliability system goal [AGREE Report, 1957, pp. 52–57; Henney, 1956, Chapter 1; Von Alven, 1964, Chapter 6]; many of the methods discussed in this section are an outgrowth of this early work. We continue to assume that there are about 7 subsystems at depth 1. Our problem is how to allocate the reliability goal among the subsystems, for which several procedures exist on which to base such an allocation early in the design process; these are listed in Table 7.2.

**TABLE 7.2  Approaches to Apportionment**

| Approach | Basis | Comments |
|---|---|---|
| Equal weighting | All subsystems should have the same reliability. | Easy first attempt. |
| Relative difficulty | Some knowledge of relative cost or difficulty to improve subsystem reliability. | Heuristic method requiring only approximate ordering of cost of difficulty. |
| Relative failure rates | Requires some knowledge of the relative subsystem failure rates. | Easier to use than the relative difficulty method. |
| Albert's method | Requires an initial estimate of the subsystem reliabilities. | A well-defined algorithm is used that is based on assumptions about the improvment-effort function. |
| Stratified optimization | Requires detailed model of the subsystem. | Discussed in Section 7.6.5. |

### 7.6.1  Equal Weighting

The simplest approach to apportionment is to assume equal subsystem reliability, $r$. In such a case, Eq. (7.11) becomes

$$R_s = P(X_1)P(X_2)\cdots P(X_7) = r^7 \tag{7.14a}$$

For the general case of $n$ independent subsystems in series,

$$R_s = r^n \tag{7.14b}$$

Solving Eq. (7.14a) for $r$ yields

$$r = (R_s)^{1/7} \tag{7.15a}$$
$$r = (R_s)^{1/n} \tag{7.15b}$$

This equal weighting apportionment is so simple that it is probably one of the first computations made in a project. System engineers typically "whip out" their calculators and record such a calculation on the back of an envelope or a piece of scrap paper during early discussions of system design.

As an example, suppose that we have a system reliability goal of 0.95, in which case Eq. (7.15a) would yield an apportioned goal of $r = 0.9927$. Of course, it is unlikely that it would be equally easy or costly to achieve the apportioned goal of 0.9927 for each of the subsystems. Thus this method gives a ballpark estimate, but not a lot of time should be spent using it in the design before a better method replaces it.

## 7.6.2 Relative Difficulty

Suppose that we have some knowledge of the subsystems and can use it in the apportionment process. Assume that we are at level 1, that we have seven subsystems to deal with, and that we know for three of the subsystems achieving a high level of reliability (e.g., the level required for equal apportionment) will be difficult. We envision that these three systems could meet their goals if they can be realized by two parallel elements. We then would have reliability expressions similar to those of Eq. (7.14b) for the four "easier" systems and a reliability expression $2r - r^2$ for the three "harder systems." The resultant expression is

$$R_s = r^4(2r - r^2)^3 \tag{7.16}$$

Solving Eq. (7.16) numerically for a system goal of 0.95 yields $r = 0.9874$. Thus the four "easier" subsystems would have a single system with a reliability of 0.9874, and the three harder systems would have two parallel systems with the same reliability. Another solution is to keep the goal of $r = 0.9927$, calculated previously for the easier subsystems. Then, the three harder systems would have to meet the goal of $0.95/0.99274 = 0.9783$. The three harder systems would have to meet a somewhat lower goal: $(2r - r^2)^3 = 0.9783$, or $r = 0.953$. Other similar solutions can easily be obtained.

The previous paragraph dealt with unequal apportionments by considering a parallel system for the three harder systems. If we assume that parallel systems are not possible at this level, we must choose a solution where the easier systems exceed a reliability of 0.9927 so that the harder systems can have a smaller reliability. For convenience, we could rewrite Eq. (7.11) in terms of unreliabilities, $r_i = 1 - u_i$, obtaining

$$R_s = (1 - u_1)(1 - u_2) \cdots (1 - u_7) \tag{7.17a}$$

Again, suppose there are four easier systems with a failure probability of $u_1 = u_2 = u_3 = u_4 = u$. The harder systems will have twice the failure probability $u_5 = u_6 = 2u$, and Eq. (7.17a) becomes

$$R_s = (1 - u)^4(1 - 2u)^3$$

that yields a 7th-order polynomial.

The easiest way to solve the polynomial is through trial and error with a calculator or by writing a simple program loop to calculate a range of values. The equal reliability solution was $r = 0.9927 = 1 - 0.0073$. If we try $r_{\text{easy}} = 0.995 = 1 - 0.005$, $r_{\text{hard}} = 0.99 = 1 - 0.01$, and substitute in Eq. (7.17a), the result is

$$0.951038 = (0.995)^4(0.99)^3 \tag{7.17b}$$

Trying some slightly larger values of $u$ results in a solution of

$$0.950079 = (0.9949)^4(0.9898)^3 \qquad (7.17c)$$

The accuracy of this method depends largely on how realistic the guesses are regarding the hard and easy systems. The method of the next section is similar, but the calculations are easier.

### 7.6.3 Relative Failure Rates

It is simpler to use knowledge about easier and harder systems during apportionment if we work with failure rates. We assume that each subsystem has a constant-failure rate $\lambda_i$ and that the reliability for each subsystem is given by

$$r_i = e^{-\lambda_i} \qquad (7.18a)$$

and substitution of Eq. (7.18a) into Eq. (7.11) yields

$$R_s = P(X_1)P(X_2)\cdots P(X_7) = e^{-\lambda_1}e^{-\lambda_2}\cdots e^{-\lambda_7} \qquad (7.18b)$$

and Eq. (7.18b) can be written as

$$R_s = e^{-\lambda_s} \qquad (7.19)$$

where

$$\lambda_s = \lambda_1 + \lambda_2 + \cdots + \lambda_7$$

Continuing with our example of the previous section, in which the goal is 0.95, the four "easier" systems have a failure rate of $\lambda$, and the three harder ones have a failure rate of $5\lambda$, Eq. (7.19) becomes

$$R_s = 0.95 = e^{-19\lambda t} \qquad (7.20)$$

Solving for $\lambda t$, we obtain $\lambda t = 0.0026996$, and the reliabilities are $e^{-0.0026996} = 0.9973$, and $e^{-5 \times 0.0026996} = 0.9865$. Thus our apportioned goals for the four easier systems are 0.9973; for the three harder systems, 0.9865. As a check, we see that $0.9973^4 \times 0.9865^3 = 0.9497$. Clearly, one can use this procedure to achieve other allocations based on some relative knowledge of the nominal failure rates of the various subsystems or on how difficult it is to achieve various failure rates.

### 7.6.4 Albert's Method

A very interesting method that results in an algorithm rather than a design procedure is known as *Albert's method* and is based on some analytical principles [Albert, 1958; Lloyd, 1977, pp. 267–271]. The procedure assumes that initially there are some estimates of what reliabilities can be achieved for the subsystems to be apportioned. In terms of our notation, we will say that $P(X_1)$, $P(X_2), \ldots, P(X_7)$ are given by some nominal values: $R_1, R_2, \ldots, R_7$. Note that we continue to speak of seven subsystems at level 1; however, this clearly can be applied to any number of subsystems. The fact that we assume nominal values for the $R_i$ implies that we have a preliminary design. However, in any large system there are many iterations in system design, and this method is quite useful even if it is not the first one attempted. Adopting the terminology of government contracting (which generally has parallels in the commercial world), we might say that the methods of Sections 7.6.1–7.6.3 are useful in formulating the request for proposal (RFQ) (the requirements) and that Albert's method is useful during the proposal preparation phase (specifications and proposed design) and during the early design phases after the contract award. A properly prepared proposal will have some early estimates of the subsystem reliabilities. Furthermore, we assume that the system specification or goal is denoted by $R_g$, and the preliminary estimates of the subsystem reliabilities yield a system reliability estimate given by

$$R_s = R_1 R_2 \cdots R_7 \qquad (7.21)$$

If the design team is lucky, $R_s > R_g$, and the first concerns about reliability are thus satisfied. In fact, one might even think about trading off some reliability for reduced cost. An experienced designer would tell us that this almost never happens and that we are dealing with the situation where $R_s < R_g$. This means that one or more of the $R_i$ values must be increased. Albert's method deals with finding which of the subsystem reliability goals must be increased and by how much so that $R_s$ is increased to the point where $R_s = R_g$.

Based on the bounds developed in Eq. (7.13), we can comment that any subsystem reliability that is less than the system goal, $R_i < R_g$, must be increased (others may also need to be increased). For convenience in developing our algorithm, we assume that the subsystems have been renumbered so that the reliabilities are in ascending order: $R_1 < R_2 < \cdots < R_7$. Thus, in the special case where $R_7 < R_g$, all the subsystem goals must be increased. In this case, Albert's method reduces to equal apportionment and Eqs. (7.14) and (7.15) hold. In the more general case, $j$ of the $i$ subsystems must have the reliability increased. Albert's method requires that all the $j$ subsystems have their reliability increased to the same value, $r$, and that the reliabilities of the $(i - j)$ subsystems remain unchanged. Thus, Eq. (7.21) becomes

$$R_g = R_1 R_2 \cdots R_j R_{j+1} \cdots R_7 \qquad (7.22)$$

where

$$R_1 = R_2 = \cdots = R_j = r \qquad (7.23)$$

Substitution of Eq. (7.23) into Eq. (7.22) yields

$$R_g = (r^j)(R_{j+1} \cdots R_7) \tag{7.24}$$

We solve Eq. (7.24) for the value of $r$ (or, more generally, $r_i$):

$$r^j = R_g/(R_{j+1} \cdots R_7) \tag{7.25a}$$
$$r = (R_g/[R_{j+1} \cdots R_7])^{1/j} \tag{7.25b}$$

Equations (7.22)–(7.25) describe Albert's method, but an important step must still be discussed: how to determine the value of $j$. Again, we turn to Eq. (7.13) to shed some light on this question. We can place a lower bound on $j$ and say that all the subsystems having reliabilities smaller than or equal to the goal, $R_i < R_g$, must be increased. It is possible that if we choose $j$ equal to this lower bound and substitute into Eq. (7.25b), the computed value of $r$ will be >1, which is clearly impossible; thus we must increase $j$ by 1 and try again. This process is repeated until the values of $r$ obtained are <1. We now have a feasible value for $j$, but we may be requiring too much "effort" to raise all the 1 through $j$ subsystems to the resulting high value of $r$. It may be better to increment $j$ by 1 (or more), reducing the value of $r$ and "spreading" this value over more subsystems. Albert showed that based on certain effort assumptions, the optimum value of $j$ is bounded from above when the value for $r$ first decreases to the point where $r < R_j$. The optimum value of $j$ is the previous value of $j$, where $r > R_j$. More succinctly, the optimum value for $j$ is the largest value for $j$, where $r > R_j$. Clearly it is not too hard to formulate a computer program for this algorithm; however, since we are assuming about seven systems and have bounded $j$ from below and above, the most efficient solution is probably done with paper, pencil, and a scientific calculator.

The reader may wonder why we have spent quite a bit of time explaining Albert's method rather than just stating it. The original exposition of the method is somewhat terse, and the notation may be confusing to some; thus the enhanced development is warranted. The remainder of this section is devoted to an example and a discussion of when this method is "optimum." The reader will note that some of the underlying philosophy behind the method can be summarized by the following principle: "The most efficient way to improve the reliability of a series structure (sometimes called a chain) is by improving the weakest links in the chain." This principle will surface a few more times in later portions of this chapter.

A simple example should clarify the procedure. Suppose that we have four subsystems with initial reliabilities $R_1 = 0.7$, $R_2 = 0.8$, $R_3 = 0.9$, and $R_4 = 0.95$, and the system reliability goal is $R_g = 0.8$. The existing estimates predict a system reliability of $R_s = 0.7 \times 0.8 \times 0.9 \times 0.95 = 0.4788$. Clearly, some or all of the subsystem goals must be raised for us to meet the system goal. Based on Eq. (7.13), we know that we must improve subsystems 1 and 2, so we begin our calculations at this point. The system reliability goal, $R_g = 0.8$, and Eq. (7.25b) yields

$$r = (R_g/[R_{j+1}\cdots R_7])^{1/j} = (0.8/0.9 \times 0.95)^{1/2} = (0.93567)^{0.5} = 0.96730 \quad (7.26)$$

Since $0.96730 > 0.9$, we continue our calculation. We now recompute for subsystems 1, 2, and 3, and Eq. (7.25b) yields

$$r = (0.8/0.95)^{1/3} = 0.9443 \quad (7.27)$$

Now, $0.9443 < 0.95$, and we choose the previous value of $j = 2$ as our optimum. As a check, we now compute the system reliability.

$$0.96730 \times 0.96730 \times 0.9 \times 0.95 = 0.7999972 = R_g$$

which equals our goal of 0.8 when rounded to one place. Thus, the conclusion from the use of Albert's method is that the apportionment goals for the four systems are $R_1 = R_2 = 0.96730$; $R_3 = 0.90$; and $R_4 = 0.95$. This solution assumes equal effort for improving the reliability of all subsystems.

The use of Albert's method produces an optimum allocation policy if the following assumptions hold [Albert, 1958; Lloyd, 1977, pp. 267–271]:

1. Each subsystem has the same effort function that governs the amount of effort required to raise the reliability of the $i$th subsystem from $R_i$ to $r_i$. This effort function is denoted by $G(R_i, r_i)$, and increased effort always increases the reliability: $G(R_i, r_i) \geq 0$.
2. The effort function $G(x, y)$ is nondecreasing in $y$ for fixed $x$, that is, given an initial value of $R_i$, it will always require more effort to increase $r_i$ to a higher value. For example,

$$G(0.35, 0.65) \leq G(0.35, 0.75)$$

The effort function $G(x, y)$ is nonincreasing in $x$ for fixed $y$, that is, given an increase to $r_i$, it will always require less effort if we start from a larger value of $R_i$. For example,

$$G(0.25, 0.65) \geq G(0.35, 0.65)$$

3. If $x \leq y \leq z$, then $G(x, y) + G(y, z) = G(x, z)$. This is a superposition (linearity) assumption that states that if we increase the reliability in two steps, the sum of the efforts for each step is the same as if we did the increase in a single step.
4. $G(0, x)$ has a derivative $h(x)$ such that $xh(x)$ is strictly increasing in ($0 < x < 1$).

The proof of the algorithm is given in Albert [1958]. If the assumptions of Albert's method are not met, the equal effort rule is probably violated, for which the methods of Sections 7.6.2 and 7.6.3 are suggested.

### 7.6.5 Stratified Optimization

In a very large system, we might consider continuing the optimization to level 2 by applying apportionment again to each of the subsystem goals. In fact, we can continue this process until we reach the LRU level and then utilize Eqs. (7.3) or (7.6) (or else improve the LRU reliability) to achieve our system design. Such decisions require some intuition and design experience on the part of the system engineers; however, the foregoing methods provide some engineering calculations to help guide intuition.

### 7.6.6 Availability Apportionment

Up until now, the discussion of apportionment has dealt entirely with system and subsystem reliabilities. Now, we discuss the question of how to proceed if system availabilities are to be apportioned. Under certain circumstances, the subsystem availabilities are essentially independent, and the system availability is given by the same formula as for the reliabilities, with the availabilities replacing the reliabilties. A discussion of availability modeling in general, and a detailed discussion of the circumstances under which such substitutions are valid appears in Shooman [1990, Appendices F]. One situation in which the availabilities are independent is where each subsystem has its own repairman (or repaircrew). This is called *repairman decoupling* in Shooman [1990, Appendices F-4 and F-5]. In the decoupled case, one can use the same system structural model that is constructed for reliability analysis to compute system availability. The steady-state availability probabilities are substituted in the model just as the reliability probabilities would be. Clearly, this is a convenient situation and is *often, but not always, approximately valid.*

Suppose, however, that the same repairman or repaircrew serves one or more subsystems. In such a case, there is the possibility that a failure will occur in subsystem $y$ while the repairman is still busy working on a repair for subsystem $x$. In such a case, a queue of repair requests develops. The queuing phenomena result in dependent coupled subsystems that can be denoted as being *repairman coupling*. When repairman coupling is significant, one should formulate a Markov model to solve for the resulting availability. Since Markov modeling for a large subsystem can be complicated, as the reader can appreciate from the analytical solutions of Chapter 3, a practical designer would be happy to use a decoupled solution even if the results were only a good approximation.

Intuition tells us that the possibility of a queue forming is a function of the ratio of repair rate to failure rate $(\lambda/\mu)$. If the repair rate is much larger than the failure rate, the approximation should be quite satisfactory. These approximations were explored extensively in Section 4.9.3, and the reader should review the results. We can explore the decoupled approximation again by considering a slightly different problem than that in Chapter 4: two series subsystems that are served by the same repairman. Returning to the results derived in Chapter 3, we can compute the exact availability using the model given in Fig. 3.16 and Eqs. (3.71a–c). This model holds for two identical elements (series, parallel, and standby). If we want the model to hold for two series subsystems, we must compute the probability that both elements are good, which is $P_{s_0}$. We can compute the steady-state solution by letting $s \to 0$ in Eqs. (3.71a–c), as was discussed in Chapter 3, and solving the resulting equations. The result is

$$A_\infty = P_{s_0} = \frac{\mu' \mu''}{\lambda \lambda' + \lambda' \mu'' + \mu' \mu''} \qquad (7.28a)$$

This result is derived in Shooman [1990, pp. 344–346]. For ordinary (not standby) two-element systems, $\lambda' = 2\lambda$ and $\mu' = \mu'' = \mu$. Substitution yields

$$A_\infty = \frac{\mu^2}{2\lambda^2 + 2\lambda\mu + \mu^2} \qquad (7.28b)$$

The approximate result is given by the probability that both elements are up, which is the product of the steady-state availability for a single element $\mu/(\lambda + \mu)$:

$$A_\infty \approx \frac{\mu}{\mu + \lambda} \cdot \frac{\mu}{\mu + \lambda} \qquad (7.29)$$

We can compare the two expressions for various values of $(\mu/\lambda)$ in Table 7.3, where we have assumed that the values of $\mu$ and $\lambda$ for the two elements are identical. From the third column in Table 7.3, we see that the ratio of the approximate unavailability $(1 - A\approx)$ to the exact unavailability $(1 - A=)$ approaches unity and is quite acceptable in all the cases shown. Of course, one might check the validity of the approximation for more complex cases; however, the results are quite encouraging, and we anticipate that the approximation will be applicable in many cases.

**TABLE 7.3  Comparison of Exact and Approximate Availability Formulas**

| Ratio $\mu/\lambda$ | Approximate Formula: Eq. (7.30), $A \approx$ | Exact Formula: Eq. (29b), $A =$ | Ratio of Unavailability: $(1 - A \approx)/(1 - A =)$ |
|---|---|---|---|
| 1 | 0.25 | 0.20 | 0.94 |
| 10 | 0.826496 | 0.819672 | 0.96 |
| 100 | 0.9802962 | 0.980199961 | 0.995 |

### 7.6.7 Nonconstant-Failure Rates

In many cases, the apportionment approaches discussed previously depend on constant-failure rates (see especially Table 7.2, third row). If the failure rates vary with time, it is possible that the optimization results will hold only over a certain time range and therefore must be recomputed for other ranges. The analyst should consider this approach if nonconstant-failure rates are significant. In most cases, detailed information on nonconstant-failure rates will not be available until late in design, and approximate methods using upper and lower bounds on failure rates or computations for different ranges assuming linear variation will be adequate.

## 7.7 OPTIMIZATION AT THE SUBSYSTEM LEVEL VIA ENUMERATION

### 7.7.1 Introduction

In the previous section, we introduced apportionment as an approximate optimization procedure at the system level. Now, we assume that we are at the subsystem level. At this point, we assume that each subsystem is at a level where we can speak of subsystem redundancy and where we can now consider exact optimization. (It is possible that in some smaller problems, the use of apportionment at the system level as a precursor is not necessary and we can begin exact optimization at this level. Also, it is possible that we are dealing with a system that is so complex that we have to apportion the subsystems into sub-subsystems—or even lower—before we can speak of redundant elements.) In all cases, we view apportionment as an approximate optimization process, which may or may not come first.

The subject of system optimization has been extensively discussed in the reliability literature [Barlow, 1965, 1975; Bellman, 1958; Messinger, 1970; Myers, 1964; Tillman, 1980] and also in more general terms [Aho, 1974; Bellman, 1957; Bierman, 1969; Cormen, 1992; Hiller, 1974]. The approach used was generally dynamic programming or greedy methods; these approaches will be briefly reviewed later in this chapter. This section will discuss a *bounded enumeration approach* [Shooman and Marshall, 1993] that the author proposes as the simplest and most practical method for redundancy optimization. We begin our development by defining the brute force approach of *exhaustive enumeration*.

### 7.7.2 Exhaustive Enumeration

This approach is straightforward, but it represents a brute force approach to the problem. Suppose we have subsystem $i$ that has five elements and we wish to improve the subsystem reliabiity to meet the apportioned subsystem goal $R_g$. If practical considerations of cost, weight, or volume limit us to choosing at most a single parallel subsystem for each of the five elements, each of the

## 352  RELIABILITY OPTIMIZATION

five subsystems has zero or one element in parallel, and the total number of possibilities is $2^5 = 32$. Given the powerful computational power of a modern personal computer, one could certainly write a computer program and evaluate all 32 possibilities in a short period of time. The designer would then choose the combination with the highest reliability or some other combination of good properties and use the complete set of possibilities as the basis of design. As previously stated, sometimes a close suboptimum solution is preferred because of risk, uncertainty, sensitivity, or other factors. Suppose we could consider at most two parallel subsystems for each of the five elements, in which case the total number of possibilities is $3^5 = 243$. This begins to approach an unwieldy number for computation and interpretation.

The actual number of computations involved in exhaustive enumeration is much larger if we do not impose a restriction such as "considering at most two parallel subsystems for each of the five elements." To illustrate, we consider the following two examples [Shooman, 1994]:

*Example 1:* The initial design of a system yields 3 subsystems at the first level of decomposition. The system reliability goal, $R_g$, is 0.9 for a given number of operating hours. The initial estimates of the subsystem reliabilities are $R_1 = 0.85$, $R_2 = 0.5$, and $R_3 = 0.3$. Parallel redundancy is to be used to improve the initial design so that it meets the reliability goal. The constraint is cost; each subsystem is assumed to cost 1 unit, and the total cost budget is 16 units.

The existing estimates predict an initial system reliability of $R_{s_0} = 0.85 \times 0.5 \times 0.3 = 0.1275$. Clearly, some or all of the subsystem reliabilities must be raised for us to meet the system goal. Lacking further analysis, we can state that the initial system costs 3 units and that 13 units are left for redundancy. Thus we can allocate 0 or 1 or 2 or any number up to 13 parallel units to subsystem 1, a similar number to subsystem 2, and a similar number to subsystem 3. An upper bound on the number of states that must be considered would therefore be $14^3 = 2,744$. Not all of these states are possible because some of them violate the weight constraint; for example, the combination of 13 parallel units for each of the 3 subsystems costs 39 units, which is clearly in excess of the 13-unit budget. However, even the actual number will be too cumbersome if not too costly in computer time to deal with. In the next section, we will show that by using the bounded enumeration technique, only 10 cases must be considered!

*Example 2:* The initial design of a system yields 5 subsystems at the first level of decomposition. The system reliability goal, $R_g$, is 0.95 for a given number of operating hours. The initial estimates of the subsystem reliabilities are $R_1 = 0.8$, $R_2 = 0.8$, $R_3 = 0.8$, $R_4 = 0.9$, and $R_5 = 0.9$. Parallel redundancy is to be used to improve the initial design so that it meets the reliability goal. The constraint is cost; the subsystems are assumed to cost 2, 2, 2, 3, and 3 units, respectively, and the total cost budget is 36 units.

The existing estimates predict an initial system reliability of $R_{s_0} = 0.8^3 \times 0.9^2 = 0.41472$. Clearly, some or all of the subsystem reliabilities must be raised for us to meet the system goal. Lacking further analysis, we can state that the initial system costs 12 units; thus we can allocate up to 24 cost units to each of the subsystems. For subsystems 1, 2, and 3, we can allocate 0 or 1 or 2 or any number up to 12 parallel units. For subsystems 4 and 5, we can allocate 0 or 1 or 2 or any number up to 8 parallel units. Thus an upper bound on the number of states that must be considered would be $13^3 \times 9^2 = 177{,}957$. Not all of these states are possible because some of them violate the cost constraint. In the next section, we will show that by using the bounded enumeration technique, only 31 cases must be considered!

Now, we begin our discussion of the significant and simple method of optimization that results when we apply bounds to constrain the enumeration process.

## 7.8 BOUNDED ENUMERATION APPROACH

### 7.8.1 Introduction

An analyst is often so charmed by the neatness and beauty of a closed-form synthesis process that they overlook the utility of an enumeration procedure. Engineering design is inherently a trial-and-error iterative procedure, and seldom are the parameters known so well that an analyst can stand behind a design and defend it as the true optimum solution to the problem. In fact, presenting a designer with a group of good designs rather than a single one is generally preferable since there may be many ancillary issues to consider in making a choice. Some of these issues are the following:

- Sensitivity to variations in the parameters that are only approximately known.
- Risk of success for certain state-of-the-art components presently under design or contemplated.
- Preferences on the part of the customer. (The old cliché about the "Golden Rule"—he who has the gold makes the rules—really does apply.)
- Conflicts between designs that yield high reliability but only moderate availability (because of repairability problems), and the converse.
- Effect of maintenance costs on the chosen solution.
- Difficulty in mathematically including multiple prioritized constraints (some independent multiple constraints are easy to deal with; these are discussed below).

Of course, the main argument against generating a family of designs and choosing among them is the effort and confusion involved in obtaining such a family. The prediction of the number of cases needed for direct enumeration in the two simple examples discussed previously are not encouraging. However, we will now show that the adoption of some simple lower- and upper-bound procedures greatly reduces the number of cases that need to be considered and results in a very practical and useful approach.

### 7.8.2 Lower Bounds

The discussion following Eqs. (7.1) and (7.2) pointed out that there is an infinite number of solutions that satisfy these equations. However, once we impose the constraint that the individual subsystems are made up of a finite number of parallel (hot or cold) systems, the problem becomes integer rather than continuous in nature, and a finite but still large number of solutions exists. Our task is to eliminate as many of the infeasible combinations as we can in a manner as simple as possible. The lower bounds on the system reliability developed in Eqs. (7.11), (7.12) and (7.13) allow us to eliminate a large number of combinations that constitute infeasible solutions. These bounds, powerful though they may be, merely state the obvious—that the reliability of a series of independent subsystems yields a product of probabilities and, since each probability has an upper bound of unity, that each subsystem reliability must equal or exceed the system goal. To be practical, it is impossible to achieve a reliability of 1 for any subsystem; thus each subsystem reliability must *exceed* the system goal. One can easily apply these bounds by enumeration or by solving a logarithmic equation.

The reliability expression for a chain of $k$ subsystems, where each subsystem is composed of $n_i$ parallel elements, is given in Eq. (7.3). If we allow all the subsystems other than subsystem $i$ to have a reliability of unity and we compare them with Eq. (7.13), we obtain

$$(1 - [1 - r_i]^{n_i}) > R_g \tag{7.30}$$

We can easily solve this equation by choosing $n_i = 1, 2, \ldots$, substituting and solving for the smallest value of $n_i$ that satisfies Eq. (7.30). A slightly more direct method is to solve Eq. (7.30) in closed form as an equality:

$$(1 - r_i)^{n_i} = 1 - R_g \tag{7.31a}$$

Taking the log of both sides of Eq. (7.31a) and solving yields

$$n_i = \log(1 - R_g)/\log(1 - r_i) \tag{7.31b}$$

The solution is the smallest integer that exceeds the value of $n_i$ computed in Eq. (7.31b).

We now show how these bounds apply to Example 1 of the last section. Solving Eq. (7.31b) for Example 1 yields

$$n_1 = \log(1 - R_g)/\log(1 - r_1)$$
$$= \log(1 - 0.9)/\log(1 - 0.85)$$
$$= 1.21 \qquad (7.32a)$$
$$n_2 = \log(1 - R_g)/\log(1 - r_2)$$
$$= \log(1 - 0.9)/\log(1 - 0.5)$$
$$= 3.32 \qquad (7.32b)$$
$$n_3 = \log(1 - R_g)/\log(1 - r_3)$$
$$= \log(1 - 0.9)/\log(1 - 0.3)$$
$$= 6.46 \qquad (7.32c)$$

Clearly, the minimum values for $n_1$, $n_2$, and $n_3$ from the preceding computations are 2, 4, and 7, respectively. Thus, these three simple computations have advanced our design from the original statement of the problem given in Fig. 7.4(a) to the *minimum system design* given in Fig. 7.4(b). The subsystem reliabilities are given by Eq. (7.33):

$$R_i = 1 - (1 - r_i)^{n_i} \qquad (7.33)$$

Substitution yields

$$R_1 = 1 - (1 - 0.85)^2 = 0.9775 \qquad (7.34a)$$
$$R_2 = 1 - (1 - 0.5)^4 = 0.9375 \qquad (7.34b)$$
$$R_3 = 1 - (1 - 0.3)^7 = 0.9176 \qquad (7.34c)$$
$$R_s = R_1 R_2 R_3 = 0.9775 \times 0.9375 \times 0.9176 = 0.84089 \qquad (7.34d)$$

The minimum system design represents the first step toward achieving an optimized system design. The reliability has been raised from 0.1275 to 0.8409, a large step toward achieving the goal of 0.9. Furthermore, only 3 cost units are left, so 0, 1, 2, and 3 are the number of units that can be added to the minimum design. An upper bound on the number of cases to be considered is $4 \times 4 \times 4 = 64$ cases, a huge decrease from the initial estimate of 2,744 cases. (This number, 64, will be further reduced once we add the upper bounds of Section 7.8.3.) In fact, because this problem is now reduced, we can easily enumerate exactly how many cases remain. If we allocate the remaining 3 units to subsystem 1, no additional units can be allocated to subsystems 2 and 3 because of the cost constraint. We could label this policy $n_1 = 5$, $n_2 = 4$, and $n_3 = 7$. However, the minimum design represents such an important initial step that we will now assume that it is always the first step in optimization and only deal with increments (deltas) added to the initial design. Thus, instead of labeling this policy (case) $n_1 = 5$, $n_2 = 4$, and $n_3 = 7$, we will call it $\Delta n_1 = 3$, $\Delta n_2 = 0$, and $\Delta n_3 = 0$, or incremental policy (3, 0, 0).

**356** RELIABILITY OPTIMIZATION

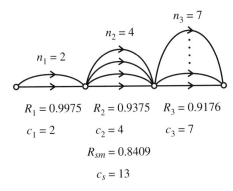

(a) Initial Problem Statement

(b) Minimum System Design

**Figure 7.4** Initial statement of Example 1 and minimum system design.

We can now apply the same minimum design approach to Example 2. Solving Eq. (7.31b) for Example 2 yields

$$n_1 = \log(1 - R_g)/\log(1 - r_1)$$
$$= \log(1 - 0.95)/\log(1 - 0.8)$$
$$= 1.86 \qquad (7.35a)$$
$$n_3 = n_2 = n_1 \qquad (7.35b)$$
$$n_4 = \log(1 - R_g)/\log(1 - r_4)$$
$$= \log(1 - 0.95)/\log(1 - 0.9)$$
$$= 1.3 \qquad (7.35c)$$
$$n_5 = n_4 \qquad (7.35d)$$

Clearly, the minimum values for $n_1$, $n_2$, $n_3$, $n_4$, and $n_5$ are all 2. The original statement of the problem and the *minimum system design* are given in Fig. 7.5. The subsystem reliabilities are given by substitution in Eq. (7.33):

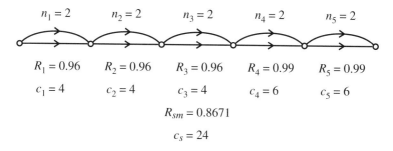

**Figure 7.5** Initial statement of Example 2 and minimum system design.

$$R_1 = 1 - (1 - 0.8)^2 = 0.96 \quad (7.36a)$$
$$R_3 = R_2 = R_1 \quad (7.36b)$$
$$R_4 = 1 - (1 - 0.9)^2 = 0.99 \quad (7.36c)$$
$$R_5 = R_4 \quad (7.36d)$$

Again, the minimum system design is a significant step toward an optimized system design. The product of the 5 parallel subsystems yields a system reliability of 0.8671. The reliability has been raised from 0.41472 to 0.8671. Since 24 cost units are consumed by the minimum design, 12 are left; these will buy up to 6 redundant elements for the first 3 elements or up to 4 for the last 2. An upper bound on the number of cases to be considered is $7 \times 7 \times 7 \times 5 \times 5 = 8{,}575$ cases—a great reduction from the initial estimate of 177,957, but much larger than the 31 cases that really must be calculated once upper bounds are added. The next section discusses how we may rapidly find the remaining cases that need to be enumerated.

### 7.8.3 Upper Bounds

In the previous section, we showed that the lower-bound procedure greatly decreased the number of cases that must be evaluated if enumeration is to be used. In this section, we show that this number is further reduced by the imposition of upper bounds that are related to the resource constraint, which has been modeled as cost. However, the procedure would be the same for another if a single constraint (such as volume or weight) were involved. The case of multiple constraints is discussed later.

We begin by discussing the rare case in which the lower bound that yields the minimum design meets or exceeds the system goal. For example, suppose that a system reliability $R_s > R_g$, can be achieved by expending 90% of the cost, $c_o$. In such a case, we wish to ask how much better can we make the system if we spend a bit more and how much we save if we are content to accept a slightly smaller $R_s$ that does not exceed $R_g$. The easiest way to formulate such a set of policies is to compute the resultant system reliabilities and costs for adding or deleting one parallel element for subsystem $X_1$, repeating the procedure for subsystem $X_2$, and so on. The design team and customer then examine this family of policies to determine which policy is to be pursued.

In the more familiar case, $R_s < R_g$, and we wish to expend all or some of the remaining resource $c_o$ to improve the system reliability to meet the desired goal. We seek a more efficient procedure for achieving an optimum system than blind enumeration. We can use the minimum solution as a lower bound and add in the resource constraint to achieve an upper bound on the solution. The resource constraint forms upper bounds on the number of additional elements that can be allocated. For Example 1, the minimum system leaves 3 cost units, which allow up to 3 additional parallel elements for each subsystem. However, each time we allocate a unit to a subsystem, we expend 1 resource unit, and the number of units available to other subsystems is also reduced by 1. We call the allocation of these additional resources the *augmentation policy*; it is the addition of the best of the augmentation policies to the minimum design that results in the optimization policy.

The use of a branching search tree (policy tree) is one way to illustrate how the upper bounds are computed and how they constrain the solution to a small number of cases. We start with the minimum system design that is the result of the lower bounds and use the remaining constraint to generate a set of augmentation policies from which we select the optimum policy or, as discussed previously, one of the suboptima close to the true optimum. We use Example 1 to illustrate the procedure. The minimum design absorbs 13 cost units, leaving 3 additional units. Thus the number of redundant elements is bounded from below by the minimum system design and from above by at most 3 additional units, yielding $2 \leq n_1 \leq 5$; $4 \leq n_2 \leq 7$; and $7 \leq n_3 \leq 10$. One can improve on these bounds by applying the upper bounds as computation of the cases proceeds. The easist way to accomplish this is to compute the minimum design and allocate the remaining resource in an incremental manner.

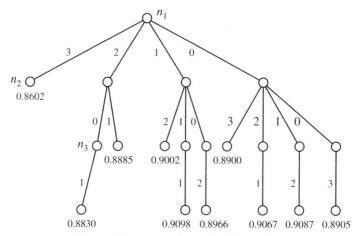

**Figure 7.6** A search tree for Example 1.

The incremental policy is implemented in the following manner. Consider the alternatives; one could expend all the 3 incremental units on element 1, for instance, resulting in an augmentation policy $\Delta n_1 = 3$, $\Delta n_2 = 0$, and $\Delta n_3 = 0$, or one could use incremental notation, resulting in an incremental policy (3, 0, 0), with $c = 16$. Computation of this policy's reliability yields $R_s = 0.8602$. Another alternative for element 1 is (2, –, –), with $c = 15$. One cost unit would be left to spend on element 2 or element 3. These two policies would be denoted as (2, 1, 0) and (2, 0, 1) using the incremental notation and yield reliabilities of 0.8885 and 0.8830. These three policies discussed above, as well as the seven other possible ones, are shown in the search tree of Fig. 7.6. Branches radiating from the start node at the top of the tree represent the number of additional components (in addition to the minimum solution) assigned to element 1. The second level displays the incremental choices for element 2, and the third level displays the incremental choices for element 3. Inspection of Fig. 7.6 shows that the maximum reliability occurs for augmentation policy (1, 1, 1), the center path in the diagram, which corresponds to a total solution of $n_1 = 2 + 1$, $n_2 = 4 + 1$, and $n_3 = 7 + 1$ at a cost of 16 and a reliability of 0.9098. Of course, the other solutions denoted by augmentation policies (0, 2, 1) and (0, 1, 2) with reliabilities 0.9068 and 0.9087 are very close; one of these could be chosen as the policy of choice based on other factors. Other possibilities are to use standby reliability for some of the systems, especially in the case of element 3, which has a large number of parallel units. In some cases, we may not be able to reach the system goal, and either the goal must be relaxed or the cost budget must be increased.

### 7.8.4 An Algorithm for Generating Augmentation Policies

The basic approach is simple: the lower-bound solution for the minimum system design is the starting point. The resources for the minimum system design

**TABLE 7.4 An Algorithm that Solves for the Minimum System Design and the Augmentation Policies for Example 1**

**procedure** (optimum reliability policy computation)
{Three subsystems: 1, 2, 3}
{Reliability of subsystems R1, R2, R3}
{Cost Constraint, C}
{Reliability Goal, RG}
**input** (R1, R2, R3, C, RG)
  **begin** {Minimum System Design}
    M1 := ceiling [log(1 − RG)/log(1 − R1)]
    M2 := ceiling [log(1 − RG)/log(1 − R2)]
    M3 := ceiling [log(1 − RG)/log(1 − R3)]
    RS := [(1 − (1 − R1)**M1)] * [(1 − (1 − R2)**M2)] * [(1 − (1 − R3)**M3)]
    **PRINT** (M1, M2, M3, RS)
  **end** {Minimum System}
  **begin** {Augmentation Policy}
    CA := C − M1 − M2 − M3
    **for** I := 0 **to** CA
      **for** J := 0 **to** CA − I
        **for** K := 0 **to** CA − I − J
          N1 := M1 + I
          N2 := M2 + J
          N3 := M3 + K
          RS := [(1 − (1 − R1)**N1)] * [(1 − (1 − R2)**N2)]
            * [(1 − (1 − R3)**N3)]
          **PRINT** (N1, N2, N3, RS)
        **end** K
      **end** J
    **end** I
  **end** {Augmentation}
**end** {Procedure}

Note: The control statements have their usual meanings: The assignment operator is denoted by := and the ceiling ($x$) function is the smallest integer that is greater than or equal to $x$. The symbol ** means "raise to the power of" and * means "multiplied by."

are subtracted from the resource budget to obtain the resources available for augmentation. All possible augmentation policies are generated for element 1, along with the concomitant reduction in augmentation resources. For each of the policies for element 1, the remaining augmentation resources are used for element 2 to form the second step of the policies. This process is continued for the rest of the elements. Since the augmentation resources quickly decrease to 0 for many of the policies, the number of combinations to be considered is greatly reduced as the process continues. Once an augmentation policy is completed, the reliability is calculated and the information is listed in a table (or on a search tree for smaller problems). A choice is then made among the policies yielding high reliabilities.

A simple algorithm is given in Table 7.4 that solves for the minimum sys-

**TABLE 7.5 Results of the Algorithm for Computing the Augmentation Policies for Example 1 (Optimum Solutions)**

| Minimum System Design | | | |
|---|---|---|---|
| M1 | M2 | M3 | RS |
| 2 | 4 | 7 | 0.8409362 |
| Optimum Policies = Minimum Design + Augmentation Policies | | | |
| N1 | N2 | N3 | RS |
| 2 | 4 | 10 | 0.8905201 |
| 2 | 5 | 9 | 0.9087401 |
| 2 | 6 | 8 | 0.9067561 |
| 2 | 7 | 7 | 0.8899909 |
| 3 | 4 | 9 | 0.896632 |
| 3 | 5 | 8 | 0.9098224 |
| 3 | 6 | 7 | 0.9002588 |
| 4 | 4 | 8 | 0.8830077 |
| 4 | 5 | 7 | 0.8885192 |
| 5 | 4 | 7 | 0.860227 |

tem design and the augmentation policies for Example 1 of Section 7.7.2. The algorithm is written in a pseudocode form similar to that given in Appendix 2 of Rosen [1999]. It generates the minimum system design (2, 4, 7) and the 10 augmentation policies. In the augmentation policies, the I-loop allocates 0 to 3 resource units to subsystem 1 and the J-loop allocates the $(3 - I)$ remaining units to subsystem 2. Once resources are allocated to subsystems 1 and 2, the amount of resources left for subsystem 3, K, is $(3 - I - J)$. Thus, the I-loop takes on the values [I = (0), (1), (2), (3)]. The J-loop takes on the values J = (3 − I), which generates the pairs [I, J = (0, 0), (0, 1), (0, 2), (0, 3), (1, 0), (1, 1), (1, 2), (2, 0), (2, 1), (3, 0)]. Lastly, the K-loop assigns the remaining variable K = [3 − (I + J)], which completes the 10 triplets [I, J, K = (0, 0, 3), (0, 1, 2), (0, 2, 1), (0, 3, 0), (1, 0, 2), (1, 1, 1), (1, 2, 0), (2, 0, 1), (2, 1, 0), (3, 0, 0)].

Execution of a program based on the algorithm in Table 7.4 enumerates the 10 augmentation policies discussed previously. The results, given in Table 7.5, agree with the search tree in Fig. 7.6.

The concept of an optimum is clearly defined mathematically, but in terms of an engineering design, a family of near-optimum solutions is preferred along with the optimum one. Since we have a simple algorithm (program), it is easy for the designer to explore such a family of solutions. Suppose we ask what reliability could be achieved if one decided to shave the cost from 16 units to 15 units. Substituting an augmentation budget of 2 instead of 3 in the loops given in Table 7.4 yields the solutions in Table 7.6. Now our comparison begins: Is the solution of (2, 5, 8), with a reliability of 0.8923632 and a cost of 15, a good substitute for the solution of (3, 5, 8), with a reliability of 0.9098224 and

**TABLE 7.6 Reliability of Various Designs for Example 1 (Optimum Solutions) In Which the Maximum Cost Is Reduced to 15 Units**

| Minimum System Design | | | |
|---|---|---|---|
| M1 | M2 | M3 | RS |
| 2 | 4 | 7 | 0.8409362 |
| Optimum Policies = Minimum Design + Augmentation Policies | | | |
| N1 | N2 | N3 | RS |
| 2 | 4 | 9 | 0.8794260 |
| 2 | 5 | 8 | 0.8923632 |
| 2 | 6 | 7 | 0.8829831 |
| 3 | 4 | 8 | 0.8804733 |
| 3 | 5 | 7 | 0.8859690 |
| 4 | 4 | 7 | 0.8598573 |

a cost of 16? Suppose we can afford a budget of 17 units. The optimum, with an increased budget of 17, achieves a reliability of 0.9265198 by using policy (3, 5, 9). Is it worth the extra cost unit to raise the reliability? Only a design review that includes the system designer, the customer, and possibly various practical considerations can be used to decide such issues.

We repeat the procedure in Table 7.6 for Example 2. An algorithm for the solution of Example 2 is given in Table 7.7. Note in the table that the loop-control end point is adjusted for the resources allocated in outer loops (e.g., **for** L := 0 **to** (CA − I − J − K)/C4), and note that the end point is divided by the item weight, C4, so that L is incremented in multiples of C4.

The results of executing a program corresponding to the algorithm of Table 7.7 are given in Table 7.8. (The running time of the program was one or two seconds on a Pentium 400 MHz personal computer.) The minimum system design requires 2 elements for each subsystem, expends 24 units of resources, and achieves a reliability of 0.8671. The 31 policies generated by the algorithm are listed in Table 7.8 in descending order of reliability.

Note that some of the policies are dominated by other policies; for example, policy 14, which is (4, 5, 3, 2, 2) and requires 36 resource units, is dominated by policy 1, which is (4, 4, 4, 2, 2) and also uses the entire 36 resource units to achieve a higher reliability. A study of the table shows that policy 1 dominates policies 2, 9, 10, 11, 12, 13, 14, 18, 19, 22, 23, and 24. Similarly, policy 15 dominates policies 25, 26, and 27. Thus out of the 31 policies, a total of 15 are dominated, leaving 16 to represent "good solutions" that should be considered.

Further inspection of Table 7.8 shows that there are many good policies that yield suboptima. For example, policy 31 satisfies the minimum requirement $RS > 0.95$ with fewer resources—30 rather than the 36 budgeted. Such calculations should result in a conference between the design leader, the management, and the customer to answer such questions as the following:

**TABLE 7.7  An Algorithm That Solves for the Minimum System Design and the Augmentation Policies for Example 2**

**procedure** (optimum reliability policy computation)
{Five subsystems: 1, 2, 3, 4, 5}
{Reliability of subsystems R1, R2, R3, R4, R5}
{Cost Constraint, C; individual costs, C1, C2, C3, C4, C5}
{Reliability Goal, RG}
**input** (R1, R2, R3, R4, R5, C, RG, C1, C2, C3, C4, C5)
  **begin** {Minimum System Design}
    M1 := ceiling [log(1 − RG)/log(1 − R1)]
    M2 := ceiling [log(1 − RG)/log(1 − R2)]
    M3 := ceiling [log(1 − RG)/log(1 − R3)]
    M4 := ceiling [log(1 − RG)/log(1 − R4)]
    M5 := ceiling [log(1 − RG)/log(1 − R5)]
    RS := [(1 − (1 − R1)\*\*M1)] \* [(1 − (1 − R2)\*\*M2)] \* [(1 − (1 − R3)\*\*M3)]
        \* [(1 − (1 − R4)\*\*M4)] \* [(1 − (1 − R5)\*\*M5)]
    CS := M1 \* C1 + M2 \* C2 + M3 \* C3 + M4 \* C4 + M5 \* C5
    **PRINT** (M1, M2, M3, M4, M5, CS, RS)
  **end** {Minimum System}
  **begin** {Augmentation Policy}
    CA := C − M1 − M2 − M3 − M4 − M5
    **for** I := 0 **to** CA/C1
      **for** J := 0 **to** (CA − I)/C2
        **for** K := 0 **to** (CA − I − J)/C3
          **for** L := 0 **to** (CA − I − J − K)/C4
            **for** M := 0 **to** (CA − I − J − K − L)/C5
              N1 := M1 + I
              N2 := M2 + J
              N3 := M3 + K
              N4 := M4 + L
              N5 := M5 + M
              RS := [(1 − (1 − R1)\*\*N1)] \* [(1 − (1 − R2)\*\*N2)]
                  \* [(1 − (1 − R3)\*\*N3)] \* [(1 − (1 − R4)\*\*N4)]
                  \* [(1 − (1 − R5)\*\*N5)]
              CS := 2 \* N1 + 2 \* N2 + 2 \* N3 + 3 \* N4 + 3 \* N5
              **PRINT** (N1, N2, N3, N4, N5, CS, RS)
            **end** M
          **end** L
        **end** K
      **end** J
    **end** I
  **end** {Augmentation}
**end** {Procedure}

1. Is a reliability of 0.9754 that uses resources of 36 units significantly better than one of 0.95677 that uses resources of 30 units?
2. What would be the cost reduction for a system that uses 30 resource units?

**TABLE 7.8 Parallel Redundancy Optimum and Suboptimum Solutions for Example 2**

| Rank | $n_1$ | $n_2$ | $n_3$ | $n_4$ | $n_5$ | Cost | Reliability |
|---|---|---|---|---|---|---|---|
| 1  | 4 | 4 | 4 | 2 | 2 | 36 | 0.97540 |
| 2  | 3 | 3 | 3 | 3 | 3 | 36 | 0.97424 |
| 3  | 3 | 3 | 4 | 2 | 3 | 35 | 0.97169 |
| 4  | 3 | 3 | 4 | 3 | 2 | 35 | 0.97169 |
| 5  | 3 | 4 | 3 | 2 | 3 | 35 | 0.97169 |
| 6  | 3 | 4 | 3 | 3 | 2 | 35 | 0.97169 |
| 7  | 4 | 3 | 3 | 2 | 3 | 35 | 0.97169 |
| 8  | 4 | 3 | 3 | 3 | 2 | 35 | 0.97169 |
| 9  | 3 | 4 | 5 | 2 | 2 | 36 | 0.97039 |
| 10 | 3 | 5 | 4 | 2 | 2 | 36 | 0.97039 |
| 11 | 4 | 3 | 5 | 2 | 2 | 36 | 0.97039 |
| 12 | 5 | 3 | 4 | 2 | 2 | 36 | 0.97039 |
| 13 | 5 | 4 | 3 | 2 | 2 | 36 | 0.97039 |
| 14 | 4 | 5 | 3 | 2 | 2 | 36 | 0.97039 |
| 15 | 4 | 3 | 4 | 2 | 2 | 34 | 0.96915 |
| 16 | 3 | 4 | 4 | 2 | 2 | 34 | 0.96915 |
| 17 | 4 | 4 | 3 | 2 | 2 | 34 | 0.96915 |
| 18 | 3 | 3 | 3 | 4 | 2 | 36 | 0.96633 |
| 19 | 3 | 3 | 3 | 2 | 4 | 36 | 0.96633 |
| 20 | 3 | 3 | 3 | 3 | 2 | 33 | 0.96546 |
| 21 | 3 | 3 | 3 | 2 | 3 | 33 | 0.96546 |
| 22 | 3 | 3 | 6 | 2 | 2 | 36 | 0.96442 |
| 23 | 6 | 3 | 3 | 2 | 2 | 36 | 0.96442 |
| 24 | 3 | 6 | 3 | 2 | 2 | 36 | 0.96442 |
| 25 | 5 | 3 | 3 | 2 | 2 | 34 | 0.96417 |
| 26 | 3 | 3 | 5 | 2 | 2 | 34 | 0.96417 |
| 27 | 3 | 5 | 3 | 2 | 2 | 34 | 0.96417 |
| 28 | 4 | 3 | 3 | 2 | 2 | 32 | 0.96294 |
| 29 | 3 | 3 | 4 | 2 | 2 | 32 | 0.96294 |
| 30 | 3 | 4 | 3 | 2 | 2 | 32 | 0.96294 |
| 31 | 3 | 3 | 3 | 2 | 2 | 30 | 0.95677 |

3. If 30 resource units are used for reliability purposes, can the additional budgeted 6 resource units be used for something else of value in the system design?

To further answer these questions, additional studies should be attempted with perhaps 32 or 34 resource units, the results of which should be used in the study. The major result demonstrated in this section is that the use of upper and lower bounds and modern, relatively fast personal computers allows a designer the luxury of computing a range of design solutions and comparing them; in general, the more complex methods discussed later in this chapter are seldom needed. Using the results of the discussion of availability in Section 7.4.6, it is easy to adapt any of the foregoing algorithms to availability apportionment by substituting probabilities in the algorithms that represent unit availabilities rather than reliabilities.

### 7.8.5 Optimization with Multiple Constraints

The preceding material in this chapter has dealt with a single constraint and has used cost to illustrate the constraint. Sometimes, however, there are multiple constraints—cost, volume, and weight, for instance. Of obvious importance is the use of these three constraints in satellites, spacecraft, and aircraft. Without loss of generality, we can assume that there are three constraints: cost, volume, and weight ($c$, $v$, and $w$) and that the constraints (given in the forthcoming equations) are similar to those of Eq. (7.4). Clearly, these constraints as well as the following equations can represent other variables and/or can be extended to more than three constraints.

$$c = \sum_{i=1}^{k} n_i c_i \qquad (7.37a)$$

$$v = \sum_{i=1}^{k} n_i v_i \qquad (7.37b)$$

$$w = \sum_{i=1}^{k} n_i w_i \qquad (7.37c)$$

Generally, optimization techniques such as dynamic programming become much more difficult when more than one variable is involved. Since we are dealing with discrete optimization and enumeration, however, the extra work of multiple constraints is modest. First of all, the computation of the minimum system design is not affected by the constraints; thus lower bounds are computed as in the case of a single constraint (cf. Section 7.6.2). Once the minimum design (lower bound) is obtained, the values are substituted in Eqs. (7.37a–c) and the remaining values of the constraints are computed for the augmentation phase. In some cases, the minimum system design exceeds one or more of the constraints and the reliabilty goal and constraints are incompatible (one can call such a situation an *ill-formulated problem*). The only recourse in the case of an ill-formulated problem is to have a high-level design review with all members of the designer's and the customer's senior management present to change the requirements so that the problem is solvable.

Assume that the minimum system design still leaves some values of all the constraints for the augmentation phase. The constraints are still used to compute the upper bounds; however, we now have more than one upper bound. For the case under discussion, we have three upper bounds—one governed by cost, one by weight, and one by volume—that result in three values of $n$: $(n_i, n'_i, n''_i)$. The bound we choose is the minimum value of the three bounds, that is, the minimum value of $(n_i, n'_i, n''_i)$ [Rice, 1999]. Computation of the augmentation policy proceeds in the same manner as discussed in Section 7.6.3; however, at each stage, three upper bounds are computed, and the minimum is used in each case. Once the augmentation policy is obtained, the system reliability is computed. If the reliability goal cannot be obtained, a high-level design review is enacted. The designer should compute beforehand some alternative designs for presentation that violate one or more of the constraints but still achieve the goals.

## 7.9 APPORTIONMENT AS AN APPROXIMATE OPTIMIZATION TECHNIQUE

The bounded solutions of the previous section led to a family of solutions that included the maximum reliability combination. The apportionment techniques discussed in Section 7.6 can be viewed as an approximate solution. In this section, we explore how these solutions compare with the optimum solutions of the previous section.

We begin by considering Example 1 given in Section 7.7.2. The system goal $R_g = 0.9$, and by using Eq. (7.15b) we obtain a goal of 0.9655 for each of the three subsystems. We can determine how many parallel components are needed for each subsystem if we use the reliability goal of 0.9655 and each subsystem reliability (0.85, 0.5, 0.3) and substitute into Eq. (7.31b). The results for the subsystem are

$$n_1 = \log(1 - R_g)/\log(1 - r_1) \tag{7.38a}$$
$$n_1 = \log(1 - 0.9655)/\log(1 - 0.85) = 1.77 \tag{7.38b}$$
$$n_2 = \log(1 - 0.9655)/\log(1 - 0.5) = 4.86 \tag{7.38c}$$
$$n_3 = \log(1 - 0.9655)/\log(1 - 0.3) = 9.44 \tag{7.38d}$$

This represents a solution of $n_1, n_2, n_3 = 2, 5, 10$, using 17 cost units and exceeding both the reliability goal and the cost budget. One could then try removing one unit from each of the subsystems to arrive at approximate solutions. If we remove one unit from subsystem 2 or 3, we obtain the 16-unit solutions $n_1, n_2, n_3 = 2, 4, 10$ and $n_1, n_2, n_3 = 2, 5, 9$, which correspond to the optimum designs given in Table 7.5 (rows 1 and 2). If we remove one unit from subsystem 1, we obtain solution $n_1, n_2, n_3 = 1, 5, 10$, corresponding to a reliability of 0.8002 that is clearly inferior to all the 10 solutions in Table 7.5.

We now consider the apportionment solutions for Example 2 given in Section 7.7.2. The system goal is $R_g = 0.95$, and by using Eq. (7.15b), we obtain a goal of 0.9898 for each of the five subsystems. We can determine how many parallel components are needed for each subsystem to meet the reliability goal of 0.9898 with each subsystem reliability (0.8, 0.8, 0.8, 0.9, 0.9). Substitution into Eq. (7.31b) yields the desired results that need to be calculated only for the subsystem values of 0.8 and 0.9; these are

$$n_1 = \log(1 - R_g)/\log(1 - r_1) \qquad (7.39a)$$
$$n_{1,2,3} = \log(1 - 0.9898)/\log(1 - 0.8) = 2.85 \qquad (7.39b)$$
$$n_{4,5} = \log(1 - 0.9898)/\log(1 - 0.9) = 1.99 \qquad (7.39c)$$

Thus rounding up represents a solution of $n_1, n_2, n_3, n_4, n_5 = 3, 3, 3, 2, 2$. Since the costs are 2, 2, 2, 3, 3 units per subsystem, respectively, this apportioned solution expends 30 cost units and yields a reliability of 0.9568. This is the last policy in Table 7.8. We conclude that the simplest equal apportionment method is a good approximation, and since it only takes a few minutes with paper, pencil, and calculator, it is a valuable check on the results of the previous section.

## 7.10 STANDBY SYSTEM OPTIMIZATION

In principle, the optimization of a standby system is the same procedure as that of a parallel system, but instead of the reliability expression for $n$ items in parallel given in Eq. (7.3), the expression given in Eqs. (7.5) and (7.6) is used. Because Eq. (7.6) is a series, the simple solution for the number of elements in the minimum system design given in Eqs. (7.30) and (7.31) is not applicable. A slightly more complicated solution for a standby system involves the evaluation of Eq. (7.6) for increasing values of $k$ until the right-hand side exceeds the reliability goal $R_g$. This is a little more complicated for paper-pencil-and-calculator computation, but the complexity increase is insignificant when a computer program is used. Another approach is to use cumulative Poisson tables or graphs to solve Eq. (7.6), the Chebyschev bound, or the normal approximation to the Poisson (these later techniques are explained and compared in Messinger [1970]). The reader should note that the techniques for a standby system also apply to a system with spares. We assume that a standby system switches in the standby component quickly enough for the system performance to not be affected. In other systems, a set of spare components is kept near an operating system that has self-diagnostics to rapidly signal the failure. If the replacement of a spare is rapid (e.g., opening a panel and putting in a printed circuit board spare), the system would not be down for a significant interval (this is essentially a standby system). If the time to switch a standby system is long enough to interrupt system performance or if the downtime in replacing a spare is significant, we must treat the system as an availability problem and formulate a Markov model.

**368**   RELIABILITY OPTIMIZATION

One can use Example 1 of Section 7.7.2 to illustrate a standby system reliability optimization [Shooman, 1994]. Since standby systems are generally more complex than parallel ones because of the failure detection and switching involved, we assume that each element costs 1.5 units rather than 1 unit. Furthermore, we equate the probability of no failures ($x = 0$) for the Poisson, $e^{-\mu}$, to the reliabilities of each unit and solve for the expected number of failures, $\mu$: $\mu_1 = \ln(0.85) = 0.1625189$; $\mu_2 = \ln(0.5) = 0.6931471$; and $\mu_3 = \ln(0.3) = 1.2039728$. Substitution into Eqs. (7.5) and (7.6) yields

$$e^{-\mu_1}(1 + \mu_1 + \mu_1^2/2! + \cdots +)$$
$$\geq 0.90 \quad \text{for two terms } 0.85(1 + 0.1625189)$$
$$= 0.9881 \geq 0.9 \tag{7.40a}$$
$$e^{-\mu_2}(1 + \mu_2 + \mu_2^2/2! + \cdots +)$$
$$\geq 0.90 \quad \text{for three terms } 0.5(1 + 0.6931471 + 0.2402264)$$
$$= 0.9667 \geq 0.9 \tag{7.40b}$$
$$e^{-\mu_3}(1 + \mu_3 + \mu_3^2/2! + \cdots +)$$
$$\geq 0.90 \quad \text{for four terms } 0.3(1 + 0.12039728 + 0.7247752 + 0.2908735)$$
$$= 0.9659 \geq 0.9 \tag{7.40c}$$

Thus, the minimum values for standby system reliability are $n_1 = 2$, $n_2 = 3$, $n_3 = 4$. The minimum system cost is $9 \times 1.5 = 13.5$ and the reliability is $0.9881 \times 0.9667 \times 0.9659 = 0.9226$. Since this exceeds the reliability goal, this is the optimum solution, and the augmentation policy phase is not needed. If augmentation had been required, we could use an algorithm similar to that of Table 7.4; however, instead of equations for M1, M2, and M3, **do while** $R \leq R_g$ loops that increment $n$ are used. Similarly, the RS:= equation becomes a product of the series expansions for the Poisson that are computed by **for** loops in a manner similar to that of Eqs. (7.40a–c). If the assumed cost of 1.5 for a standby element is accurate, and if there are no other overriding factors, the standby system would be preferred to that of any of the parallel system policies of Table 7.5 since the resource cost is less and the reliability is higher.

It is also possible to adapt the foregoing optimization techniques to an $r$-out-of-$n$ system design; see Shooman [1994, p. 946].

One should not forget that reliability can be improved by component improvement as an alternative to parallel or standby system redundancy. In general, there are extra costs involved (development and production), and typically such an improved design begins by listing all the ways in which the element can fail in the order of frequency. Design engineers then propose schemes to eliminate, mitigate, or reduce the frequency of occurrence. The design changes are made, sometimes one at a time or a few at a time, and the prototype is tested to confirm the success of the redesign. Sometimes overstress (accelerated) testing is used in such a process to demonstrate unknown failure

## 7.11 OPTIMIZATION USING A GREEDY ALGORITHM

### 7.11.1 Introduction

If one studies the optimum solutions of various optimization procedures, we find that the allocation of parallel subsystems tends to raise all the subsystems to nearly the same reliability. For example, consider the optimum solutions given in Table 7.5. The subsystem reliabilities start out (0.85, 0.5, 0.3) and the minimal system design (with 2, 4, and 7 parallel systems) yields upon substitution into Eq. (7.3), giving (0.9775, 0.9375, 0.9176), and the optimum solution of 3, 5, 8 results in reliabilities of (0.9966, 0.9688, 0.9424). This is one of the reasons that the equal apportionment approximation gave reasonably good results, leading one to a heuristic procedure for optimization. If one starts with the initial design or, better, with the *minimal system design*, one can allocate additional parallel systems to the subsystems by computing which allocation to subsystem 1, 2, or 3 will produce the largest increase in reliability. Such an allocation is made and the computations are repeated with the new parallel system added. Based on these new computations, a second parallel system is added, and the procedure is repeated until the entire resource has been expended. This procedure generates an augmentation policy that, when added to the minimal system design, generates a good policy.

### 7.11.2 Greedy Algorithm

The foregoing algorithm that makes an optimal choice at each step is often referred to as a "greedy" algorithm [Cormen, Chapter 17]. Starting with the minimum system design, we compute the increase in reliability obtained by adding a single element to each subsystem. In the case of Example 1, Fig. 7.4(b) shows that the minimum system design requires $n_1 = 2$, $n_2 = 4$, and $n_3 = 7$, yielding a reliability $R_{sm} = 0.9775 \times 0.9375 \times 0.9176 = 0.8409$. The addition of one element to $n_1$ raises the reliability of this subsystem to 0.996625 and $R_{sm}$ to 0.8573, which represents a reliability increment $\Delta R$ of 0.0164. Repeating this process for an increase in $n_2$ from 4 to 5 results in raising subsystem 2 to a reliability of 0.96875 and $R_{sm}$ to 0.8689, which represents a reliability increment $\Delta R$ of 0.0280. Increasing $n_3$ to 8 yields a subsystem reliability of 0.9424, and $R_{sm}$ increases to 0.8636, which represents a reliability increment $\Delta R$ of 0.0227. Raising subsystem 2 from 4 to 5 parallel elements produces the largest $\Delta R$; thus the first stage of the greedy algorithm yields the following:

1. Stage 0, minimum system design:

$$n_1 = 2, \quad n_2 = 4, \quad n_3 = 7,$$
$$R_{sm} = 0.9775 \times 0.9375 \times 0.9176 = 0.8409$$

2. Stage 1, add one to $n_2$:

$$n_1 = 2, \quad n_2 = 5, \quad n_3 = 7,$$
$$R_{sm} = 0.9775 \times 0.96875 \times 0.9176 = 0.86875$$

Continuing the greedy process yields the following:

1. Stage 2, add one to $n_3$:

$$n_1 = 2, \quad n_2 = 5, \quad n_3 = 8,$$
$$R_{sm} = 0.9775 \times 0.96875 \times 0.9424 = 0.8924$$

2. Stage 3, add one to $n_1$:

$$n_1 = 3, \quad n_2 = 5, \quad n_3 = 8,$$
$$R_{sm} = 0.996625 \times 0.96875 \times 0.95424 = 0.909868$$

When we compare the solution of stage 3 with Table 7.5, we see that they both have reached the same policy and the same optimum reliability (within round-off errors). The greedy algorithm always yields a good solution, but it is not always the optimum (cf. Section 7.11.4).

### 7.11.3 Unequal Weights and Multiple Constraints

There is something special about Example 1—it is that all the weights are equal. If we consider Example 2, where there are unequal weights, it may not be fair to compare the reliability increase, $\Delta R$, achieved through the adding of one additional parallel component. For example, a component with a cost of 2 should be compared to adding two components with costs of 1 each. Thus, a better procedure in implementing the greedy algorithm is to compare values of $\Delta R/\Delta C$ as the single constraint of cost. When there are multiple constraints, say, $c$, $w$, and $v$, then the comparison should be made based on some function of $c$, $w$, and $v$; $f(c, w, v)$, that is, use $\Delta R/\Delta f(c, w, v)$ as the comparison factor. One possible function to use is a linear combination of the fraction of the constraints expended. If $m$ represents the stage of the augmentation policy, then we would view the ratio $C_m/C_a$ as the fraction of the augmentation cost that has been allocated. Thus, if at the first stage we allocate 20% of the augmentation cost, then $C_m/C_a = 0.2$ and the inverse ratio $C_a/C_m = 5$. If we let $f(c, w, v) = k_1(C_m/C_a) + k_2(W_m/W_a) + k_3(V_m/V_a)$, and also let $k_1 = k_2 = k_3 = 1$, then the constraint with the most available capacity has a stronger influence. Obviously, there are many other good choices for $f(c, w, v)$.

### 7.11.4 When Is the Greedy Algorithm an Optimum?

The greedy algorithm seems like a fine approach to generating a sequence of good policies that lead to an optimum or a good semioptimum policy. The main question is when the greedy solution is an optimum and when it is a semioptimum, a question that has been studied by many [Barlow, 1965, 1975]. The geometrical model discussed in Section 7.2 can be used to explain optimization techniques. Suppose that the reliability surface has two spires: one at $x_1y_1$ that reaches a reliability peak of 0.98 and another at $x_2y_2$ that reaches a reliability peak of 0.99. If we start the greedy algorithm at the base of spire one, it is possible that it will reach a reliability maximum of 0.98 rather than 0.99. There are similarities between the greedy algorithm and the gradient algorithm for continuous functions [Hiller, 1974, p. 729]. Recent work has focused on developing a theory (called the *matroid theory*) that provides the basis for greedy algorithms [Cormen, p. 345]. However, as long as the upper and lower bounds discussed previously provide a family of solutions that includes the optimum as well as a group of good suboptimum policies, and if the computer time is modest, the use of the greedy algorithm is probably unnecessary.

### 7.11.5 Greedy Algorithm Versus Apportionment Techniques

We can understand how the apportionment algorithm reaches an approximate solution if we compare it with a greedy approximation to exact optimization. The greedy approximation adds a new redundant component at each step, which yields the highest gain in reliability on each iteration. The result is to "spread the redundancy around the system" in a bottom–up fashion. The apportionment process also spreads the redundancy about the system, but in a top–down fashion. In general, the two techniques will yield a different set of suboptima; most of the time, both will yield good solutions. The apportionment approach has a number of advantages, including the following: (a) it fits in well with the philosophy of system design, which is generally adopted by designers of large systems; (b) its computations will in general be simpler; and (c) it may provide more insight into when the suboptimal solutions it generates are good.

## 7.12 DYNAMIC PROGRAMMING

### 7.12.1 Introduction

Dynamic programming provides a different approach to optimization that requires fewer steps than exhaustive enumeration and always leads to an optimal policy. The discussion of this section is included for completeness, since the author believes that the application of lower bounds to obtain a minimum system design and the subsequent use of upper bounds to obtain an augmentation policy both require less effort but still yield the optimum. The incremental

**372**  RELIABILITY OPTIMIZATION

reliability method is a competitor to the bounding techniques, but unless one makes a careful study, it is not possible to be sure that this method does indeed yield an optimum.

Dynamic programming is based on an optimality principle established by Bellman [1957, 1958], who states, "An optimal policy has the property that whatever the initial state and initial decision are, the remaining decisions must constitute an optimal policy with regard to the state resulting from the first decision." Clearly, this is a high-level principle that can apply to a large variety of situations. A large number of examples that describe how dynamic programming can be applied to various situations appear in Hiller [1974, Chapter 6]. The best way to understand its application to reliability optimization is to apply it to a problem.

### 7.12.2 Dynamic Programming Example

The following example used to illustrate dynamic programming is a modification of Example 1 of Section 7.7.2.

***Example 3 (Modification of Example 1):*** The initial design of a system yields 3 subsystems at the first level of decomposition. The system reliability goal, $R_g$, is 0.8 for a given number of operating hours. The initial estimates of the subsystem reliabilities are $R_1 = 0.85$, $R_2 = 0.5$, and $R_3 = 0.3$. Parallel redundancy is to be used to improve the initial design so that it meets the reliability goal. The constraint is cost; subsystem 1 is assumed to cost 2 units and subsystems 2 and 3 to cost 1 unit each. The total cost budget is 16 units.

### 7.12.3 Minimum System Design

Dynamic programming can deal with the optimization problem as stated. However, one should take advantage of the minimum system design procedures (lower bounds) to reduce the size of the problem. Thus the minimum design is computed and dynamic programming is used to solve for the augmentation policy. The minimum system design is computed in a manner similar to that of Eqs. (7.35a–d).

$$n_1 = \log(1 - 0.8)/(1 - 0.85) = 0.848 \tag{7.41a}$$
$$n_2 = \log(1 - 0.8)/(1 - 0.5) = 2.322 \tag{7.41b}$$
$$n_3 = \log(1 - 0.8)/(1 - 0.3) = 4.512 \tag{7.41c}$$

Thus the minimum system design consists of one subsystem 1, three subsystem 2, and five subsystem 3. The cost of the minimum design is $C = 1 \times 2 + 3 \times 1 + 5 \times 1 = 10$, and the cost available for the augmentation policy is $\Delta C = 16 - 10 = 6$. The reliability of the minimum system design is

$$R_{sm} = [1 - (1 - 0.85)] \times [1 - (1 - 0.5)^3] \times [1 - (1 - 0.3)^5]$$
$$= (0.85) \times (0.875) \times (0.83193) = 0.6187 \tag{7.42}$$

### 7.12.4 Use of Dynamic Programming to Compute the Augmentation Policy

Thus we now wish to use dynamic programming to determine what is the best augmentation policy from which to raise the reliability 0.6187 to 0.8 by expending the remaining six cost units. Dynamic programming for this problem consists of two phases: I and II. Phase I is used to construct a series of tables that correspond to the best solution for cost allocation for various combinations of the subsystems. The first table considers the first subsystem only. The second table corresponds to the best cost allocation for the first and second subsystem; its construction uses information from the first table. A third table is constructed that gives the best allocation for the third system based on the second table, which depicts the best allocation for the first and second subsystems. For phase II, certain features of the three tables of phase I are combined to construct a new table that displays cost allocation per subsystem and also the resulting reliability optimization. A "backtracking" solution-procedure is used to compute the optimal policy for the cost constraint. The solution procedure automatically allows backtracking to compute optimization solutions for smaller cost constraints. The description of these procedures will be clearer when they are applied to Example 3.

We begin our discussion by constructing the first table of phase I of the solution, which is found in Table 7.9 and labeled as "Table 1." The first column in Table 1 is the amount of cost constraint allocated to subsystem 1. The maximum allocation is 6 cost units; the minimum, 0 cost units. The increments between 0 and 6 are sized to be equal to the greatest common divisor of all the subsystem costs, $gcd(C_1, C_2, C_3)$. In the case of Example 3, this is the $gcd(2, 1, 1) = 1$. Thus the first column in Table 1 comprises the cost allocations 0, 1, 2, 3, 4, 5, and 6.

The details of constructing Table 1 are as follows:

1. Consider the bottom line of Table 1. This table considers only the allocation of cost to buy additional parallel units for subsystem 1. If 0 cost is allocated to subsystem 1 for the augmentation policy, then no additional components can be allocated above the single subsystem of the minimal system design, and the optimal reliability is the same as the minimum system design—that is, 0.6187.
2. Because subsystem 1 costs 2 units, no additional units can be purchased with 1 cost unit, and the solution is the same as the 0 cost allocation.
3. If we increase the cost allocation to 2 units, we can allocate 1 additional unit to subsystem 1 for a total cost of 2, from which the reliability

**TABLE 7.9 Phase I Constraint Allocation Tables**

Table 1: Allocation Table for Subsystem 1

| $\Delta$Cost Constraint | $\Delta n_1$ Allocation | Optimum Reliability |
|---|---|---|
| 6 | 3 | 0.7276 |
| 5 | 2 | 0.7255 |
| 4 | 2 | 0.7255 |
| 3 | 1 | 0.7116 |
| 2 | 1 | 0.7116 |
| 1 | 0 | 0.6187 |
| 0 | 0 | 0.6187 |

Table 2: Allocation Table for Subsystems 1 and 2

| $\Delta$Cost Constraint | $\Delta n_2$ Allocation | Optimum Reliability |
|---|---|---|
| 6 | 4 | 0.8068 |
| 5 | 3 | 0.8063 |
| 4 | 2 | 0.7878 |
| 3 | 1 | 0.7624 |
| 2 | 0 | 0.7116 |
| 1 | 1 | 0.6629 |
| 0 | 0 | 0.6187 |

Table 3: Allocation Table for Subsystems 1, 2, and 3

| $\Delta$Cost Constraint | $\Delta n_3$ Allocation | Optimum Reliability |
|---|---|---|
| 6 | 2 | 0.8689 |
| 5 | 2 | 0.8409 |
| 4 | 1 | 0.8086 |
| 3 | 0 | 0.7624 |
| 2 | 0 | 0.7116 |
| 1 | 0 | 0.6629 |
| 0 | 0 | 0.6187 |

becomes $R_s = [1 - (1 - 0.85)^2] \times (0.875) \times (0.83193) = (0.9775) \times (0.875) \times (0.83193) = 0.7116$. This solution also holds for 3 cost units.

4. For 4 and 5 cost units, there can be an allocation of two subsystem 1 units, from which the reliability becomes $R_s = [1 - (1 - 0.85)^3] \times (0.875) \times (0.83193) = (0.996625) \times (0.875) \times (0.83193) = 0.7255$.

5. Lastly, for an allocation of 6 units, the total number of subsystem 1 units is $1 + 3$, from which the reliability becomes $R_s = [1 - (1 - 0.85)^4] \times (0.875) \times (0.83193) = (0.999493) \times (0.875) \times (0.83193) = 0.7276$.

The details of constructing Table 2 are as follows:

1. Consider the bottom line of Table 2. If there are 0 cost units allocated, then there can be no additional parallel elements for subsystem 2 and none for subsystem 1. Thus the reliability is the same as the bottom line in Table 1—that is, 0.6187.

2. If there is 1 cost unit allocated in Table 2, we can allocate 1 additional unit to subsystem 2 or we can consult Table 1 to see the result of allocating 1 cost unit to subsystem 1 instead. Since subsystem 1 is 2 cost units, there is no gain obtained by the allocation of 1 cost unit to subsystem 1. Therefore, the optimum is to allocate 1 additional element to subsystem 2 (for a total of 4), from which the reliability is $R_s = (0.85) \times [1 - (1 - 0.5)^4] \times (0.83193) = (0.85) \times (0.9375) \times (0.83193) = 0.6629$. Note that this is actually the optimum for subsystems 1 and 2, which is the meaning of Table 2.

3. For a cost allocation of 2, we have three choices: (a) 1 additional element for $\Delta n_2$ and 0 for $\Delta n_1$; (b) two additional elements for $\Delta n_2$ and 0 for $\Delta n_1$; and (c) 0 additional elements for $\Delta n_2$ and 1 for $\Delta n_1$. Clearly, choice (b) is superior to choice (a), so for the optimum policy we need to compare choices (b) and (c). Note that for choice (c), we can obtain the achieved reliability by reading the appropriate row in Table 1, which indicates $R_s = 0.7116$. For choice (2), we obtain $R_s = (0.85) \times [1 - (1 - 0.5)^5] \times (0.83193) = (0.85) \times (0.96875) \times (0.83193) = 0.6850$. Thus choice (b) is superior, and we allocate 0 elements to $\Delta n_2$.

4. In the case of a cost constraint of 3, there are again three choices: (a) 1 additional element for $\Delta n_2$ and 1 for $\Delta n_1$; (b) 2 additional elements for $\Delta n_2$ and 0 for $\Delta n_1$; and (c) 3 additional elements for $\Delta n_2$ and 0 for $\Delta n_1$. Clearly, choice (c) is superior to choice (b). To compare choice (c) with choice (a), we say that it is a comparison of 1 additional element for $n_2$ and 1 for $n_1$ versus 1 additional element for $n_2$ along with 2 more additional elements for $n_2$. However, we already showed that 1 for $n_1$ is better that 2 for $n_2$; therefore, choice (a) is superior, from which the reliability is $R_s = [1 - (1 - 0.85)^2] \times [1 - (1 - 0.5)^4] \times [1 - (1 - 0.3)^5 = (0.9775) \times (0.9375) \times (0.83193) = 0.7624$.

5. For the case of 4 units of incremental cost, there are again three choices: (a) 0 additional elements for $\Delta n_2$ and 2 for $\Delta n_1$; (b) 2 additional elements for $\Delta n_2$ and 1 for $\Delta n_1$; and (c) 4 additional elements for $\Delta n_2$ and 0 for $\Delta n_1$. From Table 1, we see that choice (a) yields a reliability of 0.7255, which is smaller than the previous allocation in Table 2; thus we should consider choice (b) or choice (c). The reliability for these two choices is given by $R_s = [1 - (1 - 0.85)^2] \times [1 - (1 - 0.5)^5] \times [1 - (1 - 0.3)^5] = (0.9775) \times (0.96875) \times (0.83193) = 0.7878$, and $R_s = [1 - (1 - 0.85)^1] \times [1 - (1 - 0.5)^6] \times [1 - (1 - 0.3)^5] = (0.85) \times (0.9844) \times (0.83193) = 0.6961]$. Thus choice (b) is superior.

6. For the case of 5 cost units, one choice is 2 units for $n_1$ and 1 for $n_2$, which gives a reliability of $R_s = [1 - (1 - 0.85)^3] \times [1 - (1 - 0.5)^5] \times [1$

**376**  RELIABILITY OPTIMIZATION

$- (1 - 0.3)^5] = (0.996625) \times (0.96875) \times (0.83193) = 0.8032$. Another choice is 1 unit of $n_1$ and 3 units of $n_2$, with a reliability of $R_s = [1 - (1 - 0.85)^2] \times [1 - (1 - 0.5)^6] \times [1 - (1 - 0.3)^5] = (0.9775) \times (0.9844) \times (0.83193) = 0.8063$. The remaining choice is 0 units of $n_1$ and 5 units of $n_2$, with a reliability of $R_s = [1 - (1 - 0.85)^1] \times [1 - (1 - 0.5)^8] \times [1 - (1 - 0.3)^5] = (0.85) \times (0.9961) \times (0.83193) = 0.7044$. The second choice is the best.

7. Lastly, a cost increment of 6 units allows a policy of 3 units for $n_1$ and 0 units for $n_2$, which gives a reliability of $R_s = [1 - (1 - 0.85)^4] \times [1 - (1 - 0.5)^3] \times [1 - (1 - 0.3)^5] = (0.99949) \times (0.875) \times (0.83193) = 0.72757$, which is not an improvement from the previous policy. Another choice is 2 units for $n_1$ and 2 units for $n_2$, which gives a reliability of $R_s = [1 - (1 - 0.85)^3] \times [1 - (1 - 0.5)^5] \times [1 - (1 - 0.3)^5] = (0.996625) \times (0.96875) \times (0.83193) = 0.8032$; because this is less than the previous policy, it is not optimum. The remaining choice is 1 unit for $n_1$ and 4 units for $n_2$, which gives a reliability of $R_s = [1 - (1 - 0.85)^2] \times [1 - (1 - 0.5)^7] \times [1 - (1 - 0.3)^5] = (0.9775) \times (0.99219) \times (0.83193) = 0.80686$.

The details of constructing Table 3 are as follows:

1. If there is 0 weight allocation for the incremental policy, the minimum system prevails as is shown in the last row of Table 3. If 1 unit of cost is available, it could be added to element 3 to give a reliability of $R_s = [1 - (1 - 0.85)^1] \times [1 - (1 - 0.5)^3] \times [1 - (1 - 0.3)^6] = (0.85) \times (0.875) \times (0.8824) = 0.6562$. Inspection of Table 2 shows that this is inferior to allocating the single cost unit to subsystems 1 and 2. Thus the entry from Table 2 is inserted in Table 3.

2. If 2 cost units are available, they can be used for two additional subsystem 3 units to give a reliability of $R_s = [1 - (1 - 0.85)^1] \times [1 - (1 - 0.5)^3] \times [1 - (1 - 0.3)^7] = (0.85) \times (0.875) \times (0.9176) = 0.6825$. Another choice is 1 unit for subsystem 3 and the optimum 1-unit cost from Table 2, which is one additional subsystem 2 that gives a reliability of $R_s = [1 - (1 - 0.85)^1] \times [1 - (1 - 0.5)^4] \times [1 - (1 - 0.3)^6] = (0.85) \times (0.9375) \times (0.8824) = 0.7032$. The last possible choice is 0 cost for subsystem 3. Table 2 shows that all the weight is allocated to subsystem 1, which achieves a reliability of 0.7116. This solution is entered in Table 3.

3. For the cost increment of 3 units, one choice is to allocate all of this to subsystem 3 to give a reliability of $R_s = [1 - (1 - 0.85)^1] \times [1 - (1 - 0.5)^3] \times [1 - (1 - 0.3)^8] = (0.85) \times (0.875) \times (0.9424) = 0.70091$. If 2 cost units are allocated to subsystem 2, then Table 2 shows that we should allocate the remaining cost to purchase an additional subsystem 1, from which the reliability becomes $R_s = [1 - (1 - 0.85)^2] \times [1 - (1 - 0.5)^3] \times [1 - (1 - 0.3)^6] = (0.9775) \times (0.875) \times (0.8824) = 0.7547$. The

remaining choice is to allocate 0 cost to subsystem 3 and use the solution for 3 cost units from Table 2, which uses one additional subsystem 1 and one additional subsystem 2 and gives the best reliability of 0.7624. For the case of 4 cost units, all can be allocated to subsystem 3 to give a reliability of $R_s = [1 - (1 - 0.85)^1] \times [1 - (1 - 0.5)^3] \times [1 - (1 - 0.3)^9] = (0.85) \times (0.9375) \times (0.99974) = 0.7967$. Allocating 3 cost units to subsystem 3 plus the remaining unit to subsystem 2 gives a reliability of $R_s = [1 - (1 - 0.85)^1] \times [1 - (1 - 0.5)^4] \times [1 - (1 - 0.3)^8] = (0.85) \times (0.9375) \times (0.9424) = 0.7509$. By allocating 2 cost units to subsystem 3, Table 2 reveals that the remaining 2 cost units should be allocated to subsystem 1 to give a reliability of $R_s = [1 - (1 - 0.85)^2] \times [1 - (1 - 0.5)^3] \times [1 - (1 - 0.3)^7] = (0.9775) \times (0.875) \times (0.9176) = 0.7848$. Allocating 1 cost unit to subsystem 3 and, from Table 2, 1 additional unit for subsystems 1 and 2 gives a reliability of $R_s = [1 - (1 - 0.85)^2] \times [1 - (1 - 0.5)^4] \times [1 - (1 - 0.3)^6] = (0.9775) \times (0.9375) \times (0.8824) = 0.8086$. Allocating 0 cost units to subsystem 3 and, from Table 2, 1 additional unit for subsystem 1 as well as 2 additional units for subsystem 2 gives a reliability of $R_s = [1 - (1 - 0.85)^2] \times [1 - (1 - 0.5)^5] \times [1 - (1 - 0.3)^5] = (0.9775) \times (0.96875) \times (0.8319) = 0.7878$.

4. Similar computations yield the allocations for the 5 and 6 cost units shown in Table 3.

We now describe Phase II of dynamic programming: the backtracking procedure. This procedure is merely a reorganization of the information contained in the phase I tables so that an optimum policy can be easily chosen. In a "shorthand" way, the cost allocated to each subsystem defines the policy because dividing the cost by the cost per element yields the number of elements.

The optimum policy for a cost constraint of 6 units is found by starting at the point in the optimum reliability column of Table 7.10 that corresponds to a cost constraint of 6 ($c = 6$). The optimum reliability is 0.8689; to the immediate left, we see that for this policy, 2 cost units ($\Delta n_3 = 2$) have been allocated to $ss_3$ (subsystem 3)—leaving 4 units available. If we look in the allocation to $ss_2$ for a 4-unit cost constraint, we see that 2 cost units are used; thus ($\Delta n_2 = 2$). This leaves 2 cost units for the first subsystem, which means ($\Delta n_1 = 1$). The augmentation policy is therefore (1, 2, 2); when added to the minimum system design (1, 3, 5), it yields the optimal policy (2, 5, 7). The circles and lines in Table 7.10 connect the backtracking steps. A feature of the dynamic programming solution is that it gives the optimal solution for all constraint values below the maximum. For example, suppose that we wanted the solution for 4 cost units. By backtracking, we have 1 unit for $n_3$, 1 unit for $n_2$, and 1 unit for $n_1$. The policy, together with the minimal system design, is (2, 4, 6), which achieves a reliability of 0.8086. For additional examples of dynamic programming, see Messinger [1970].

**TABLE 7.10  Phase II: Backtracking Table for Reliability Augmentation**

| Cost Constraint | Allocation to $ss_1$ Cost | Allocation to $ss_2$ Cost | Allocation to $ss_3$ Cost | Optimum Reliability |
|---|---|---|---|---|
| 6 | 6 | 4 | 2 | 0.8689 |
| 5 | 4 | 3 | 2 | 0.8409 |
| 4 | 4 | 2 | 1 | 0.8086 |
| 3 | 2 | 1 | 0 | 0.7624 |
| 2 | 2 | 0 | 0 | 0.7116 |
| 1 | 0 | 1 | 0 | 0.6629 |
| 0 | 0 | 0 | 0 | 0.6187 |

Note: Solid line is the policy for a cost constraint of 6; dashed line is the policy for a cost constraint of 4.

### 7.12.5  Use of a Bounded Approach to Check Dynamic Programming Solution

To check the results of the dynamic programming solution to Example 3, a slightly revised version of the algorithm in Table 7.4 is written for Example 3 and an associated program was run for reliability values that exceed 0.8. The program generated 14 solutions with costs of 14, 15, and 16 units. The optimum reliabilities for each of these cost constraints are given in Table 7.11. The results given in Table 7.11 are, of course, identical with those obtained by backtracking in Table 7.10.

**TABLE 7.11  Computation of Optimum Reliability for Example 3 Using the Minimum System Design and Augmentation Policy**

| Minimum System Design | | | | |
|---|---|---|---|---|
| $n_1$ | $n_2$ | $n_3$ | | |
| 1 | 3 | 5 | | |
| Augmentation Policy | | | | |
| $n_1$ | $n_2$ | $n_3$ | $C$ | $R_s$ |
| 2 | 5 | 7 | 16 | 0.8689674 |
| 2 | 4 | 7 | 15 | 0.8409362 |
| 2 | 4 | 6 | 14 | 0.808592 |

## 7.13 CONCLUSION

The methods of this chapter provide a means of implementing component reliability at the lowest level possible in most systems: the line-replaceable unit (LRU) level. However, in some cases the designer may not wish to implement strict component redundancy. For example, if a major portion of a system is already available because it was used in a previous design, then it may be most cost-effective to use this as a fixed subsystem. If the reliability is too low, we merely place additional copies of this subsystem in parallel rather than delve within the design to provide a lower-level redundancy. A similar case occurs when portions of a design are being implemented by using existing very high level integrated circuits.

Optimizing a design is a difficult problem for many reasons. Designers often rush to meet schedule and costs and look for feasible solutions that meet the performance requirements; thus reliability may be treated as an afterthought. This approach seldom leads to a design with optimum reliability—much less a good suboptimal design. The methods outlined in this chapter provide the designer with many tools to rapidly generate a family of good optimum and suboptimum system designs. This provides guidance when choices must be made rapidly and conflicting design constraints must be satisfied.

## REFERENCES

AGREE Report. Reliability of Electronic Equipment. Washington, DC: Advisory Group on Reliability of Electronic Equipment, Office of the Assistant Secretary of Defense, U.S. Government Printing Office (GPO), June 4, 1957.

Aho, A. V. et al. *The Design and Analysis of Computer Algorithms.* Addison-Wesley, Reading, MA, 1974.

Albert, A. A Measure of the Effort Required to Increase Reliability. Technical Report No. 43. Stanford, CA: Stanford University, Applied Mathematics and Statistics Laboratory, November 5, 1958.

Barlow, R. E., and F. Proschan. *Mathematical Theory of Reliability.* Wiley, New York, 1965, ch. 6.

Barlow, R. E., and F. Proschan. *Statistical Theory of Reliability and Life Testing—Probability Models.* Holt, Rinehart and Winston, New York, 1975, ch. 7.

Bartlett, J. *Bartlett's Familiar Quotations.* Little, Brown, and Co., Boston, 1968.

Bellman, R. *Dynamic Programming.* Princeton University Press, Princeton, NJ, 1957.

Bellman, R., and S. Dreyfus. Dynamic Programming and the Reliability of Multi-Component Devices. *Operations Research* 6, 2 (1958): 200–206.

Bierman, H. et al. *Quantitative Analysis for Business Decisions.* Richard D. Irwin, Homewood, IL, 1969, ch. 23.

Blischke, W. R., and D. N. Prabhakar. *Reliability: Modeling, Prediction, and Optimization.* Wiley, New York, 2000.

Claudio, M. et al. A Cellular, Evolutionary Approach Applied to Reliability Optimization of Complex Systems. *Proceedings Annual Reliability and Maintainability Symposium*, 2000. IEEE, New York, NY, pp. 210–215.

Coit, D. W., and A. E. Smith, Reliability Optimization of Series-Parallel Systems Using a Genetic Algorithm. *IEEE Transactions on Reliability* 45 (1996): 254–260.

Cormen, T. H., C. E. Leiserson, and R. L. Riverset. *Introduction to Algorithms*. McGraw-Hill, New York, 1992.

Fragola, J. R., and J. F. Spahn. Economic Models for Satellite System Effectiveness. *Proceedings Annual Reliability and Maintainability Symposium*, 1973. IEEE, New York, NY, pp. 167–176.

Gilbert, G. A. *Air Traffic Control: The Uncrowded Sky*. Smithsonian Institution Press, Washington, DC, 1973.

Henney, K. et al. *Reliability Factors for Ground Electronic Equipment*. McGraw-Hill, New York, 1956.

Hiller, S. F., and G. J. Lieberman. *Operations Research*. Holden-Day, San Francisco, 1974, ch. 6.

Lloyd, D. K., and M. Lipow. *Reliability: Management, Methods, and Mathematics*. Prentice-Hall, Englewood Cliffs, NJ, 1962.

Lloyd, D. K., and M. Lipow. *Reliability: Management, Methods, and Mathematics*, 2d ed. ASQC, Milwaukee, WI, 1977.

Mancino, V. J. et al. Reliability Considerations for Communications Satellites. *Proceedings Annual Reliability and Maintainability Symposium*, 1986. IEEE, New York, NY, pp. 389–396.

Marseguerra, M., and E. Zio. System Design Optimization by Genetic Algorithms. *Proceedings Annual Reliability and Maintainability Symposium*, 2000. IEEE, New York, NY, pp. 222–227.

Messinger, M. and M. L. Shooman. Techniques for Optimum Spares Allocation: A Tutorial Review. *IEEE Transactions on Reliability* 19, 4 (November 1970).

Mettas, A. Reliability Allocation and Optimization for Complex Systems. *Proceedings Annual Reliability and Maintainability Symposium*, 2000. IEEE, New York, NY, pp. 216–221.

Miller, G. A. The Magical Number Seven, Plus or Minus Two: Some Limits on our Capacity for Processing Information. *The Psychological Review* 63, 2 (March 1956): p. 81.

Myers, R. H. et al. (eds.). *Reliability Engineering for Electronic Systems*. Wiley, New York, 1964.

Oliveto, F. E. An Algorithm to Partition the Operational Availability Parts of an Optimal Provisioning Strategy. *Proceedings Annual Reliability and Maintainability Symposium*, January 1999. IEEE, NY, pp. 310–316.

Rice, W. F., C. R. Cassady, and T. R. Wise. Simplifying the Solution of Redundancy Allocation Problems. *Proceedings Annual Reliability and Maintainability Symposium*, January 1999. IEEE, New York, NY.

Rosen, K. H. *Discrete Mathematics and its Applications*. McGraw-Hill, New York, 1999.

Shooman, M. L. *Software Engineering, Design, Reliability, and Management*. McGraw-Hill, New York, 1983.

Shooman, M. L. *Probabilistic Reliability: An Engineering Approach*, 2d ed. Krieger, Melbourne, FL, 1990.

Shooman, M. L., and C. Marshall. A Mathematical Formulation of Reliability Optimized Design. *Proceedings IFIP 16th Conference on System Modeling and Optimization*, Compiegne, France, July 5–9, 1993. (Also published in Henry, J., and J.-P. Yvon. System Modeling and Optimization. *Lecture Notes in Control and Information Sciences 197*. Springer-Verlag, London, 1994, pp. 941–950.)

Tillman, F. A., C. L. Hwang, and W. Kuo. Optimization Techniques for System Reliability with Redundancy—A Review. *IEEE Transactions on Reliability* 26, 1977, 3, pp. 148–155.

Tillman, F. A., C. L. Hwang, and W. Kuo. *Optimization of Systems Reliability*. Marcel Dekker, New York, 1980.

Von Alven, W. H., and ARINC Research Corporation Staff. *Reliability Engineering*. Prentice-Hall, Englewood Cliffs, NJ, 1964.

## PROBLEMS

**7.1.** Why was Eq. (7.3) written in terms of probability of failure rather than probability of success?

**7.2.** There have been many studies of *software* errors that relate the number of errors to the number of interfaces. Suppose that many of the hardware design errors are also related to the number of interfaces. What can you say about the complexity of the design and the number of errors based on the results of Sections 7.5.2 and 7.5.3?

**7.3.** Repeat the apportionment example of Section 7.6.1 for reliability goals of 0.90 and 0.99.

**7.4.** Repeat the apportionment example of Section 7.6.2 for reliability goals of 0.90 and 0.99.

**7.5.** Repeat the apportionment example of Section 7.6.3 for reliability goals of 0.90 and 0.99.

**7.6.** Repeat the apportionment example of Section 7.6.4 for reliability goals of 0.90 and 0.99.

**7.7.** Comment on the results of problems 7.3–7.6 with respect to the difficulty of the computations, how close the results agree, and which results you think are the most realistic.

**7.8.** Derive Eqs. (7.28a, b) by formulating a Markov model and solving the associated equations.

**7.9.** Suppose the reliability goal for Example 1 of Section 7.7.2 is 0.95 and compute the minimum system design. Repeat for a reliability goal of 0.99.

**382** RELIABILITY OPTIMIZATION

**7.10.** Repeat problem 7.9 for Example 2.

**7.11.** Write a computer program corresponding to the algorithm of Table 7.4 and verify the results of Tables 7.5 and 7.6.

**7.12.** Change the program of problem 7.11 so that it prints out the results in descending order of reliability.

**7.13.** Repeat problem 7.11 for the algorithm of Table 7.7.

**7.14.** Repeat problem 7.12 for Example 2.

**7.15.** Rewrite the algorithm of Table 7.4 to include volume and weight constraints as discussed in Section 7.8.5.

**7.16.** Repeat problem 7.15 for the algorithm of Table 7.7.

**7.17.** Compare the results of apportionment with the optimum system design for Example 1 where the reliability goals are 0.95 and 0.99, as was done in Section 7.9.

**7.18.** Compare the results of apportionment with the optimum system design for Example 2 where the reliability goals are 0.95 and 0.99, as was done in Section 7.9.

**7.19.** Repeat problem 7.9 for Example 3 given in Section 7.12.2, with reliability goals of 0.85, 0.90, and 0.95.

**7.20.** Write a computer program to solve for the minimum system design and the augmentation policy of Example 3 of Section 7.12.2.

**7.21.** Modify the algorithm of Table 7.4 for the case of standby systems as discussed in Section 7.10.

**7.22.** Repeat problem 7.21 for Example 2.

**7.23.** Repeat problem 7.21 for Example 3.

**7.24.** Write a general program for bounded optimization that includes the following features: (a) the input of the number of series subsystems and their reliability, cost, weight, and volume; and (b) the input of the system reliability goals and the cost, weight, and volume constraints. Then, specify which subsystems will use parallel and which will use standby redundancy. Policies are to be printed in descending reliability order.

**7.25.** Write a program to solve the greedy algorithm of Section 7.11.2.

**7.26.** Use the greedy algorithm of Section 7.11.2 to solve for the optimum for Example 1.

**7.27.** Repeat problem 7.26 for Example 2.

**7.28.** Repeat problem 7.26 for Example 3.

**7.29.** Repeat problem 7.26 for the multiple constraints discussed in Section 7.11.3.

**7.30.** Write a program to solve for the dynamic programming algorithm of Section 7.12 and verify Tables 7.9 and 7.10.

**7.31.** A satellite communication system design is discussed in the article by Mancino [1986]. The structure is essentially a series system with many components paralleled to increase the reliability. Practical failure rates are given for the system. Use the article to redesign the system by incorporating optimum design principles, making any reasonable assumptions. How does your design compare with that in the article? How was the design in the article achieved?

**7.32.** Starting with a component that has a failure rate of $\lambda$, compare two different ways of improving reliability: (a) by placing a second component in parallel and (b) by improving the reliability of a single component by high-quality design. What is the reduced equivalent failure rate of (a)? Comment on what you think the cost would be to achieve the same reductions if (b) is used.

**7.33.** Starting with a component that has a failure rate of $\lambda$, compare two different ways of improving reliability: (a) by placing a second component in parallel and (b) by placing a second component in standby. What is the reduced equivalent failure rate of (a)? Of (b)? Comment on what you think the comparative cost would be to achieve the same reductions if (b) is used.

**7.34.** Choose a project with which you are familiar. Decompose the structure as was done in Fig. 7.3; then discuss.

# APPENDIX A

# SUMMARY OF PROBABILITY THEORY*

## A1 INTRODUCTION

Several of the analytical techniques discussed in this text are based on probability theory. Many readers have an adequate background in probability and need only refer to this appendix for notation and brief review. However, some readers may not have studied probability, and this appendix should serve as a brief and concise introduction for them. If additional explanation is required, an introductory probability text should be consulted [Meeker, 1998; Mendenhall, 1990; Stone, 1996; Wadsworth and Bryan, 1960].

## A2 PROBABILITY THEORY

"Probability had its beginnings in the 17th century when the Chevalier de Méré, supposedly an ardent gambler, became puzzled over how to divide the winnings in a game of chance. He consulted the French mathematician Blaise Pascal (1623–1662), who in turn wrote about this matter to Pierre Fermat (1601–1665); it is this correspondence which is generally considered the origin of modern probability theory" [Freund, 1962]. In the 18th century Karl Gauss (1777–1855) and Pierre Laplace (1749–1827) further developed probability theory and applied it to fields other than games of chance.

Today, probability theory is viewed in three different ways: the a pri-

---

*This appendix is largely extracted from Appendix A of *Software Engineering: Design, Reliability, and Management*, by M. L. Shooman, McGraw-Hill, New York, 1983.

ori (equally-likely-events) approach, the relative-frequency approach, and the axiomatic definition [Papoulis, 1965]. Intuitively we state that the probability of obtaining the number 2 on a single roll of a die is $\frac{1}{6}$. Assuming each of the six faces is equally likely and that there is one favorable outcome, we merely take the ratio. This is a convenient approach; however, it fails in the case of a loaded die, where all events are not equally likely, and also in the case of compound events, where the definition of "equally likely" is not at all obvious. The relative-frequency approach begins with a discussion of an experiment such as the rolling of a die. The experiment is repeated $n$ times (or $n$ *identical* dice are all rolled at the same time in *identical* fashion). If $n_2$ represents the number of times that two dots face up, then the ratio $n_2/n$ is said to approach the probability of rolling a 2 as $n$ approaches infinity. The requirement that the experiment be repeated an infinite number of times and that the probability be defined as the limit of the frequency ratio can cause theoretical problems in some situations unless stated with care. The newest and most generally accepted approach is to base probability theory on three fundamental axioms. The entire theory is built in a deductive manner on these axioms in much the same way plane geometry is developed in an axiomatic manner. This approach has the advantage that if it is followed carefully, there are no loopholes, and all properties are well defined. As with any other theory or abstract model, the engineering usefulness of the technique is measured by how well it describes problems in the physical world. In order to evaluate the parameters in the axiomatic model one may perform an experiment and utilize the relative-frequency interpretation or evoke a hypothesis on the basis of equally likely events. In fact a good portion of mathematical statistics is devoted to sophisticated techniques for determining probability values from an experiment.

The axiomatic approach begins with a statement of the three fundamental axioms of probability:

1. The probability that an event $A$ occurs is a number between zero and unity:

$$0 \leq P(A) \leq 1 \tag{A1}$$

2. The probability of a certain event (also called the *entire sample space* or the *universal set*) is unity:

$$P(S) = 1 \tag{A2}$$

3. The probability of the *union* (also called sum) of *two disjoint* (also called *mutually exclusive*) events is the sum of the probabilities:

$$P(A_1 + A_2) = P(A_1) + P(A_2) \tag{A3}$$

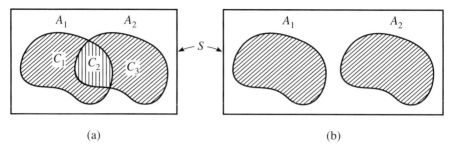

**Figure A1** Venn diagram illustrating the union of sets $A_1$ and $A_2$: (a) ordinary sets; (b) disjoint sets.

## A3 SET THEORY

### A3.1 Definitions

Since axiomatic probability is based on set theory, we shall discuss briefly a few concepts of sets. The same concept often appears in set theory and in probability theory, with different notation and nomenclature being used for the same ideas. A *set* is simply a collection or enumeration of objects. The order in which the objects of the set are enumerated is not significant. Typical sets are the numbers 1, 2, 3, all 100 cars in a parking lot, and the 52 cards in a deck. Each item in the collection is an *element* of the set. Thus, in the examples given there are 3, 100, and 52 elements, respectively. Each set (except the trivial one composed of only one element) contains a number of *subsets*. The subsets are defined by a smaller number of elements selected from the original set. To be more specific one first defines the largest set of any interest in the problem and calls this the *universal set U*. The universal set contains all possible elements in the problem. Thus, a universal set of $n$ elements has a maximum of $2^n$ distinct subsets. The universal set might be all cars in the United States, all red convertibles in New York, or all cars in the parking lot. This is a chosen collection which is fixed throughout a problem. In probability theory, the type of sets one is interested in consists of those which can, at least in theory, be viewed as outcomes of an experiment. These sets are generally called *events*. When the concept of universal set is used in probability theory, the term *sample space S* is generally applied. It is often convenient to associate a geometric picture, called a *Venn diagram*, with these ideas of sample space and event (or set and subset), and the sample space is represented by a rectangle (see Fig. A1).

### A3.2 Axiomatic Probability

With the above background one can discuss intelligently the meaning of probability axioms 1 and 2 given in Eqs. (A1) and (A2). Equation (A1) implies that the probability of an event $A$ is a *positive number* between zero and one. From the relative-frequency interpretation we know that the probability of a certain event is unity and the probability of an impossible event is zero. All other events have probabilities between zero and unity. In Eq. (A2) we let the event A be the entire sample space S, and not too surprisingly we find that this is a certain event. This is true because we say that S occurs if at least one element of S occurs.

### A3.3 Union and Intersection

The union of sets $A_1$ and $A_2$ is a third set $B$. Set $B$ contains all the elements which are in set $A_1$ or in set $A_2$ or in both sets $A_1$ and $A_2$. Symbolically,

$$B = A_1 \cup A_2 \quad \text{or} \quad B = A_1 + A_2 \qquad (A4)$$

The $\cup$ notation is more common in mathematical work, whereas the $+$ notation is commonly used in applied work. The union operation is most easily explained in terms of the Venn diagram of Fig. A1(a). Set $A_1$ is composed of disjoint subsets $C_1$ and $C_2$ and set $A_2$ of disjoint subsets $C_2$ and $C_3$. Subset $C_2$ represents points common to $A_1$ and $A_2$, whereas $C_1$ represents points in $A_1$ but not in $A_2$, and $C_3$ represents points that are in $A_2$ but not $A_1$. When the two sets have no common elements, the areas do not overlap [Fig. A1(b)], and they are said to be *disjoint* or *mutually exclusive*.

The intersection of events $A_1$ and $A_2$ is defined as a third set $D$ which is composed of all elements which are contained in both $A_1$ and $A_2$. The notation is:

$$D = A_1 \cap A_2 \quad \text{or} \quad D = A_1 A_2 \quad \text{or} \quad D = A_1 \cdot A_2 \qquad (A5)$$

As before, the former is more common in mathematical literature and the latter more common in applied work. In Fig. A1(a), $A_1 A_2 = C_2$, and in Fig. A1(b), $A_1 A_2 = \phi$. If two sets are disjoint, they contain no common elements, and their intersection is a set with no elements called a *null set*, $\phi$. $P(\phi) = 0$.

### A3.4 Probability of a Disjoint Union

We can now interpret the third probability axiom given in Eq. (A3) in terms of a card-deck example. The events in the sample space are disjoint and (using the notation $S_3 \equiv$ three of spades, etc.),

$$P(\text{spades}) = P(S_1 + S_2 + \cdots + S_Q + S_K)$$

Since all events are disjoint,

$$P(\text{spades}) = P(S_1) + P(S_2) + \cdots + P(S_Q) + P(S_K) \qquad (A6)$$

From the equally-likely-events hypothesis one would expect that for a fair deck (without nicks, spots, bumps, torn corners, or other marking) the probability of drawing a spade is given by:

$$P(\text{spades}) = \tfrac{1}{52} + \tfrac{1}{52} + \cdots + \tfrac{1}{52} + \tfrac{1}{52} = \tfrac{13}{52} = \tfrac{1}{4}$$

## A4 COMBINATORIAL PROPERTIES

### A4.1 Complement

The complement of set $A$, written as $\overline{A}$, is another set $B$. The notation $A'$ is sometimes used for complement, and both notations will be used interchangeably in this book. Set $B = \overline{A}$ is composed of all the elements of the universal set which are not in set $A$. (The term $A$ *not* is often used in engineering circles instead of $A$ *complement*.) By definition the union of $A$ and $\overline{A}$ is the universal set.

$$A + \overline{A} = U \tag{A7}$$

Applying axioms 2 and 3 from Eqs. (A3) and (A2) to Eq. (A7) yields

$$P(A + \overline{A}) = P(A) + P(\overline{A}) = P(S) = 1$$

This is valid since $A$ and $\overline{A}$ are obviously disjoint events (we have substituted the notation $S$ for $U$, since the former is more common in probability work). Because probabilities are merely numbers, the above algebraic equation can be written in three ways:

$$\begin{aligned} P(A) + P(\overline{A}) &= 1 \\ P(A) &= 1 - P(\overline{A}) \\ P(\overline{A}) &= 1 - P(A) \end{aligned} \tag{A8}$$

There is considerable similarity between the logic operations presented above and the digital logic of Section C1.

### A4.2 Probability of a Union

Perhaps the first basic relationship to be deduced is the probability of a union of two events which are not mutually exclusive. We begin by extending the axiom of Eq. (A3) to three or more events. Assuming that event $A_2$ is the union of two other disjoint events $B_1 + B_2$, we obtain

$$A_2 = B_1 + B_2$$
$$P(A_2 + A_2) = P(A_1) + P(B_1 + B_2) = P(A_1) + P(B_1) + P(B_2)$$

By successive application of this stratagem of splitting events into unions of other mutually exclusive events, we obtain the general result by induction

## COMBINATORIAL PROPERTIES

$$P(A_1 + A_2 + \cdots + A_n) = P(A_1) + P(A_2) + \cdots + P(A_n) \quad \text{for disjoint } A\text{'s}$$

(A9)

If we consider the case of two events $A_1$ and $A_2$ which are not disjoint, we can divide each event into the union of two subevents. This is most easily discussed with reference to the Venn diagram shown in Fig. A1(a). The event (set) $A_1$ is divided into those elements (1) which are contained in $A_1$ and not in $A_2$, $C_1$ and (2) which are common to $A_1$ and $A_2$, $C_2$. Then $A_1 = C_1 + C_2$. Similarly we define $A_2 = C_3 + C_2$. We have now broken $A_1$ and $A_2$ into disjoint events and can apply Eq. (A9):

$$P(A_1 + A_2) = P(C_1 + C_2 + C_2 + C_3) = P[C_1 + C_3 + (C_2 + C_2)]$$

By definition, the union of $C_2$ with itself is $C_2$; therefore

$$P(A_1 + A_2) = P(C_1 + C_2 + C_3) = P(C_1) + P(C_2) + P(C_3)$$

We can manipulate this result into a more useful form if we add and subtract the number $P(C_2)$ and apply Eq. (A3) in reverse

$$\begin{aligned} P(A_1 + A_2) &= [P(C_1) + P(C_2)] + [P(C_2) + P(C_3)] - P(C_2) \\ &= P(A_1) + P(A_2) - P(A_1 A_2) \end{aligned}$$

(A10)

Thus, when events $A_1$ and $A_2$ are not disjoint, we must subtract the probability of the union of $A_1$ and $A_2$ from the sum of the probabilities. Note that Eq. (A10) reduces to Eq. (A3) if events $A_1$ and $A_2$ are disjoint since $P(A_1 A_2) = 0$ for disjoint events.

Equation (A10) can be extended to apply to three or more events:

$$P(A_1 + A_2 + \cdots + A_n)$$

$$= [P(A_1) + P(A_2) + \cdots + P(A_n)] \qquad \leftarrow \binom{n}{1} = n \text{ terms}$$

$$- \left[ P(A_1 A_2) + P(A_1 A_3) + \cdots + P\left(\underset{i \neq j}{A_i A_j}\right) \right] \qquad \leftarrow \binom{n}{2} \text{ terms}$$

$$+ \left[ P(A_1 A_2 A_3) + P(A_1 A_2 A_4) + \cdots + P\left(\underset{i \neq j \neq k}{A_i A_j A_k}\right) \right] \leftarrow \binom{n}{3} \text{ terms}$$

$$\cdots \cdots \cdots \cdots \cdots \cdots \cdots \cdots \cdots \cdots \cdots \cdots$$

$$(-1)^{n-1}[P(A_1 A_2 \cdots A_n)] \qquad \leftarrow \binom{n}{n} = 1 \text{ term}$$

(A11)

The complete expansion of Eq. (A11) involves $(2^n - 1)$ terms.

## A4.3 Conditional Probabilities and Independence

It is important to study in more detail the probability of an intersection of two events, that is, $P(A_1 A_2)$. We are especially interested in how $P(A_1 A_2)$ is related to $P(A_1)$ and $P(A_2)$.

Before proceeding further we must define conditional probability and introduce a new notation. Suppose we want the probability of obtaining the four of clubs on one draw from a deck of cards. The answer is of course 1/52, which can be written: $P(C_4) = 1/52$. Let us change the problem so it reads: What is the probability of drawing the four of clubs *given that a club is drawn*? The answer is 1/13.

In such a situation we call the probability statement a *conditional probability*. The notation $P(C_4|C) = 1/13$ is used to represent the conditional probability of drawing a four of clubs given that a club is drawn. We read $P(A_2|A_1)$ as the probability of $A_2$ occurring conditioned on the previous occurrence of $A_1$, or more simply as the probability of $A_2$ given $A_1$.

$$P(A_1 A_2) = P(A_1) P(A_2|A_1) \quad \text{(A12a)}$$

$$P(A_1 A_2) = P(A_2) P(A_1|A_2) \quad \text{(A12b)}$$

Intuition tells us that there must be many cases in which

$$P(A_2|A_1) = P(A_2)$$

In other words, the probability of occurrence of event $A_2$ is independent of the occurrence of event $A_1$. From Eq. (A12a) we see that this implies $P(A_1 A_2) = P(A_1) P(A_2)$, and this latter result in turn implies

$$P(A_1|A_2) = P(A_1)$$

Thus we define independence by any one of the three equivalent relations

$$P(A_1 A_2) = P(A_1) P(A_2) \quad \text{(A13a)}$$

or

$$P(A_1|A_2) = P(A_1) \quad \text{(A13b)}$$

or

$$P(A_2|A_1) = P(A_2) \quad \text{(A13c)}$$

Conditional probabilities are sometimes called *dependent probabilities*.

One can define conditional probabilities for three events by splitting event

$B$ into the intersection of events $A_2$ and $A_3$. Then letting $A = A_1$ and $B = A_2A_3$, we have

$$P(AB) = P(A)P(B|A) = P(A_1)P(A_2A_3|A_1)$$
$$= P(A_1)P(A_2|A_1)P(A_3|A_1A_2)$$

Successive application of this technique leads to the general result

$$P(A_1A_2 \cdots A_n) = P(A_1)P(A_2|A_1)P(A_3|A_1A_2) \cdots$$
$$P(A_n|A_1A_2 \cdots A_{n-1}) \tag{A14}$$

Thus, the probability of the union of $n$ terms is expressed as the joint product of one independent probability and $n - 1$ dependent probabilities.

## A5 DISCRETE RANDOM VARIABLES

### A5.1 Density Function

We can define **x** as a random variable if we associate each value of **x** with an element in event $A$ defined on sample space $S$. If the random variable **x** assumes a finite number of values, then **x** is called a *discrete random variable*. In the case of a discrete random variable, we associate with each value of **x** a number $x_i$ and a probability of occurrence $P(x_i)$. We could describe the probabilities associated with the random variable by a table of values, but it is easier to write a formula that permits calculation of $P(x_i)$ by substitution of the appropriate value of $x_i$. Such a formula is called a *probability function* for the random variable **x**. More exactly, we use the notation $f(x)$ to mean a discrete probability density function associated with the discrete random variable **x**. (The reason for the inclusion of the word "density" will be clear once the parallel development for continuous random variables is completed.) Thus,

$$P(\mathbf{x} = x_i) = P(x_i) = f(x_i) \tag{A15}$$

In general we use the sequence of positive integers $0, 1, 2, \ldots, n$ to represent the subscripts of the $n + 1$ discrete values of **x**. Thus, the random variable is denoted by **x** and particular values of the random variable by $x_1, x_2, \ldots, x_n$. If the random variable under consideration is a nonnumerical quantity, e.g., the colors of the spectrum (red, orange, yellow, green, blue, indigo, violet), then the colors (or other quantity) would first be coded by associating a number 1 to 7 with each. If the random variable **x** is defined over the entire sample space $S$, $P(A)$ is given by

$$P(A) = \sum_{\substack{\text{for all } x_i \\ \text{values which} \\ \text{are elements of } A}} P(x_i) = \sum_{\substack{\text{for all } x_i \\ \text{in } A}} f(x_i) \tag{A16}$$

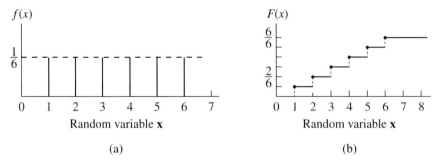

**Figure A2** (a) Line diagram depicting the discrete density function for the throw of one die; (b) step diagram depicting the discrete distribution function for the density function given in (a).

The probability of the sample space is

$$P(S) = \sum_{\substack{\text{over all} \\ i}} f(x_i) = 1 \tag{A17}$$

As an example of the above concepts we shall consider the throw of one die. The random variable **x** is the number of spots which face up on any throw. The domain of the random variable is $x = 1, 2, 3, 4, 5, 6$. Using the equally-likely-events hypothesis, we conclude that

$$P(x=1) = P(x=2) = \cdots = \tfrac{1}{6}$$

Thus, $f(x) = 1/6$, a constant density function. This can also be depicted graphically as in Fig. A2(a). The probability of an even roll is

$$P(\text{even}) = \sum_{i=2,4,6} f(x_i) = \tfrac{1}{6} + \tfrac{1}{6} + \tfrac{1}{6} = \tfrac{1}{2}$$

## A5.2  Distribution Function

It is often convenient to deal with another, related function rather than the density function itself. The *distribution function* is defined in terms of the probability that $\mathbf{x} \leq x$:

$$P(\mathbf{x} \leq x) \equiv F(x) = \sum_{\mathbf{x} \leq x} f(x) \tag{A18}$$

The distribution function is a cumulative probability and is often called the *cumulative distribution function*. The analytical form of $F(x)$ for the example in Fig. A2 is

$$F(x) = \frac{x}{6} \quad \text{for } 1 \le x \le 6 \tag{A19}$$

Equation (A19) related $F(x)$ to $f(x)$ by a process of summation. One can write an inverse relation[1] defining $f(x)$ in terms of the difference between two values of $F(x)$

$$f(x) = F(x^+) - F(x^-) \tag{A20}$$

In other words, $f(x)$ is equal to the value of the discontinuity at $x$ in the step diagram of $F(x)$. There are a few basic properties of density and distribution functions which are of importance: (1) since $f(x)$ is a probability, $0 \le f(x) \le 1$; (2) because $P(S) = 1$,

$$\sum_{\text{all } x} f(x) = 1$$

### A5.3 Binomial Distribution

Many discrete probability models are used in applications, the foremost being the binomial distribution and the Poisson distribution. The *binomial distribution* (sometimes called the *Bernoulli distribution*) applies to a situation in which an event can either occur or not occur (the more common terms are *success* or *failure*, a legacy from the days when probability theory centered around games of chance). The terms success and failure, of course, are ideally suited to reliability applications. The probability of success on any one trial is $p$, and that of failure is $1 - p$. The number of independent trials is denoted by $n$, and the number of successes by $r$. Thus, the probability of $r$ successes in $n$ trials with the probability of one success being $p$ is

$$B(r; n, p) = \binom{n}{r} p^r (1-p)^{n-r} \quad \text{for } r = 0, 1, 2, \ldots, n \tag{A21}$$

where

$$\binom{n}{r} = n!/r!(n-r)! \equiv \text{number of combinations of } n \text{ things taken } r \text{ at a time}$$

A number of line diagrams for the binomial density function[2] are given in Fig. A3. In Fig. A3 the number of trials is fixed at nine, and the probability of success on each trial is changed from 0.2 to 0.5 to 0.8. Intuition tells us

---

[1]The notations $F(x^+)$ and $F(x^-)$ mean the limits approached from the right and left, respectively.
[2]We use the notation $B(r; n, p)$ rather than the conventional and less descriptive notation $f(x)$.

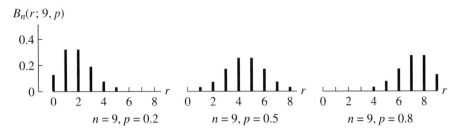

**Figure A3** Binomial density function for fixed $n$. (Adapted from Wadsworth and Bryan [1960].)

that the most probable number of successes is $np$, which is 1.8, 4.5, and 7.2, respectively. (It is shown in Section A7 that intuition has predicted the mean value.)

**Example 1:** Clearly, we could use the binomial distribution to predict the probability of twice obtaining a 3, in six throws of a die:

$$r = 2 \quad n = 6 \quad p = \tfrac{1}{6}$$

$$B(2; 6, \tfrac{1}{6}) = \binom{6}{2} (\tfrac{1}{6})^2 (1 - \tfrac{1}{6})^{6-2} = 15 \times 0.0131 = 0.196$$

**Example 2:** We can also use the binomial distribution to evaluate the probability of picking three aces on ten draws with replacement from a deck; however, if we do not replace the drawn cards after each pick, the binomial model will no longer hold, since the parameter $p$ will change with each draw. The binomial distribution does not hold when draws are made without replacement, because the trials are no longer independent. The proper distribution to use in such a case is the hypergeometric distribution [Freeman, 1963, pp. 113–120; Wadsworth and Bryan, 1960, p. 59].

$$H(k; j, n, N) = \frac{\binom{n}{k} \binom{N-n}{j-k}}{\binom{N}{j}} \quad \text{(A21a)}$$

where

$k$ = the number of successes
$j$ = the number of trials
$n$ = the finite number of possible successes
$N$ = the finite number of possible trials

## A5.4 Poisson Distribution

Another discrete distribution of great importance is the *Poisson distribution*, which can be derived in a number of ways. One derivation will be outlined in this section (see Shooman [1990], pp. 37–42), and a second derivation in Section A8. If $p$ is very small and $n$ is very large, the binomial density, Eq. (A21), takes on a special limiting form, which is the Poisson law of probability. Starting with Eq. (A21), we let $np$, the most probable number of occurrences, be some number $\mu$

$$\mu = np \qquad \therefore p = \frac{\mu}{n}$$

$$B\left(r; n, \frac{\mu}{n}\right) = \frac{n!}{r!(n-r)!} \left(\frac{\mu}{n}\right)^r \left(1 - \frac{\mu}{n}\right)^{n-r}$$

The limiting form called the Poisson distribution is

$$f(r; \mu) = \frac{\mu^r e^{-\mu}}{r!} \tag{A22}$$

The Poisson distribution can be written in a second form, which is very useful for our purposes. If we are interested in events which occur in time, we can define the rate of occurrence as the constant $\lambda$ = occurrences per unit time; thus $\mu = \lambda t$. Substitution yields the alternative form of the Poisson distribution:

$$f(r; \lambda, t) = \frac{(\lambda t)^r e^{-\lambda t}}{r!} \tag{A23}$$

Line diagrams for the Poisson density function given in Eq. (A22) are shown in Fig. A4 for various values of $\mu$. Note that the peak of the distribution is near $\mu$ and that symmetry about the peak begins to develop for larger values of $\mu$.

## A6 CONTINUOUS RANDOM VARIABLES

### A6.1 Density and Distribution Functions

The preceding section introduced the concept of a discrete random variable and its associated density and distribution functions. A similar development will be pursued in this section for continuous variables. Examples of some continuous random variables are the length of a manufactured part, the failure time of a system, and the value of a circuit resistance. In each of these examples there is no reason to believe that the random variable takes on discrete values. On the contrary, the variable is continuous over some range of definition. In a manner analogous to the development of the discrete variable, we define a continuous

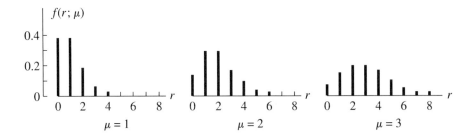
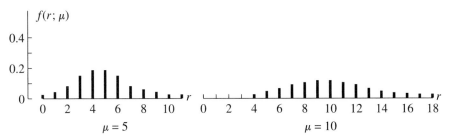

**Figure A4** Poisson density function for several values of $\mu$.

density function and a continuous distribution function. We shall start with the cumulative distribution function.

The cumulative distribution function for the discrete case was defined in Eq. (A18) as a summation. If the spacings between the discrete values of the random variable **x** are $\Delta x$ and we let $\Delta x \to 0$, then the discrete variable becomes a continuous variable, and the summation becomes an integration. Thus, the cumulative distribution function of a continuous random variable is given by

$$F(x) = \int_{\text{over the domain of } \mathbf{x}} f(x)\, dx \tag{A24}$$

If we let **x** take on all values between points $a$ and $b$

$$P(\mathbf{x} \leq x) = F(x) = \int_a^x f(x)\, dx \quad \text{for} \quad a < \mathbf{x} \leq b \tag{A25}$$

The density function $f(x)$ is given by the derivative of the distribution function. This is easily seen from Eq. (A25) and the fact that the derivative of the integral of a function is the function itself.

$$\frac{dF(x)}{dx} = f(x) \tag{A26}$$

The probability that **x** lies in an interval $x < \mathbf{x} < x + dx$ is given by

$$P(x < \mathbf{x} < x + dx) = P(\mathbf{x} \le x + dx) - P(x \le \mathbf{x})$$

$$= \int_a^{x+dx} f(x)\,dx - \int_a^x f(x)\,dx = \int_x^{x+dx} f(x)\,dx$$

$$= F(x + dx) - F(x) \qquad (A27)$$

It is easy to see from Eq. (A27) that if $F(x)$ is continuous and we let $dx \to 0$, $P(\mathbf{x} = x)$ is zero. Thus, when we deal with continuous probability, it makes sense to talk of the probability that **x** is within an interval rather than at one point. In fact since the $P(\mathbf{x} = x)$ is zero, the numerical value is the same in the continuous case whether the interval is open or closed since

$$P(a \le \mathbf{x} \le b) = P(a < \mathbf{x} < b) = P(a \le \mathbf{x} < b) = P(a < \mathbf{x} \le b)$$

Thus, the density function $f(x)$ is truly a density, and like any other density function it has a value only when integrated over some finite interval. The basic properties of density and distribution functions previously discussed in the discrete case hold in the continuous case. At the lower limit of **x** we have $F(a) = 0$, and at the upper limit $F(b) = 1$. These two statements, coupled with Eq. (A27), lead to $\int_a^b f(x)\,dx = 1$. Since $f(x)$ is a probability, $f(x)$ is nonnegative, and $F(x)$, its integral, is a nondecreasing function.

## A6.2 Rectangular Distribution

The simplest continuous variable distribution is the uniform or rectangular distribution shown in Fig. A5(a). The two parameters of this distribution are the limits $a$ and $b$. This model predicts a uniform probability of occurrence in any interval

$$P(x < \mathbf{x} \le x + \Delta x) = \Delta x (b - a)^{-1}$$

between $a$ and $b$.

## A6.3 Exponential Distribution

Another simple continuous variable distribution is the exponential distribution. The exponential density function is

$$f(x) = \lambda e^{-\lambda x} \qquad 0 < \mathbf{x} \le +\infty \qquad (A28)$$

which is sketched in Fig. A5(b). This distribution recurs time and time again in reliability work. The exponential is the distribution of the time to failure $t$ for a great number of electronic-system parts. The parameter $\lambda$ is constant and

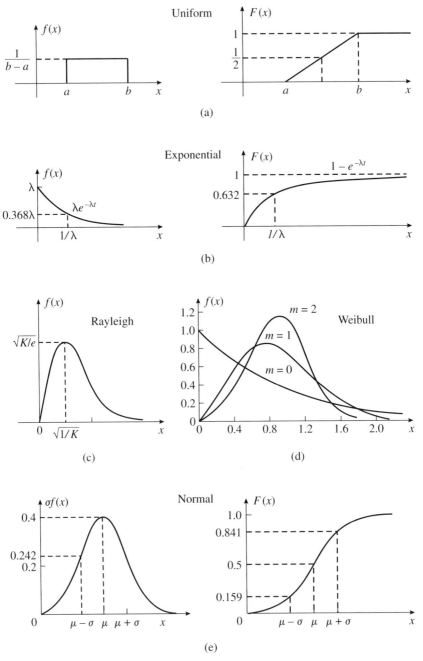

**Figure A5** Various continuous variable probability distributions: (a) uniform distribution; (b) exponential distribution; (c) Rayleigh distribution; (d) Weibull distribution; and (e) normal distribution.

is called the *conditional failure rate* with the units fractional failures per hour. The distribution function yields the failure probability and $1 - F(t)$ the success probability. Specifically, the probability of no failure (success) in the interval $0 - t$ is given by

$$P_s(t_1) = 1 - F(t_1) = e^{-\lambda t_1}$$

## A6.4  Rayleigh Distribution

Another single-parameter density function of considerable importance is the *Rayleigh distribution*, which is given as

$$f(x) = Kxe^{-Kx^2/2} \qquad 0 < x \leq +\infty \tag{A29}$$

and for the distribution function,

$$F(x) = 1 - e^{-Kx^2/2} \tag{A30}$$

The density function is sketched in Fig. A5(c). The Rayleigh distribution finds application in noise problems in communication systems and in reliability work. Whereas the exponential distribution holds for time to failure of a component with a constant conditional failure rate $\lambda$, the Rayleigh distribution holds for a component with a linearly increasing conditional failure rate $Kt$. The probability of success of such a unit is

$$P_s(t) = 1 - F(t) = e^{-Kt^2/2}$$

## A6.5  Weibull Distribution

Both the exponential and the Rayleigh distributions are single-parameter distributions which can be represented as special cases of a more general two-parameter distribution called the *Weibull distribution*. The density and distribution functions for the Weibull are

$$f(x) = Kx^m e^{-Kx^{m+1}/(m+1)} \qquad F(x) = 1 - e^{-Kx^{m+1}/(m+1)} \tag{A31}$$

This family of functions is sketched for several values of $m$ in Fig. A5(d). When $m = 0$, the distribution becomes exponential, and when $m = 1$, a Rayleigh distribution is obtained. The parameter $m$ determines the shape of the distribution, and parameter $K$ is a scale-change parameter.

## A6.6 Normal Distribution

The best-known two-parameter distribution is the *normal*, or *Gaussian*, distribution. This distribution is very often a good fit for the size of manufactured parts, the size of a living organism, or the magnitude of certain electric signals. It can be shown that when a random variable is the sum of many other random variables, the variable will have a normal distribution in most cases.

The density function for the normal distribution is written as

$$f(x) = \frac{1}{\sigma\sqrt{2\pi}} e^{-x^2/2\sigma^2} \qquad -\infty < \mathbf{x} < +\infty$$

This function has a peak of $1/\sigma\sqrt{2\pi}$ at $x = 0$ and falls off symmetrically on either side of zero. The rate of falloff and the height of the peak at $x = 0$ are determined by the parameter $\sigma$, which is called the *standard deviation*. In general one deals with a random variable $\mathbf{x}$ which is spread about some value such that the peak of the distribution is not at zero. In this case one shifts the horizontal scale of the normal distribution so that the peak occurs at $x = \mu$

$$f(x) = \frac{1}{\sigma\sqrt{2\pi}} e^{-(x-\mu)^2/2\sigma^2} \qquad (A32)$$

The effect of changing $\sigma$ is as follows: a large value of $\sigma$ means a low, broad curve and a small value of $\sigma$ a thin, high curve. A change in $\mu$ merely slides the curve along the $x$ axis.

The distribution function is given by

$$F(x) = \frac{1}{\sigma\sqrt{2\pi}} \int_{-\infty}^{x} e^{-(\xi-\mu)^2/2\sigma^2} d\xi \qquad (A32a)$$

where $\xi$ is a dummy variable of integration. The shapes of the normal density and distribution functions are shown in Fig. A5(e). The distribution function given in Eq. (A32) is left in integral form since the result cannot be expressed in closed form. This causes no particular difficulty, since $f(x)$ and $F(x)$ have been extensively tabulated and approximate expansion formulas are readily available [Abramovitz and Stegun, 1972, pp. 931–936, Section 26.2]. In tabulating the integral of Eq. (A32) it is generally convenient to introduce the change of variables $\mathbf{t} = (\mathbf{x} - \mu)/\sigma$, which shifts the distribution back to the origin and normalizes the $x$ axis in terms of $\sigma$ units.

The area under the $f(x)$ curve between $a$ and $b$ is of interest since it represents the probability that $\mathbf{x}$ is within the interval $a < \mathbf{x} \le b$. The areas for $-1 < \mathbf{t} \le +1$, $-2 < \mathbf{t} \le +2$, and $-3 < \mathbf{t} \le +3$ are shown in Fig. A6 along with a short table of areas between $-\infty$ and $t$.

The normal distribution can also be used as a limiting form for many other distributions. The binomial distribution approaches the normal distribution for large $n$.

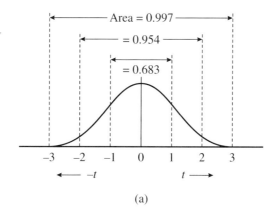

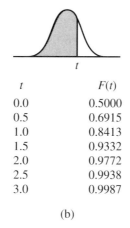

| $t$ | $F(t)$ |
|---|---|
| 0.0 | 0.5000 |
| 0.5 | 0.6915 |
| 1.0 | 0.8413 |
| 1.5 | 0.9332 |
| 2.0 | 0.9772 |
| 2.5 | 0.9938 |
| 3.0 | 0.9987 |

(b)

**Figure A6** Area under the normal curve: (a) for $-1 < \mathbf{t} \leq 1$, $-2 < \mathbf{t} \leq 2$, and $-3 < \mathbf{t} \leq 3$; and (b) between $-\infty$ and various values of $t$.

## A7 MOMENTS

The density or distribution function of a random variable contains all the information about the variable, i.e., the entire story. Sometimes the entire story of the random variable is not necessary, and an excerpt which sufficiently characterizes the distribution is sufficient. In such a case one computes a few moments (generally two) for the distribution and uses them to delineate the salient features. The moments are weighted integrals of the density function which describe various geometrical properties of the density function.

### A7.1 Expected Value

It is easy to express the various moments of a probability distribution in terms of an operator called the *expected value*. The expected value of the continuous random variable **x** defined over the range $a < \mathbf{x} \leq b$ with density function $f(x)$ is

$$E(\mathbf{x}) = \int_a^b x f(x)\, dx \tag{A33}$$

For a discrete random variable **x** taking on values $x = x_1, x_2, \ldots, x_n$ the expected value is defined in terms of a summation

$$E(\mathbf{x}) = \sum_{i=1}^{n} x_i f(x_i) \tag{A34}$$

## A7.2  Moments

To be more general one defines an entire set of moments. The $n$th moment of the random variable **x** computed about the origin and defined over the range $a < \mathbf{x} \leq b$ is given by

$$m_r = \int_{-\infty}^{+\infty} x^r f(x)\, dx \tag{A35}$$

The zero-order moment $m_0$ is the area under the density function, which is, of course, unity. The first-order moment is simply the expected value, which is called the *mean* and is given the symbol $\mu$

$$m_1 = E(\mathbf{x}) = \mu \tag{A36}$$

The origin moments for a discrete random variable which takes on the values $x_1, x_2, \ldots, x_n$ are given by

$$m_r = \sum_{i=1}^{n} x_i^r f(x_i) \tag{A37}$$

It is often of importance to compute moments about the mean rather than the origin. The set of moments about the mean are defined as follows:
For continuous random variables:

$$m_r' = E[(\mathbf{x} - \mu)^r] = \int_{-\infty}^{+\infty} (x - \mu)^r f(x)\, dx \tag{A38}$$

For discrete random variables:

**TABLE A1  Mean and Variance for Several Distributions**

| Distribution | $E(x)$ | var $x$ |
|---|---|---|
| Binomial | $np$ | $np(1-p)$ |
| Poisson | $\mu$ | $\mu$ |
| Exponential | $\dfrac{1}{\lambda}$ | $\dfrac{1}{\lambda^2}$ |
| Rayleigh | $\sqrt{\dfrac{\pi}{2K}}$ | $\dfrac{0.4292}{K}$ |
| Weibull | $\left(\dfrac{K}{m+1}\right)^{1-\varepsilon}\Gamma(\varepsilon)$ | $\left(\dfrac{K}{m+1}\right)^{1-\delta}\Gamma(\delta) - [E(x)]^2$ |
|  | $\varepsilon \equiv \dfrac{m+2}{m+1}$ | $\delta \equiv \dfrac{m+3}{m+1}$ |
|  | $\Gamma \equiv$ the gamma function |  |
| Normal | $\mu$ | $\sigma^2$ |

$$m'_r = E[(\mathbf{x}-\mu)^r] = \sum_{i=1}^{n}(x_i-\mu)^r f(x_i) \tag{A39}$$

The second moment about the mean, $m'_2 = \int_{-\infty}^{+\infty}(x-\mu)^2 f(x)\,dx$, is called the *variance* of $\mathbf{x}$, var $\mathbf{x}$, and is a measure of the sum of the squares of the deviations from $\mu$. Generally this is expressed in terms of the standard deviation $\sigma = \sqrt{\text{var } \mathbf{x}}$. One can easily express var $\mathbf{x}$ and $\sigma$ in terms of the expected-value operator:

$$\sigma^2 = \text{var } \mathbf{x} = E(\mathbf{x}^2) - \mu^2$$

The means and variances of the distributions discussed in Section A5 are given in Table A1.

## A8  MARKOV MODELS

### A8.1  Properties

There are basically four kinds of Markov probability models, one of which plays a central role in reliability. Markov models are functions of two random variables: the state of the system $\mathbf{x}$ and the time of observation $\mathbf{t}$. The four kinds of models arise because both $\mathbf{x}$ and $\mathbf{t}$ may be either discrete or continuous random variables, resulting in four combinations. As a simple example of the concepts of state and time of observation, we visualize a shoe box with two interior partitions which divide the box into three interior compartments labeled 1, 2, and 3. A Ping-Pong ball is placed into one of these compartments, and the box is periodically tapped on the bottom, causing the Ping-Pong ball to jump up and fall back into one of the three compartments. (For the moment we neglect the possibility that it falls out onto the floor.) The states of the system are the three compartments in the box. The time of observation is immediately after each rap, when the ball has fallen back into one of the compartments. Since we specified that the raps occur periodically, the model is discrete in both state and time. This sort of model is generally called a *Markov chain model* or a *discrete-state discrete-time model*. When the raps at the bottom of the box occur continuously, the model becomes a discrete-state continuous-time model, called a *Markov process*. If we remove the partitions and call the long axis of the box the $x$ axis, we can visualize a continuum of states from $x = -l/2$ to $x = +l/2$. If the ball is coated with rubber cement, it will stick wherever it hits when it falls back into the box. In this manner we can visualize the other two types of models, which involve a continuous-state variable. We shall be concerned only with the discrete-state continuous-time model, the Markov process.

Any Markov model is defined by a set of probabilities $p_{ij}$ which define the probability of transition from any state $i$ to any state $j$. If in the discrete-state case we make our box compartments and the partitions equal in size, all the transition probabilities should be equal. (In the general case, where each compartment is of different size, the transition probabilities are unequal.) One of the most important features of any Markov model is that the transition probability $p_{ij}$ depends only on states $i$ and $j$ and is completely independent of all past states except the last one, state $i$. This seems reasonable in terms of our shoe-box model since transitions are really dependent only on the height of the wall between adjacent compartments $i$ and $j$ and the area of the compartments and not on the sequence of states the ball has occupied before arriving in state $i$. Before delving further into the properties of Markov processes, an example of great importance, the Poisson process, will be discussed.

## A8.2 Poisson Process

In Section A5.4 the Poisson distribution was introduced as a limiting form of the binomial distribution. In this section we shall derive the Poisson distribution as the governing probability law for a Poisson process, a particular kind of Markov process. In a Poisson process we are interested in the number of occurrences in time, the probability of each occurrence in a small time $\Delta t$ being a constant which is the parameter of the process. Examples of Poisson processes are the number of atoms transmuting as a function of time in the radioactive decay of a substance, the number of noise pulses as a function of time in certain types of electric systems, and the number of failures for a group of components operating in a standby mode or in an instantaneous-replacement situation. The occurrences are discrete, and time is continuous; therefore this is a discrete-state continuous-time model. The basic assumptions which are necessary in deriving a Poisson process model are as follows:

1. The probability that a transition occurs from the state of $n$ occurrences to the state of $n + 1$ occurrences in time $\Delta t$ as $\lambda \Delta t$. The parameter $\lambda$ is a constant and has the dimensions of occurrences per unit time. The occurrences are irreversible, which means that the number of occurrences can never decrease with time.
2. Each occurrence is independent of all other occurrences.
3. The transition probability of two or more occurrences in interval $\Delta t$ is negligible. Another way of saying this is to make use of the independence-of-occurrence property and write the probability of two occurrences in interval $\Delta t$ as the product of the probability of each occurrence, that is, $(\lambda \Delta t)(\lambda \Delta t)$. This is obviously an infinitesimal of second order for $\Delta t$ small and can be neglected.

We wish to solve for the probability of $n$ occurrences in time $t$, and to that end we set up a system of difference equations representing the state probabilities and transition probabilities. The probability of $n$ occurrences having taken place by time $t$ is denoted by

$$P(x = n, t) \equiv P_n(t)$$

For the case of zero occurrences at time $t + \Delta t$ we write the following difference equation:

$$P_0(t + \Delta t) = (1 - \lambda \Delta t) P_0(t) \tag{A40}$$

which says that the probability of zero occurrences at time $t + \Delta t$ is $P_0(t + \Delta t)$. This probability is given by the probability of zero occurrences at time $t$, $P_0(t)$, multiplied by the probability of no occurrences in interval $\Delta t$, $1 - \lambda \Delta t$. For the case of one occurrence at time $t + \Delta t$ we write

$$P_1(t + \Delta t) = (\lambda \Delta t) P_0(t) + (1 - \lambda \Delta t) P_1(t) \tag{A41}$$

The probability of one occurrence at $t + \Delta t$, $P_1(t + \Delta t)$, can arise in two ways: (1) either there was no occurrence at time $t$, $P_0(t)$, and one happened in the interval $\Delta t$ (with probability $\lambda \Delta t$), or (2) there had already been one occurrence at time $t$, $P_1(t)$, and no additional ones came along in the time interval $\Delta t$ (probability $1 - \lambda \Delta t$). It is clear that Eq. (A41) can be generalized, yielding

$$P_n(t + \Delta t) = (\lambda \Delta t) P_{n-1}(t) + (1 - \lambda \Delta t) P_n(t) \qquad \text{for } n = 1, 2, \ldots \tag{A42}$$

The difference equations (A40) and (A41) really describe a discrete-time system, since time is divided into intervals $\Delta t$, but by taking limits as $\Delta t \to 0$ we obtain a set of differential equations which truly describe the continuous-time Poisson process. Rearranging Eq. (A40) and taking the limit of both sides of the equation at $\Delta t \to 0$ leads to

$$\lim_{\Delta t \to 0} \frac{P_0(t + \Delta t) - P_0(t)}{\Delta t} = \lim_{\Delta t \to 0} -\lambda P_0(t)$$

By definition the left-hand side of the equation is the time derivative of $P_0(t)$ and the right-hand side is independent of $\Delta t$; therefore

$$\frac{dP_0(t)}{dt} = \dot{P}_0(t) = -\lambda P_0(t) \tag{A43}$$

Similarly for Eq. (A41)

$$\lim_{\Delta t \to 0} \frac{P_n(t + \Delta t) - P_n(t)}{\Delta t} = \lim_{\Delta t \to 0} \lambda P_{n-1}(t) - \lim_{\Delta t \to 0} \lambda P_n(t)$$

$$\frac{dP_n(t)}{dt} = \dot{P}_n(t) = \lambda P_{n-1}(t) - \lambda P_n(t)$$

$$\text{for } n = 1, 2, \ldots, n \tag{A44}$$

Equations (A43) and (A44) are a complete set of differential equations which, together with a set of initial conditions, describe the process. If there are no occurrences at the start of the problem, $t = 0$, $n = 0$, and

$$P_0(0) = 1, P_1(0) = P_2(0) = \cdots = P_n(0) = 0$$

Solution of this set of equations can be performed in several ways: classical differential-equation techniques, Laplace transforms, matrix methods, etc. In this section we shall solve them using the classical technique of undetermined coefficients. Substituting a solution of the form $Ae^{st}$ gives $s = -\lambda$, and substituting the initial condition $P_0(0) = 1$ gives

$$P_0(t) = e^{-\lambda t} \tag{A45}$$

For $n = 1$, Eq. (A42) becomes

$$\dot{P}_1(t) = \lambda P_0(t) - \lambda P_1(t)$$

Substitution from Eq. (A45) and rearrangement yields

$$\dot{P}_1(t) + \lambda P_1(t) = \lambda e^{-\lambda t}$$

The homogeneous portion of this equation is the same as that for $P_0(t)$. The particular solution is of the form $Bte^{-\lambda t}$. Substituting yields $B = \lambda$, and using the initial condition $P_1(0) = 0 = A$ gives

$$P_1(t) = \lambda t e^{-\lambda t} \tag{A46}$$

**TABLE A2  A Transition Matrix**

| Initial States | | Final States | | | | |
|---|---|---|---|---|---|---|
| | | $s_0(t+\Delta t)$ | $s_1(t+\Delta t)$ | $s_2(t+\Delta t)$ | $\cdots$ | $s_n(t+\Delta t)$ |
| $s_0(t)$ | $n=0$ | $p_{00}$ | $p_{01}$ | $p_{02}$ | $\cdots$ | $p_{0n}$ |
| $s_1(t)$ | $n=1$ | $p_{10}$ | $p_{11}$ | $p_{12}$ | $\cdots$ | $p_{1n}$ |
| $s_2(t)$ | $n=2$ | $p_{20}$ | $p_{21}$ | $p_{22}$ | $\cdots$ | $p_{2n}$ |
| $\cdots$ | | | | | | |
| $s_n(t)$ | $n=n$ | $p_{n0}$ | $p_{n1}$ | $p_{n2}$ | $\cdots$ | $p_{nn}$ |

It should be clear that solving for $P_n(t)$ for $n = 2, 3, \ldots$ will generate the Poisson probability law given in Eq. (A23). (Note: $m \equiv r$.)

Thus, the Poisson process has been shown to be a special type of Markov process which can be derived from the three basic postulates with no mention of the binomial distribution. We can give another important interpretation to $P_0(t)$. If we let $t_0$ be the time of the first occurrence, then $P_0(t)$ is the probability of no occurrences:

$$P_0(t) \equiv P(t < t_0) = 1 - P(t_0 < t)$$

Thus, $1 - P_0(t)$ is a cumulative distribution function for the random variable **t,** the time of occurrence. The density function for time of first occurrence is obtained by differentiation:

$$f(t) = \frac{d}{dt}(1 - e^{-\lambda t}) = \lambda e^{-\lambda t} \qquad (A47)$$

This means that the time of first occurrence is exponentially distributed. Since each occurrence is independent of all others, it also means that the time between any two occurrences is exponentially distributed.

## A8.3  Transition Matrix

Returning to some of the basic properties of Markov processes, we find that we can specify the process by a set of differential equations and their associated initial conditions. Because of the basic Markov assumption that only the last state is involved in determining the probabilities, we always obtain a set of first-order differential equations. The constants in these equations can be specified by constructing a transition-probability matrix.[3] The rows represent the probability of being in any state $A$ at time $t$ and the columns the probability of being in state $B$ at time $t + \Delta t$. The former are called *initial states* and the latter *final states*. An example is given in Table A2 for a process with $n$

---

[3] In Appendix B6–B8 a flowgraph model for a Markov process will be developed which parallels the use of the transition matrix. The flowgraph model is popular in engineering analysis.

**TABLE A3** The First Five Rows and Columns of the Transition Matrix for a Poisson Process

|        | $s_0(t+\Delta t)$ | $s_1(t+\Delta t)$ | $s_2(t+\Delta t)$ | $s_3(t+\Delta t)$ | $s_4(t+\Delta t)$ |
|--------|---|---|---|---|---|
| $s_0(t)$ | $1-\lambda\Delta t$ | $\lambda\Delta t$ | 0 | 0 | 0 |
| $s_1(t)$ | 0 | $1-\lambda\Delta t$ | $\lambda\Delta t$ | 0 | 0 |
| $s_2(t)$ | 0 | 0 | $1-\lambda\Delta t$ | $\lambda\Delta t$ | 0 |
| $s_3(t)$ | 0 | 0 | 0 | $1-\lambda\Delta t$ | $\lambda\Delta t$ |
| $s_4(t)$ | 0 | 0 | 0 | 0 | $1-\lambda\Delta t$ |

+ 1 discrete states. The transition probability $p_{ij}$ is the probability that in time $\Delta t$ the system will undergo a transition from initial state $i$ to final state $j$. Of course $p_{ii}$, a term on the main diagonal, is the probability that the system will remain in the same state during one transition. The sum of the $p_{ij}$ terms in any row must be unity, since this is the sum of all possible transition probabilities. In the case of a Poisson process, there is an infinite number of states. The transition matrix for the first five terms of a Poisson process is given in Table A3. Inspection of the Poisson example reveals that the difference equations[4] for the system can be obtained simply. The procedure is to equate the probability of any final state at the top of each column to the product of the transition probabilities in that column and the initial probabilities in the row. Specifically, for the transition matrix given in Table A2,

$$P_{s_0}(t+\Delta t) = p_{00}P_{s_0}(t) + p_{10}P_{s_1}(t) + \cdots + p_{n0}P_{s_n}(t)$$

If the $p_{ij}$ terms are all independent of time and depend only on constants and $\Delta t$, the process is called *homogeneous*. For a homogeneous process, the resulting differential equations have constant coefficients, and the solutions are of the form $e^{-rt}$ or $t^n e^{-rt}$. If for a homogeneous process the final value of the probability of being in any state is independent of the initial conditions, the process is called *ergodic*. A finite-state homogeneous process is ergodic if every state can be reached from any other state with positive probability. Whenever it is not possible to reach any other state from some particular state, the latter state is called an *absorbing state*. Returning to the partitioned shoebox example of Section A8.1, if we allow the ball to hop completely out of the box onto the floor, the floor forms a fourth state, which is absorbing. In a transition matrix any column $j$ having only a single entry ($p_{ij}$ along the main diagonal) is an absorbing state.

---

[4]The differential equations are obtained by taking the limit of the difference equations as $\Delta t \to 0$.

# REFERENCES

Abramovitz, M., and I. Stegun (eds.). *Handbook of Mathematical Functions*. Washington, DC: National Bureau of Standards, U.S. Government Printing Office (GPO), 1972.

Freeman, H. *Introduction to Statistical Inference*. Addison-Wesley, Reading, MA, 1963.

Freund, J. E. *Mathematical Statistics*. Prentice-Hall, Englewood Cliffs, NJ, 1962.

Meeker, W. Q., and L. A. Escobar. *Statistical Methods for Reliability Data*. Wiley, New York, 1998.

Mendenhall, W. et al. *Mathematical Statistics with Applications*, 4th ed. PWS-Kent, Boston, 1990.

Papoulis, A. *Probability, Random Variables, and Stochastic Processes*. McGraw-Hill, New York, 1965.

Shooman, M. L. *Probabilistic Reliability: An Engineering Approach*. McGraw-Hill, New York, 1968.

Shooman, M. L. *Software Engineering: Design, Reliability, and Management*. McGraw-Hill, New York, 1983.

Shooman, M. L. *Probabilistic Reliability: An Engineering Approach*, 2d ed. Krieger, Melbourne, FL, 1990.

Stone, C. J. *A Course in Probability and Statistics*. Wadsworth Publishers, New York, 1996.

Wadsworth, G. P., and J. G. Bryan. *Introduction to Probability and Random Variables*. McGraw-Hill, New York, 1960.

# PROBLEMS

Note: Problems A1–A5 are taken from Shooman [1990].

**A1.** The following two theorems are known as De Morgan's theorems:

$$\overline{A + B + C} = \overline{A}\,\overline{B}\,\overline{C}$$
$$\overline{ABC} = \overline{A} + \overline{B} + \overline{C}$$

Prove these two theorems using a Venn diagram. Do these theorems hold for more than three events? Explain.

**A2.** We wish to compute the probability of winning on the first roll of a pair of dice by throwing a seven or an eleven.
   (a) Define a sample space for the sum of the two dice.
   (b) Delineate the favorable and unfavorable outcomes.
   (c) Compute the probability of winning and losing.
   (d) List any assumptions you made in this problem.

**A3.** Suppose a resistor has a resistance $R$ with mean of $100\ \Omega$ and a tolerance of 5%, i.e., variation of $5\ \Omega$.
  - **(a)** If the resistance values are normally distributed with $\mu = 100\ \Omega$ and $\sigma = 5\ \Omega$, sketch $f(R)$.
  - **(b)** Assume that the resistance values have a Rayleigh distribution. If the peak is to occur at $100\ \Omega$, what is the value of $K$? Plot the Rayleigh distribution on the same graph as the normal distribution of part (a).

**A4.** A certain resistor has a nominal value (mean) of $100\ \Omega$.
  - **(a)** Assume a normal distribution and compute the value of $\sigma$ if we wish $P(95 < \mathbf{R} < 105) = 0.95$.
  - **(b)** Repeat part (a) assuming a Weibull distribution and specify the values of $K$ and $m$.
  - **(c)** Plot the density function for parts (a) and (b) on the same graph paper.

**A5.** Let a component have a good, a fair, and a bad state. Assume the transition probabilities of failure are from good to fair, $\lambda_{gf}\Delta t$; from good to bad, $\lambda_{gb}\Delta t$; and from fair to bad, $\lambda_{fb}\Delta t$.
  - **(a)** Formulate a Markov model.
  - **(b)** Compute the probabilities of being in any state.

# APPENDIX B

# SUMMARY OF RELIABILITY THEORY*

## B1 INTRODUCTION

### B1.1 History

Since its beginnings following World War II, reliability theory has grown into an engineering science in its own right. (The early development is discussed in Chapter 1 of Shooman [1990].) Much of the initial theory, engineering, and management techniques centered about hardware; however, human and procedural elements of a system were often included. Since the late 1960s the term *software reliability* has become popular, and now reliability theory refers to both *software* and *hardware reliability*.

### B1.2 Summary of the Approach

The conventional approach to reliability is to decompose the system into smaller subsystems and units. Then by the use of combinatorial reliability, the system probability of success is expressed in terms of the probabilities of success of the elements. Then by the use of failure rate models, the element probabilities of success are computed. These two concepts are combined to calculate the system reliability.

When reliability or availability of repairable systems is the appropriate fig-

---

*Parts of this appendix have been abstracted from Appendix B of *Software Engineering: Design, Reliability, and Management*, by M. L. Shooman, McGraw-Hill, New York, 1983; and also *Probabilistic Reliability: An Engineering Approach*, 2d ed., by M. L. Shooman, Krieger, 1990.

ure of merit, Markov models are generally used to compute the associated probabilities.

Often a proposed system does not meet its reliability specifications, and various techniques of reliability improvement are utilized to improve the predicted reliability of the design.

Readers desiring more detail are referred to Shooman [1990] and the references cited in that text.

## B1.3 Purpose of This Appendix

This appendix was written to serve several purposes. The prime reason is to provide additional background for those techniques and principles of reliabilty theory which are used in the software reliability models developed in Chapter 5. A second purpose is to expose software engineers who are not familiar with reliability theory to some of the main methods and techniques. This is especially important since many discussions of software reliability end up discussing how much of "hardware reliability theory" is applicable to software. This author feels the correct answer is "some"; however, the only way to really appreciate this answer is to learn something about reliability.

The third purpose is to allow readers who are software engineers to talk with and understand hardware reliability engineers. If a reliability and quality control (R&QC) engineer handles the hardware reliability estimates and the software engineer generates software reliability estimates, they must meet at the interface. Even if the R&QC engineer computes reliability estimates for both the hardware and the software, it is still necessary for the software engineer to work with him or her and provide information as well as roughly evaluate the thoroughness and quality of the software effort.

## B2 COMBINATORIAL RELIABILITY

### B2.1 Introduction

In performing the reliability analysis of a complex system, it is almost impossible to treat the system in its entirety. The logical approach is to decompose the system into functional entities composed of units, subsystems, or components. Each entity is assumed to have two states, one good and one bad. The subdivision generates a block-diagram or fault-tree description of system operation. Models are then formulated to fit this logical structure, and the calculus of probability is used to compute the system reliability in terms of the subdivision reliabilities. Series and parallel structures often occur, and their reliability can be described very simply. In many cases the structure is of a more complicated nature, and more general techniques are needed.

The formulation of a structural-reliability model can be difficult in a large, sophisticated system and requires much approximation and judgment. This is best done by a system engineer or someone closely associated with one who knows the system operation thoroughly.

## B2.2 Series Configuration

The simplest and perhaps most common structure in reliability analysis is the *series configuration*. In the series case the *functional* operation of the system depends on the proper operation of all system components. A series string of Christmas tree lights is an obvious example. The word *functional* must be stressed, since the electrical or mechanical configuration of the circuit may differ from the logical structure.

A series reliability configuration will be portrayed by the block-diagram representation shown in Fig. B1(a), or the reliability graph shown in Fig. B1(b). In either case, a single path from cause to effect is created. Failure of any component is represented by removal of the component, which interrupts the path and thereby causes the system to fail.

The system shown in Fig. B1 is divided into $n$ series–connected units. This system can represent $n$ parts in an electronic amplifier, the $n$ subsystems in an aircraft autopilot, or the $n$ operations necessary to place a satellite in orbit. The event signifying the success of the $n$th unit will be $x_n$, and $\bar{x}_n$ will represent the failure of the $n$th unit. The probability that unit $n$ is successful will be $P(x_n)$, and the probability that unit $n$ fails will be $P(\bar{x}_n)$. The probability of system success is denoted by $P_s$. In keeping with the definition of reliability, $P_s \equiv R$, where $R$ stands for the system reliability. The probability of system failure is

$$P_f = 1 - P_s$$

Since the series configuration requires that all units operate successfully for system success, the event representing system success is the intersection of $x_1, x_2, \ldots, x_n$. The probability of this event is given by

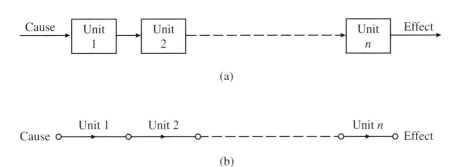

**Figure B1** Series reliability configuration: (a) reliability block diagram (RBD); (b) reliability graph.

**414** SUMMARY OF RELIABILITY THEORY

$$R = P_s = P(x_1 x_2 x_3 \cdots x_n) \tag{B1}$$

Expansion of Eq. (B1) yields

$$P_s = P(x_1)P(x_2|x_1)P(x_3|x_1 x_2) \cdots P(x_n|x_1 x_2 \cdots x_{n-1}) \tag{B2}$$

The expression appearing in Eq. (B2) contains conditional probabilities, which must be evaluated with care. For example, $P(x_3|x_1 x_2)$ is the probability of success of unit 3 evaluated under the condition that units 1 and 2 are operating. In the case where the power dissipation from units 1 and 2 affects the temperature of unit 3 and thereby its failure rate, a conditional probability is involved. If the units do not interact, the failures are independent, and Eq. (B2) simplifies to

$$P_s = P(x_1)P(x_2)P(x_3) \cdots P(x_n) \tag{B3}$$

The reliability of a series system is always smaller than the smallest reliability of the set of components. The lowest relability in the series is often referred to as "the weakest link in the chain." An alternative approach is to compute the probability of failure. The system fails if *any* of the units fail, and therefore we have a union of events

$$P_f = P(\bar{x}_1 + \bar{x}_2 + \bar{x}_3 + \cdots + \bar{x}_n) \tag{B4}$$

Expansion of Eq. (B4) yields

$$\begin{aligned} P_f = &[P(\bar{x}_1) + P(\bar{x}_2) + P(\bar{x}_3) + \cdots + P(\bar{x}_n)] \\ &- [P(\bar{x}_1 \bar{x}_2) + P(\bar{x}_1 \bar{x}_3) + \cdots + P(\bar{x}_i \bar{x}_j)] \\ & \quad\quad\quad\quad\quad\quad\quad\quad\quad\quad\quad\quad\quad\quad i \neq j \\ &+ \cdots + (-1)^{n-1}[P(\bar{x}_1 \bar{x}_2 \cdots \bar{x}_n)] \end{aligned} \tag{B5}$$

Since

$$P_s = 1 - P_f \tag{B6}$$

the probability of system success becomes

$$\begin{aligned} P_s = &1 - P(\bar{x}_1) - P(\bar{x}_2) - P(\bar{x}_3) - \cdots - P(\bar{x}_n) + P(\bar{x}_1)P(\bar{x}_2|\bar{x}_1) \\ &+ P(\bar{x}_1)P(\bar{x}_3|\bar{x}_1) + \cdots + P(\bar{x}_i)P(\bar{x}_i|\bar{x}_j) \\ &\quad\quad\quad\quad\quad\quad\quad\quad\quad\quad\quad i \neq j \\ &- \cdots + (-1)^n P(\bar{x}_1)P(\bar{x}_2|\bar{x}_1) \cdots P(\bar{x}_n|\bar{x}_1 \cdots \bar{x}_{n-1}) \end{aligned} \tag{B7}$$

The reliability expression in Eq. (B7) is equivalent to that in Eq. (B2) but is much more difficult to evaluate because of the many terms involved. Equation

(B7) also involves conditional probabilities; for example, $P(\bar{x}_3|\bar{x}_1\bar{x}_2)$ is the probability that unit 3 will fail given the fact that units 1 and 2 have failed. In the case of independence $P(\bar{x}_3|\bar{x}_1\bar{x}_2)$ becomes $P(\bar{x}_3)$, and the other conditional probability terms in Eq. (B7) simplify, yielding

$$P_s = 1 - P(\bar{x}_1) - P(\bar{x}_2) - P(\bar{x}_3) - \cdots - P(\bar{x}_n)$$
$$+ P(\bar{x}_1)P(\bar{x}_2) + P(\bar{x}_1)P(\bar{x}_3) + \cdots + \underset{i \neq j}{P(\bar{x}_i)P(\bar{x}_j)}$$
$$- \cdots + (-1)^n P(\bar{x}_1)P(\bar{x}_2)\cdots P(\bar{x}_n) \tag{B8}$$

Equation (B8) is still more complex than Eq. (B3). It is interesting to note that the reliability of any particular configuration may be computed by considering either the probability of success or the probability of failure. In a very complex structure both approaches may be used at different stages of the computation.

### B2.3 Parallel Configuration

In many systems several signal paths perform the same operation. If the system configuration is such that failure of one or more paths still allows the remaining path or paths to perform properly, the system can be represented by a parallel model.

A block diagram and reliability graph for a parallel system are shown in Fig. B2. There are $n$ paths connecting input to output, and all units must fail in order to interrupt all the paths. This is sometimes called a redundant configuration.

In a parallel configuration the system is successful if any one of the parallel channels is successful. The probability of success is given by the probability of the union of the $n$ successful events.

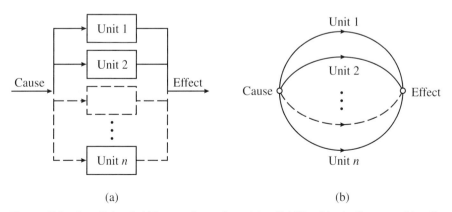

**Figure B2** Parallel reliability configuration: (a) reliability block diagram; (b) reliability graph.

$$P_s = P(x_1 + x_2 + x_3 + \cdots + x_n) \tag{B9}$$

Expansion of Eq. (B9) yields

$$P_s = [P(x_1) + P(x_2) + P(x_3) + \cdots + P(x_n)]$$
$$- [P(x_1 x_2) + P(x_1 x_3) + \cdots + P(x_i x_j)]_{i \neq j}$$
$$+ \cdots + (-1)^{n-1} P(x_1 x_2 \cdots x_n) \tag{B10}$$

The conditional probabilties which occur in Eq. (B10) when the *intersection terms* are expanded must be interpreted properly, as in the previous section [see Eq. (B7)]. A simpler formula can be developed in the parallel case if one deals with the probability of system failure. System failure occurs if *all* the system units fail, yielding the probability of their intersection.

$$P_f = P(\bar{x}_1 \bar{x}_2 \bar{x}_3 \cdots \bar{x}_n) \tag{B11}$$

where

$$P_s = 1 - P_f \tag{B12}$$

Substitution of Eq. (B11) into Eq. (B12) and expansion yields

$$P_s = 1 - P(\bar{x}_1) P(\bar{x}_2 | \bar{x}_1) P(\bar{x}_3 | \bar{x}_1 \bar{x}_2) \cdots P(\bar{x}_n | \bar{x}_1 \bar{x}_2 \cdots \bar{x}_{n-1}) \tag{B13}$$

If the unit failures are independent, Eq. (B13) simplifies to

$$P_s = 1 - P(\bar{x}_1) P(\bar{x}_2) \cdots P(\bar{x}_n) \tag{B14}$$

## B2.4  An $r$-out-of-$n$ Configuration

In many problems the system operates if $r$ out of $n$ units function, e.g., a bridge supported by $n$ cables, $r$ of which are necessary to support the maximum load. If each of the $n$ units is identical, the probabiality of exactly $r$ successes out of $n$ units is given by Eq. (A21)

$$B(r; n, p) = \binom{n}{r} p^r (1-p)^{n-r} \quad \text{for } r = 0, 1, 2 \cdots n \tag{B15}$$

where $p$ is the probability of success of any unit. The system will succeed if $r$, $r+1 \cdots n-1$, or $n$ units succeed. The probability of system success is given by

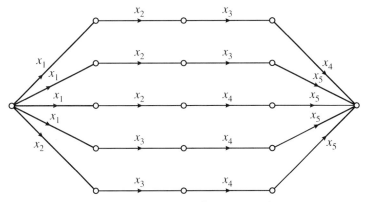

**Figure B3** Reliability graph for a 4-out-of-5 system.

$$P_s = \sum_{k=r}^{n} \binom{n}{k} p^k (1-p)^{n-k} \tag{B16}$$

If the units all differ, Eqs. (B15) and (B16) no longer hold, and one is faced with the explicit enumeration of all possible successful combinations. One can draw a reliability graph as an aid. The graph will have $\binom{n}{r}$ parallel paths. Each parallel path will contain $r$ different elements, corresponding to one of the combinations of $n$ things $r$ at a time. Such a graph for a four-out-of-five system is given in Fig. B3. The system succeeds if *any* path succeeds. Each path success depends on the success of four elements:

$$P_s = P(x_1 x_2 x_3 x_4 + x_1 x_2 x_3 x_5 + x_1 x_2 x_4 x_5 + x_1 x_3 x_4 x_5 + x_2 x_3 x_4 x_5) \tag{B17}$$

Expanding Eq. (B17) involves simplification of redundant terms. For example, the term $P[(x_1 x_2 x_3 x_4)(x_1 x_2 x_3 x_5)]$ becomes by definition $P(x_1 x_2 x_3 x_4 x_5)$. Thus, the equation simplifies to

$$P_s = P(x_1 x_2 x_3 x_4) + P(x_1 x_2 x_3 x_5) + P(x_1 x_2 x_4 x_5) + P(x_1 x_3 x_4 x_5)$$
$$+ P(x_2 x_3 x_4 x_5) - 4P(x_1 x_2 x_3 x_4 x_5) \tag{B18}$$

It is easy to check Eq. (B18). For independent, identical elements Eq. (B18) gives $P_s = 5p^4 - 4p^5$. From Eq. (B16) we obtain

$$P_s = \sum_{k=4}^{5} \binom{5}{k} p^k (1-p)^{5-k} = \binom{5}{4} p^4 (1-p)^1 + \binom{5}{5} p^5 (1-p)^0$$
$$= 5p^4 - 4p^5$$

## B2.5 Fault-Tree Analysis

Fault-tree analysis (FTA) is an application of deductive logic to produce a failure- or fault-oriented pictorial diagram, which allows one to analyze system safety and reliability. Various failure modes that can contribute to a specified undesirable event are organized deductively and represented pictorially.

First the top undesired event is defined and drawn. Below this, secondary undesired events are drawn. These secondary undesired events include the potential hazards and failures that are immediate causes of the top event. Below each of these subevents are drawn second-level events, which are the immediate causes of the subevents. The process is continued until basic events are reached (often called *elementary faults*). Since the diagram branches out and there are more events at each lower level, it resembles an inverted tree. The treelike structure of the diagram illustrates the various critical paths of subevents leading to the occurrence of the top undesired event. A fault tree for an auto braking system example is given in Section B5, Fig. B13.

Both FTAs and RBDs are useful for both qualitative and quantitative analyses:

1. They force the analyst to actively seek out failure events (success events) in a deductive manner.
2. They provide a visual display of how a system can fail, and thus aid understanding of the system by persons other than the designer.
3. They point out critical aspects of systems failure (system success).
4. They provide a systematic basis for quantitative analysis of reliability.

Often in a difficult practical problem one utilizes other techniques to decompose the system prior to effecting either an RBD or an FTA.

## B2.6 Failure Mode and Effect Analysis

Failure mode and effect analysis (FMEA) is a systematic procedure for identifying the modes of failures and for evaluating their consequences. It is a tabular procedure which considers hazards in terms of single-event chains and their consequences. The FMEA is generally performed on the basis of limited design information during the early stages of design and is periodically updated to reflect changes in design and improved knowledge of the system. The basic questions which must be answered by the analyst in performing an FMEA are:

1. How can each component or subsystem fail? (What is the failure mode?)
2. What cause might produce this failure? (What is the failure mechanism?)
3. What are the effects of each failure if it does occur?

Once the FMEA is completed, it assists the analyst in:

1. Selecting, during initial stages, various design alternatives with high reliability and high safety potential
2. Ensuring that all possible failure modes, and their effects on operational success of the system, have been taken into account
3. Identifying potential failures and the magnitude of their effects on the system
4. Developing testing and checkout methods
5. Providing a basis for qualitative reliability, availability, and safety analysis
6. Providing input data for construction of RBD and FTA models
7. Providing a basis for establishing corrective measures
8. Performing an objective evaluation of design requirements related to redundancy, failure detection systems, and fail-safe character

An FMEA for the auto braking example is given in Section B5, Table B3.

## B2.7 Cut-Set and Tie-Set Methods

A very efficient general method for computing the reliability of any system not containing dependent failures can be developed from the properties of the reliability graph. The reliability graph consists of a set of branches which represent the $n$ elements. There must be at least $n$ branches in the graph, but there can be more if the same branch must be repeated in more than one path (see Fig. B3). The probability of element success is written above each branch. The nodes of the graph tie the branches together and form the structure. A path has already been defined, but a better definition can be given in terms of graph theory. The term *tie set*, rather than path, is common in graph nomenclature. A tie set is a group of branches which forms a connection between input and output when traversed in the arrow direction. We shall primarily be concerned with *minimal* tie sets, which are those containing a minimum number of elements. If no node is traversed more than once in tracing out a tie set, the tie set is minimal. If a system has $i$ minimal tie sets denoted by $T_1, T_2, \ldots, T_i$, then the system has a connection between input and output if at least one tie set is intact. The system reliability is thus given by

$$R = P(T_1 + T_2 + \cdots + T_i) \tag{B19}$$

One can define a *cut set* of a graph as a set of branches which interrupts all connections between input and output when removed from the graph. The minimal cut sets are a group of distinct cut sets containing a minimum number of terms. All system failures can be represented by the removal of at least one minimal cut set from the graph. The probability of system failure is, therefore,

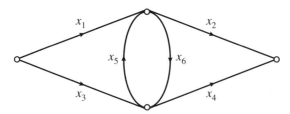

**Figure B4** Reliability graph for a six-element system.

given by the probability that at least one minimal cut set fails. If we let $C_1$, $C_2, \ldots, C_j$ represent the $j$ minimal cut sets and $\overline{C}_j$ the failure of the $j$th cut set, the system reliability is given by

$$P_f = P(\overline{C}_1 + \overline{C}_2 + \cdots + \overline{C}_j)$$
$$R = 1 - P_f = 1 - P(\overline{C}_1 + \overline{C}_2 + \cdots + \overline{C}_j) \tag{B20}$$

As an example of the application of cut-set and tie-set analysis we consider the graph given in Fig. B4. The following combinations of branches are *some* of the several tie sets of the system:

$$T_1 = x_1 x_2 \qquad T_2 = x_3 x_4 \qquad T_3 = x_1 x_6 x_4 \qquad T_4 = x_3 x_5 x_2 \qquad T_5 = x_1 x_6 x_5 x_2$$

Tie sets $T_1$, $T_2$, $T_3$, and $T_4$ are minimal tie sets. Tie set $T_5$ is nonminimal since the top node is encountered twice in traversing the graph. From Eq. (B19)

$$R = P(T_1 + T_2 + T_3 + T_4) = P(x_1 x_2 + x_3 x_4 + x_1 x_6 x_4 + x_3 x_5 x_2) \tag{B21}$$

Similarly we may list *some* of the several cut sets of the structure

$$C_1 = x_1 x_3 \qquad C_2 = x_2 x_4 \qquad C_3 = x_1 x_5 x_3 \qquad C_4 = x_1 x_5 x_4$$
$$C_5 = x_3 x_6 x_1 \qquad C_6 = x_3 x_6 x_2$$

Cut sets $C_1$, $C_2$, $C_4$, and $C_6$ are minimal. Cut sets $C_3$ and $C_5$ are nonminimal since they are both contained in cut set $C_1$. Using Eq. (B20),

$$R = 1 - P(\overline{C}_1 + \overline{C}_2 + \overline{C}_4 + \overline{C}_6) = 1 - P(\overline{x}_1 \overline{x}_3 + \overline{x}_2 \overline{x}_4 + \overline{x}_1 \overline{x}_5 \overline{x}_4 + \overline{x}_3 \overline{x}_6 \overline{x}_2) \tag{B22}$$

In a large problem there will be many cut sets and tie sets, and although Eqs. (B19) and (B20) are easily formulated, the expansion of either equation is a formidable task. (If there are $n$ events in a union, the expansion of the probability of the union involves $2^n - 1$ terms.) Several approximations which are useful in simplifying the computations are discussed in Messinger and Shooman [1967] and in Shooman [1990, p. 138].

## B3 FAILURE-RATE MODELS

### B3.1 Introduction

The previous section has shown how one constructs various combinatorial reliability models which express system reliability in terms of element reliability. This section introduces several different failure models for the system elements. These element failure models are related to life-test results and failure-rate data via probability theory.

The first step in constructing a failure model is to locate test data or plan a test on parts substantially the same as those to be used. From these data the part failure rate is computed and graphed. On the basis of the graph, any physical failure information, engineering judgment, and sometimes statistical tests, a failure-rate model is chosen. The parameters of the model are estimated from the graph or computed using the statistical principles of estimation, which are developed in Section A9. This section discusses the treatment of the data and the choice of a model.

The emphasis is on simple models, which are easy to work with and contain one or two parameters. This simplifies the problems of interpretation and parameter determination. Also in most cases the data are not abundant enough and the test conditions are not sufficiently descriptive of the proposed usage to warrant more complex models.

### B3.2 Treatment of Failure Data

Part failure data are generally obtained from either of two sources: the failure times of various items in a population placed on a life test, or repair reports listing operating hours of replaced parts in equipment already in field use. Experience has shown that a very good way to prevent these data is to compute and plot either the failure density function or the hazard rate as a function of time.

The data we are dealing with are a sequence of times to failure, but the failure density function and the hazard rate are continuous variables. We first compute a piecewise-continuous failure density function and hazard rate from the data.

We begin by *defining* piecewise-continuous failure density and hazard-rate functions in terms of the data. It can be shown that these discrete functions approach the continuous functions in the limit as the number of data becomes large and the interval between failure times approaches zero. Assume that our data describe a set of $N$ items placed in operation at time $t=0$. As time progresses, items fail, and at any time $t$ the number of survivors is $n(t)$. The data density function (also called empirical density function) defined over the time interval $t_i < t \leq t_i + \Delta t_i$ is given by the ratio of the number of failures occurring in the interval to the *size of the original population*, divided by the length of the time interval[1]

---

[1] In general a sequence of time intervals $t_0 < t \leq t_0 + \Delta t_0$, $t_1 < t \leq t_1 + \Delta t_1$, etc., is defined, where $t_1 = t_0 + \Delta t_0$, $t_2 = t_1 + \Delta t_1$, etc.

**TABLE B1  Failure Data for 10 Hypothetical Electronic Components**

| Failure Number | Operating Time, h |
|---|---|
| 1 | 8 |
| 2 | 20 |
| 3 | 34 |
| 4 | 46 |
| 5 | 63 |
| 6 | 86 |
| 7 | 111 |
| 8 | 141 |
| 9 | 186 |
| 10 | 266 |

$$f_d(t) = \frac{[n(t_i) - n(t_i + \Delta t_i)]/N}{\Delta t_i} \quad \text{for } t_i < t \leq t_i + \Delta t_i \quad (B23)$$

Similarly, the data hazard rate[2] over the interval $t_i < t \leq t_i + \Delta t_i$ is defined as the ratio of the number of failures occurring in the time interval to the *number of survivors at the beginning of the time interval*, divided by the length of the time interval.

$$z_d(t) = \frac{[n(t_i) - n(t_i + \Delta t_i)]/n(t_i)}{\Delta t_i} \quad \text{for } t_i < t \leq t_i + \Delta t_i \quad (B24)$$

The failure density function $f_d(t)$ is a measure of the *overall speed* at which failures are occurring, whereas the hazard rate $z_d(t)$ is a measure of the *instantaneous speed* of failure. Since the numerators of both Eqs. (B23) and (B24) are dimensionless, both $f_d(t)$ and $z_d(t)$ have the dimensions of inverse time (generally the time unit is hours).

The failure data for a life test run on a group of 10 hypothetical electronic components are given in Table B1. The computation of $f_d(t)$ and $z_d(t)$ from the data appears in Table B2.

The time intervals $\Delta t_i$ were chosen as the times between failure, and the first time interval $t_0$ started at the origin; that is, $t_0 = 0$. The remaining time intervals $t_i$ coincided with the failure times. In each case the failure was assumed to have occurred just before the end of the interval. Two alternate procedures are possible. The failure could have been assumed to occur just after the time interval closed, or the beginning of each interval $t_i$ could have been defined as the midpoint between failures. In this book we shall consistently use the first method, which is illustrated in Table B2.

---

[2]Hazard rate is sometimes called hazard or failure rate.

FAILURE-RATE MODELS    423

**TABLE B2  Computation of Data Failure Density and Data Hazard Rate**

| Time Interval, $h$ | Failure Density per Hour, $f_d(t)(\times 10^{-2})$ | Hazard Rate per Hour, $z_d(t)(\times 10^{-2})$ |
|---|---|---|
| 0–8     | $\dfrac{1}{10 \times 8} = 1.25$  | $\dfrac{1}{10 \times 8} = 1.25$ |
| 8–20    | $\dfrac{1}{10 \times 12} = 0.84$ | $\dfrac{1}{9 \times 12} = 0.93$ |
| 20–34   | $\dfrac{1}{10 \times 14} = 0.72$ | $\dfrac{1}{8 \times 14} = 0.96$ |
| 34–46   | $\dfrac{1}{10 \times 12} = 0.84$ | $\dfrac{1}{7 \times 12} = 1.19$ |
| 46–63   | $\dfrac{1}{10 \times 17} = 0.59$ | $\dfrac{1}{6 \times 17} = 0.98$ |
| 63–86   | $\dfrac{1}{10 \times 23} = 0.44$ | $\dfrac{1}{5 \times 23} = 0.87$ |
| 86–111  | $\dfrac{1}{10 \times 25} = 0.40$ | $\dfrac{1}{4 \times 25} = 1.00$ |
| 111–141 | $\dfrac{1}{10 \times 30} = 0.33$ | $\dfrac{1}{3 \times 30} = 1.11$ |
| 141–186 | $\dfrac{1}{10 \times 45} = 0.22$ | $\dfrac{1}{2 \times 45} = 1.11$ |
| 186–266 | $\dfrac{1}{10 \times 80} = 0.13$ | $\dfrac{1}{1 \times 80} = 1.25$ |

Since $f_d(t)$ is a density function, we can *define* a data failure distribution function and a data success distribution function by

$$F_d(t) = \int_0^t f_d(\xi)\, d\xi \tag{B25a}$$

$$R_d(t) = 1 - F_d(t) = 1 - \int_0^t f_d(\xi)\, d\xi \tag{B25b}$$

where $\xi$ is just a dummy variable of integration. Since the $f_d(t)$ curve is a piecewise-continuous function consisting of a sum of step functions, its integral is a piecewise-continuous function made up of a sum of ramp functions.

The functions $F_d(t)$ and $R_d(t)$ are computed for the preceding example by the appropriate integration of Fig. B5(a) and are given in Fig. B5(c) and (d). By inspection of Eqs. (B23) and (B25b) we see that

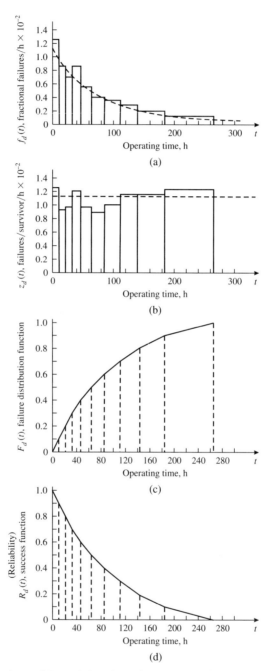

**Figure B5** Density and hazard functions for the data of Table B1. (a) Data failure density functions; (b) data hazard rate; (c) data failure distribution function; (d) data success function.

$$R_d(t_i) = \frac{n(t_i)}{N} \tag{B26}$$

In the example given in Table B1, only 10 items were on test, and the computations were easily made. If many items are tested, the computation intervals $\Delta t_i$ cannot be chosen as the times between failures since the computations become too lengthy. The solution is to divide the same scale into several equally spaced intervals. Statisticians call these *class intervals*, and the midpoint of the interval is called a *class mark*. Graphical diagrams such as Fig. B5(a) and (b) are called *histograms*.

## B3.3 Failure Modes and Handbook Failure Data

After plotting and examining failure data for several years, people began to recognize several modes of failure. Early in the lifetime of equipment or a part, there are a number of failures due to initial weakness or defects; poor insulation, weak parts, bad assembly, poor fits, etc. During the middle period of equipment operation fewer failures take place, and it is difficult to determine their cause. In general they seem to occur when the environmental stresses exceed the design strengths of the part or equipment. It is difficult to predict the environmental-stress amplitudes or the part strengths as deterministic functions of time; thus the middle-life failures are often called *random failures*.[3] As the item reaches old age, things begin to deteriorate, and many failures occur. This failure region is quite naturally called the *wear-out region*. Typical $f(t)$ and $z(t)$ curves[4] illustrating these three modes of behavior are shown in Fig. B6. The early failures, also called *initial failures* or *infant mortality*,[5] appear as decreasing $z(t)$ and $f(t)$ functions. The random-failure, or constant-hazard-rate, mode is characterized by an approximately constant $z(t)$ and a companion $f(t)$ which is approximately exponential. In the wear-out or rising-failure-rate region, the $z(t)$ function increases whereas $f(t)$ has a humped appearance.

It is clear that it is easier to distinguish the various failure modes by inspection of the $z(t)$ curve than it is from the appearance of the $f(t)$ function. This is one of the major reasons why hazard rate is introduced. Because of the monotonic nature of $F(t)$ and $R(t)$ these functions are even less useful in distinguishing failure modes.

The curve of Fig. B6(b) has been discussed by many of the early writers on the subject of reliability [Carhart, 1953] and is often called the *bathtub curve* because of its shape. The fact that such a hazard curve occurs for many types of equipment

---

[3]Actually all the failures are random; thus a term such as *unclassifiable as to cause* would be more correct.
[4]We are now referring to continuous hazard and failure density functions, which represent the limiting forms of $f_d(t)$ and $z_d(t)$ as discussed.
[5]Some of the terms, as well as the concept of hazard, have been borrowed from those used by actuaries, who deal with life insurance statistics.

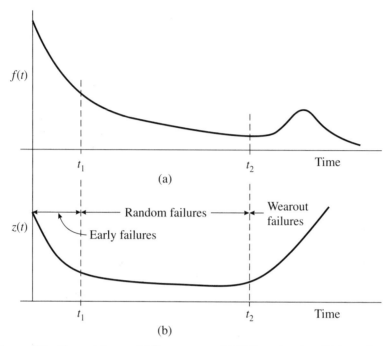

**Figure B6** General form of failure curves. (a) Failure density; (b) hazard rate.

has been verified by experience. Also when failed components have been dismantled to determine the reasons for failure, the conclusions have again justified the hypothesis of three failure modes. In fact most manufacturers of high-reliability components now subject their products to an initial burn-in period of $t_i$ hours to eliminate the initial failure region shown in Fig. B6. At the onset of wearout at time $t_2$, the hazard rate begins to increase rapidly, and it is wise to replace the item after $t_2$ hours of operation. Thus, if the bathtub curve were really a universal model, one would pretest components for $t_1$ hours, place the survivors in use for an additional $t_2 - t_1$ hours, and then replace them with fresh pretested components. This would reduce the effective hazard rate and improve the probability of survival if burn-in and replacement are feasible. Unfortunately, many types of equipment have a continuously decreasing or continuously increasing hazard and therefore behave differently. It often happens that electronic components have a constant hazard and mechanical components a wear-out characteristic. Unfortunately, even though reliability theory is 4 to 5 decades old, not enough comparative analysis has been performed on different types of hazard models and failure data to make a definitive statement as to which models are best for all types of components.

Many failure data on parts and components have been recorded since the beginning of formal interest in reliability in the early 1950s. Large industrial organizations such as Radio Corporation of America, General Electric Company, Motorola, etc., published handbooks of part-failure-rate data compiled from life-test and field-failure data. These data and other information were compiled into an evolving series of part-failure-rate handbooks: MIL-HDBK-217, 217A, 217B, 217C, 217D, 217E, 217F Government Printing Office, Washington, DC. Another voluminous failure data handbook is *Failure Rate Data Handbook* (FARADA), published by the GIDEP program, Government Industrial Data Exchange Program, Naval Fleet Missile Systems, Corona, CA. The FARADA handbook includes such vital information as the *number on test*, the *number of failures*, and some details on the *source* of the data and the *environment*. This information allows one to use engineering judgments in selecting failure rates from this reference. Many practitioners now use failure databases compiled by various telecommunications companies in the United States and worldwide (see Shooman [1990, Appendix K], and Section D2.1 of this book).

In practice, failure rates for various components are determined from handbook data, field failure-rate data, or life test data provided by component manufacturers. A large fraction of the components used in modern systems are microelectronics circuits. Many are analog in nature; however, even more are digital integrated circuits (ICs). One reason that modern electronic equipment is very reliable is because these ICs have a very low failure rate. Furthermore, the failure rate of ICs increases only slowly as their complexity increases. Analysis of past failure-rate data allows one to develop a simple model for the failure rate of digital integrated circuits, which is very useful for initial reliability comparisons of various designs.

A curve showing the failure rate per gate versus gate complexity for digital integrated IC is given in Fig. B7, which was adapted from Siewiorek [1982]. The light solid and light dotted lines in the figure as well as the dark and light circles represent data plotted from various sources. The heavy solid lines were fitted to the data by the author. Note that the slopes are all approximately parallel and that the reliability improved from 1965 to 1975 to 1985.

The heavy lines are based on the assumption that the failure rate increases as the square root of the number of gates:

$$\lambda_b = C \times (g)^{1/2} \tag{B27a}$$

where

$\lambda_b$ = is the base failure rate
$C$ = is a constant
$g$ = is the number of gates in the equivalent circuit for the chip

Others use a different failure-rate model, where $\lambda_{b'} \sim (g)^a$ and $0.1 < a < 0.3$ [Healey, 2001]. If we express the failure rates per gate, $\lambda_{b'}$, we obtain from Eq. (B27a):

$$\lambda_{b'} = \lambda_b/g = C/g^{1/2} \tag{B27b}$$

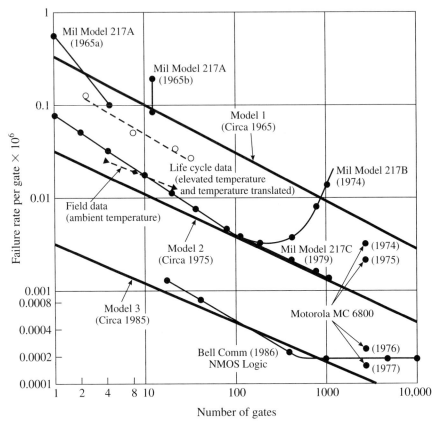

**Figure B7** Failure rate per gate as a function of chip complexity for bipolar technology. (Adapted from Siewiorek [1982, p. 10, Fig. 1–5]).

The values of $C$, determined by fitting the heavy curve to the data for 1965, 1975, and 1985, are $C = 0.32$, $0.04$, and $0.004$. This indicates that IC reliability has improved by about a factor of 10 for each decade. More details of this model are given in Shooman [1990, pp. 641–644].

We now propose a *hypothetical* explanation for why the failure rate should be proportional to the square root of the number of gates if we assume that most of the IC failures occur from *electromigration*—a process that produces projections of conducting material growing out of the various areas carrying current on an IC chip. These projections grow as a function of current and time. If a projection touches another projection or current-carrying area on the chip, then shorting may occur resulting in a failure of the IC. (Another possible result of electromigration is the forming of voids, creating open circuits, or unacceptable changes in line/contact resistances. Note that newer technologies now in use are less prone to electromigration failure modes [Bernstein, 2001].)

We can model the time to failure in the following manner: Let $s$ (mm) be the average spacing between current-carrying elements on the chip and let $v$ (mm/hr) be the average speed at which the projections elongate. The average time to failure, $t_f$ (hr), is then proportional to $s/v$. In general, $t_f$, $s$, and $v$ will be random variables; however, we can characterize $t_f$ by its expected value, which is generally called the mean time to failure (MTTF), yielding

$$\text{MTTF} = K_1 s/v \tag{B27c}$$

where $K_1$ is a proportionality constant computed by taking the expected value of the distribution for the ratio of the random variables $s$ and $v$.

For a fixed chip area, $A$, we assume that the horizontal spacing $s_h$ is inversely proportional to the number of gates across the width of the chip. Similarly, the vertical spacing, $s_v$, is inversely proportional to the number of gates along the length of the chip. Assuming a square chip and $g$ total gates, the number of gates across the width (gates along the length) will be equal to $\sqrt{g}$ and $s_h = s_v = s$, which is given by

$$s = \frac{K_2}{\sqrt{g}} \tag{B27d}$$

where $K_2$ is a proportionality constant, which is a function of the fabrication techniques used for the IC.

A substitution of Eq. (B27d) into Eq. (B27c) yields

$$\text{MTTF} = \frac{K_1 K_2}{v \sqrt{g}} \tag{B27e}$$

The simplest failure-rate model is the constant failure-rate model, where $\lambda$ is a constant with units failures/hr. For this model, it can be shown that $\lambda = 1/\text{MTTF}$. Thus

$$\lambda = \frac{v \sqrt{g}}{K_1 K_2} \tag{B27f}$$

This development therefore leads to a failure rate that is proportional to the square root of the number of gates on the chip. The author emphasizes that this is a *hypothesis*, not a proven explanation.

Another *hypothesis* for newer chip technologies relates initial yield and failure rate to residual chip defects. Thus $\lambda \sim$ Area $\sim g^1$ [Bernstein, 2001].

## B3.4 Reliability in Terms of Hazard Rate and Failure Density

In the previous section, various functions associated with failure data were defined and computed for the data given in the examples. These functions were $z_d(t)$, $f_d(t)$, $F_d(t)$, and $R_d(t)$. In this section we begin by defining two random variables and deriving in a careful manner the basic definitions and relations between the theoretical hazard, failure density function, failure distribution function, and reliability function.

The random variable **t** is defined as the failure time of the item in question.[6] Thus, the probability of failure as a function of time is given as

$$P(\mathbf{t} \leq t) = F(t) \tag{B28}$$

which is simply the definition of the failure distribution function. We can define the reliability, which is a probability of success in terms of $F(t)$, as

$$R(t) = P_s(t) = 1 - F(t) = P(\mathbf{t} \geq t) \tag{B29}$$

The failure density function is of course given by

$$\frac{dF(t)}{dt} = f(t) \tag{B30}$$

We now consider a population of $N$ items with the same failure-time distribution. The items fail independently with probability of failure given by $F(t) = 1 - R(t)$ and probability of success given by $R(t)$. If the random variable $\mathbf{N}(t)$ represents the number of units surviving at time $t$, then $\mathbf{N}(t)$ has a binomial distribution with $p = R(t)$. Therefore,

$$P[\mathbf{N}(t) = n] = B[n; N, R(t)] = \frac{N!}{n!(N-n)!} [R(t)]^n [1 - R(t)]^{N-n}$$

$$n = 0, 1, \ldots, N \tag{B31}$$

The number of units operating at any time $t$ is a random variable and is not fixed; however, we can compute the expected value $\mathbf{N}(t)$. From Table A1 we see that the expected value of a random variable with a binomial distribution is given by $NR(t)$ and leads to

$$n(t) \equiv E[N(t)] = NR(t) \tag{B32}$$

Solving for the reliability yields

$$R(t) = \frac{n(t)}{N} \tag{B33}$$

Thus, the reliability at time $t$ is the average fraction of surviving units at time $t$. This verifies Eq. (B27), which was obtained as a consequence of the definition of $f_d(t)$. From Eq. (B29) we obtain

---

[6]In some problems, a more appropriate random variable is the number of miles, cycles, etc. The results are analogous.

## FAILURE-RATE MODELS

$$F(t) = 1 - \frac{n(t)}{N} = \frac{N - n(t)}{N} \tag{B34}$$

and from Eq. (B30)

$$f(t) = \frac{dF(t)}{dt} = -\frac{1}{N}\frac{dn(t)}{dt}$$

$$f(t) \equiv \lim_{\Delta t \to 0} \frac{n(t) - n(t + \Delta t)}{N \Delta t} \tag{B35}$$

Thus, we see that Eq. (B23) is valid, and as $N$ becomes large and $\Delta t_i$ becomes small, Eq. (B23) approaches Eq. (B35) in the limit. From Eq. (B34) we see that $F(t)$ is the average fraction of units having failed between 0 and time $t$, and Eq. (B35) states that $f(t)$ is the rate of change of $F(t)$, or its slope. From Eq. (B35) we see that the failure density function $f(t)$ is *normalized* in terms of the size of the original population $N$. In many cases it is more informative to normalize with respect to $n(t)$, the number of survivors. Thus, we define the hazard rate as

$$z(t) \equiv -\lim_{\Delta t \to 0} \frac{n(t) - n(t + \Delta t)}{n(t) \Delta t} \tag{B36}$$

The definition of $z(t)$ in Eq. (B36) of course agrees with the definition of $z_d(t)$ in Eq. (B24). We can relate $z(t)$ and $f(t)$ using Eqs. (B35) and (B36):

$$z(t) = -\lim_{\Delta t \to 0} \frac{n(t) - n(t + \Delta t)}{\Delta t} \frac{1}{n(t)} = Nf(t) \frac{1}{n(t)}$$

Substitution of Eq. (B33) yields

$$z(t) = \frac{f(t)}{R(t)} \tag{B37}$$

We now wish to relate $R(t)$ to $f(t)$ and to $z(t)$. From Eqs. (B29) and (B30) we see that

$$R(t) = 1 - F(t)$$

$$= 1 - \int_0^t f(\xi)\, d\xi \tag{B38}$$

where $\xi$ is merely a dummy variable. Substituting into Eq. (B37) from Eqs. (B35) and (B33), we obtain

## SUMMARY OF RELIABILITY THEORY

$$z(t) = -\frac{1}{N}\frac{dn(t)}{dt}\frac{N}{n(t)} = -\frac{d}{dt}\ln n(t)$$

Solving the differential equation yields:

$$\ln n(t) = -\int_0^t z(\xi)\, d\xi + c$$

where $\xi$ is a dummy variable and $c$ is the constant of integration. Taking the antilog of both sides of the equation gives:

$$n(t) = e^c \exp\left[-\int_0^t z(\xi)\, d\xi\right]$$

Inserting initial conditions

$$n(0) = N = e^c$$

gives

$$n(t) = N \exp\left[-\int_0^t z(\xi)\, d\xi\right]$$

Substitution of Eq. (B33) completes the derivation

$$R(t) = \exp\left[-\int_0^t z(\xi)\, d\xi\right] \tag{B39}$$

Equations (B35) and (B36) serve to define the failure density function and the hazard rate, and Eqs. (B37) to (B39) relate $R(t)$ to $f(t)$ and $z(t)$.[7]

### B3.5 Hazard Models

On first consideration it might appear that if failure data and graphs such as Fig. B5(a–d) are available, there is no need for a mathematical model. However, in drawing conclusions from test data on the behavior of other, similar components it is necessary to fit the failure data with a mathematical model. The discussion will start with several simple models and gradually progress to the more involved problem of how to choose a general model which fits all cases through adjustment of constants.

---

[7]An alternative derivation of these expressions is given in Shooman [1990, p. 183].

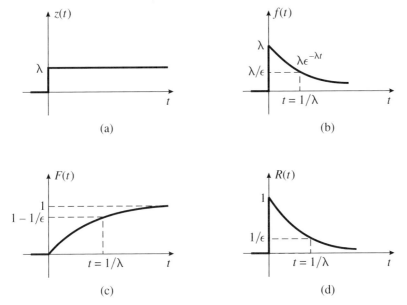

**Figure B8** Constant-hazard model: (a) constant hazard; (b) decaying exponential density function; (c) rising exponential distribution function; and (d) decaying exponential reliability function.

*Constant Hazard.* For a good many years, reliability analysis was almost wholly concerned with constant-hazard rates. Indeed many data have been accumulated, like those in Fig. B5(b), which indicate that a constant-hazard model is appropriate in many cases.

If a constant-hazard rate $z(t) = \lambda$ is assumed, the time integral is given by $\int_0^t \lambda \, d\xi = \lambda t$. Substitution in Eqs. (B37) to (B39) yields

$$z(t) = \lambda \qquad (B40)$$
$$f(t) = \lambda e^{-\lambda t} \qquad (B41)$$
$$R(t) = e^{-\lambda t} = 1 - F(t) \qquad (B42)$$

The four functions $z(t), f(t), F(t)$, and $R(t)$ are sketched in Fig. B8. A constant-hazard rate implies an exponential density function and an exponential reliability function.

The constant-hazard model forbids any deterioration in time of strength or soundness of the items in the population. Thus, if $\lambda = 0.1$ per hour, we can expect 10 failures in a population of 100 items during the first hour of operation and the same number of failures between the thousandth and thousand and first hours of operation in a population of 100 items that have already survived 1,000 hours. A simple hazard model that admits deterioration in time, i.e., wear, is one in which the failure rate increases with time.

**434** SUMMARY OF RELIABILITY THEORY

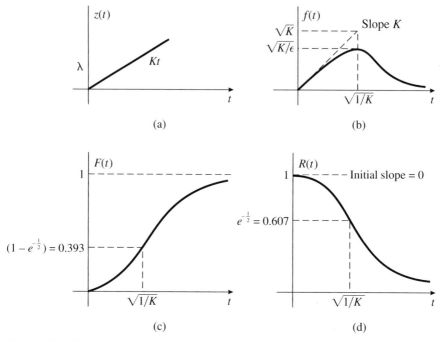

**Figure B9** Linearly increasing hazard: (a) linearly increasing hazard; (b) Rayleigh density function; (c) Rayleigh distribution function; (d) Rayleigh reliability function.

Sometimes a test is conducted for $N$ parts for $T$ hours and no parts fail. The total number of test hours is $H = NT$, but the number of failures is zero. Thus one is tempted to estimate $\lambda$ by $0/H$, which is incorrect. A better procedure is to say that the failure rate is less than if one failure occurred, $\lambda < 1/H$. More advanced statistical techniques suggest that $\lambda \approx (1/3)/H$ [Welker, 1974].

***Linearly Increasing Hazard.*** When wear or deterioration is present, the hazard will increase as time passes. The simplest increasing-hazard model that can be postulated is one in which the hazard increases linearly with time. Assuming that $z(t) = Kt$ for $t \geq 0$ yields

$$z(t) = Kt \tag{B43}$$
$$f(t) = Kte^{-Kt^2/2} \tag{B44}$$
$$R(t) = e^{-Kt^2/2} \tag{B45}$$

These functions are sketched in Fig. B9. The density function of Eq. (B44) is a Rayleigh density function.

***The Weibull Model.*** In many cases, the $z(t)$ curve cannot be approximated by

FAILURE-RATE MODELS    435

a straight line, and the previously discussed models fail. In order to fit various $z(t)$ curvatures, it is useful to investigate a hazard model of the form

$$z(t) = Kt^m \quad \text{for } m > -1 \tag{B46}$$

This form of model was discussed in detail in a paper by Weibull [1951] and is generally called a *Weibull model*. The associated density and reliability functions are

$$f(t) = Kt^m e^{-Kt^{m+1}/(m+1)} \tag{B47}$$

$$R(t) = e^{-Kt^{m+1}/(m+1)} \tag{B48}$$

By appropriate choice of the two parameters $K$ and $m$, a wide range of hazard curves can be approximated. The various functions obtained for typical values of $m$ are shown in Fig. B10. For fixed values of $m$, a change in the parameter $K$ merely changes the vertical amplitude of the $z(t)$ curve; thus, $z(t)/K$ is plotted versus time. Changing $K$ produces a time-scale effect on the $R(t)$ function; therefore, time is normalized so that $\tau^{m+1} = [K/(m+1)]t^{m+1}$. The amplitude of the hazard curve affects the time scale of the reliability function; consequently, the parameter $K$ is often called the *scale parameter*. The parameter $m$ obviously affects the shape of all the reliability functions shown and is consequently called the *shape parameter*. The curves $m = 0$ and $m = 1$ are constant-hazard and linearly-increasing-hazard models, respectively. It is clear from inspection of Fig. B10 that a wide variety of models is possible by appropriate selection of $K$ and $m$. The drawback is, of course, that this is a two-parameter model, which means a greater difficulty in sketching the results and increased difficulty in estimating the parameters. A three-parameter Weibull model can be formulated by replacing $t$ by $t - t_0$, where $t_0$ is called the *location parameter*.

## B3.6 Mean Time To Failure

It is often convenient to characterize a failure model or set of failure data by a single parameter. One generally uses the mean time *to* failure or the mean time *between* failures for this purpose. If we have life-test information on a population of $n$ items with failure times $t_1, t_2, \ldots, t_n$, then the MTTF[8] is defined by the following equation [see also Eq. (A34)]:

$$\text{MTTF} = \frac{1}{n} \sum_{i=1}^{n} t_i \tag{B49}$$

---

[8]Sometimes the term *mean time between failures* (MTBF) is used interchangeably with the term MTTF; however, strictly speaking, the MTBF has meaning only when one is discussing a renewal situation, where there is repair or replacement. See Shooman [1990, Section 6.10].

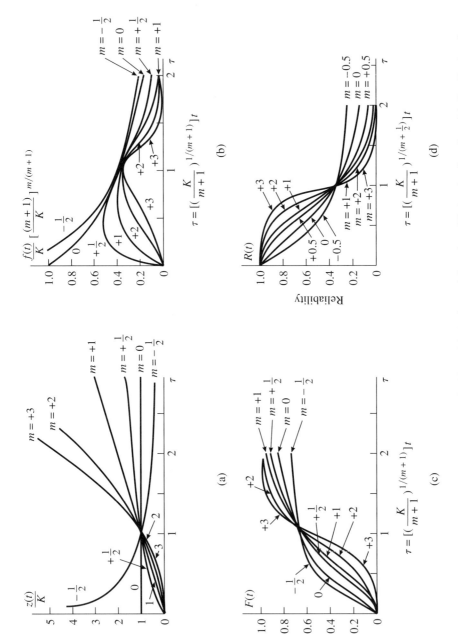

**Figure B10** Reliability functions for the Weibull model: (a) hazard function; (b) density function; (c) distribution function; (d) reliability function.

If one is discussing a hazard model, the MTTF for the probability distribution defined by the model is given by Eq. (A33) as

$$\text{MTTF} = E(t) = \int_0^\infty t f(t)\, dt \tag{B50}$$

In a single-parameter distribution, specification of the MTTF fixes the parameter. In a multiple-parameter distribution, fixing the MTTF only places one constraint on the model parameters.

One can express Eq. (B50) by a simpler computational expression involving the reliability function:[9]

$$\text{MTTF} = \int_0^\infty R(t)\, dt \tag{B51}$$

As an example of the use of Eq. (B51) the MTTF for several different hazards will be computed. For a single component with a constant hazard:

$$\text{MTTF} = \int_0^\infty e^{-\lambda t}\, dt = \left.\frac{e^{-\lambda t}}{-\lambda}\right|_0^\infty = \frac{1}{\lambda} \tag{B52}$$

For a linearly increasing hazard:

$$\text{MTTF} = \int_0^\infty e^{-Kt^2/2}\, dt = \frac{\Gamma(\tfrac{1}{2})}{2\sqrt{K/2}} = \sqrt{\frac{\pi}{2K}} \tag{B53}$$

For a Weibull distribution:

$$\text{MTTF} = \int_0^\infty e^{-Kt^{(m+1)}/(m+1)}\, dt = \frac{\Gamma[(m+2)/(m+1)]}{[K/(m+1)]^{1/(m+1)}} \tag{B54}$$

In Eq. (B52) the MTTF is simply the reciprocal of the hazard, whereas in Eq. (B53) it varies as the reciprocal of the square root of the hazard slope. In Eq. (B54) the relationship between MTTF, $K$, and $m$ is more complex (see Table A1).

In many cases we assume an exponential density (constant hazard) for simplicity, and for this case we frequently hear the statement "the MTBF is the reciprocal of the failure rate." The reader should not forget the assumptions necessary for this statement to hold.

---

[9]For the derivation, see Shooman [1990, p. 197].

## B4 SYSTEM RELIABILITY

### B4.1 Introduction

The previous two sections have divided reliability into two distinct phases: a formulation of the reliability structure of the problem using combinatorial reliability, and a computation of the element probabilities in terms of hazard models. This section unites these two approaches to obtain the reliability function for the system.

When the element probabilities are independent, computations are straightforward. The only real difficulties encountered here are the complexity of the calculations in large problems.

### B4.2 The Series Configuration

The *series configuration*, also called a *chain structure*, is the most common reliability model and the simplest. Any system in which the system success depends on the success of all its components is a series reliability configuration. Unfortunately for the reliability analyst (but fortunately for the user of the product or device), not all systems have this simple structure.

A series configuration of $n$ items is shown in Fig. B11(a). The reliability of this structure is given by

$$R(t) = P(x_1, x_2, \ldots, x_n) = P(x_1)P(x_2|x_1)P(x_3|x_1x_2) \cdots P(x_n|x_1x_2\cdots x_{n-1}) \tag{B55}$$

If the $n$ items $x_1, x_2, \ldots, x_n$ are independent, then

$$R(t) = P(x_1)P(x_2)\cdots P(x_n) = \prod_{i=1}^{n} P(x_i) \tag{B56}$$

If each component exhibits a constant hazard, then the appropriate component model is $e^{-\lambda_i t}$, and Eq. (B56) becomes

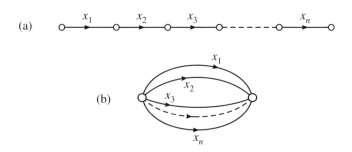

**Figure B11** Series (a) and parallel (b) reliability configurations.

$$R(t) = \prod_{i=1}^{n} e^{-\lambda_i t} = \exp\left(-\sum_{i=1}^{n} \lambda_i t\right) \tag{B57}$$

Equation (B57) is the most commonly used and the most elementary system reliability formula. In practice this formula is often misused (probably because it is so simple and does work well in many situations, people have become overconfident). The following assumptions must be true if Eq. (B57) is to hold for a system:

1. The system reliability configuration must truly be a series one.
2. The components must be independent.
3. The components must be governed by a constant-hazard model.

If assumptions 1 and 2 hold but the components have linearly increasing hazards $z_i(t) = K_i t$, Eq. (B56) then becomes

$$R(t) = \prod_{i=1}^{n} e^{-K_i t^2/2} = \exp\left(-\sum_{i=1}^{n} \frac{K_i t^2}{2}\right) \tag{B58}$$

If $p$ components have a constant hazard and $n - p$ components a linearly increasing hazard, the reliability becomes

$$R(t) = \left(\prod_{i=1}^{p} e^{-\lambda_i t}\right)\left(\prod_{i=p+1}^{n} e^{-\lambda_i t^2/2}\right)$$

$$= \exp\left(-\sum_{i=1}^{p} \lambda_i t\right) \exp\left(-\sum_{i=p+1}^{n} \frac{K_i t^2}{2}\right) \tag{B59}$$

In some cases no simple composite formula exists, and the reliability must be expressed as a product of $n$ terms. For example, suppose each component is governed by the Weibull distribution, $z(t) = K_i t^{m_i}$. If $m$ and $K$ are different for each component,

$$R(t) = \prod_{i=1}^{n} \exp\left(\frac{-K_i t^{m_i+1}}{m_i + 1}\right) = \exp\left(-\sum_{i=1}^{n} \frac{K_i t^{m_i+1}}{m_i + 1}\right) \tag{B60}$$

The series reliability structure serves as a lower-bound configuration. To illustrate this principle we pose a hypothetical problem. Given a collection of $n$ elements, from the reliability standpoint what is the worst possible reliability structure they can assume? The intuitive answer, of course, is a series structure. (A proof is given in Shooman [1990, p. 205]; see also Section B6.4 of this book.)

## B4.3 The Parallel Configuration

If a system of $n$ elements can function properly when only one of the elements is good, a parallel configuration is indicated. A parallel configuration of $n$ items is shown in Fig. B11(b). The reliability expression for a parallel system may be expressed in terms of the probability of success of each component or, more conveniently, in terms of the probability of failure

$$R(t) = P(x_1 + x_2 + \cdots + x_n) = 1 - P(\bar{x}_1 \bar{x}_2 \cdots \bar{x}_n) \tag{B61}$$

In the case of constant-hazard components, $P_f = P(\bar{x}_i) = 1 - e^{-\lambda_i t}$, and Eq. (B61) becomes

$$R(t) = 1 - \prod_{i=1}^{n} (1 - e^{-\lambda_i t}) \tag{B62}$$

In the case of linearly increasing hazard, the expression becomes

$$R(t) = 1 - \prod_{i=1}^{n} (1 - e^{-K_i t^2/2}) \tag{B63}$$

In the general case, the system reliability function is

$$R(t) = 1 - \prod_{i=1}^{n} (1 - e^{-Z_i(t)}) \tag{B64}$$

where $Z_i(t) \equiv \int_0^t z(\xi)\, d\xi$.

In order to permit grouping of terms in Eq. (B64) to simplify computation and/or interpretation, the equation must be expanded. The expansion results in

$$R(t) = (e^{-Z_1} + e^{-Z_2} + \cdots + e^{-Z_n}) - (e^{-(Z_1+Z_2)} + e^{-(Z_1+Z_3)} + \cdots) \\ + (e^{-(Z_1+Z_2+Z_3)} + e^{-(Z_1+Z_2+Z_4)} + \cdots) - \cdots e^{-(Z_1+Z_2+Z_3+\cdots+Z_n)} \tag{B65}$$

Note that the signs of the terms in parentheses alternate and that in the first set of parentheses, the exponents are all the Zs taken singly; in the second, all the sums of Zs taken two at a time; and in the last term, the sum of all the Zs. The $r$th parentheses in Eq. (B65) contain $n!/[r!(n-r)!]$ terms.

Just as the series configuration served as a lower-bound structure, the parallel model can be thought of as an upper-bound structure.

If we have a system of $n$ elements with information on each element reliability but little or no information on their interconnection, we can bound the reliability function from below by Eq. (B56) and from above by Eq. (B64). We would in general expect these bounds to be quite loose; however, they do provide some information even when we are grossly ignorant of the system structure.

## B4.4 An *r*-out-of-*n* Structure

Another simple structure which serves as a useful model for many reliability problems is an *r*-out-of-*n* structure. Such a model represents a system of $n$ components in which $r$ of the $n$ items must be good for the system to succeed. Of course $r$ is less than $n$. Two simple examples of an *r*-out-of-*n* system are (1) a piece of stranded wire with $n$ strands in which at least $r$ are necessary to pass the required current and (2) a battery composed of $n$ series cells of $E$ volts each where the minimum voltage for system operation[10] is $rE$.

We may formulate a structural model for an *r*-out-of-*n* system, but it is simpler to use the binomial distribution if applicable. The binomial distribution can be used only when the $n$ components are independent and identical. If the components differ or are dependent, the structural-model approach must be used.[11] Success of exactly $r$ out of $n$ identical, independent items is given by

$$B(r:n) = \binom{n}{r} p^r (1-p)^{n-r} \tag{B66}$$

where $r:n$ stands for $r$ out of $n$, and the success of at least $r$ out of $n$ items is given by

$$P_s = \sum_{k=r}^{n} B(k:n) \tag{B67}$$

For constant-hazard components Eq. (B66) becomes

$$R(t) = \sum_{k=r}^{n} \binom{n}{k} e^{-k\lambda t} (1 - e^{-\lambda t})^{n-k} \tag{B68}$$

Similarly for linearly increasing or Weibull components, the reliability functions are

---

[10]Actually when one cell of a series of $n$ cells fails, the voltage of the string does not become $(n-1)E$ unless a special circuit arrangement is used. Such a circuit is discussed in Shooman [1990, p. 2.9].
[11]The reader should refer to the example given in Eq. (B17) and Fig. B3.

$$R(t) = \left[ \sum_{k=r}^{n} \binom{n}{k} e^{-kKt^2/2} \right] (1 - e^{-Kt^2/2})^{n-k} \tag{B69}$$

$$R(t) = \left[ \sum_{k=r}^{n} \binom{n}{k} e^{-kKt^{m+1}/(m+1)} \right] (1 - e^{-Kt^{m+1}/(m+1)})^{n-k} \tag{B70}$$

It is of interest to note that for $r = 1$, the structure becomes a parallel system and for $r = n$ the structure becomes a series system. Thus, in a sense series and parallel systems are subclasses of an $r$-out-of-$n$ structure.

## B5 ILLUSTRATIVE EXAMPLE OF SIMPLIFIED AUTO DRUM BRAKES

### B5.1 Introduction

The preceding sections have attempted to summarize the pertinent aspects of reliability theory and show the reader how analysis can be performed. This section illustrates via an example how the theory can be applied.

The example chosen for this section is actually a safety analysis. In the case of the automobile, the only difference between a reliability and a safety analysis is in the choice of subsystems included in the analysis. In the case of safety, we concentrate on the subsystems whose failure could cause injury to the occupants, other passengers, pedestrians, etc. In the case of reliability analysis, we would include all subsystems whose failure either makes the auto inoperative or necessitates a repair (depending on our definition of success).

### B5.2 The Brake System

The example considers the braking system of a typical older auto, without power brakes or antilock brakes and excluding the parking (emergency) brake and the dash warning light. An analysis at the detailed (piece-part) level is a difficult task. The major subsystems in a typical braking system may contain several hundred parts and assemblies.

The major subsystems and approximate parts counts are: pressure differential valve (8 parts), self-adjusting drum brakes (4 × 15 parts), wheel cylinder (4 × 9 parts), tubing, brackets, and connectors (50 parts), dual master cylinder (22 parts), and master cylinder installation parts (20 parts). Frequently, because of lack of data, analysis is not carried out at this piece-part level. Even with scanty data, an analysis is still important, since often it will show obvious weaknesses of such a braking system which can be improved when redesigning it. In such a case, redesign is warranted, based on engineering judgment, even without statistics on frequencies of the failure modes. The example will be performed

ILLUSTRATIVE EXAMPLE OF SIMPLIFIED AUTO DRUM BRAKES   443

**TABLE B3  A Simplified Braking System FMECA**

| Failure Mode | Mechanism | Safety Criticality | Comments |
|---|---|---|---|
| M1: Differential valve failure (1/2 system) | Leakage or Blockage | Medium | Reduces braking efficiency |
| M2: Differential valve failure (total system) | Leakage affecting both front and back systems | High | Loss of both systems |
| M3: Master cylinder failure (1/2 system) | Leakage or blockage | Medium | Reduces braking efficiency |
| M4: Master cylinder failure (total system) | Leakage (front and back) | High | Loss of both systems |
| M5: Drum brakes self-adjusting | Leakage or blockage of one assembly | Medium | Unbalance of brakes causes erratic behavior |
| M6: Tubing, brackets, and connectors (1/2 system) | Leakage or blockage | Medium | Reduces braking efficiency |
| M7: Pedal and linkage | Broken or jammed | High | Loss of both systems |

at a higher level, and will group together all failure mechanisms which cause the particular failure mode in question.

### B5.3  Failure Modes, Effects, and Criticality Analysis

An FMECA[12] for the simplified braking system is given in Table B3. Inspection of the table shows that the modes which most seriously affect safety are modes M2, M4, and M7, and the design should be scrutinized in a design review to assure that the probability of occurrence for these modes is minimized.

### B5.4  Structural Model

The next step in analysis would be the construction of an SBD or an FT.[13] Assume that a safety failure occurs if modes M2, M4, or M7 occur singly and

---

[12]Sometimes a column is added to an FMEA analysis which discusses (evaluates) the severity or *criticality* of the failure mode. In such a case, the analysis is called an FMECA.
[13]Safety block diagram or fault tree.

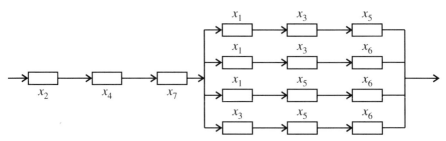

**Figure B12** Safety block diagram for simplified brake example. The notation $x_i$ means a failure mode for element $i$ does *not* occur, indicating success of element $i$.

modes M1, M3, M5, and M6 occur in pairs. (Actually, the paired modes must affect both front and rear systems to constitute a failure; but approximations are made here to simplify the analysis.) Based on the above assumptions, the SBD and FT given in Figs. B12 and B13, respectively, are obtained.

### B5.5 Probability Equations

Given either the SBD of Fig. B12 or the FT of Fig. B13, an equation can be written for the probability of safety (or unsafety), using probability principles. Computer programs are used in complex cases; this simplified case can be written:

$P_s \equiv$ probability of a safe condition from Fig. B12
$= P[(x_2 x_4 x_7)(x_1 x_3 x_5 + x_1 x_3 x_6 + x_1 x_5 x_6 + x_3 x_5 x_6)]$
$P_u \equiv$ probability of an unsafe condition from Fig. B13
$= P[(\bar{x}_2 + \bar{x}_4 + \bar{x}_7) + (\bar{x}_1 \bar{x}_3 + \bar{x}_1 \bar{x}_5 + \bar{x}_1 \bar{x}_6 + \bar{x}_3 \bar{x}_5 + \bar{x}_3 \bar{x}_6 + \bar{x}_5 \bar{x}_6)]$

An analysis in more depth would require more detail (and more input data). The choice of how much decomposition to lower levels of detail is required in an analysis is often determined by data availability. To continue the analysis, failure data on the modes M1, M2, ..., M6 is required. If, for M3, the failure rate were constant and equal to $\lambda_3$ failures per mile, then the possibility of mode M3 occurring or not occurring in $M$ miles would be:

$$P\begin{pmatrix} \text{mode M3 does not occur} \\ \text{in } M \text{ miles} \end{pmatrix} = P(X_3) = e^{-\lambda_3 M}$$

$$P\begin{pmatrix} \text{mode M3 does occur in} \\ M \text{ miles} \end{pmatrix} = P(\bar{X}_3) = 1 - e^{-\lambda_3 M}$$

To complete the analysis, the failure-rate data $\lambda_i$ are substituted into the

# ILLUSTRATIVE EXAMPLE OF SIMPLIFIED AUTO DRUM BRAKES

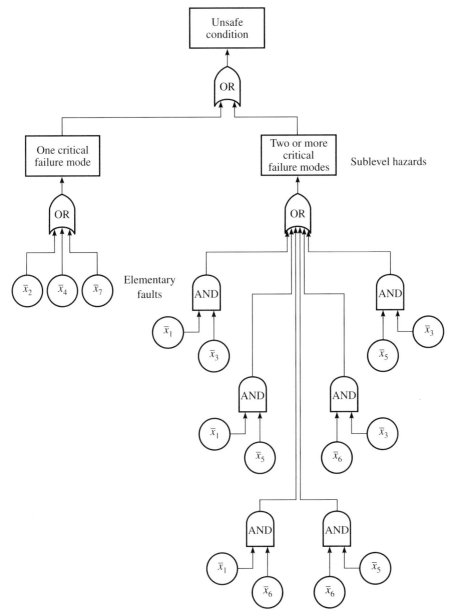

**Figure B13** Safety fault tree for simplified brake example. Presence of any $\bar{x}_i$ means failure mode $i$ *does* occur, indicating failure of element $i$.

equations to determine $P(X_i)$ or $P(\overline{X}_i)$; then these terms are substituted into the equations for $P_s$ or $P_u$; and last, $P_s$ or $P_u$ is substituted into a system safety equation along with probabilities for all subsystems that affect safety. For more advanced FT concepts, see Dugan [1996].

### B5.6  Summary

The safety analysis consists of:

1. Decomposing the system into subsystems or piece-parts.
2. Drawing a safety block diagram (SBD) or fault tree (FT) (computer programs are available for this purpose). See Appendix D.
3. Computation of the probability of safety or unsafety from the SBD or FT (computer programs are also available for this purpose).
4. Determining the failure rates of each component element. This is a data collection and estimation problem.
5. Substitution of failure rates into expression of step 3 (also done by computer programs).

## B6  MARKOV RELIABILITY AND AVAILABILITY MODELS

### B6.1  Introduction

Dependent failures, repair, or standby operation complicates the direct calculation of element reliabilities. In this section we shall discuss three different approaches to reliability computations for systems involving such computations. The first technique is the use of Markov models, which works well and has much appeal as long as the failure hazards $z(t)$ and repair hazards $w(t)$ are constant. When $z(t)$ and $w(t)$ become time-dependent, the method breaks down, except in a few special cases. (See Shooman [1990, pp. 348–359].) A second method, using joint density functions, and a third method, using convolution-like integrations, are more difficult to set up, but they are still valid when $z(t)$ or $w(t)$ is time-dependent.[14]

Some of the Markov modeling programs discussed in Appendix D deal with nonconstant hazards. In many cases, there is a paucity of failure-rate data; both constant-failure rates and -repair rates are used by default.

### B6.2  Markov Models

The basic properties of Markov models have already been discussed in Section A8. In this section we shall briefly review some of the assumptions necessary

---

[14]See Shooman [1990, Section 5.8].

for formulation of a Markov model and show how it can be used to make reliability computations.

In order to formulate a Markov model (to be more precise we are talking about continuous-time and discrete-state models) we must first define all the mutually exclusive states of the system. For example, in a system composed of a single nonrepairable element $x_1$ there are two possible states: $s_0 = x_1$, in which the element is good, and $s_1 = \bar{x}_1$, in which the element is bad. The states of the system at $t = 0$ are called the *initial states*, and those representing a final or equilibrium state are called *final states*. The set of Markov state equations describes the probabilistic transitions from the initial to the final states.

The transition probabilities must obey the following two rules:

1. The probability of transition in time $\Delta t$ from one state to another is given by $z(t)\Delta t$, where $z(t)$ is the hazard associated with the two states in question. If all the $z_i(t)$'s are constant, $z_i(t) = \lambda_i$, and the model is called *homogeneous*. If any hazards are time functions, the model is called *nonhomogeneous*.

2. The probabilities of more than one transition in time $\Delta t$ are infinitesimals of a higher order and can be neglected.

For the example under discussion the state-transition equations can be formulated using the above rules. The probability of being in state $s_0$ at time $t + \Delta t$ is written $P_{s_0}(t + \Delta t)$. This is given by the probability that the system is in state $s_0$ at time $t$, $P_{s_0}(t)$, times the probability of *no* failure in time $\Delta t$, $1 - z(t)\Delta t$, plus the probability of being in state $s_1$ at time $t$, $P_{s_1}(t)$, times the probability of repair in time $\Delta t$, which equals zero.

The resulting equation is

$$P_{s_0}(t + \Delta t) = [1 - z(t)\Delta t]P_{s_0}(t) + 0P_{s_1}(t) \tag{B71}$$

Similarly, the probability of being in state $s_1$ at $t + \Delta t$ is given by

$$P_{s_1}(t + \Delta t) = [z(t)\Delta t]P_{s_0}(t) + 1P_{s_1}(t) \tag{B72}$$

The transition probability $z(t)\Delta t$ is the probability of failure (change from state $s_0$ to $s_1$), and the probability of remaining in state $s_1$ is unity.[15] One can summarize the transition Eqs. (B71) and (B72) by writing the transition matrix given in Table B4. Note that it is a property of transition matrices that its rows must sum to unity. Rearrangement of Eqs. (B71) and (B72) yields

---

[15]Conventionally, state $s_1$ would be called an absorbing state since transitions out of the state are not permitted.

**TABLE B4   State Transition Matrix for a Single Element**

| Initial States | Final States | |
|---|---|---|
| | $s_0$ | $s_1$ |
| $s_0$ | $1 - z(t)\Delta t$ | $z(t)\Delta t$ |
| $s_1$ | 0 | 1 |

$$\frac{P_{s_0}(t + \Delta t) - P_{s_0}(t)}{\Delta t} = -z(t)P_{s_0}(t)$$

$$\frac{P_{s_1}(t + \Delta t) - P_{s_1}(t)}{\Delta t} = z(t)P_{s_0}(t)$$

Passing to a limit as $\Delta t$ becomes small, we obtain

$$\frac{dP_{s_0}(t)}{dt} + z(t)P_{s_0}(t) = 0 \tag{B73}$$

$$\frac{dP_{s_1}(t)}{dt} = z(t)P_{s_0}(t) \tag{B74}$$

Equations (B73) and (B74) can be solved in conjunction with the appropriate initial conditions for $P_{s_0}(t)$ and $P_{s_1}(t)$, the probabilities of ending up in state $s_0$ or state $s_1$, respectively. The most common initial condition is that the system is good at $t = 0$, that is, $P_{s_0}(t = 0) = 1$ and $P_{s_1}(t = 0) = 0$. Equations (B73) and (B74) are simple first-order linear differential equations which are easily solved by classical theory. Equation (B73) is homogeneous (no driving function), and separation of variables yields

$$\frac{dP_{s_0}(t)}{P_{s_0}(t)} = -z(t) \, dt$$

$$\ln P_{s_0}(t) = -\int_0^t z(\xi) \, d\xi + C_1$$

$$P_{s_0}(t) = \exp\left[-\int_0^t z(\xi) \, d\xi + C_1\right] = C_2 \exp\left[-\int_0^t z(\xi) \, d\xi\right] \tag{B75}$$

Inserting the initial condition $P_{s_0}(t = 0) = 1$,

$$P_{s_0}(t=0) = 1 = C_2 e^{-0}$$
$$\therefore C_2 = 1$$

and one obtains the familiar reliability function

$$R(t) = P_{s_0}(t) = \exp\left[-\int_0^t z(\xi)\,d\xi\right] \tag{B76}$$

Formal solution of Eq. (B76) proceeds in a similar manner.

$$P_{s_1}(t) = 1 - \exp\left[\int_0^t z(\xi)\,d\xi\right] \tag{B77}$$

Of course a formal solution of Eq. (B74) is not necessary to obtain Eq. (B77), since it is possible to recognize at the outset that

$$P_{s_0}(t) + P_{s_1}(t) = 1.$$

The role played by the initial conditions is clearly evident from Eq. (B75). Since $C_2 = P_{s_0}(0)$, if the system was initially bad, $P_{s_0}(t) = 0$, and $R(t) = 0$. If there is a fifty-fifty chance that the system is good at $t = 0$, then $P_{s_0}(t) = \frac{1}{2}$, and

$$R(t) = \frac{1}{2}\exp\left[-\int_0^t z(\xi)\,d\xi\right]$$

This method of computing the system reliability function yields the same results, of course, as the techniques of Sections B3 to B5. Even in a single-element problem it generates a more general model. The initial condition allows one to include the probability of initial failure before the system in question is energized.

## B6.3 Markov Graphs

It is often easier to characterize Markov models by a graph composed of nodes representing system states and branches labeled with transition probabilities. Such a Markov graph for the problem described by Eqs. (B73) and (B74) or Table B4 is given in Fig. B14. Note that the sum of transition probabilities for the branches leaving each node must be unity. Treating the nodes as signal sources and the transition probabilities as transmission coefficients, we can write Eqs. (B73) and (B74) by inspection. Thus, the probability of being at any node at time $t + \Delta t$ is the sum of all signals arriving at that node. All other

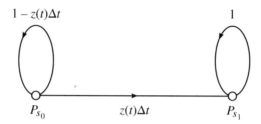

**Figure B14** Markov graph for a single nonrepairable element.

nodes are considered probability sources at time $t$, and all transition probabilities serve as transmission gains. A simple algorithm for writing Eqs. (B73) and (B74) by inspection is to equate the derivative of the probability at any node to the sum of the transmissions coming into the node. Any unity gain factors of the self-loops must first be set to zero, and the $\Delta t$ factors are dropped from the branch gains. Referring to Fig. B14, the self-loop on $P_{s_1}$ disappears, and the equation becomes $\dot{P}_{s_1} = zP_{s_0}$. At node $P_{s_0}$ the self-loop gain becomes $-z$, and the equation is $\dot{P}_{s_0} = -zP_{s_0}$. The same algorithm holds at each node for more complex graphs.

### B6.4 Example—A Two-Element Model

One can illustrate dependent failures,[16] standby operation, and repair by discussing a two-element system. For simplicity repair is ignored at first. If a two-element system consisting of elements $x_1$ and $x_2$ is considered, there are four system states: $s_0 = x_1 x_2$, $s_1 = \bar{x}_1 x_2$, $s_2 = x_1 \bar{x}_2$, and $s_3 = \bar{x}_1 \bar{x}_2$. The state transition matrix is given in Table B5 and the Markov graph in Fig. B15.

The probability expressions for these equations can be written by inspection, using the algorithms previously stated.

$$\frac{dP_{s_0}(t)}{dt} = -[z_{01}(t) + z_{02}(t)]P_{s_0}(t) \tag{B78}$$

$$\frac{dP_{s_1}(t)}{dt} = -[z_{13}(t)]P_{s_1}(t) + [z_{01}(t)]P_{s_0}(t) \tag{B79}$$

$$\frac{dP_{s_2}(t)}{dt} = -[z_{23}(t)]P_{s_2}(t) + [z_{02}(t)]P_{s_0}(t) \tag{B80}$$

$$\frac{dP_{s_3}(t)}{dt} = [z_{13}(t)]P_{s_1}(t) + [z_{23}(t)]P_{s_2}(t) \tag{B81}$$

---

[16]For dependent failures, see Shooman [1990, p. 235].

**TABLE B5  State Transition Matrix for Two Elements**

| Initial States | | Final States | | | |
|---|---|---|---|---|---|
| | | $s_0$ | $s_1$ | $s_2$ | $s_3$ |
| Zero failures | $s_0$ | $1 - [z_{01}(t) + z_{02}(t)]\Delta t$ | $z_{01}(t)\Delta t$ | $z_{02}(t)\Delta t$ | 0 |
| One failure | $s_1$ | 0 | $1 - [z_{13}(t)]\Delta t$ | 0 | $z_{13}(t)\Delta t$ |
| | $s_2$ | 0 | 0 | $1 - [z_{23}(t)]\Delta t$ | $z_{23}(t)\Delta t$ |
| Two failures | $s_3$ | 0 | 0 | 0 | 1 |

The initial conditions associated with this set of equations are $P_{s_0}(0)$, $P_{s_1}(0)$, $P_{s_2}(0)$, $P_{s_3}(0)$.

It is difficult to solve these equations for a general hazard function $z(t)$, but if the hazards are specified, the solution is quite simple. If all the hazards are constant, $z_{01}(t) = \lambda_1$, $z_{02}(t) = \lambda_2$, $z_{13}(t) = \lambda_3$, and $z_{23}(t) = \lambda_4$. The solutions are

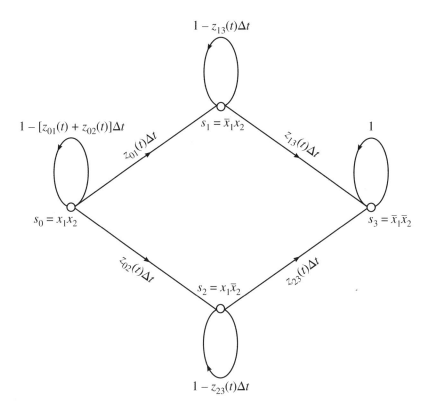

**Figure B15**  Markov graph for two distinct nonrepairable elements.

$$P_{s_0}(t) = e^{-(\lambda_1 + \lambda_2)t} \tag{B82}$$

$$P_{s_1}(t) = \frac{\lambda_1}{\lambda_1 + \lambda_2 - \lambda_3} (e^{-\lambda_3 t} - e^{-(\lambda_1 + \lambda_2)t}) \tag{B83}$$

$$P_{s_2}(t) = \frac{\lambda_2}{\lambda_1 + \lambda_2 - \lambda_4} (e^{-\lambda_4 t} - e^{-(\lambda_1 + \lambda_2)t}) \tag{B84}$$

$$P_{s_3}(t) = 1 - [P_{s_0}(t) + P_{s_1}(t) + P_{s_2}(t)] \tag{B85}$$

where

$$P_{s_0}(0) = 1 \quad \text{and} \quad P_{s_1}(0) = P_{s_2}(0) = P_{s_3}(0) = 0$$

Note that we have not as yet had to say anything about the configuration of the system, but only have had to specify the number of elements and the transition probabilities. Thus, when we solve for $P_{s_0}$, $P_{s_1}$, $P_{s_2}$, we have essentially solved for all possible two-element system configurations. In a two-element system, formulation of the reliability expressions in terms of $P_{s_0}$, $P_{s_1}$, and $P_{s_2}$ is trivial, but in a more complex problem we can always formulate the expression using the tools of Sections B3 to B5.

For a series system, the only state representing success is no failures; that is, $P_{s_0}(t)$. Therefore

$$R(t) = P_{s_0}(t) = e^{-(\lambda_1 + \lambda_2)t} \tag{B86}$$

If the two elements are in parallel, one failure can be tolerated, and there are three successful states, $P_{s_0}(t)$, $P_{s_1}(t)$, $P_{s_2}(t)$. Since the states are mutually exclusive,

$$\begin{aligned} R(t) = P_{s_0}(t) + P_{s_1}(t) + P_{s_2}(t) &= e^{-(\lambda_1 + \lambda_2)t} \\ &+ \frac{\lambda_1}{\lambda_1 + \lambda_2 - \lambda_3} (e^{-\lambda_3 t} - e^{-(\lambda_1 + \lambda_2)t}) \\ &+ \frac{\lambda_2}{\lambda_1 + \lambda_2 - \lambda_4} (e^{-\lambda_4 t} - e^{-(\lambda_1 + \lambda_2)t}) \end{aligned} \tag{B87}$$

It is easy to see why a series configuration of $n$ components has the poorest reliability and why a parallel configuration has the best. The only successful state for a series system is where all components are good; thus, $R(t) = P_{s_0}(t)$. In the case of a parallel system, all states except the one in which all components have failed are good, and $R(t) = P_{s_0}(t) + P_{s_1}(t) + P_{s_2}(t)$. It is clear that any other system configuration falls somewhere in between.

## B6.5 Model Complexity

The complexity of a Markov model depends on the number of system states. In general we obtain for an $m$-state problem a system of $m$ first-order differential equations. The number of states is given in terms of the number of components $n$ as

$$m = \binom{n}{0} + \binom{n}{1} + \binom{n}{2} + \cdots + \binom{n}{n} = 2^n$$

Thus, our two-element model has 4 states, and a four-element model 16 states. This means that an $n$-component system may require a solution of as many as $2^n$ first-order differential equations. In many cases we are interested in fewer states. Suppose we want to know only how many failed items are present in each state and not which items have failed. This would mean a model with $n + 1$ states rather than $2^n$, which represents a tremendous saving. To illustrate how such simplifications affect the Markov graph we consider the collapsed flowgraph shown in Fig. B16 for the example given in Fig. B15. Collapsing the flowgraph is equivalent to the restriction $P_{s_1'}(t) = P_{s_1}(t) + P_{s_2}(t)$ applied to Eqs. (B78) to (B81). Note that this can collapse the flowgraph only if $z_{13} = z_{23}$; however, $z_{01}$ and $z_{02}$ need not be equal. These results are obvious if Eqs. (B79) and (B80) are added.

Markov graphs for a system with repair are shown in Fig. B17(a) and (b). The graph in Fig. B17(a) is a general model, and that of Fig. B17(b) is a collapsed model.

The system equations can be written for Fig. B17(a) by inspection using the algorithm previously discussed.

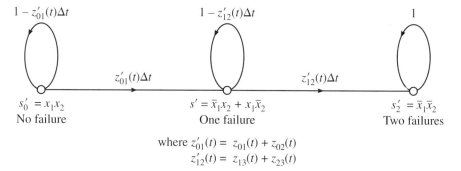

**Figure B16** Collapsed Markov graph corresponding to Fig. B15.

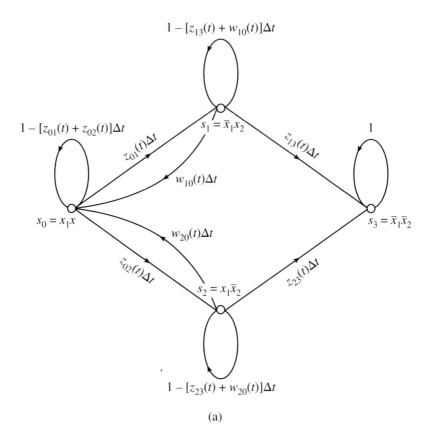
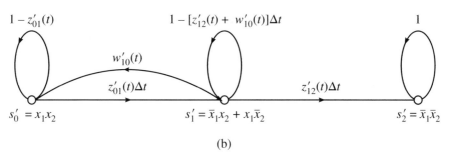

**Figure B17** Markov graphs for a system with repair: (a) general model; (b) collapsed model.

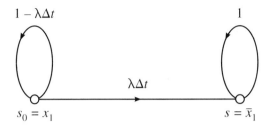

**Figure B18** Markov graph for the *reliability* of a single component with repair.

$$\dot{P}_{s_0} = -(z_{01} + z_{02})P_{s_0} + w_{10}P_{s_1} + w_{20}P_{s_2}$$
$$\dot{P}_{s_1} = -(z_{13} + w_{10})P_{s_1} + z_{01}P_{s_0}$$
$$\dot{P}_{s_2} = -(z_{23} + w_{20})P_{s_2} + z_{02}P_{s_0}$$
$$\dot{P}_{s_3} = z_{13}P_{s_1} + z_{23}P_{s_2} \tag{B88}$$

Similarly for Fig. B18(b)

$$\dot{P}_{s'_0} = -z'_{01}P_{s'_0} + w'_{10}P_{s'_1}$$
$$\dot{P}_{s'_1} = -(z'_{12} + w'_{10})P_{s'_1} + z'_{01}P_{s'_0}$$
$$\dot{P}_{s'_2} = z'_{12}P_{s'_1} \tag{B89}$$

The solution to Eqs. (B88) and (B89) for various values of the $z$s and $w$s will be deferred until the next section.

## B7  REPAIRABLE SYSTEMS

### B7.1  Introduction

In general, whenever the average repair cost in time and money of a piece of equipment is a fraction of the initial equipment cost, one considers system repair. If such a system can be rapidly returned to service, the effect of the failure is minimized. Obvious examples are such equipment as a television set, an automobile, or a radar installation. In such a system the time between failures, repair time, number of failures in an interval, and percentage of operating time in an interval are figures of merit which must be considered along with the system reliability. Of course, in some systems, such as those involving life support, surveillance, or safety, any failure is probably catastrophic, and repair is of no avail.

## B7.2 Availability Function

In order to describe the beneficial features of repair in a system that tolerates shutdown times, a new system function called *availability* is introduced. The availability function $A(t)$ is defined as the probability that the system is operating *at time t*. By contrast, the reliability function $R(t)$ is the probability that the system has operated *over the interval 0 to t*. Thus, if $A(250) = 0.95$, then if 100 such systems are operated for 250 hours on the average, 95 will be operative when 250 hours is reached and 5 will be undergoing various stages of repairs. The availability function contains no information on how many (if any) failure-repair cycles have occurred prior to 250 hours. On the other hand, if $R(250) = 0.95$, then if 100 such systems are operated for 250 hours, on the average, 95 will have operated without failure for 250 hours and 5 will have failed at some time within this interval. It is immaterial in which stage of the first or subsequent failure-repair cycles the five failed systems are. Obviously the requirement that $R(250) = 0.95$ is much more stringent than the requirement that $A(250) = 0.95$. Thus, in general, $R(t) \leq A(t)$.

If a *single unit* has no repair capability, then by definition $A(t) = R(t)$. If we allow repair, then $R(t)$ does not change, but $A(t)$ becomes greater than $R(t)$. The same conclusions hold for a *chain structure*. The situation changes for any system involving more than one tie set, i.e., systems with inherent or purposely introduced *redundancy*. In such a case, repair can beneficially alter both the $R(t)$ and $A(t)$ functions. This is best illustrated by a simple system composed of two parallel units. If a system consists of components $A$ and $B$ in parallel and no repairs are permitted, the system fails when both $A$ and $B$ have failed. In a repairable system if $A$ fails, unit $B$ continues to operate, and the system survives. Meanwhile, a repairer begins repair of unit $A$. If the repairer restores $A$ to usefulness before $B$ fails, the system continues to operate. The second component failure might be unit $B$, or unit $A$ might fail the second time in a row. In either case there is no system failure as long as the repair time is shorter than the time between failures. In the long run, at some time a lengthy repair will be started and will be in progress when the alternate unit fails, causing system failure. It is clear that repair will improve system reliability in such a system. It seems intuitive that the increase in reliability will be a function of the mean time to repair divided by the MTTF.

To summarize, in a series system, repair will not affect the reliability expression; however, for a complete description of system operation we shall have to include measures of repair time and time between failures. If the system structure has any parallel paths, repair will improve reliability, and repair time and time between failures will be of importance. In some systems, e.g., an unmanned space vehicle, repair may be impossible or impractical.[17]

---

[17]Technology is rapidly reaching the point where repair of an orbiting space vehicle is practical. The Hubble Space Telescope has already been repaired twice.

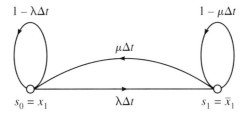

**Figure B19** Markov graph for the *availability* of a single component with repair.

## B7.3 Reliability and Availability of Repairable Systems

As long as the failure and repair density functions are exponential, i.e., constant-hazard, we can structure Markov repair models, as done in the previous section. The reliability and availability models will differ, and we must exercise great care in assigning absorbing states in a reliability model for a repairable system.

The reliability of a single component $x_1$ with constant failure hazard $\lambda$ and constant repair hazard $\mu$ can be derived easily using a Markov model. The Markov graph is given in Fig. B18 and the differential equations and reliability function in Eqs. (B90) and (B91).

$$\dot{P}_{s_0} + \lambda P_{s_0} = 0$$
$$\dot{P}_{s_1} = \lambda P_{s_0} \tag{B90}$$
$$P_{s_0}(0) = 1 \quad P_{s_1}(0) = 0$$
$$R(t) = P_{s_0}(t) = 1 - P_{s_1}(t) = e^{-\lambda t} \tag{B91}$$

Note that repair in no way influenced the reliability computation. Element failure $\bar{x}_1$ is an absorbing state, and once it is reached, the system never returns to $x_1$.

If we wish to study the availability, we must make a different Markov graph. State $\bar{x}_1$ is no longer an absorbing state, since we now allow transitions from state $\bar{x}_1$ back to state $x_1$. The Markov graph is given in Fig. B19 and the differential equations and state probabilities in Eqs. (B92) and (B93). The corresponding differential equations are

$$\dot{P}_{s_0} + \lambda P_{s_0} = \mu P_{s_1} \quad \dot{P}_{s_1} + \mu P_{s_1} = \lambda P_{s_0}$$
$$\dot{P}_{s_0}(0) = 1 \quad P_{s_1}(0) = 0 \tag{B92}$$

Solution yields the probabilities

## 458 SUMMARY OF RELIABILITY THEORY

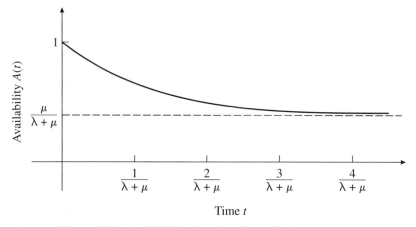

**Figure B20** Availability function for a single component.

$$P_{s_0}(t) = \frac{\mu}{\lambda + \mu} + \frac{\lambda}{\lambda + \mu} e^{-(\lambda + \mu)t}$$

$$P_{s_1}(t) = \frac{\lambda}{\lambda + \mu} - \frac{\lambda}{\lambda + \mu} e^{-(\lambda + \mu)t} \qquad (B93)$$

By definition, the availability is the probability that the system is good, $P_{s_0}(t)$:

$$A(t) = P_{s_0}(t) = \frac{\mu}{\lambda + \mu} + \frac{\lambda}{\lambda + \mu} e^{-(\lambda + \mu)t} \qquad (B94)$$

The availability function given in Eq. (B94) is plotted in Fig. B20.

### B7.4 Steady-State Availability

An important difference between $A(t)$ and $R(t)$ is their steady-state behavior. As $t$ becomes large, all reliability functions approach zero, whereas availability functions reach some steady-state value. For the single component the steady-state availability

$$A_{ss}(t) = \lim_{t \to \infty} A(t) = \mu/(\lambda + \mu) \qquad (B95a)$$

In the normal case, the mean repair time $1/\mu$ is much smaller than the time to failure $1/\lambda$, and we can expand the steady-state availability in a series and approximate by truncation:

$$A_{ss}(t) = A(\infty) = \frac{1}{1+\lambda/\mu} = 1 - \frac{\lambda}{\mu} + \frac{\lambda^2}{2\mu^2} + \cdots \approx 1 - \frac{\lambda}{\mu} \qquad \text{(B95b)}$$

The transient part of the availability function decays to zero fairly rapidly. The time at which the transient term is negligible with respect to the steady-state term depends on $\lambda$ and $\mu$. As an upper bound we know that the term $e^{-\alpha t} \leq 0.02$ for $t > 4/\alpha$; therefore, we can state that the transient term is over before $t = 4/(\lambda + \mu)$. If $\mu > \lambda$, the transient is over before $t = 4/\mu$. The interaction between reliability and availability specifications is easily seen in the following example.

Suppose a system is to be designed to have a reliability of greater than 0.90 over 1,000 hours and a minimum availability of 0.99 over that period. The reliability specification yields

$$R(t) = e^{-\lambda t} \geq 0.90 \qquad 0 < t < 1,000$$
$$e^{-1,000\lambda} \approx 1 - 10^3\lambda = 0.90 \qquad \lambda \geq 10^{-4}$$

Assuming $A(\infty)$ for the minimum value of the availability, Eq. (B95) yields

$$A(\infty) = 1 - \frac{\lambda}{\mu} = 0.99$$
$$\mu = 100\lambda = 10^{-2}$$

Thus, we use a component with an MTTF of $10^4$ hours, a little over 1 year, and a mean repair time of 100 hours (about 4 days). The probability of any failure within 1,000 hours (about 6 weeks) is less than 10%. Furthermore, the probability that the system is down and under repair at any chosen time between $t = 0$ and $t = 10^3$ hours is less than 1%. Now to check the approximations. The transient phase of the availability function lasts for $4/(10^{-2} + 10^{-4}) \approx 400$ hours; thus the availability will be somewhat greater than 0.99 for 400 hours and then settle down at 0.99 for the remaining 600 hours. Since $\mu$ is $100\lambda$, the approximation of Eq. (B95) is valid. Also since $\lambda t = 10^{-4} \times 10^3 = 10^{-1}$, the two-term series expansion of the exponential is also satisfactory.

The availability function has been defined as a probability function, just as the reliability function was. There is another statistical interpretation which sheds some light on the concept. Suppose that a large number of system operating hours are accumulated. This can be done either by operating one system for a long time, so that many failure and repair cycles are obtained and recorded, or by operating a large number of identical systems (an ensemble) for a shorter period of time and combining the data. If the ratio of cumulative operating time to total test time is computed, it approaches $A(\infty)$ as $t \to \infty$. Actually the data taken during the transient period of availability should be discarded to avoid any distortions. In fact if one wished to compute the transient phase of availability from experimental data, one would be forced to use

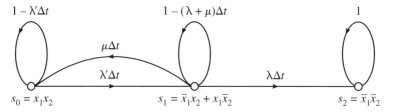

**Figure B21** Markov *reliability* model for two identical parallel elements and one repairer.

a very large number of systems over a short period of time. In analyzing the data one would break up the time scale into many small intervals and compute the ratio of cumulative operating time over the intervals divided by the length of the interval. [See Eq. (3.80).]

In a two-element nonseries system, the reliability function as well as the availability function is influenced by system repairs. The Markov reliability and availability graphs for systems with two components are given in Figs. B21, B22, and B23, and their solution is discussed in Shooman [1990, p. 345].

### B7.5 Computation of Steady-State Availability

When only the steady-state availability is of interest, a simplified computational procedure can be used. In the steady state, all the state probabilities should approach a constant; therefore, setting the derivatives to zero yields the following:

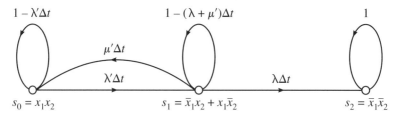

**Figure B22** Markov *reliability* model for two identical parallel elements and $k$ repairers.

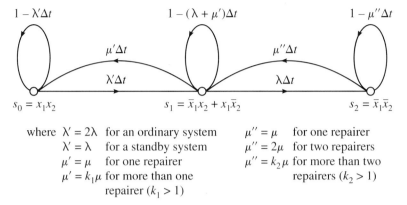

where  $\lambda' = 2\lambda$  for an ordinary system
       $\lambda' = \lambda$  for a standby system
       $\mu' = \mu$  for one repairer
       $\mu' = k_1\mu$  for more than one repairer ($k_1 > 1$)

       $\mu'' = \mu$  for one repairer
       $\mu'' = 2\mu$  for two repairers
       $\mu'' = k_2\mu$  for more than two repairers ($k_2 > 1$)

**Figure B23** Markov *availability* graph for two identical parallel elements and $k$ repairers.

$$\dot{P}_{s_0}(t) = \dot{P}_{s_1}(t) = \dot{P}_{s_2}(t) = 0$$

This set of equations cannot be solved for the steady-state probabilities, since their determinant is zero. Any two of these equations can, however, be combined with the identity

$$P_{s_0}(\infty) + P_{s_1}(\infty) + P_{s_2}(\infty) = 1 \tag{B96}$$

to yield a solution. A simpler method for computing steady-state availability using Laplace transforms is discussed in Chapter 4.

So far we have discussed reliability and availability computations only in one- and two-element systems. Obviously we could set up and solve Markov models for a larger number of elements, but the complexity of the problem increases very rapidly as $n$, the number of elements, increases. (If the elements are distinct, it goes as $2^n$, and if identical, as $n + 1$.)

## B8  LAPLACE TRANSFORM SOLUTIONS OF MARKOV MODELS

The formulation of a Markov model always leads to a set of first-order differential equations. For simple models, these equations are easily solved by using conventional differential equation theory. As we add more components, however, the model becomes more complex, and when repair is added to the model, the equations become coupled, making the solution more difficult. The easiest approach to the solution of such equations is through the use of Laplace transforms. In addition, the Laplace transform method provides a particularly simple method of calculating the mean time to failure (MTTF), the steady-state availability ($A_{ss}$), and the initial behavior of the reliability or availability functions: $R(t \to 0)$ or $A(t \to 0)$.

## B8.1 Laplace Transforms

One can appreciate the simplification that the Laplace transform provides in the solution of differential equations by analogy to the use of logarithms for simplifying certain numerical computations. In the predigital computer era, accurate computations to many decimal places for expressions such as $\alpha = (A \times B)/(C \times D)$ depended on lengthy hand computations, cumbersome mechanical calculators, or logarithms. Using logarithms, the $\log(\alpha) = \log(A) + \log(B) - \log(C) - \log(D)$. Thus multiplication and division are reduced to addition and subtraction of logarithms and $\alpha$ is determined by taking the antilogarithm of $\log(\alpha)$. Of course, for high accuracy, log tables with many digits of precision are required; such tables were calculated by mathematicians during the Depression of the 1930s as part of Franklin D. Roosevelt's New Deal programs used for creating jobs. Logarithm tables up to 10 or 16 digits appear in Abramowitz [1972, pp. 95–113]. The concept is to use logarithms to convert multiplication and division to simpler addition and subtraction and recover the answer by taking antilogarithms. The analogous procedure is to use Laplace transforms to convert *differential* equations to *algebraic* equations, solve the simpler algebraic equations, and use inverse Laplace transforms to recover the answer. The Laplace transform will now be introduced as an aid in the solution of differential equations. The Laplace transform of a time function $f(t)$ is defined by the integral[18]

$$\mathcal{L}\{f(t)\} = F(s) = \{f(t)\}^* = f^*(s) = \int_0^\infty f(t)e^{-st}\,dt \qquad \text{(B97)}$$

Four equivalent sets of notation for the Laplace transform are given in Eq. (B97). The first two are the most common, but they will not always be used since the symbol $F(s)$ causes confusion when we take the Laplace transform of both a density and a distribution function in the same equation. The third and fourth notation will be used whenever confusion might arise. The asterisk or the change in argument from $t$ to $s$ (or both) symbolizes the change from the time domain to the transform domain. The utility of the Laplace transform is that it reduces ordinary constant-coefficient linear differential equations to algebraic equations in $s$ which are easily solved. The solution in terms of $s$ is then converted back to a time function by an inverse-transform procedure. Sometimes the notation $\mathcal{L}^{-1}$ is used to denote the inverse-transform procedure; thus one could write $\mathcal{L}^{-1}\{F(s)\} = f(t)$. A pictorial representation of the Laplace transform solution of a differential equation is given in Fig. B24.

---

[18]The Laplace transform is defined only over the range of $s$ for which this integral exists.

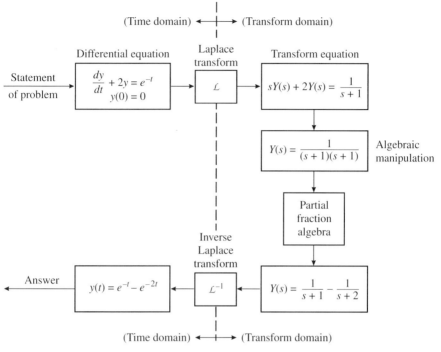

**Figure B24** The solution of a differential equation by Laplace transform techniques.

Since only a few basic transform techniques will be needed in this book, this discussion will be brief and will not touch on the broad aspects of the method. The Laplace transforms of three important functions follow.

***Example 1:*** For the exponential functions $f(t) = e^{-at}$

$$\mathcal{L}\{f(t)\} = f^*(s) = \int_0^\infty e^{-at} e^{-st}\, dt = \int_0^\infty e^{-(a+s)t}\, dt$$

$$= \left. \frac{-e^{-(a+s)t}}{s+a} \right|_0^\infty$$

$$= \frac{1}{s+a} \quad \text{for } s > -a \qquad (B98)$$

The restriction $s > -a$ is necessary in order that the integral not diverge.

***Example 2:*** Similarly, for the cosine function

$$f(t) = \cos at = \frac{e^{ia} + e^{-ia}}{2}$$

$$f^*(s) = \left\{\frac{e^{ia}}{2}\right\}^* + \left\{\frac{e^{-ia}}{2}\right\}^* = \frac{\frac{1}{2}}{s+ia} + \frac{\frac{1}{2}}{s-ia} = \frac{s}{s^2 + a^2} \quad \text{for } s > 0$$

(B99)

Note that in the above computation two properties were used. The Laplace transform is an integral, and the integral of a sum of two time functions is the sum of the integrals; thus, the transform of the sum is the sum of transforms. This is referred to as the *superposition property*. Also, the result of Eq. (B98) was used for each term in Eq. (B99).

**Example 3:** As a third example we consider the unit step function $u_{-1}(t)$ and the constant unity. When $f(t) = 1$ or $f(t) = u_{-1}(t)$

$$f^*(s) = \int_0^\infty 1 e^{-st}\, dt = \left.\frac{e^{-st}}{-s}\right|_0^\infty = \frac{1}{s} \quad \text{for } s > 0 \qquad (B100)$$

Note that although a step function and a constant are different functions, their Laplace transforms are the same, since over the region from $0 < t < +\infty$ they have the same value. Thus, the Laplace transform holds only for positive $t$. We can view a step as the limit of an exponential as we increase the time constant, $1/a \to \infty$. Equation (B100) could therefore be obtained from Eq. (B98) by letting $a \to 0$. The transforms for several time functions of interest are given in Table B6.

In order to solve differential equations with Laplace transform techniques we must compute the transform of a derivative. This can be done directly from Eq. (B97) using integration by parts

$$\mathcal{L}\left\{\frac{df(t)}{dt}\right\} = \left\{\frac{df(t)}{dt}\right\}^* = \int_0^\infty \frac{df(t)}{dt} e^{-st}\, dt = \int_0^\infty e^{-st}\, df(t)$$

Letting $dv = df(t)$ and $u = e^{-st}$ and integrating by parts,

$$\mathcal{L}\left\{\frac{df(t)}{dt}\right\} = \left. e^{-st}f(t)\right|_0^\infty + s\int_0^\infty f(t) e^{-st}\, dt$$

We first discuss the evaluation of the $e^{-st}f(t)$ term at its upper and lower limits. Since the Laplace transform is defined only for functions $f(t)$ which build up more slowly then $e^{-st}$ decay [see footnote to Eq. (B97)], $\lim_{t \to \infty} e^{-st}f(t) = 0$. At

**TABLE B6  A Short Table of Laplace Transforms**

| No. | $f(t)$ | $\{f(t)\}^* = f^*(s) = \mathcal{L}\{f(t)\} = F(s)$ |
|---|---|---|
| 1 | $u_0(t)$ | $1$ |
| 2 | $u_{-1}(t)$ | $\dfrac{1}{s}$ |
| 3 | $u_{-2}(t)$ | $\dfrac{1}{s^2}$ |
| 4 | $e^{-at}$ | $\dfrac{1}{s+a}$ |
| 5 | $\dfrac{1}{(n-1)!}t^{n-1}e^{-at}$ | $\dfrac{1}{(s+a)^n}$ |
| 6 | $\sin at$ | $\dfrac{a}{s^2+a^2}$ |
| 7 | $\cos at$ | $\dfrac{s}{s^2+a^2}$ |
| 8 | $e^{-bt}\sin at$ | $\dfrac{a}{(s+b)^2+a^2}$ |
| 9 | $e^{-bt}\cos at$ | $\dfrac{s+b}{(s+b)^2+a^2}$ |
| 10 | $Ae^{-at} + Be^{-bt}$ | $\dfrac{(A+B)s + Ab + Ba}{(s+a)(s+b)}$ |

Note: The functions $u_0(t), u_{-1}(t), \ldots, u_{-n}(t)$ are called the singularity functions, where $u_{-n}(t)$ is the derivative of $u_{-(n+1)}(t)$. The unit step, $u_{-1}(t)$, was already defined as 0 for $-\infty \le t < 0$ and 1 for $0 \le t \le +\infty$. The unit ramp, $u_{-2}(t)$, is the integral of the step function (of course, the step is the derivative of the ramp). The function $u_0(t)$ is the unit-impulse function in which the amplitude is the derivative of the step function and is 0 everywhere except $t = 0$, where it is infinite. The area of the impulse is unity at $t = 0$, as it must be since the step is the integral of the impulse.

the lower limit we obtain the initial value of the function $f(0)$.[19] The integral is of course the Laplace transform itself

$$\left\{\frac{df(t)}{dt}\right\}^* = sf^*(s) - f(0) \tag{B101}$$

By letting $g(t) = d^n f(t)/dt^n$ it is easy to generate a recursion relationship

---

[19]The notation $f(0)$ means the value of the function at $t = 0$. If singularity functions occur at $t = 0$, we must use care and write $f(0^-)$, which is the limit as 0 is approaches from the left.

$$\left\{\frac{dg(t)}{dt}\right\}^* = \left\{\frac{d^{n+1}f(t)}{dt^{n+1}}\right\}^* = s\left\{\frac{d^n f(t)}{dt^n}\right\}^* - f^n(0) \tag{B102}$$

for the second derivation

$$\left\{\frac{d^2 f(t)}{dt^2}\right\}^* = s^2 f^*(s) - s f(0) - \dot{f}(0) \tag{B103}$$

Using the information discussed, we can solve the homogeneous differential equation

$$\frac{d^2 y}{dt^2} + 5\frac{dy}{dt} + 6y = 0 \qquad \begin{array}{l} y(0) = 0 \\ \dot{y}(0) = 1 \end{array}$$

Taking the transform of each term, we have

$$[s^2 y^*(s) - sy(0) - \dot{y}(0)] + 5[sy^*(s) - y(0)] + 6y^*(s) = 0$$
$$[s^2 y^*(s) - 1] + 5[sy^*(s)] + 6y^*(s) = 0$$
$$(s^2 + 5s + 6)y^*(s) = 1$$
$$y^*(s) = \frac{1}{(s+2)(s+3)}$$

Using transform 10 in Table B6,

$$A + B = 0 \qquad 3A + 2B = 1$$
$$A = +1 \qquad B = -1$$
$$y(t) = e^{-2t} - e^{-3t}$$

Suppose we add the driving function $e^{-4t}$ to the above example, that is,

$$\left\{\frac{d^2 y}{dt^2} + 5\frac{dy}{dt} + 6y\right\}^* = \{e^{-4t}\}^*$$

$$(s+2)(s+3)y^*(s) - 1 = \frac{1}{s+4}$$

$$y^*(s) = \frac{s+5}{(s+4)(s+2)(s+3)}$$

No transform for this function exists in the table, but we can use partial-fraction algebra to reduce this to known results

$$\frac{s+5}{(s+4)(s+2)(s+3)} = \frac{\frac{1}{2}}{s+4} + \frac{\frac{3}{2}}{s+2} + \frac{-2}{s+3}$$

Thus, each term represents an exponential, and

$$y(t) = \tfrac{1}{2} e^{-4t} + \tfrac{3}{2} e^{-2t} - 2 e^{-3t}$$

The $e^{-4t}$ term represents the *particular solution* (driving function), and the $e^{-2t}$ and $e^{-3t}$ terms represent the *homogeneous solution* (natural response).

The partial-fraction-expansion coefficients can be found by conventional means or by the following shortcut formula

$$f(s) = \frac{N(s)}{D(s)} = \frac{N(s)}{\prod_{i=1}^{n}(s+r_i)} = \frac{A_1}{s+r_1} + \frac{A_2}{s+r_2} + \cdots + \frac{A_n}{s+r_n}$$

where

$$A_i = \left[ \frac{N(s)}{D(s)} (s+r_i) \right]_{s=-r_i} \tag{B104}$$

For the above example

$$A_1 = \left[ \frac{s+5}{(s+4)(s+2)(s+3)} (s+4) \right]_{s=-4} = \frac{1}{2}$$

$$A_2 = \left[ \frac{s+5}{(s+4)(s+2)(s+3)} (s+2) \right]_{s=-2} = \frac{3}{2}$$

$$A_3 = \left[ \frac{s+5}{(s+4)(s+2)(s+3)} (s+3) \right]_{s=-3} = -2$$

The derivation of Eq. (B104) as well as a similar one for the case of repeated roots can be found in any text on Laplace transforms.

We have already discussed two Laplace transform theorems, superposition and derivative property. Some additional ones useful in solving Markov models appear in Table B7.

The first and second theorems have already been discussed. The third theorem is simply the integral equivalent of the differentiation theorems. The convolution theorem is important since it describes the time-domain equivalent of a product of two Laplace transforms. Theorems 6 and 7 are useful in computing the initial and final behavior of reliability functions.

**TABLE B7  A Short Table of Laplace Transform Theorems**

| No. | Operation | $f(t)$ | $\mathcal{L}\{f(t)\} = F(s)$ |
|---|---|---|---|
| 1 | Linearity (superposition) property | $a_1 f_1(t) + a_2 f_2(t)$ | $a_1 F_1(s) + a_2 F_2(s)$ |
| 2 | Differentiation theorems | $\dfrac{df(t)}{dt}$ | $sF(s) - f(0)$ |
|  |  | $\dfrac{d^2 f(t)}{dt^2}$ | $s^2 F(s) - sf(0) - \dot{f}(0)$ |
|  |  | $\dfrac{d^n f(t)}{dt^n}$ | $s\mathcal{L}\left\{\dfrac{d^{n-1} f(t)}{dt^{n-1}}\right\}$ $- f^{n-1}(0)$ |
| 3 | Integral theorems | $\displaystyle\int_0^t f(t)\,dt$ | $\dfrac{F(s)}{s}$ |
|  |  | $\displaystyle\int_{-\infty}^t f(t)\,dt$ | $\dfrac{F(s)}{s} + \dfrac{\displaystyle\int_{-\infty}^0 f(t)\,dt}{s}$ |
| 4 | Convolution theorem | $\displaystyle\int_0^t f_1(\tau) f_2(t-\tau)\,d\tau$ | $F_1(s) F_2(s)$ |
| 5 | Multiplication-by-$t$ property | $t f(t)$ | $-\dfrac{dF(s)}{ds}$ |
| 6 | Initial-value theorem | $\lim\limits_{t \to 0} f(t)$ | $\lim\limits_{s \to \infty} sF(s)$ |
| 7 | Final-value theorem | $\lim\limits_{t \to \infty} f(t)$ | $\lim\limits_{s \to 0} sF(s)$ |

Note: The function $sF(s)$ is a ratio of polynomials in problems we shall consider. The roots of the denominator polynomial are called *poles*. We cannot apply the initial- and final-value theorems if *any* pole of $sF(s)$ has *a zero or positive real part*. The statement is conventionally worded: the initial- and final-value theorems hold only provided that all poles of $sF(s)$ lie in the left half of the $s$ plane.

## B8.2  MTTF from Laplace Transforms

The MTTF can also be computed from the Laplace transform of $R(t)$. [See Eq. (B51).]

Another form in terms of $R^*(s)$—an alternate notation for $\mathcal{L}\{R(\tau)\}$—is obtained by considering $\int_0^t R(\tau)\,d\tau$. Using Theorem 3 of Table B7, we obtain

$$\mathcal{L}\left\{\int_0^t R(\tau)\,d\tau\right\} = \frac{R^*(s)}{s} \qquad \text{(B105)}$$

However,

$$\text{MTTF} = \lim_{t \to \infty} \int_0^t R(\tau)\, d\tau$$

Using Theorem 7 of Table B12,

$$\text{MTTF} = \lim_{t \to \infty} \int_0^t R(\tau)\, d\tau = \lim_{s \to 0} s\mathcal{L}\left\{\int_0^t R(\tau)\, d\tau\right\} = \lim_{s \to 0} s\, \frac{R^*(s)}{s}$$

Thus,

$$\text{MTTR} = \lim_{s \to 0} R^*(s) \qquad (B106)$$

The above formula is extremely useful for computing the MTTF for a Markov model.

### B8.3 Time-Series Approximations from Laplace Transforms

One of the objectives of the MTTF computations of the previous section is to simplify the algebra involved in obtaining time functions from the transformed function. In most practical cases, the transform expression (ratio of two polynomials in $s$) has a denominator that is of second, third, or higher order. The solution requires the factoring of the polynomial (generally requiring numerical methods) and subsequent partial-fraction expansion. Calculating the MTTF from $F(s)$ by taking the limit as given in Eq. (B106), is simple; however, it provides only partial information. In Section 3.4.1, we discussed approximating the system-time function in the high-reliability region by the leading terms in the Taylor-series expansion. If this is our objective, and if we have the Laplace transform, we can find the Taylor-series coefficients simply and directly without first finding the time function.

Any function $f(t)$, whose various derivatives exist, can be expanded in a Taylor series:

$$f(t) = f(0) + f'(0)t/1! + f''(0)t^2/2! + f'''(0)t^3/3! + \cdots \qquad (B107)$$

Where $f'''(0)$ is the third time derivative of $f(t)$ evaluated at $t = 0$, and similarly for $f(0)$, $f'$, and so on.

Note that the derivatives of $f(t)$ always exist for a reliability function that is a linear combination of exponential terms, since all derivatives exist for an exponential function. We can rewrite this equation in terms of a set of constant $K_i$, which stand for the time derivatives, and obtain

$$f(t) = K_0 + K_1 t/1! + K_2 t^2/2! + \cdots + \sum_{n=0}^{\infty} K_n t^n/n! \quad (B108)$$

If we take $\mathcal{L}\{f(t)\}$ for the function in Eq. (B108), we obtain a series of simple Laplace transforms for the right-hand side of the equation. These transforms can easily be obtained from entry no. 5 in Table B6 by setting $a = 0$, yielding

$$\mathcal{L}\{f(t)\} = F(s) = \sum_{n=0}^{\infty} \frac{K_n}{s^{n+1}} \quad (B109)$$

Knowing that $f(t)$ in Eq. (B108) is a reliability function, we know that $R(0) = 1$; thus $K_0 = 1$. We can easily manipulate $F(s)$ into the series form given by Eq. (B109) by the simple process of long division. Of course, this method presupposes that we have the transform, which is generally true if we are solving a Markov model. If we already have the time function, it is probably easier to use the expansions discussed in Section 3.4.1 than to first compute the transform and use this approach. The following example illustrates the method.

Let us suppose that in the process of solving a Markov model we obtain the following Laplace transform of a reliability function:[20]

$$R(s) = \frac{s + \lambda + \lambda' + \mu'}{s^2 + [\lambda + \lambda' + \mu']s + \lambda\lambda'} \quad (B110)$$

Performing long division of the numerator and denominator polynomials, we obtain

$$s^2 + [\lambda + \lambda' + \mu']s + \lambda\lambda' \overline{\Big)\, s + [\lambda + \lambda' + \mu']} \quad \overline{\frac{1}{s} - \frac{\lambda\lambda'}{s^3} + \frac{\lambda\lambda'[\lambda + \lambda' + \mu']}{s^4} + \cdots} \quad (B111)$$

Thus, by using Table B6, entry no. 5 to obtain the inverse transform, that is, the expression for $R(t)$ that corresponds to $R(s)$ given in Eq. (B111),

$$R(t) \approx 1 - \frac{\lambda\lambda' t^2}{2} + \frac{\lambda\lambda'[\lambda + \lambda' + \mu']t^3}{6} + \cdots \quad (B112)$$

For a parallel system, $\lambda' = 2\lambda$ and $\mu' = \mu$, and substitution in Eq. (B112) yields

---

[20]This example is actually the Laplace transform of the reliability of two parallel elements with repair. For hot standby, $\lambda' = 2\lambda$, and for cold standby, $\lambda' = \lambda$. See Eqs. (3.65a, b).

$$R(t) \approx 1 - \lambda^2 t^2 + \frac{\lambda^2[3\lambda + \mu]t^3}{3} + \cdots \qquad (B113)$$

For a standby system, $\lambda' = \lambda$ and $\mu' = \mu$, and substitution in Eq. (B112) yields

$$R(t) \approx 1 - \frac{\lambda^2 t^2}{2} + \frac{\lambda^2[2\lambda + \mu]t^3}{6} + \cdots \qquad (B114)$$

Comparing Eq. (B113) with (B114), we see from the coefficients of the $t^2$ term that the standby system is superior.

## REFERENCES

Abramowitz, M., and I. A. Stegun (eds.). *Handbook of Mathematical Functions with Formulas, Graphs, and Mathematical Tables*. Washington, DC: National Bureau of Standards, U.S. Government Printing Office (GPO), 1972.

Bernstein, J. Private communication. University of Maryland, October 2001.

Carhart, R. R. A Survey of the Current Status of the Reliability Problem. Rand Corporation No. RM-1131, August 14, 1953.

Chang, C. Y., and S. M. Size. *VLSI Technology*. McGraw-Hill, New York, 1996, pp. 668–672.

*Defect and Fault-Tolerance in VLSI Systems*. International Symposium. IEEE Computer Society Press, New York, 1997.

Dugan, J. B. Software System Analysis Using Fault Trees. In *Handbook of Software Reliability Engineering*, Michael R. Lyu (ed.). McGraw-Hill, New York, 1996, ch. 15.

Dugan, J. B., S. J. Bavuso, and M. A. Boyd. Dynamic Fault-Tree Models for Fault-Tolerant Computer Systems. *IEEE Transactions on Reliability* 41, 3 (June 1992): 363–376.

Elsayed, A. *Reliability Engineering*. Addison-Wesley, Reading, MA, 1996.

Grace, K. Jr. Approximate System Availability Models. *Proceedings 1969 Annual Symposium on Reliability*, January 1969. IEEE, New York, NY, pp. 146–152.

Grace, K. Jr. Repair Queuing Models for System Availability. *Proceedings 1970 Annual Symposium on Reliability*, February 1970. IEEE, New York, NY, pp. 331–336.

Haldar, A., and S. Manadevan. *Probability, Reliability, and Statistical Methods in Engineering Design*. Wiley, New York, 2000.

Healey, J. T. Private communication. Telcordia Technologies, October 2001.

Jensen, F. *Electronic Component Reliability*. Wiley, New York, 1995, app. J, p. 344.

Leemis, L. M. *Reliability: Probabilistic Models and Statistical Methods*. Prentice-Hall, Englewood Cliffs, NJ, 1994.

Meshkat, L., J. B. Dugan, and J. D. Andrews. Analysis of Safety Systems with On-Demand and Dynamic Failure Modes. *Proceedings Annual Reliability and Maintainability Symposium*, 2000. IEEE, New York, NY, pp. 14–21.

Messinger, M., and M. L. Shooman. Approximations for Complex Structures. *Proceedings 1967 Annual Symposium on Reliability*. IEEE, New York, NY.

Mok, Y.-L., and L.-M. Ten. A Review of Plastic-Encapsulated-Microcircuit Reliability-Prediction Models. *Proceedings Annual Reliability and Maintainability Symposium*, 2000. IEEE, New York, NY, pp. 200–205.

Ohring, M. *Reliability and Failure of Electronic Materials and Devices*. Academic Press, New York, 1998, p. 340.

Pecht, M. G. (ed.). *Product Reliability, Maintainability, and Supportability Handbook*. CRC Press (www.crcpub.com), Boca Raton, FL, 1995.

Pham, H., and S. Upadhyaya. Reliability Analysis of a Class of Fault-Tolerant Systems (Digital Data Communications). *Proceedings Annual Reliability and Maintainability Symposium*, 1989. IEEE, New York, NY, pp. 114–118.

Pukiate, J., and P. Pukite. *Modeling for Reliability Analysis*. IEEE Press, New York, 1998.

Shooman, M. L. *Probabilistic Reliability: An Engineering Approach*. McGraw-Hill, New York, 1968.

Shooman, M. L. *Software Engineering: Design, Reliability, and Management*. McGraw-Hill, New York, 1983.

Shooman, M. L. *Probabilistic Reliability: An Engineering Approach*, 2d ed. Krieger, Melbourne, FL, 1990.

Siewiorek, D. P., and R. S. Swarz. *Reliable Computer Systems Design and Evaluation*. The Digital Press, New Bedford, MA, 1982.

Size, S. M. (ed.). *VLSI Technology*, 2d ed. McGraw-Hill, New York, 1988, pp. 409–414.

Weibull, W. A Statistical Distribution Function of Wide Application. *Journal of Applied Mechanics* 18 (1951): 293–297.

Welker, E. L., and M. Lipow. Estimating the Exponential Failure Rate from Data with No Failure Events. *Proceedings Annual Reliability and Maintainability Symposium*, 1974. IEEE, New York, NY, pp. 420–427.

## PROBLEMS

Note: Problems B1–B4, B6, and B10–B12 are taken from Shooman [1990].

**B1.** A series system is composed of $n$ identical independent components. The component probability of success is $p_c$ and $q_c = 1 - p_c$.

  (a) Show that if $q_c \ll 1$, the system reliability $R$ is approximately given by $R \approx 1 - nq_c$.

  (b) If the system has 10 components and $R$ must be 0.99, how good must the components be?

**B2.** A parallel system is composed of 10 identical independent components. If the system reliability $R$ must be 0.99, what is the minimum component reliability?

**B3.** A 10-element system is constructed of independent identical components so that 5 out of the 10 elements are necessary for system success. If the system reliability $R$ must be 0.99, how good must the components be?

**B4.** Draw reliability graphs for the following three reliability block diagrams. Note: The probabilities of system success for independent identical units are given for each part—(a), (b), and (c)—of Fig. P1.

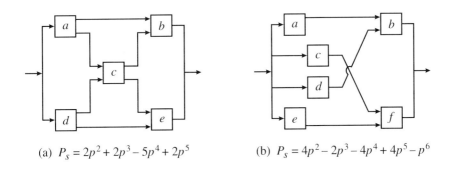

(a) $P_S = 2p^2 + 2p^3 - 5p^4 + 2p^5$

(b) $P_S = 4p^2 - 2p^3 - 4p^4 + 4p^5 - p^6$

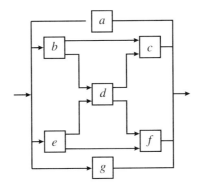

(c) $P_S = 2p + p^2 - 2p^3 - 7p^4 + 14p^5 - 9p^6 + 2p^2$

**Figure P1**

**B5.** Formulate a fault-tree model for the systems given in problem B4.

**B6.** Find all the minimal tie sets and cut sets for the three systems in problem B4.

**B7.** Solve problems B2 and B3.

**B8.** Assume numerical values for the axes of Fig. B6, and explain the cost trade-offs of burn-in and replacement.

**B9.** Check the MTTF computation given in Eq. (B54).

**B10.** A communication system is composed of a fixed-frequency transmitter $T_1$ and a fixed-frequency receiver $R_1$. The fixed frequency is $f_1$ and both receiver and transmitter have constant hazards $\lambda$. In order to improve the reliability, a second receiver and transmitter operating on frequency $f_2$ are used to provide a redundant channel. Both channels are identical, except for frequency. Construct a reliability diagram for the system and write the reliability function. In order to improve reliability, a tuning unit is added to each receiver so that it can operate at frequency $f_1$ or $f_2$. The hazard for each tuning unit is given by $\lambda'$. Draw the new reliability diagram and write the reliability function. Sketch the reliability of the improved system and the original two-channel system. Assume that $\lambda' = 0.1\lambda$ and repeat for $\lambda' = 10\lambda$. (Use series approximations, if necessary.)

**B11.** Solve for the reliability expression for a three-element standby system using a Markov model. All elements are independent identical units with constant-hazard $\lambda$.

**B12.** For a single component with repair, $R(t)$ and $A(t)$ are given by Eqs. (B91) and (B94). If we specify that $R(t_1) = 0.9$,
(a) What can you say about $A(t_1)$?
(b) How are $\lambda$ and $\mu$ constrained if $A(t_1) \geq 0.99$?

# APPENDIX C

# REVIEW OF ARCHITECTURE FUNDAMENTALS

## C1 INTRODUCTION TO COMPUTER ARCHITECTURE

Most readers of this book probably have an electrical engineering or computer science background and are familiar with the material presented in this appendix; thus they can skip it altogether or thumb through it as a refresher. However, some readers may have a background in mathematics, operations research, or some similar field; for them, this appendix will serve as a concise background. The reader is referred to the following references for more detailed information: Hill, 1981; Kohavi, 1978; Mano, 1995; Roth, 1995; Shiva, 1988; Wakerly, 2001.

### C1.1 Number Systems

Computers are constructed from switching elements that are two-state devices; thus it is common to utilize the binary number system (base 2) for computer computation, design of arithmetic algorithms, and construction of computer hardware. A number, $N$, written in radix (base), $r$, takes on the general polynomial form.

$$N = a_n r^n + a_{n-1} r^{n-1} + \cdots + a_1 r^1 + a_0 r^0 \,.\, a_{-1} r^{-1} + a_{-2} r^{-2} + \cdots$$

$\longleftarrow$ whole number portion $\longrightarrow$ | $\longleftarrow$ fraction portion $\longrightarrow$

radix point

(C1)

# REVIEW OF ARCHITECTURE FUNDAMENTALS

Each number system has $r$ distinct digits; for example, in binary, the two digits 0 and 1; in decimal, the ten digits 0, 1, 2, 3, 4, 5, 6, 7, 8, and 9.

One can convert from one number system to another using these basic definitions. As an example, consider the conversion of a base-2 number to a base-10 number:

$$(11110101)_2 = 2^7 + 2^6 + 2^5 + 2^4 + 2^2 + 2^0$$
$$= 128 + 64 + 32 + 16 + 4 + 1 = (245)_{10}$$

Note that parentheses and base subscripts are commonly used to clarify the notation when one is discussing two or more number systems and conversions. Similarly, one can convert from base 10 to base 2 by extracting the largest powers of 2 that are contained in the base-10 number. Conversion of $(245)_{10}$ to $(?)_2$ proceeds as follows:

| 245 | 117 | 53 | 21 | 5 | 5 | 1 | 1 |
|---|---|---|---|---|---|---|---|
| −128 | −64 | −32 | −16 | −8 | −4 | −2 | −1 |
| 117 | 53 | 21 | 5 | X | 1 | X | 0 |
| Yes | Yes | Yes | Yes | No | Yes | No | Yes |

Thus the subtraction process shows that 245 base 10 contains 2 to the powers 7, 6, 5, 4, 2, and 0 (the Yes's), but not $2^3 = 8$ or $2^1 = 2$ (the No's) that yields the binary number 11110101. The references give many simpler algorithms for conversion.

The first twenty numbers in the decimal, binary, octal (base 8), hexadecimal (base 16; commonly called Hex), and the binary-coded decimal (BCD) systems are given in Table C1. Since the Hex number system is base 16, we need sixteen digit symbols. Clearly, the first ten are the digits 0–9 and the remaining six are generally represented by the first six letters of the English alphabet—A, B, ... , F.

Note from Table C1 that it is easy to convert from binary to octal. One divides the binary number into groups of three digits and writes the octal numbers (0 to 7) that correspond to the group of three digits. Similarly, one can convert from binary to Hex by grouping four digits at a time and using a similar process. Reverse conversions involve expanding each octal digit into three binary digits or each Hex digit into four binary digits. The BCD number system uses four binary digits to represent the digital numerals from 1 to 9. This is emphasized in Table C1 by the vertical bar used for separating the binary digits into groups of four. The advantage of the BCD system is that each decimal digit can be converted by repetition of the same circuit; thus designs based on BCD numbers are highly modular.

**TABLE C1 Number Systems Commonly Used in Computer and Digital Circuit Design**

| Decimal | Binary | Octal | Hex | BCD |
| --- | --- | --- | --- | --- |
| 0  | 00000 | 00 | 00 | 0000\|0000 |
| 1  | 00001 | 01 | 01 | 0000\|0001 |
| 2  | 00010 | 02 | 02 | 0000\|0010 |
| 3  | 00011 | 03 | 03 | 0000\|0011 |
| 4  | 00100 | 04 | 04 | 0000\|0100 |
| 5  | 00101 | 05 | 05 | 0000\|0101 |
| 6  | 00110 | 06 | 06 | 0000\|0110 |
| 7  | 00111 | 07 | 07 | 0000\|0111 |
| 8  | 01000 | 10 | 08 | 0000\|1000 |
| 9  | 01001 | 11 | 09 | 0000\|1001 |
| 10 | 01010 | 12 | $0A$ | 0001\|0000 |
| 11 | 01011 | 13 | $0B$ | 0001\|0001 |
| 12 | 01100 | 14 | $0C$ | 0001\|0010 |
| 13 | 01101 | 15 | $0D$ | 0001\|0011 |
| 14 | 01110 | 16 | $0E$ | 0001\|0100 |
| 15 | 01111 | 17 | $0F$ | 0001\|0101 |
| 16 | 10000 | 20 | 10 | 0001\|0110 |
| 17 | 10001 | 21 | 11 | 0001\|0111 |
| 18 | 10010 | 22 | 12 | 0001\|1000 |
| 19 | 10011 | 23 | 13 | 0001\|1001 |
| 20 | 10100 | 24 | 14 | 0010\|0000 |

## C1.2 Arithmetic in Binary

One can discuss at length arithmetic in various bases; however, the algorithms can become quite detailed, especially when one considers both positive and negative numbers. In the binary number system, the rules for positive numbers are quite simple.

1. The sum of any two binary digits (0 + 0, 0 + 1, 1 + 0, 1 + 1) is 0 if the digits are the same (0 + 0, 1 + 1) and 1 if the digits differ (0 + 1, 1 + 0). A carry to the next digit is only generated when both digits are 1.
2. The difference of any two binary digits (0 − 0, 0 − 1, 1 − 0, 1 − 1) is 0 if the digits are the same (0 − 0, 1 − 1) and 1 if the digits differ (0 − 1, 1 − 0). A borrow to the next digit is only generated in the case 0 − 1.
3. To multiply two binary numbers, we treat the process just like decimal multiplication, forming partial products that are shifted left once each time we shift to another bit of the multiplier (number on the bottom). The partial products are either 0 or a replica of the multiplicand (number on the top); then they are added using binary addition.
4. Long division of two binary numbers proceeds as in the decimal number

system, and each trial divisor is subtracted from the number on the top using binary subtraction.

## C2 LOGIC GATES, SYMBOLS, AND INTEGRATED CIRCUITS

Digital-logic elements used in circuits have evolved over the years. The earliest realization of logic elements (logic gates) were relays with multiple contacts, which were soon replaced by vacuum tube switches and vacuum tube diode circuits. Vacuum tubes were in turn replaced by transistors and semiconductor diodes, and, finally, by integrated circuits. Modern-day digital circuits (often called *chips*) are composed of various integrated circuits; some, such as microprocessors and memory systems, are quite complex, whereas others are simple-logic circuits. We will discuss the simple-logic circuits since many more complex circuits can be viewed as interconnection of the simple circuits. Such logic circuits realize simple-logic functions such as union, intersection, compliment, and so forth (see Appendix A3 for a definition of these logic operations based on set theory that applies to both digital logic and probability theory). The inputs to the logic gates are called *switching variables* (represented by letters $A$, $B$, $x$, $y$, etc.), and the output is a switching function, $f(x, y)$. The union of two switching variables $A$ and $B$ is written as $f(A, B) = A + B$, $(A \cup B)$, and is called an *OR function*; the associated logic gate is an *OR gate*. Similarly, an intersection of $A$, $B$ is written as $f(A, B) = A \cdot B$, $(A \cap B)$, and is called an *AND gate*. The complement of $A$ is given by $\overline{A}$ or $A'$ and is called a *NOT gate* or an *inverter*. Note the symbols $\overline{A}$ and $A'$ are used interchangeably in this text. The logic symbols for these gates are given in Fig. C1. These three logic gates as

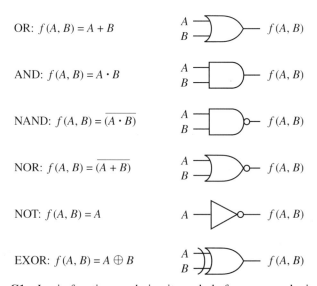

**Figure C1** Logic functions and circuit symbols for common logic gates.

well as the others given in Fig. C1 are discussed further in the next section. The complement can be denoted by a NOT gate or by a small circle shown at the output of the logic gate (cf. Fig. C2).

## C3  BOOLEAN ALGEBRA AND SWITCHING FUNCTIONS

One can define a logic function in terms of its variables, a mapping connecting the values that the variables assume, and the resulting value of the logic function. For example, if we have two switching variables, $x$ and $y$, we can write the general form of a two-variable switching function as $f(x,y)$. This is similar to the definition of a function in calculus; however, the variables $x$ and $y$ are discrete and only take on the values 0 and 1. Thus we can define the switching function mapping in terms of the 4 combinations (00, 01, 10, 11) of the variables in tabular form. Such a table is called a *truth table*; truth tables for the 6 functions given in Fig. C1 are shown in Table C2. The OR function is 1 whenever $x$ or $y$ or both are 1; the AND function is 1 only when both $x$ and $y$ are 1. The NAND function is the complement of the AND; the NOR function is the complement of the OR. Although the EXOR function is like an OR function, it excludes the case where both $x$ and $y$ are 1. The EXOR function is 1 whenever the inputs $x$ and $y$ disagree. There is another function that is sometimes defined as the complement of the exclusive OR function; this is called the *coincidence function*, which is 1 whenever $x$ and $y$ agree. Note that there is an alternate way to denote the NOR and NAND functions shown in Fig. C2(a) and (b). A circuit for implementing an EXOR function and the logic symbol are shown in Fig. C2(c).

In constructing the truth tables given in Table C2, we assumed that the properties of the complement, union, and intersection of 1s and 0s given in Table C3 hold. A more basic treatment [Hill, 1981] develops these relationships from the principles of Boolean algebra. However, we will assume that the properties of Table C3, as well as the basic Boolean algebra identities given in Table C4, have been proven.

**TABLE C2  Truth Tables for the Six Functions in Fig. C1**

| NOT (Inverter) Function | | OR, AND, NAND, NOR, EXOR, Functions | | | | | | |
|---|---|---|---|---|---|---|---|---|
| | | | | | | $f(x,y)$ | | |
| $x$ | $f(x) = \bar{x}$ | $x$ | $y$ | $x+y$ | $x \cdot y$ | $\overline{x \cdot y}$ | $\overline{x+y}$ | $x \oplus y$ |
| 0 | 1 | 0 | 0 | 0 | 0 | 1 | 1 | 0 |
| 1 | 0 | 0 | 1 | 1 | 0 | 1 | 0 | 1 |
| | | 1 | 0 | 1 | 0 | 1 | 0 | 1 |
| | | 1 | 1 | 1 | 1 | 0 | 0 | 0 |

480  REVIEW OF ARCHITECTURE FUNDAMENTALS

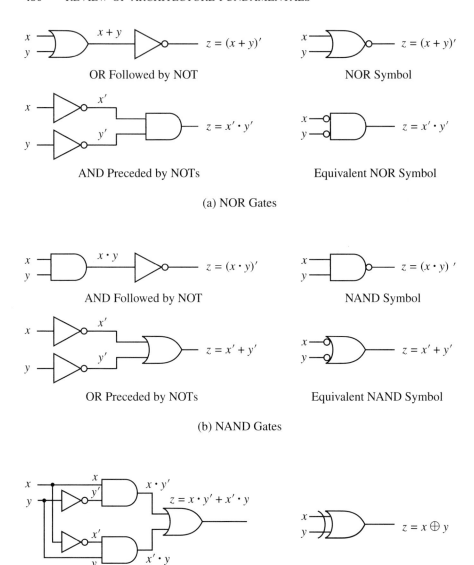

**Figure C2** Equivalent forms for logic functions; equivalent symbols/circuits for (a) NOR gates, (b) NAND gates, and (c) EXOR gates.

**TABLE C3  Properties of 1 and 0 in Boolean Algebra**

| $\bar{0} = 1$ | $0 + 0 = 0$ | $0 \cdot 0 = 0$ |
| $\bar{1} = 0$ | $0 + 1 = 1$ | $0 \cdot 1 = 0$ |
|               | $1 + 0 = 0$ | $1 \cdot 0 = 0$ |
|               | $1 + 1 = 1$ | $1 \cdot 1 = 1$ |

The identities given in Table C3 and C4 can be used to manipulate Boolean expressions. For example, consider the following expression:

$$\overline{x \cdot y \cdot z} = ? \qquad \text{let } a = x \cdot y \tag{C1}$$

Substituting $a$ and applying identity {17} (one of DeMorgan's laws),

$$\overline{a \cdot z} = \bar{a} + \bar{z}$$

Now, substituting again for $a$:

$$\overline{x \cdot y \cdot z} = \overline{x \cdot y} + \bar{z}$$

Again, applying identity {17}, we obtain

$$\overline{x \cdot y \cdot z} = \bar{x} + \bar{y} + \bar{z} \tag{C2}$$

Thus we have proved that one of DeMorgan's laws applies to three variables. (One can show that both of DeMorgan's laws apply to $n$ variables.)

Consider another example: We wish to simplify the expression, that is, obtain an equivalent expression with fewer terms, fewer variables, or both. Note that in the second form of Eq. (C3), the "dots" indicating multiplication of Boolean variables have been omitted for brevity, as is usually done.

$$f(x, y, z) = \bar{x} \cdot y \cdot z + \bar{x} \cdot y \cdot \bar{z} + x \cdot z = \bar{x}yz + \bar{x}y\bar{z} + xz \tag{C3}$$

Applying identity {14} to the first two terms, one obtains

$$f(x, y, z) = \bar{x}y(z + \bar{z}) + xz$$

From identity {5} and {6}, one obtains

$$f(x, y, z) = \bar{x}y(1) + xz = \bar{x}y + xz \tag{C4}$$

The result of our Boolean algebraic manipulation is that Eq. (C4) has two terms rather than the three in the original function given in Eq. (C3), and both

**TABLE C4  Boolean Algebra Identities**

| Complement | Union | Intersection | Commutative, Association | Distributive, DeMorgan's {16,17} |
|---|---|---|---|---|
| {1} $\bar{\bar{x}} = x$ | {2} $x + 0 = x$ | {6} $x \cdot 1 = x$ | {10} $x + y = y + x$ | {14} $x \cdot (y + z) = x \cdot y + x \cdot z$ |
| | {3} $x + 1 = 1$ | {7} $x \cdot 0 = 0$ | {11} $x \cdot y = y \cdot x$ | {15} $x + y \cdot z = (x + y) \cdot (x + z)$ |
| | {4} $x + x = x$ | {8} $x \cdot x = x$ | {12} $x + (y + z) = (x + y) + z$ | {16} $\overline{x + y} = \bar{x} \cdot \bar{y}$ |
| | {5} $x + \bar{x} = 1$ | {9} $1 \cdot 1 = 1$ | {13} $x \cdot (y \cdot z) = (x \cdot y) \cdot z$ | {17} $\overline{x \cdot y} = \bar{x} + \bar{y}$ |

terms in Eq. (C4) contain only two variables (often called *literals*). Thus the manipulation has transformed the switching function into an equivalent simpler form, which would result in a simpler circuit if one tries to build a digital-circuit realization of this function (see Section C5).

Any switching function may be written in one of two standard (canonical) forms: the sum-of-products (SOP) form and the product-of-sums (POS) form. Either form holds for $n$ variables, but for simplicity we will illustrate by considering a switching function of three variables in SOP form. The standard SOP form is as follows:

$$f(x,y,z) = [\bar{x}\bar{y}\bar{z} + \bar{x}\bar{y}z + \bar{x}y\bar{z} + \bar{x}y\bar{z} + \bar{x}yz + x\bar{y}z + xy\bar{z} + xyz] \quad (C5)$$

Various combinations of the 8 terms appear in the brackets. All possible combinations of these 8 terms represent a three-variable function, which includes the degenerate cases of no terms (a null circuit), all terms (always unity), the 8 functions that contain 1 term each, the 28 functions with 2 terms each, and so on, for a total of 256 possible functions. As an example, consider the switching function composed of first, second, and eighth terms in the bracket of Eq. (C5):

$$f(x,y,z) = [\bar{x}\bar{y}\bar{z} + \bar{x}\bar{y}z + xyz] \quad (C6)$$

The number of different switching functions, $N$, of 3 variables can be computed as the number of combinations of 8 terms taken 0 at a time plus the number of combinations of 8 terms taken 1 at a time plus the number of combinations of 8 terms taken 2 at a time, and so on. One can show that the sum of this series is given by $2^8 = 256$: Expand the binomial $(a+b)^n$ using the binomial expansion and then let $a = b = 1$; the expression reduces to the series of combinations discussed previously. In general, if there are $k$ variables, there are $2^k$ terms within the SOP bracket [cf. Eq. (C5)], and $N$ is given by

$$N = 2^{2^k} \quad (C7)$$

The 8 terms inside the bracket in the SOP form are called *minterms*, and an inspection of the example given in Eq. (C6) suggests a simplified form of notation in terms of binary numbers:

$$f(x,y,z) = [\bar{x}\bar{y}\bar{z} + \bar{x}\bar{y}z + xyz] = [000 + 001 + 111] = \sum m(0,1,7) \quad (C8)$$

One would say that the switching function is in SOP form and contains minterms 0, 1, 7, and one can write the SOP form directly from a truth table by including minterms corresponding to each row for which the function is a 1. For example, the EXOR function given in Table C2 is given by

$$f(x,y) = [01 + 10] = \sum m(1,2) = \bar{x}y + x\bar{y} \tag{C9}$$

The POS form is similar to the SOP form and is illustrated as follows for 3 variables:

$$f(x,y,z) = [(\bar{x} + \bar{y} + \bar{z}) \cdot (\bar{x} + \bar{y} + z) \cdot (\bar{x} + y + \bar{z}) \cdot (x + \bar{y} + \bar{z})$$
$$\cdot (\bar{x} + y + z) \cdot (x + \bar{y} + z) \cdot (x + y + \bar{z}) \cdot (x + y + z) \tag{C10}$$

As with the SOP form, various combinations of the 8 terms appear in the brackets.

The number $N$ is the same as with SOP, as given by Eq. (C7). The terms in the bracket of Eq. (C10) are called *maxterms*; the notation is similar to that illustrated in Eqs. (C8) and (C9), except that a capital $M$ is used and that instead of the summation symbol, a product symbol is used. One can write the POS form from the truth table in a manner similar to that of the SOP, but a maxterm is included for each row of the truth table where the function is a 0 and all variables are complemented. As an example, consider the EXOR function of Table C2, which is given in SOP form in Eq. (C9). The POS form is given by

$$g(x,y) = \prod M(0,3) = (\bar{x} + \bar{y}) \cdot (x + y)$$

Complementing all the variables, we obtain the POS form:

$$g(x,y) = (x + y) \cdot (\bar{x} + \bar{y}) \tag{C11}$$

One can show that Eqs. (C9) and (C11) are the same by expanding Eq. (C11) and simplifying

$$g(x,y) = (x + y) \cdot (\bar{x} + \bar{y}) = x\bar{x} + x\bar{y} + \bar{x}y + y\bar{y} \tag{C12}$$

The first and the last terms go to 0, and we have the same expression as Eq. (C9).

## C4 SWITCHING FUNCTION SIMPLIFICATION

### C4.1 Introduction

Digital-logic design begins by formulating the switching function and then drawing a logic circuit that implements the design. Sometimes it is possible to write the logic function in an equivalent but simpler form to lead to a simpler logic circuit.

The basis of logic simplification is when the union of two logic functions occurs where the two functions are identical; however, one contains a logic variable, the other contains its complement. For example,

$$f(x,y) = xyz \cdot xy\bar{z} = xy(z + \bar{z}) = (xy)(1) = xy \qquad (C13)$$

One can describe an algebraic simplification process as successive applications of the above simplification along with identities {5} and {8} of Table C4 to simplify a logic function. However, it is easier to define a graphical process called the *Karnaugh map (K map) simplification*.

## C4.2  K Map Simplification

The K map method is very useful in simplifying logic functions. It begins by constructing a "special matrix," in which a pair of horizontal or vertical adjacent cells represents variable simplification; such a pair means that one variable drops out. The elementary logic terms (minterms) that make up the logic expression are entered in the map as ones (zeroes are entered in the other squares), and adjacencies that signify logic simplification are identified by inspection. For *two variables*, $f(x, y)$, we use a square map as shown in Table C5. A similar rectangular map for three variables is shown in Table C6, and a larger square map for four variables is shown in Table C7. Note that in the three- and four-variable maps, the columns and rows are ordered 00, 01, 11, 10 to provide the "touching" property, not 00, 01, 10, 11 as *blind intuition* might suggest to do.

The way one proceeds with the K map method is to expand the function to be minimized to include all variables; oftentimes, it is convenient to convert to the "binary notation." (The terms in the expanded function are generally called minterms.) Consider the three examples each given in Tables C8, C9, and C10. Note the expansion, binary notation, and the shorthand notation in

**TABLE C5   Two-Variable K Map, $f(x, y)$**

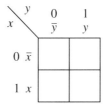

Rules:
(a) Horizontal or vertical touching of two "1" cells means that one variable drops out—the one that appears as the variable and its complement.
(b) All four cells are "1" cells, meaning both (two) variables drop out; the function becomes unity.
(c) Diagonal touching does not count.

**TABLE C6    Three-Variable K Map, $f(x, y, z)$**

| x \ yz | 00 $\bar{y}\bar{z}$ | 01 $\bar{y}z$ | 11 $yz$ | 10 $y\bar{z}$ |
|---|---|---|---|---|
| 0  $\bar{x}$ | | | | |
| 1  $x$ | | | | |

Rules:
(a) Horizontal or vertical touching of two "1" cells means that one variable drops out—the one that appears as the variable and its complement.
(b) Four adjacent cells—four across or down, or four in a square—mean two variables drop out; the function becomes a single variable.
(c) Diagonal touching does not count.
(d) All eight cells mean three variables drop out; the function becomes unity.


**TABLE C7    Four-Variable K Map, $f(w, x, y, z)$**

| wx \ yz | 00 $\bar{y}\bar{z}$ | 01 $\bar{y}z$ | 11 $yz$ | 10 $y\bar{z}$ |
|---|---|---|---|---|
| 00  $\bar{w}\bar{x}$ | | | | |
| 01  $w\bar{x}$ | | | | |
| 11  $wx$ | | | | |
| 10  $w\bar{x}$ | | | | |

Rules:
(a) Horizontal or vertical touching of two "1" cells means that one variable drops out—the one that appears as the variable and its complement.
(b) Four adjacent cells—four across or down, or four in a square—mean two variables drop out; the function becomes two variables.
(c) Eight adjacent cells—two adjacent rows or columns of four across or down—mean three variables drop out; the function becomes a single variable.
(d) Diagonal touching does not count.
(e) All sixteen cells mean four variables drop out; the function becomes unity.

### TABLE C8  Two-Variable K Map Simplification, $f(x, y)$

|   | $y$ = 0 ($\bar{y}$) | $y$ = 1 ($y$) |
|---|---|---|
| $\bar{x}$ (0) | (1 | 1) |
| $x$ (1) | 0 | (1) |

Rules:
(a) $f(x, y) = x' + xy$ expands to $x'(y + y') + xy = x'y + x'y' + xy$. The shorthand notations become

$$00 + 01 + 11 = \sum m(0, 1, 3).$$
$$\longleftarrow \text{minterms} \longrightarrow$$

Three minterms shown in the map all appear as (ones) in cells 0, 1, 3 → 00, 01, 11.

(b) The two touching horizontal terms (00, 01) are circled to show that they combine. The literal $y$ drops out since it appears as $y$ and $y'$ in the terms, yielding $x'$.

(c) The two touching vertical terms (01, 11) are circled to show that they combine. The literal $x$ drops out since it appears as $x$ and $x'$ in the terms, yielding $y$.

(d) The simplified expression is $x' + y$. Note that the minterm 01 was used twice in the simplification, which is legitimate because $f(x, y) = f(x, y) + f(x, y)$.

### TABLE C9  Three-Variable K Map Simplification, $f(x, y, z)$

|   | $yz$ = 00 ($\bar{y}\bar{z}$) | 01 ($\bar{y}z$) | 11 ($yz$) | 10 ($y\bar{z}$) |
|---|---|---|---|---|
| $\bar{x}$ (0) | (1 | (1) | 1 | 0 |
| $x$ (1) | 0 | (1 | 1) | 0 |

Rules:
(a) $f(x, y, z) = 000 + 001 + 011 + 101 + 111 = \sum m(0, 1, 3, 5, 7)$. Note the simplified minterm notation.

(b) One always uses the largest groupings first. The four adjacent cells in the center of the map are grouped, eliminating $y$ and $x$ and yielding $z$. This grouping is said to "cover" these four minterms. However, minterm 0 remains. One can take minterm 0 by itself, but further simplification occurs if we group 0 and 1, thereby eliminating $z$ and yielding $x'y'$.

(c) Simplified function $f(x, y, z) = z + x'y'$.

**TABLE C10  Four-Variable K Map Simplification, $f(w, x, y, z)$**

|    wx \ yz    | 00 $\bar{y}\bar{z}$ | 01 $\bar{y}z$ | 11 $yz$ | 10 $y\bar{z}$ |
|---|---|---|---|---|
| 00 $\bar{w}\bar{x}$ | (1 | 1 | 1) | 0 |
| 01 $\bar{w}x$ | 0 | (1 | 1) | 0 |
| 11 $wx$ | 0 | 0 | 0 | (1) |
| 10 $w\bar{x}$ | (1 | 0 | 0 | 0 |

Rules:
(a) $f(w, x, y, z) = \sum m(0, 1, 3, 5, 7, 8, 14)$. Note the simplified minterm notation.
(b) As in the three-variable example, first the four adjacent cells (0001, 0011, 0101, and 0111) are grouped, thereby eliminating $y$ and $x$ and yielding $w'z$.
(c) We can group 0000 with 0001 as in the three-variable example, but a better move is to cover 1000 so that it touches only one other cell (0000). Thus grouping these two yields $x'y'z'$. Note that the top and bottom edges of all maps "touch," as do the right and left edges. This means the four-variable "square" is mappable on the surface of a torus, as are two- and three-variable K maps.
(d) All ones are covered except for 1110. Unfortunately, this minterm does not touch any others (no diagonals are allowed) and must be included without simplification as $wxyz'$.
(e) The resulting function is $f(w, x, y, z) = w'z = x'y'z + wxyz'$.

terms of the "sigma notation" shown in the examples. For convenience, the prime notation is sometimes used to represent complement. In the four-variable example shown in Table C10, we can visualize the map as a surface, and since the top edge of the map "touches" the bottom edge, we can view the map as a cylinder. Furthermore, the left edge and right edge touch so the ends of the cylinder are joined, forming a torus (a donut shape). In addition, the four corners form a grouping, and sometimes there is more than one distinct grouping, leading to equivalent and different groupings of the same complexity.

A K map for five variables $v, w, x, y, z$ can be viewed as two four-variable maps: one suspended above the other, with $v = 1$ on the top plane and $v = 0$ on the bottom plane, and one where adjacency also holds for cells above and below each other. Maps for six or more variables involve a "stack" of four-variable maps, which become very complex. Fortunately, they are not often needed, and another method—the Quine–McCluskey (QM) method, involving a series of tables—can be used. The QM method becomes complicated; however, computer program implementation exists for large problems (see Hill [1981], Kohavi [1978], Mano [1995], Roth [1995], Shiva [1988], and Wakerly

[2001]). The physical problem is sometimes such that a particular minterm or minterms cannot exist; thus we do not care whether they exist or not. Such terms, called *don't-cares*, are entered as $d$ in the K map and are treated as ones if they help the simplification and as zeroes if they are of no aid. The cells in the map that are not ones (minterms) are called *maxterms*; these are generally labeled as zeroes. The zeroes can also be grouped to yield the simplified function called the product-of-sums form (POS) mentioned previously. The application of don't-cares and the POS form are treated in the problems at the end of this appendix and also in the references.

## C5 COMBINATORIAL CIRCUITS

### C5.1 Circuit Realizations: SOP

Once the logic functions are minimized, the designer produces circuits that realize the logic function. The resulting minimizations from grouping the ones in the K map produce a union of intersection terms (in common engineering terms, a sum of products, or SOP). The product terms are developed using AND gates, and their outputs feed into an OR gate. If complements of the variables are needed as inputs to the AND gates, inverters are required. Thus any SOP form can be realized using only {AND, OR, NOT} gates; this set of logic functions is called a *complete set*. There are other combinations of logic gates that also form a complete set (e.g., NAND gates or NOR gates). Three examples of SOP circuits are shown in Fig. C3.

### C5.2 Circuit Realizations: POS

As was discussed in the previous section, one can group zeroes in the K map and obtain a POS design. The resulting circuit is similar to those in Fig. C3; however, instead of a multiple input OR preceded by a number of AND gates, the circuit is a group of OR gates followed by an AND gate. For examples, the reader is referred to the problems at the end of this appendix.

### C5.3 NAND and NOR Realizations

Up until now, we have discussed the circuit realizations that all involved the complete set of {AND, OR, NOT} gates. Sometimes, it is more convenient or simpler to deal with other types of logic gates. It turns out that two other complete sets of logic gates are often used: {NAND} and {NOR} gates. We state without proof (see the problems at the end of this appendix) that each of the AND and OR gates in the SOP form can be replaced with a NAND gate. Furthermore, if one does not wish to use an inverter, a two-input NAND gate with inputs tied together can suffice. Similarly, in a POS design, we can replace OR and AND gates by NOR gates and the inverter by a two-input NOR with inputs tied together. With a little more effort, one can also use NAND gates

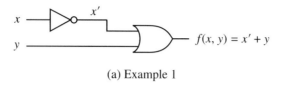

(a) Example 1

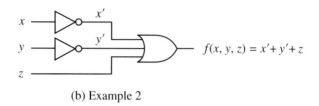

(b) Example 2

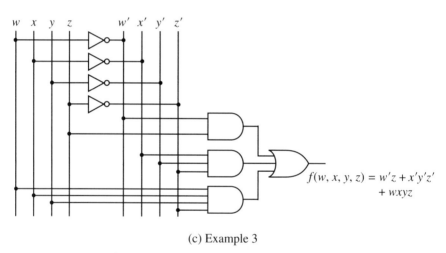

(c) Example 3

**Figure C3** Three examples of SOP circuits.

for POS designs and NOR gates for SOP designs. For more details, see the following references: Hill [1981], Kohavi [1978], Mano [1995], Roth [1995], Shiva [1988], Wakerly [2001].

## C5.4 EXOR

The standard OR gate has an output equal to one whenever either of the inputs has an input of one or *both* inputs are one. An exclusive OR function (EXOR) has an output equal to one whenever either of the inputs has an input of one but *not* when both inputs are one. The switching function can be written as $f(x, y) = xy' + x'y$. This function has a special logic symbol written as $f(x, y) = x \oplus y$. The EXOR function occurs frequently in coding theory and in other application areas. It is easy to recognize in a K map by its "checkerboard" pattern of ones and zeroes. In shorthand notation, $f(x, y) = 10 + 01 = \sum m(1, 2)$; for three and four variables, $f(x, y, z) = xy'z' + x'yz' + x'y'z + xyz = 100 + 010 + 001 + 111 = \sum m(0, 1, 2, 4, 7)$; $f(w, x, y, z) = wx'y'z' + w'xy'z' + w'x'yz' + w'x'y'z + wxyz' + wxy'z + wx'yz + w'xyz = 1000 + 0100 + 0010 + 0001 + 1110 + 1101 + 1011 + 0111 = \sum m(0, 1, 2, 4, 7, 8, 11, 13, 14)$. Note that one of the properties of the EXOR function is that in the binary notation, each minterm has an *odd number of ones*. The function is used extensively in Chapter 2 on coding.

## C5.5 IC Chips

Several courses in electrical engineering curricula study in detail the features of integrated circuits (also called ICs or "chips"); however, for our purposes, we need to know a few facts. Integrated circuits come in a wide variety of device types or families that vary in switching speed, power consumption, immunity to noise and cost, and other factors. We can illustrate some of the differences by focusing on two families: the transistor–transistor logic (TTL) logic family, which is the most common and least expensive family, and the complementary metal oxide silicon (CMOS, or "sea moss") family, which is used extensively for low-power, portable (i.e., battery and solar cell–powered) applications such as calculators. Switching delays range from about 3 to 20 nanoseconds (billionths of a second), the quiescent power dissipation range from about 0.0025 to 10 milliwatts, and a cost—depending on the complexity of the circuit and the quantity purchased—ranging from 10¢ to $2. Within each logic family there are subcategories such as the low-power Schottky subfamily, which are TTL (LSTTL) circuits with a lower-than-normal power usage, and fast Schottky TTL (FAST TTL) circuits that switch faster than regular TTL circuits. The reader should consult a recent Motorola or Texas Instruments databook and the current state of technology for more details.

The kinds of available logic gate packages depend strongly on the number of pins in the package. The simpler ICs come in standard 14- or 16-pin packages approximately 20 × 6 × 5 mm with 7 or 8 pins 5 mm long on each side. A typical 14-pin package has 2 pins devoted to power (typically, $V_{cc}$ = 5 volts and ground, which are generally pins 14 and 7); thus 12 pins are available for input and output signals. For an inverter, there is 1 input and 1 output pin; thus 6 devices come in a standard package, which is called a HEX inverter. For a two-input gate (AND, OR, NAND, NOR), 2 input pins and 1 output pin are required; thus 4 gates can be placed in a package, which is called a QUAD two-input gate. Similarly, a three-input gate has 3 gates per package and is called a TRIPLE three-input gate. A DUAL four-input gate has 2 gates per package (with 2 unused pins). The biggest standard-size gate has a single 13-input gate. See Table C11 for typical TTL gates.

Of course, complex integrated circuits such as memories and microproces-

**TABLE C11** Some Examples of Typical TTL Logic Gates.
[Reprinted with permission of ON Semiconductor; Motorola, 1992.]

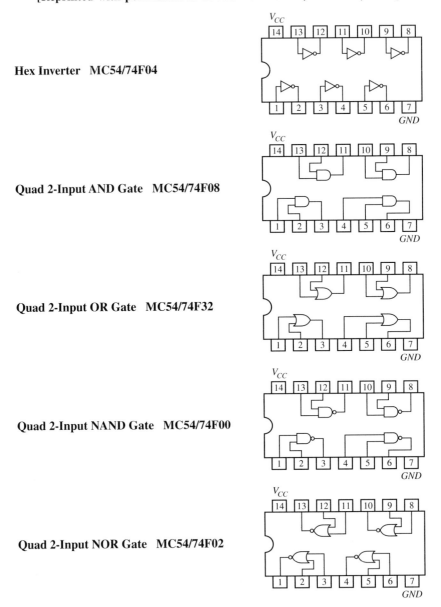

sors may come in large packages and have 50–100 pins. One can consult an Intel databook for typical examples of the present state of the art of logic for larger ICs.

### TABLE C11 (Continued)

**Triple 3-Input NAND Gate  MC54/74F10**

**Dual 4-Input AND Gate  MC54/74F21**

**13-Input NAND Gate  SN54/74LS133**

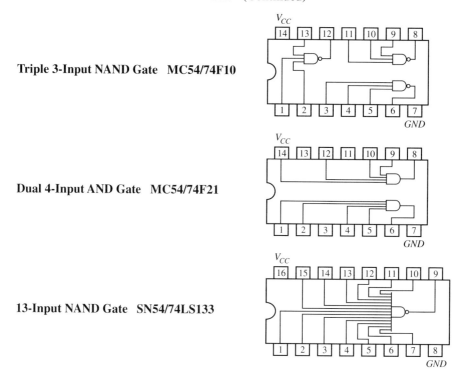

## C6  COMMON CIRCUITS: PARITY-BIT GENERATORS AND DECODERS

### C6.1  Introduction

The short discussion of digital integrated circuits in the preceding section may have left the reader with the notion that there are only small IC packages, such as QUAD two-input AND gates (called small-scale integration, or SSI) and large-memory and microprocessor chips (called large-scale integrated circuits, or LSI, or very large scale integrated circuits, or VLSI). Such is not the case, however, and IC designers have been active for several decades producing medium-scale integrated (MSI) circuits. The MSI devices are available for a large range of practical functions that could be built out of ICs but at a greater cost and size. In essence, it is easier and cheaper to do the wiring and constructing on the IC chip rather than externally. Formally, we can classify the scale of ICs in terms of the number of gates in their equivalent circuits: $1 < \text{SSI} < 20$; $20 < \text{MSI} < 200$; $200 < \text{LSI} < 200{,}000$; and $\text{VLSI} > 200{,}000$ (some say $\text{VLSI} > 500{,}000$). We will discuss two MSI devices: a parity-bit generator and a decoder, both of which were used in Chapter 2.

## C6.2 A Parity-Bit Generator

In Chapter 2, we discussed the use of a parity-bit code to help detect simple errors in transmission of digital words. The most common scheme is to add check bit to the word so that all the words have an odd number of ones. After transmission, we can count the number of ones in the word; if the count is even, we know that one error has occurred (actually, an odd number of errors), and we signal "transmission error." Sometimes we set the parity bit so that the number of ones in the word is an even number; we call this even parity. From our discussion in Section C5.4, we see that EXOR gates could be used to accomplish both the generation of a parity bit and the checking of parity for the transmitted word. Rather than use a group of AND gates (as shown in Fig. 2.1) or a "tree" of EXOR gates (shown in Fig. 2.2), one can use an MSI device—the SN74180, a 9-bit odd/even parity generator/checker (see Fig. 2.4). This circuit and the similar (newer and faster) 74LS280 shown in Fig. 2.7 and repeated in Fig. C4 can be used to compute parity-bit generation or checking for up to 9 bits. By studying Fig. C4(a) and (b), one can see that inputs 8–13 and 1, 2, 4 (note that pin 3 is not used; this is denoted as NC = no connection) labeled $A$, $B$, $C$, $D$, $E$, $F$, $G$, $H$, $I$ are used for up to nine inputs. Outputs 5 and 6 yield the EXOR function of all the inputs and its complement; these are labeled by equivalent wording of even or odd parity. Suppose that one wishes to check the parity of a 16-bit word (labeled as bits 0–15) in a computer circuit. One can use two 74LS280 chips in cascade. Bits 0–7 go into inputs $A$–$H$ of chip 1 and bits 8–15 go into inputs $A$–$H$ of chip 2. The output of chip 1 goes into input $I$ of chip 2. When an input is not used (e.g., input $I$ of chip 1), it can be left unconnected in some logic families, but in others it serves as an "antenna" that picks up stray inputs that may interfere with operation. The safest course of operation is to connect an unused input to +5 volts or ground, depending on the logic family and the function of the input. Unused inputs for a 74LS280 are connected to ground. The details of Fig. C4(b) are best left to an electrical engineer who has studied IC design.

## C6.3 A Decoder

Another MSI circuit used in Chapter 2 was a decoder. A *decoder* (sometimes called a *demultiplexer*) converts the binary representation of $n$ variables into $2^n$ outputs. One can liken the functional operation of the decoder to a selector switch such as the one found on the ventilation systems in many automobiles: Rotation of the knob switches from "off" to "interior air circulation" to "vent air conditioner" to "floor air conditioner" to "outside air vent" and so on. In a decoder, the binary input is analogous to the number of clicks of switch rotation; which one of the $2^n$ outputs selected is analogous to the air circulation function selected.

For example, the 74LS138 3-to-8 decoder (shown in Fig. 2.7 and repeated in Fig. C5) converts the three digits of the input variable $A_2A_1A_0$ (pins 1, 2,

## COMMON CIRCUITS: PARITY-BIT GENERATORS AND DECODERS  495

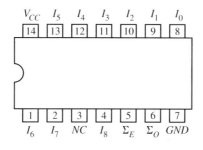

(a) Diagram

**Logic Diagram**

(b) Functional Block Diagram

**Figure C4**  A 74LS280 9-bit odd/even parity generator/checker. [Reprinted with permission of ON Semiconductor; Motorola, 1992.]

### 1-of-8 Decoder/Demultiplexer
### MC54/74F138

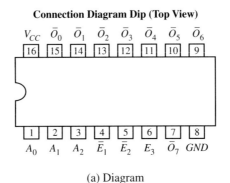

(a) Diagram

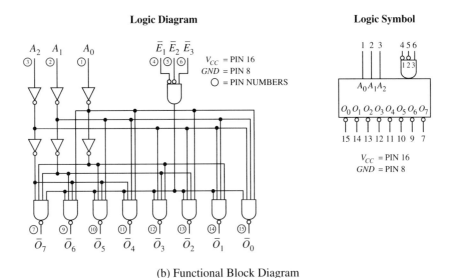

(b) Functional Block Diagram

**Figure C5** A 74LS138A 1-of-8 decoder/demultiplexer (also called a 3-to-8 decoder). [Reprinted with permission of ON Semiconductor; Motorola, 1992.]

3) into the eight possible output combinations (pins 15–7): $000 = \overline{A}\,\overline{B}\,\overline{C}$, $001 = \overline{A}\,\overline{B}\,C, \ldots, ABC$, which represent the outputs $O_0$ to $O_7$. The small circle at the outputs in the logic diagram indicates that this is the complement of the desired output. It is fairly common for integrated circuits to produce the desired output, its complement, or both signals. Similarly, the inputs to various integrated circuits may call for the desired signal or its complement. The designer keeps track of which signals are needed, alternates complementary outputs and

inputs (which cancel), or occasionally uses inverters where needed. There is an additional set of three inputs (4, 5, 6; $G1$, $G2A$, $G2B$), which are called *enable inputs*. These inputs are designed for expansion of certain ICs in groups to work with larger combinations of inputs and outputs and also serve to connect or disconnect all the inputs or outputs. (Note that output of the enable AND gate is shown in the logic diagram to be an input to all the output gates so it can switch all the outputs on or off.) For example, one can design a 4-to-16 decoder by using two 74LS138 3-to-8 decoders. One bank of 8 outputs is handled by the first decoder; the other bank of 8 outputs is handled by the second decoder. The enable inputs are configured so that they switch on or off the appropriate bank. Thus, by inserting the extra variable (fourth input) into the $G1$ input of one decoder and the variable into the $G2A$ input of the other decoder that complements the variable, the extra variable switches between the two banks. The enable input AND gate has an output if $G1 = 1$, $G2A = 0$, and $G2B = 0$. Thus, if any enable input is not used, it must be connected to 1 (5 volts) if it is $G1$ or connected to 0 (grounded) if it is the $G2A$ or $G2B$ input. In Fig. 2.7, we only need a 3-to-8 decoder; thus none of the enable inputs are used, and $G1$ is connected to +5 volts and $G2A$, $G2B$ are grounded (0 volts). Note that the $G1$ connection to +5 volts also uses a resistor to limit the current into the input to protect it when switching occurs and that the decoder requires a 16-pin package.

## C7  FLIP-FLOPS

Computers and digital circuits are composed of three broad classes of facilities: the computational units called *central processing units* (CPUs) (frequently microprocessors); the memory units (flip-flops, electronic memory, and disk, card, tape, CD, and other media); and input/output facilities (keyboards, monitors, communication connections, etc.). The fastest of the memory storage units are flip-flops (FFs), which are individually connected or connected in banks called *registers*. Registers, discussed briefly in Chapter 2, are storage devices made up of cascaded single-bit FF storage circuits with switching time delays of several nanoseconds (about the same as logic gates).

In addition, there are single-input FFs; among them is the trigger or toggle FF ($T$ FF). When the input $T$ is 0, the output $Q$ (and its complement $Q'$) holds (stores) its previous state (either a 0 or 1). When $T = 1$, the values of $Q$ and $Q'$ flip from their previous states to the complement (0 or 1). There is also a delay FF ($D$ FF), which stores as output whatever the $D$ input is after a switching delay. Both the $T$ and $D$ FF are single-input devices. A symbolic diagram of a $T$ FF is shown in Fig. C6 along with a *state table*. A state table is similar to a truth table, but it contains an additional column (serving like an additional input) that is the previous output. Note the first line of the state table reveals that if the previous storage state $Q_n$ is 0 and there is no $T$ input (0 input), then the new state $Q_{n+1}$ is the same as the old—that is, 0. The second line in the

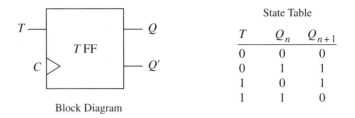

(a) A Single Input T FF

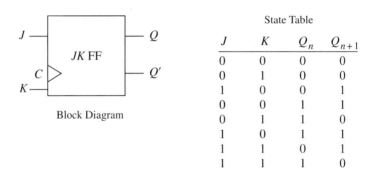

(b) A 2-Input JK FF

**Figure C6** Diagrams of various flip-flops. [Reprinted with permission of ON Semiconductor; Motorola, 1992.]

state table also reveals that with no input, the stored state does not change and the stored value of 1 remains stored. In the last two lines of the state table, the input of 1 changes the stored state from 0 to 1 or 1 to 0. In a large circuit with many interconnected FFs, there may be unwanted signals that propagate after one or more FFs switch. It is desirable to have all the FFs synchronized so that they initiate switching simultaneously in a short instant of time less than the switching time. The $C$ input (clock input) accomplishes this synchronization. A timing signal, which is a pulse of short duration (less than the switching time), is fed to the $C$ input that "opens" the $T$ input and allows it to switch the $FF$ if $T = 1$. (The inputs $T$ and $C$ are essentially fed to a two-input AND gate, and the output of the gate is the switching signal.) Other synchronization circuitry and features for clearing a stored value or setting a stored value (reset to 0 or preset to unity) are included in commercial FFs but need not be discussed here.

The second part of Fig. C6 shows a two-input $JK$ FF. A 1 signal on the $J$ and a 0 signal on the $K$ input set the output $Q$ to 1 regardless of its previous storage value (see rows 3 and 6 in the state table). A 1 signal on the $K$ and a 0 signal on the $J$ input set the output $Q$ to 0 regardless of its previous storage value (see rows 2 and 5 in the state table). As a way to remember the function,

**SN54/74LS73A**

**Logic Diagram (Each Flip-Flop)**

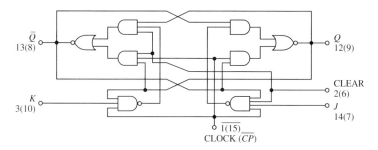

**Logic Symbol**

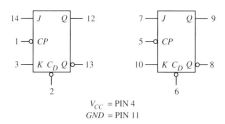

(c) A 74LS73A *JK* FF

**Figure C6 (Continued)**

one can think of *J* as "jump" and *K* as "kill." If both inputs *J* and *K* have a 0 signal, then nothing changes in the output (see rows 1 and 4 in the state table). If both the *J* and *K* inputs are simultaneously 1, then the FF behaves like a *T* FF, that is, the stored state flips its value (see rows 7 and 8 in the state table). There is also another kind of two-input FF called a reset (*R*) and set (*S*) FF (or a reset–set FF) that behaves like a *JK* FF, except the $S = R = 1$ condition is not allowed. (For more details, see the references [Hill, 1981; Kohavi, 1978; Mano, 1995; Roth, 1995; Shiva, 1988; Wakerly, 2001].) Some designers consider a *JK* FF a basic design element since it is easily connected to behave like a *T* or *D* FF. See the problems at the end of this appendix for more details.

The 74LS73A *JK* FF shown in Fig. C6(c) is essentially the same as the *JK* FF just discussed, but with the following modifications: (a) It is dual—that is, two devices are fit into a 14-pin package; (b) it is negative edge–triggered—that is, the clock-pulse input opens the *J* and *K* inputs when the pulse falls from 1 to 0; and (c) a 0 signal on the CD-input sets (clears) the *Q* output to 0 (and, of course, the *Q'* output becomes 1).

A simple application of a *T* FF is in a digital–electronic elevator control system. When you are outside the elevator and push the up or down button for calling the elevator to a floor and release the button, the signal remains stored. An implementation is to connect a power source and a switch in series and feed the signal to the $T$ input of a $T$ FF (and also into the $C$ input). The call signal will be stored. One FF is needed for the up and one for the down inputs. Once the elevator reaches your floor, a floor switch can feed another voltage into the $T$ input to switch the call state back to 0. Actually, the circuit must know if the elevator is traveling up or down to know which call signal to clear (this would require another $T$ FF storage element).

## C8 STORAGE REGISTERS

Flip-flops can serve as single-bit storage registers; however, they are generally organized inside of MSI circuits called registers, such as the 74F195 device (a 4-bit shift/storage register in a 16-pin package; see Fig. C7). Up to four bits of data can be stored in the register, and there is also shift-right function. If we wish to store a 16-bit word, we use a cascade of four such devices. Other storage registers provide more bits of storage (in packages with more pins), both shift-right and shift-left operation, and other functions. Generic block diagrams of storage/shift registers are shown in Figs. 2.10 and 2.11.

At the heart of this storage register are four reset–set ($RS$) FFs that behave like $JK$ FFs, where $R$ is like $K$ and $S$ is like $J$. The clock-pulse ($CP$) input is for a clock pulse that feeds all four FFs. The MR′ input is used to reset (clear) all the FF outputs—$Q_0$, $Q_1$, $Q_2$, $Q_3$—instantly to 0. For convenience, the complement of the $Q_3$ output is supplied. Parallel input of data is provided (as with most shift/storage registers) via inputs $D_0$, $D_1$, $D_2$, $D_3$. We can liken parallel loading of the register to four soldiers facing four adjacent doors of a barracks ($D_0$, $D_1$, $D_2$, $D_3$). When the corporal beats the drum (the $CP$), the four doors open and the soldiers enter the barracks and stand to attention inside ($Q_0$, $Q_1$, $Q_2$, $Q_3$), after which the doors close. The analogy for serial loading is that the four soldiers are in single file facing a single door ($J$, $K'$). When the corporal beats the drum (the $CP$ and the shift pulse, or PE), the single door opens, and the first soldier enters the barracks and stands to attention inside ($Q_0$). As the sound of the drum dissipates, the door closes and the first soldier comes to attention ($Q_0$). At the next drumbeat, the soldier inside makes a right turn, takes a step forward, makes a left turn, and stands to attention (the soldier shifts right from position $Q_0$ to $Q_1$); then the door opens and the second soldier steps inside ($Q_0$). As the sound of the drum dissipates, the door closes and the second soldier stands to attention ($Q_0$) next to the first who is already standing to attention ($Q_1$). The process repeats itself for two more drumbeats until all four soldiers are in the barracks standing to attention. If there is an attack, the burglar sounds the alarm ($MR'$), causing all four soldiers to run from the barracks to arm themselves and leave the barracks empty (reset to 0).

Applications of storage/shift registers are shown in Figs. 2.9, 2.10, and 2.11;

## 4-Bit Parallel Access-Shift Register
## MC74F195

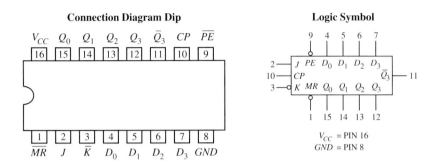

**Figure C7** Diagrams of a 74F195 shift/storage register. [Reprinted with permission of ON Semiconductor; Motorola, 1992.]

they are also discussed in the problems at the end of this appendix. Some shift/storage registers provide additional facilities, such as both shift-right and shift-left capability. The reader is referred to the following references for more details: Motorola, 1992; Shiva, 1988; and Wakerly, 2001.

## REFERENCES

Hill, F. J., and G. R. Peterson. *Introduction to Switching Theory and Logical Design*, 3d ed. Wiley, New York, 1981.

## 502  REVIEW OF ARCHITECTURE FUNDAMENTALS

Kohavi, Z. *Switching and Automata Theory*, 2d ed. McGraw-Hill, New York, 1978.

Mano, M. M. *Digital Design*, 2d ed. Prentice-Hall, Englewood Cliffs, NJ, 1995.

Motorola Logic Integrated Circuits Division. FAST and LS TTL Data, 5th ed., DL121/D, rev. 5. Motorola Literature Distribution, Phoenix, AZ, 1992.

Roth, C. H. Jr. *Fundamentals of Logic Design*, 4th ed. PWS Publishing, Boston, 1995.

Shiva, S. G. *Introduction to Logic Design*. Scott Foresman and Company, Glenview, IL, 1988.

Tinder, R. F. *Engineering Digital Design*. Academic Press, San Diego, CA, 2000.

Wakerly, J. F. *Digital Design Principles and Practice*, 3d ed. Prentice-Hall, Englewood Cliffs, NJ, 1994.

Wakerly, J. F. Digital Design Principles 2.1. Student CD-ROM Package. Prentice-Hall, Englewood Cliffs, NJ, 2001.

## PROBLEMS

**C1.** Convert the following base-2 number to base 10, base 8, base 16: $(1011110101)_2$.

**C2.** As a check, convert the base-10, base-8, and base-16 numbers obtained back to base 2.

**C3.** Expand Table C1 for numbers between 20 and 30.

**C4.** Prove the identities of Table C4 by substituting all possible combinations of ones and zeroes for the variables and computing the results.

**C5.** Add the following base-2 numbers and check the results by converting to base 10: $(1011110101)_2 + (11110101)_2 = ?$

**C6.** Repeat problem C5 for the subtraction problem: $(1011110101)_2 - (11110101)_2 = ?$

**C7.** Repeat the problems in Tables C8–C10 using the POS form. To use the POS form, draw a K map and enter ones for the maxterms (note the minterms are ones in the SOP map and the maxterms are zeroes). Proceed to minimize the K map as with the POS form and write the minimum function in SOP form. Then complement all variables.

**C8.** Which method is better for the problems of Tables C8–C10: the SOP method shown or the POS method of problem C7?

**C9.** Consider the three examples that are each given in Tables C8–C10. Assume that you can replace one of the zeroes in each K map. For each problem in the tables, choose the best location to insert a don't-care ($d$) and decide whether it should become a 1 or a 0 for best minimization. Explain your reasoning. Minimize the function and draw an SOP circuit for the minimized function.

**C10.** Repeat problem C9 for a POS design. (Hint: Study the solution of problem C1 to learn how to do a POS design.)

**C11.** Prove that the NAND gate is a complete set. (Hint: Since we know that the set of AND, OR, NOT gates is a complete set, we can perform our proof by showing that we can construct an AND, an OR, and a NOT gate from one or more NAND gates.

**C12.** Repeat problem C11 and show that a NOR gate is a complete set.

**C13.** Draw a circuit for the examples of Tables C8–C10 in the SOP form for using only NAND gates.

**C14.** Draw a circuit for the examples of Tables C8–C10 in the POS form for using only NOR gates.

**C15.** Consider the four-variable K map given in Table C7. Fill the entire map with a "checkerboard pattern" of ones and zeroes, starting with a 0 in the top left-hand corner. Minimize the function and draw an SOP circuit using AND, OR, NOT gates. Repeat the circuit using EXOR gates; then compare the two circuits.

**C16.** Repeat problem C15 for a "checkerboard pattern," starting with a 1 in the top left-hand corner.

**C17.** Draw a diagram to show how a 74LS280 IC can be used to check an 8-bit word for even parity and for odd parity.

**C18.** Repeat problem C17 for a 16-bit word.

**C19.** Show how to connect two 74LS138 3-to-8 decoders to implement a 4-to-16 decoder.

**C20.** Start with a *JK* FF and show how the two inputs can be connected to operate like a *T* FF. Explain.

**C21.** Start with a *JK* FF and show how the two inputs can be connected to operate like a *D* FF. (Hint: You will need an inverter to take the complement of one of the inputs.) What will the delay time of the *D* FF be?

**C22.** Fill in the details of the elevator-floor-button-control system outlined in Section C7 and draw the circuit diagram. Explain the operation.

**C23.** Use a 74F195 storage/shift register to design the storage application shown in Fig. 2.9. Explain how to connect the inputs and outputs of the 74F195.

**C24.** Repeat problem C23 for the application shown in Fig. 2.10.

**C25.** Repeat problem C23 for the application shown in Fig. 2.11.

# APPENDIX D

# PROGRAMS FOR RELIABILITY MODELING AND ANALYSIS

## D1 INTRODUCTION

Analysis is the theme of this book; indeed, Chapters 1–7 stressed both exact and approximate analytical approaches. However, it is clear that a large, practical system will involve computer analysis. Thus the focus of this appendix is to briefly discuss a sampling of the many available computer programs and point the reader to references that discuss such programs in more detail. In all cases, the intent is to provide a smooth transition from analysis to computation.

Analysis programs are important for many reasons, including the following:

1. to replace laborious computation;
2. to model complex effects that are difficult to solve analytically;
3. to solve a system that is so large or complex that it is intractable, even with approximations and simplifications;
4. to provide a graphic- and text-based record of the system model under consideration; and
5. to document parameter values and computations based on the model.

The reader should not underestimate the utility of a well-thought-out computer program to aid in documentation to satisfy reasons (4) and (5) of the preceding list. A program might be used for documentation even if all the computations are done analytically.

In the early days of reliability analysis, the size of a computational program, the speed of the computer, and the size of memory were of prime importance.

To put this in perspective, the reader should consider the title, subtitle, and publication date of the following article written by Schmidt and Busch, two engineers with the GE Controls Department, in *The Electronic Engineer*: "An Electronic Digital Slide Rule—If This Hand-Sized Calculator Ever Becomes Commercial, the Conventional Slide Rule Will Become Another Museum Piece" [1968]. The authors were speaking of the forerunner of the scientific calculator now available in any stationery store or drugstore for $12 (as low as $8 during fall back-to-school sales). The first pocket-sized commercial scientific calculator, introduced in the early 1970s, was the Hewlett-Packard HP-35; it sold for about $400! To the author's knowledge, the first comprehensive reliability computation program (one that models repair via solution of the analytical equations) was the GEM Markov modeling program developed by the Naval Applied Sciences Lab in the late 1960s [Orbach, 1968]. This program used the supercomputer of the day—the CDC 6600, and complex problem solutions ran one-half to one hour.

Computer solutions of reliability or availability models fall into a number of major classes. All the approaches to be discussed in this appendix are practical because of the great power and memory size of modern desktop and laptop computers. The main choice hinges on ease of use and cost of the program. The least desirable approach is to write a custom analysis program in some modern version of a standard computer language, such as C or FORTRAN. This certainly works, but unless someone within your organization has already developed such a program, the overhead cost is too great.

The next choice is to formulate a set of equations for the analysis and use one of the standard mathematical solution tools such as Mathematica [1999], Mathcad [1995], Matlab [1992], Macsyma [Ralston, 1976], and Maple [Ellis, 1992] to solve the equations. All these systems are powerful and can solve logic equations for combinatorial reliability or availability expressions or differential equations for Markov model solutions. The choice should be based on cost, ease of use, familiarity, availability within your organization, availability of "readable" manuals describing how the program is used in reliability or availability modeling, and other practical factors.

Another class of reliability analysis programs is a Monte Carlo solution (see Rubinstein [1981] and Shooman [1990]). Such a simulation approach is very flexible and allows one to model highly complex behaviors. However, solution requires the generation of random values for the times to failure, times to repair, and other parameters for each "run" of the simulation program. The program must be repeated $N$ times, and the probabilities must be estimated from the ratios of the number of favorable outcomes divided by $N$. As $N \to \infty$, these ratios approach the true probabilities. The main limitation of such an approach is the size of $N$ required and how long one must wait for the running of $N$ simulations. At one time, simulation required supercomputers and long running times. The method, invented by von Neuman and Ulam, was used initially to solve complex nuclear calculations during the Manhattan Project at Los Alamos Laboratories in New Mexico and went under the code name

"Monte Carlo" for secrecy. (Of course, Monte Carlo evoked the image of the games of chance in the casinos of the famous city in Monaco.) Many simulation programs are written in the language SIMSCRIPT (and its successors), developed by the RAND corporation in the early 1960s and implemented on early IBM computers [Sammet, 1969, p. 657]. We again comment that the power and speed of modern desktop and laptop computers makes the Monte Carlo method practical for many solutions that previously required prohibitive running times. For further details, see Shooman [1990, Section 5.10.4].

The methods introduced in the preceding paragraphs are discussed in the remainder of this chapter. The next section, however, focuses on the customized commercial programs commonly thought of as reliability and availability modeling programs. All such programs start with a model building phase that is based on interactive or tabular input or, in more modern cases, an interactive graphical editor. The next step is to choose from available component density functions, including databases for some components, or to provide input of failure-rate data; the program should then formulate the equations for the model without user assistance. The next step is solution of the model equations, which only requires information from the user regarding the points in time at which reliability or availability values are required. The final phase is the output section, which provides tabular and graphical output in addition to model documentation as selected by the user. Most of the programs to be discussed run on Windows '95, '98, or 2000 and later versions. Some of these programs have alternate versions that can be run on a Macintosh- or UNIX-based operating system.

## D2  VARIOUS TYPES OF RELIABILITY AND AVAILABILITY PROGRAMS

### D2.1  Part-Count Models

Reliability and availability programs have been developed by universities, government agencies, military contractors, and various commercial organizations. Such programs generally can be grouped under a number of headings. The simplest of such programs are those that implement a so-called *part-count* model. Such a model assumes that all parts are vital to system operation; thus they are all in series in a reliability sense. For such a model, the system failure rate is the sum of the part failure rates, so programs of this type are essentially large databases of part and component failure rates. The analyst starts with a parts list, identifies each component, and enters the environmental parameters; the program computes the failure rates based on the database and environmental and other adjustment factors, or else the user inputs failure-rate parameters. One of the most popular failure-rate databases was that contained in the military handbook MIL-HDBK-217A, B, C, D, E, and F, published from 1962 to 1992. Thus many of the earlier part-count programs used these handbooks as their databases and were sometimes called MIL-HDBK-217 programs. Newer programs frequently use data collected by the telecommunications industry [Bellcore, 1997; Shooman, 1990, p. 643].

### D2.2  Reliability Block Diagram Models

The part-count method assumes that all components are in series. This is often not the case, however, and an improved model is needed for reliability estimation. The simpler reliability block diagram (RBD) models consider elements in various series and parallel combinations. The more general RBD programs utilize cut-set or tie-set algorithms, such as those discussed in Section B2.7. Sometimes, such models are combined with a failure-rate database; in other cases, the analyst must input the set of required failure rates to the program. The programs generally include a plotting option that graphs the system $R$ versus $t$. Generally, the program allows one to deal with discrete probabilities at some point in time (often called *demand probabilities*). If the system contains repair and the system is decoupled, then discrete availabilities can be used in the model to compute steady-state availability.

### D2.3  Reliability Fault Tree Models

The fault tree (FT) method discussed in Section B2.5 and shown in Fig. B13 introduces an analysis technique that is a competitor to the RBD method. System analysts and designers who focus on system success paths often favor the RBD method. Many feel that it is easier to list modes of failure and build a system model from these modes; this results in an FT model. Also, those who perform safety analysis often claim that the FT method is more intuitive. In any event, however, the two methods are mathematically equivalent [Shooman, 1970]. One can describe the RBD method as a probability of success viewpoint and the FT method as a probability of failure viewpoint. The classic FT modeling method is described in McCormick [1981, Chapter 8]; the recent work on FT modeling can be found in Dugan [1996]. The exact analytical solution of the FT and the RBD methods can both be based on cut sets or tie sets, and approximations are frequently incorporated. Cut sets are generally used because they represent failure combinations that have small probability in reliable systems, and the omission of a cut set in error only has a small effect on the computation of reliability or availability.

### D2.4  Markov Models

If repair is involved in a system, the components are decoupled, and steady-state availability solutions are satisfactory, the availability probabilities can then be substituted into an RBD or an FT model. If the components are not decoupled, discrete-state and continuous-time Markov models would be the conventional approaches. Such Markov models result in a set of differential equations. The following are five approaches to solving such models:

1. Solve the differential equations and obtain a closed-form time function that can be plotted.
2. Solve the differential equations and obtain a numerical solution that can be plotted.
3. Use Laplace transforms to help solve the differential equations and obtain a closed-form time function that can be plotted.
4. Use Laplace transforms to help solve the differential equations and obtain a numerical solution that can be plotted.
5. Solve only for the steady-state values that reduce the differential equations to a set of algebraic equations that can be more easily solved for an algebraic solution.

The analysis techniques discussed in Chapters 3 and 4 are based on a combination of these five approaches. In many cases, the analytical approach is all that is required. For large, complex problems, however, a computer program may be required to check analytical approximations, to obtain a solution in intractable complex cases, and to document the model solution. The various Markov modeling programs allow one or more of these approaches, and the simpler ones only use approach (5) for steady state. The more comprehensive programs include graphical input programs for constructing a Markov state model to define the problem and provide facilities for printing a copy of the Markov diagram to document the model. Sometimes the Markov model is built around a simulation program that allows great flexibility in modeling the repair process. A discussion of the use of one program to perform Markov modeling is given in Section D5.

### D2.5 Mathematical Software Systems: Mathcad, Mathematica, and Maple

The vast power of the digital computer stimulated two areas in numerical computation. The most obvious was the consolidation and improvement of the many algorithms that were available for computing roots of polynomials, solving systems of algebraic equations and differential equations, and so on. In addition, much research was done on the symbolic solution of expressions; for example, integration, differentiation, the closed-form solution of differential equations, and the factoring of polynomials. These developments culminated in a wide variety of mathematical packages that are very helpful in many analysis and solution tasks. Some of the leading programs are Mathematica [1999], Mathcad [1995], Matlab [1992], Macsyma [Ralston, 1976], and Maple [Ellis, 1992]. There is a great amount of overlap among these programs, for which reason the initial comparison should be based on whether the program supports symbolic manipulation. The choice of which program to use may be based on specific features as well as availability and prior experience with a particular program at work. However, if the program must be acquired, its flexibility and

**TABLE D1  Information on Mathematical Programs**

| Product Name | Company Name and Address | Telephone Number | Web Address |
|---|---|---|---|
| Mathematica | Wolfram Research, Inc.<br>100 Trade Center Drive<br>Champaign, IL 61820 | (217) 398-0700 | www.wolfram.com |
| Mathcad | MathSoft, Inc.<br>101 Main Street<br>Cambridge, MA 02142 | (617) 577-1017 | www.mathsoft.com |
| Matlab | The MathWorks<br>3 Apple Hill Drive<br>Natick, MA 01760 | (508) 647-7000 | www.mathworks.com |
| Macsyma | Symbolics Technology, Inc. | — | www.symbolics.com |
| Maple | Waterloo Maple, Inc. | (800) 267-6583 | www.maplesoft.com |

ease of use in addition to one's confidence in its accuracy and validity should all be the major deciding factors. The use of Maple to check analytical solutions is discussed in Section D5. A good discussion of some of these programs and their origins appears in Ralston [1976, pp. 35–46].

At this point, a table comparing the features and prices of such programs as well as the availability of test copies is in order. However, because the factors change rapidly and all of this information is available on the Web, readers are urged to contact the manufacturers to make their own comparison. To facilitate such a search, contact information for the programs is provided in Table D1.

## D2.6  Fault-Tolerant Computing Programs

In the mid-1970s, researchers in the fault-tolerant computer field began to develop specialized reliability and availability programs. These programs incorporated a general reliability computation program and added some features of special interest to the fault-tolerant field, such as coverage and transient faults. Some of the first such programs were developed by Professor Algirdas Avizienis and his students at UCLA (ARIES '76 and ARIES '82) [Markam, 1982]. Several other programs (ASSIST, CARE, HARP, and SHURE) were developed soon after ARIES by various researchers at NASA's Langley Research Center. (For a description of HARP and CARE, see Bavuso [1988]; for ASSIST and SHURE, see Johnson [1988].) Some of the more recent fault-tolerant programs (e.g., SHARPE) were developed by Professor Kishor Trivedi and his students at Duke University [Sahner, 1987, 1995].

**510** PROGRAMS FOR RELIABILITY MODELING AND ANALYSIS

## D2.7 Risk Analysis Programs

Risk analysis programs represent another class of large, comprehensive reliability and availability analysis programs. The term *risk analysis* generally implies that the analysis includes the consequences of failure [Shooman, 1990, Appendix J]. The impetus of such programs was the first probabilistic risk analysis for a nuclear power reactor performed for the U.S. Nuclear Regulatory Commission (NRC) by a team of experts lead by Professor Neils Rasmussen of MIT [McCormick, 1981, p. 240; NRC, 1975]. Risk analysis generally includes a final stage that predicts the probability of various classes of failures (accidents, calamities) and the result of such accidents (property loss, injuries, deaths). In addition, the team that conducted the NRC study (known in familiar terms as Wash 1400) found that it was difficult to include all the possible effects with such a large, complex system as a nuclear power reactor by using reliability block diagrams and fault trees. New techniques called *event trees* and *event sequence diagrams* were developed to help in the analysis, along with fault tree methods [McCormick, 1981, p. 193]. Such programs have been developed for analysis of nuclear reactors and other similar risk situations [McCormick, Chapter 10].

Several risk analysis programs have been evolved over the past few decades, namely: SAPHIRE [Long, 1999]; RISKMAN [Wakefield]; NUPRA, REBECCA, and CAFTA/ETA [Smith]. Presently, NASA Headquarters is developing a comprehensive risk analysis program called QRAS for its space projects [Safie, 1998; Shooman, 2000]. The analyst must judge whether one of these risk programs is suitable for fault-tolerant studies.

## D2.8 Software Reliability Programs

Software reliability models and the supporting programs differ from those of hardware reliability. Many programs have been developed by researchers and companies; however, three multimodel programs exist: SMERFS, CASRE, and SoRel. The best description and comparison of software reliability modeling programs is in Appendix A of the *Handbook of Software Reliability Engineering* [Stark, 1996]. The SMERFS program was developed in 1983 by the U.S. Naval Surface Warfare Center in Dahlgren, Virginia. In 1991, the LAAS Laboratory at the National Center for Scientific Research in Toulouse, France developed SoRel. The Jet Propulsion Laboratory developed CASRE in 1993. For further details, see Stark [1996].

## D3 TESTING PROGRAMS

The development of large, comprehensive reliability and availability programs requires the technical skills of reliability analysts who know probability and reliability methods and those of skilled programmers who can translate the

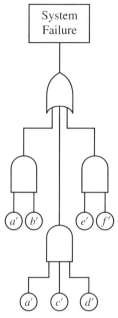

**Figure D1** Fault tree for testing reliability programs.

algorithms into code. Seldom do these skills reside in the same people; generally, the reliability analysts explain the algorithms and the programmers code the program. There is often insufficient coordination and review between the two groups, resulting in user-unfriendly interfaces or errors in the algorithms. The user should not expect a polished program such as a commercial word processor or a spreadsheet program, where the millions of users eventually report all the *major* bugs that are fixed in later releases.

The user should test any new program by comparing the solutions with those obtained with prior programs. The author has found that some of these programs do not make an exact computation when elementary events are repeated in a fault tree. Instead, they use an approximation that is generally (but not always) valid. One should test any program to see if it properly computes the fault tree given in Fig. D1.

The example given in Fig. D1 fails if $a$ and $b$ fail, if $a$ and $c$ and $d$ fail, or if $e$ and $f$ fail; thus the cut sets for the example are $a'b'$, $a'c'd'$, $e'f'$. The correct reliability expression for the probability of failure is given by Eq. (D1a).

$$P_f = P(a'b' + a'c'd' + e'f') = P(a'b') + P(a'c'd') + P(e'f')$$
$$- P(a'b'a'c'd') - P(a'b'e'f') - P(a'c'd'e'f')$$
$$+ P(a'b'a'c'd'e'f') \qquad \text{(D1a)}$$

The intersection of two like elements obeys the following logic law: $x \cdot x = x$. Thus Eq. (D1a) becomes

$$P_f = P(a'b' + a'c'd' + e'f') = P(a'b') + P(a'c'd') + P(e'f')$$
$$- P(a'b'c'd') - P(a'b'e'f') - P(a'c'd'e'f')$$
$$+ P(a'b'c'd'e'f') \qquad \text{(D1b)}$$

If all the elements are independent and have a probability of failure of $q$, that is, $P(a') = P(b') = P(c') = P(d') = P(e') = P(f') = q$, the probability of failure becomes

$$P_f = 2q^2 + q^3 - 2q^4 - q^5 + q^6 \qquad \text{(D1c)}$$

Many programs do not perform the step given in Eq. (D1b); rather, they expand Eq. (D1a) as

$$P_f = 2q^2 + q^3 - q^4 - 2q^5 + q^7 \qquad \text{(D1d)}$$

Equations (D1c) and (D1d) have the same first two terms, but the following three terms differ. If $q$ is small, the higher-order powers of $q$ are negligible, and the two expressions give approximately the same numerical result. If $q$ is not small, however, the two expressions differ. If $q = 0.2$, Eq. (D1c) gives $P_f = 0.084544$, and Eq. (D1d) yields 0.0857728. Larger values of $q$ result in even a larger difference—thus *caveat emptor!* ("let the buyer beware").

## D4 PARTIAL LIST OF RELIABILITY AND AVAILABILITY PROGRAMS

Many reliability and availability programs exist, all varying greatly in their ease of use, facilities provided, and cost. The purchaser should be wary of the basic validity of some of these programs, as was discussed in the preceding section. The contact information provided in Table D2 should allow users to conduct their own search and comparison via the Web.

Additional reliability and availability software advertisements can be found in the end pages of the *Proceedings Annual Reliability and Maintainability Symposium*. Sometimes, specialized reliability programs appear in the literature, such as the NAVY TIGER program [Luetjen, 1982], which was designed to analyze reliability and availability of naval ships and incorporates certain preventive maintenance programs used by the U.S. Navy.

TABLE D2  Information on Reliability and Availability Programs

| Product Name | Company Name and Address | Telephone Number | Web Address |
|---|---|---|---|
| Gidep | Government–Industry Data Exchange Program<br>P.O. Box 8000<br>Corona, CA 91718 | (909) 273-4677 | www.gidep.corona.navy.mil |
| Item | RAMS Software Tools<br>2030 Main Street<br>Suite 1130<br>Irvine, CA 92614 | (049) 260-0900 | www.itemsoft.com |
| Prism | Reliability Analysis Center<br>201 Mill Street<br>Rome, NY 13440 | (888) RAC-USER | http://rac.ittri.org |
| Meadep | SoHaR, Inc.<br>8421 Wilshire Boulevard<br>Beverly Hills, CA 90211 | (323) 653-4717 | www.sohar.com |
| ASENT<br>RAM Commander<br>Isograph Direct<br>Relex | Raytheon<br>—<br>—<br>Relex Software Corporation<br>540 Pellis Road<br>Greensburg, PA 15601 | (972) 575-6172<br>(800) 292-4519<br>—<br>(724) 836-8800 | http://asent.raytheon.com<br>www.isograph.com<br>www.relexsoftware.com |
| — | ReliaSoft Corporation<br>115 South Sherwood Village Drive<br>Suite 103<br>Tucson, AZ 85710 | (888) 886-0410 | www.Reliasoft.com |
| CARE Modules | BQR Reliability Engineering Ltd.<br>7 Bialik Street<br>P.O. Box 208<br>Rishon–LeZion 75101, Israel | (972) 3-966-3569 | www.bqr.com |
| Markov I | Decision System Associates<br>4244 Jefferson Avenue<br>Woodside, CA 94062 | (415) 851-7591<br>or<br>(415) 369-0501 | — |

## D5  AN EXAMPLE OF COMPUTER ANALYSIS

As part of a consulting assignment, the author was asked to derive a closed-form analytical solution for a spacecraft system with one on-line element and two different standby elements with dormancy failure rates. By dormancy failure rates, one means that the two standby elements have small but nonzero failure rates while on standby. A full Markov model for the three elements would require eight states, resulting in eight differential equations. Normally, one would use a numerical solution; however, the company staff for whom the author was consulting wished to include the solution in a proposal and felt that a closed-form solution would be more impressive and that it had to be checked for validity. (Errors had been found in previous company derivations). Assuming that the two standby elements had identical on-line and standby failure rates allowed a reduction to a six-state model. Formulation of the six equations, computing the Laplace transforms, and checking the resulting pencil-and-paper equations and solutions took the author a day while he worked with one of the company's engineers.

To check the results, the six basic differential equations were submitted in algebraic form to the Maple symbolic equation program, and an algebraic solution was requested. The first four of the state probabilities were easily checked, but the fifth equation took about half a page in printed form and was difficult to check. The Maple program provided a factoring function; when it was asked to factor the equation, another form was printed. Careful checking showed that the second form and the pencil-and-paper solution were both identical. The last (sixth) equation was the most complex, for which the Maple solution produced an algebraic form with many terms that covered more than a page. Even after using the Maple factoring function, it was not possible to show that the two equations were identical. As an alternative, the numerical values of the failure rates were substituted into the pencil-and-paper solution and numerical values were obtained. Failure rates were substituted into the Maple equations, and the program was asked for numerical solutions of the differential equations. These numerical solutions were identical (within round-off error to many decimal places) and easily checked.

There are several lessons to be learned from this discussion. The Maple symbolic equation program is very useful in checking solutions. However, as problems become larger, numerical solutions may be required, though it is possible that newer versions of Maple or some of the other symbolic programs may be easier to use with large problems. Checking an analytical solution is a good way of ensuring the accuracy of your results. Even in a very large problem, it is common to make a simplified model that could be checked in this way. Because of potential errors in modeling or in computational programs, it is wise to check all results in two ways: (a) by using two different modeling programs, or (b) by using an analytical solution (sometimes an approximate solution) as well as a modeling program.

# REFERENCES

Bavuso, S. J. A User's View of CARE III. *Proceedings Annual Reliability and Maintainability Symposium*, January 1984. IEEE, New York, NY, pp. 382–389.

Bavuso, S. J., and A. Martensen. A Fourth Generation Reliability Predictor. *Proceedings Annual Reliability and Maintainability Symposium*, 1988. IEEE, New York, NY, pp. 11–16.

Bavuso, S. J. et al. CARE III Hands-On Demonstration and Tutorial. NASA Technical Memorandum 85811. Langley Research Center, Hampton, VA, May 1984.

Bavuso, S. J. et al. CARE III Model Overview and User's Guide. NASA Technical Memorandum 85810. Langley Research Center, Hampton, VA, June 1984. [Updated, NASA Technical Memorandum 86404, April 1985.]

Bavuso, S. J. et al. Analysis of Typical Fault-Tolerant Architectures Using HARP. *IEEE Transactions on Reliability* 36, 2 (June 1987).

Bellcore. Reliability Prediction Procedure for Electronic Equipment. TR-NWT-000332, Issue 6, 1997.

Bryant, L. A., and J. J. Stiffler. CARE III, Version 6 Enhancements. NASA Contractor Report 177963. Langley Research Center, Hampton, VA, November 1985.

Butler, R. W. An Abstract Language for Specifying Markov Reliability Models. *IEEE Transactions on Reliability* 35, 5 (December 1986).

Butler, R. W., and P. H. Stevenson. The PAWS and STEM Reliability Analysis Program. NASA Technical Memorandum 100572. Langley Research Center, Hampton, VA, March 1988.

Butler, R. W., and A. L. Martensen. The FTC Fault Tree Program. Draft, NASA Technical Memorandum. Langley Research Center, Hampton, VA, December 1988.

Butler, R. W., and A. L. White. SHURE Reliability Analysis—Program and Mathematics. NASA Technical Paper 2764. Langley Research Center, Hampton, VA, March 1988.

Dugan, J. B. "Software System Analysis Using Fault Trees." In *Handbook of Software Reliability Engineering*, M. R. Lyu (ed.). McGraw-Hill, New York, 1996.

Ellis, W. Jr. et al. *Maple V Flight Manual*. Brooks/Cole Division of Wadsworth Publishers, Pacific Grove, CA, 1992.

Gedam, S. G., and S. T. Beaudet. Monte Carlo Simulation Using Excel Spreadsheet for Predicting Reliability of a Complex System. *Proceedings Annual Reliability and Maintainability Symposium*, 2000. IEEE, New York, NY, pp. 188–193.

Hayhurst, K. J. Testing of Reliability—Analysis Tools. *Proceedings Annual Reliability and Maintainability Symposium*, 1989. IEEE, New York, NY, pp. 487–490.

Huff, D. S. The Prophet™ Risk Management Toolset. *Proceedings Annual Reliability and Maintainability Symposium*, 1999. IEEE, New York, NY, pp. 426–431.

Johnson, S. C. ASSIST User's Manual. NASA Technical Memorandum 87735. Langley Research Center, Hampton, VA, August 1986.

Johnson, S. C. Reliability Analysis of Large, Complex Systems Using ASSIST. *Eighth Digital Avionics Systems Conference*, AIAA/IEEE, San Jose, CA, October 1988.

Johnson, S. C., and R. W. Butler. Automated Generation of Reliability Models. *Proceedings Annual Reliability and Maintainability Symposium, 1988. IEEE, New York, NY, pp. 17–25.

Laviron, A. SESCAF: Sequential Complex Systems Are Analyzed with ESCAF Through an Add-On Option. *IEEE Transactions on Reliability* (August 1985).

Laviron, A. et al. ESCAF—A New and Cheap System for Complex Reliability Analysis and Computation. *IEEE Transactions on Reliability* 31 4 (October 1982).

Long, S. M. Current Status of the SAPHIRE Models for ASP Evaluations. *Probabilistic Safety Assessment and Management, PSAM 4*, A. Mosleh and R. A. Bari (eds.). Springer-Verlag, New York, 1998, pp. 1195–1199.

Luetjen, P., and P. Hartman. Simulation with the Restricted Erlang Distribution. *Proceedings Annual Reliability and Maintainability Symposium*, 1982. IEEE, New York, NY, pp. 233–237.

Markam, S., A. Avizienis, and G. Grusas. ARIES '82 Users' Guide. Technical Report No. CSD-820830/UCLA-ENG-8262. University of California at Los Angeles (UCLA), Computer Science Department, August 1982.

MathSoft, Inc. Users' Guide Mathcad. MathSoft, Inc., Cambridge, MA, 1995.

Math Works, Inc. The Student Edition of MATHLAB. Prentice-Hall, Englewood Cliffs, NJ, 1992.

McCormick, N. *Reliability and Risk Analysis*. Academic Press, New York, 1981.

Mulvihill, R. J., and Safie, F. M. Application of the NASA Risk Assessment Tool to the Evaluation of the Space Shuttle External Tank Re-Welding Process. *Proceedings Annual Reliability and Maintainability Symposium*, 2000. IEEE, New York, NY, pp. 364–369.

National Aeronautics and Space Administration (NASA). Practical Reliability—Volume II Computation. NASA Contractor Report, NASA CR-1127. Research Triangle Institute, August 1968.

Ng, Y. W., and A. A. Avizienis. ARIES—An Automated Reliability Estimation System for Redundant Digital Structures. *Proceedings Annual Reliability and Maintainability Symposium*, 1977. IEEE, New York, NY, pp. 108–113.

Nuclear Regulatory Commission (NRC). Reactor Safety Study—An Assessment of Accident Risks in U.S. Commercial Nuclear Power Plants. Report Wash 1400, NUCREG 75/014, 1995.

Orbach, S. The Generalized Effectiveness Methodology (GEM) Analysis Program. Lab. Project 920-71-1, SF 013-14-03, Task 1604, Progress Report 1. U.S. Naval Applied Science Lab., Brooklyn, NY, May 8, 1968.

Ralston, A. *Encyclopedia of Computer Science*. Van Nostrand Reinhold, New York, 1976.

Rubinstein, R. Y. *Simulation and the Monte Carlo Method*. Wiley, New York, 1981.

Safie, F. M. An Overview of Quantitative Risk Assessment of Space Shuttle Propulsion Elements. *Probabilistic Safety Assessment and Management, PSAM 4*, A. Mosleh and R. A. Bari (eds.). Springer-Verlag, New York, 1998, pp. 425–430.

Safie, F. M. NASA New Approach [QRAS Risk Tool] for Evaluating Risk Reduction Due To Space Shuttle Upgrades. *Proceedings Annual Reliability and Maintainability Symposium*, 2000. IEEE, New York, NY, pp. 288–291.

Sahner, R. A., and K. S. Trivedi. Reliability Modeling Using SHARPE. *IEEE Transactions on Reliability* 36, 2 (June 1987): 186–193.

Sahner, R., K. S. Trivedi, and A. Puliafito. *Performance and Reliability Analysis of Computer Systems: An Example-Based Approach Using the SHARPE Software Package*. Kluwer Academic Publishers, Boston, MA, 1995.

Sammet, J. E. *Programming Languages: History and Fundamentals*. Prentice-Hall, Englewood Cliffs, NJ, 1969.

Schmidt, H., and D. Busch. An Electronic Digital Slide Rule—If This Hand-Sized Calculator Ever Becomes Commercial, the Conventional Slide Rule Will Become Another Museum Piece. *The Electronic Engineer* (July 1968).

Shooman, M. L. The Equivalence of Reliability Diagrams and Fault Tree Analysis. *IEEE Transactions on Reliability* (May 1970): 74–75.

Shooman, M. L. *Probabilistic Reliability: An Engineering Approach*, 2d ed. Krieger, Melbourne, FL, 1990.

Shooman, M. L. and four others participated in a review of QRAS for NASA in late 1999 and early 2000.

Smith, D. CAFTA, SAIC Facilities Group (http://fsg.saic.com).

Stark, G. *Software Reliability Tools, Handbook of Software Reliability Engineering*, M. R. Lyu (ed.). McGraw-Hill, New York, 1995, app. A.

Trivedi, K. S., and R. M. Geist. A Tutorial on the CARE III Approach to Reliabiity Modeling. NASA Contractor Report 3488. Langley Research Center, Hampton, VA, 1981.

Trivedi, K. S. et al. HARP: The Hybrid Automated Reliability Predictor: Introduction and Guide for Users. NASA, Langley Research Center, Hampton, VA, September 1986.

Trivedi, K. S. et al. HARP Programmer's Maintenance Manual. NASA, Langley Research Center, Hampton, VA, April 1988.

Turconi, G., and E. Di Perma. A Design Tool for Fault-Tolerant Systems. *Proceedings Annual Reliability and Maintainability Symposium*, 2000. IEEE, New York, NY, pp. 317–326.

Vesley, W. E. et al. PREP and KITT: Computer Codes for the Automatic Evaluation of a Fault Tree. Idaho Nuclear Corporation, Report for the U.S. Atomic Energy Commission, No. IN-1349, August 1970.

Wakefield. RISKMAN (Wakefield@plg.com).

White, A. Motivating the SURE Bounds. *Proceedings Annual Reliability and Maintainability Symposium*, 1989. IEEE, New York, NY, pp. 277–282.

Wolfram, S. *The Mathematica Book*, 4th ed. Cambridge University Press, New York, 1999.

## PROBLEMS

**D1.** Search the Web for reliability and availability analysis programs. Make a table comparing the type of program, the platforms supported, and the cost.

**D2.** Use a reliability analysis program to compute the reliability for the first three systems in Table 7.8 and check the reliability.

**D3.** Use a symbolic modeling program to check Eq. (3.56).

**D4.** Use a Markov modeling program to check the results given in Eq. (3.58).

**D5.** Use a fault tree program to solve the model of Fig. D1 to see if the results agree with Eqs. (D1c) or (d).

# NAME INDEX

Numbers in parentheses that follow page references indicate "Problems."

Abramovitz, M., 400, 462
Advanced Hardware Architectures (AHA), 74
AGREE Report, 342
Aho, A. V., 337, 351
Ahuja, V., 320
AIAA/ANSI, 259, 261
Aktouf, C., 23
Albert, 345, 346, 348, 349
Anderson, T., 25, 131, 135
Arazi, B. A., 34, 65, 71, 72
ARINC, 89
Ascher, 111, 112
Aversky, D. R., 23
Avizienis, A., 23, 263, 265, 509, 516

Baker, W. A., 131
Barlow, R. E., 331, 351
Bartlett, J., 337
Bavuso, S. J., 509
Bazovsky, I., 110
Bell, C. G., 25, 146
Bell, T., 25
Bellman, R., 351, 372
Bernstein, J., 428
Bhargava, B. K., 275
Bierman, H., 333, 351
Billings, C. W., 226
Bloch, G. S, 23

Boehm, B., 207
Booch, G., 207
Bouricius, 115, 117
Braun, E., 25
Breuer, M. A, 23.
Brooks, F. P., 207, 211
Burks, A. W., 25
Buzen, J. P. 119, 120

Calabro, S. R., 186
Carhart, R. R., 425
Carter, W. C. 115, 117
Cassady, C. R., 366
Chen, L., 263, 265
Christian, F., 23
Clark, R., 25
Colbourn, C. J., 284, 294, 309, 320
*Computer Magazine*, 23, 24
Cormen, T. H., 213, 314, 320, 337, 351, 369, 371
Cramer, H., 203

Dacin, M., 24
Daves, D. W., 24
Dierker, P. F., 312, 314, 317
Ditlea, S., 5, 25
Dougherty, E. M., 24, 202
Dugan, J. B., 117, 284, 446, 507

**519**

**520** NAME INDEX

Echte, K., 24
Ellis, W., 505, 508

Farr, W., 258
Fault-Tolerant Computing Symposium, 24
Federal Aviation Administration, 25
Fisher, L. M., 25
Fowler, M., 207
Fragola, J. R., 24, 25, 202, 331
Frank, H., 284, 301, 317, 320
Freeman, H., 394
Friedman, A. D., 23
Friedman, M. B., 25, 120
Friedman, W. F. (*see* Clark, R.)
Frisch, I. T., 284, 317
Fujiwara, E., 25, 77

Gibson, G., 24, 119, 120, 126
Gilbert, G. A., 340
Giloth, P. K., 26
Goldstine, H. H., 25
Graybill, F. A., 261
Greiner, H. de Meer, 23
Grisamore, 158, 159

Hafner, K., 26
Hall, H. S., 214
Hamming, R. W., 31
Hawicska, A., 24
Healey, J. T., 427
Hennessy, J. L., 136
Henney, K., 342
Hill, F. J., 475, 479, 488, 490
Hiller, S. F., 333, 351, 371, 372
Hoel, P. G., 255
Hopper, G., 226
Hwang, C. L., 351

Iaciofano, C., 26

Jelinski, Z., 258, 259
Jia, S., 283
Johnson, B. W., 24, 117.
Johnson, G., 26
Johnson, S. C., 509

Kanellakis, P. C., 24
Kaplan, G., 24
Kaufman, L. M. 117
Karnaugh, M., 35
Katz, R., 24, 26
Keller, T. W., 269

Kershenbaum, A., 287, 308, 309, 310, 314, 318, 320, 321
Knight, J. C., 265
Knight, S. R., 214
Knox-Seith, 154, 155, 157
Kohavi, Z., 475, 488, 490
Kuo, W., 351

Lala, P. K., 24, 263
Larsen, J. K., 24
Lee, P. A., 23, 24
Leveson, N. G., 265
Lewis, P. H., 26
Lipow, M., 333, 346, 348, 434
Littlewood, B., 259
Lloyd, D. K., 333, 346, 348
Long, S. M., 510
Luetjen, P., 512
Lyu, M. R., 24, 258, 261

Mancino, V. J., 331, 383
Mann, C. C., 5, 26
Mano, M. M., 475, 488, 490
Markam, S., 509
Markoff, J., 26
Marshall, 351, 369
Massaglia, P., 120
McAllister, D. F., 189, 191, 192, 194, 268
McCormick, N., 24, 507, 510
McDonald, S., 25
Meeker, W. Q., 384
Mendenhall, W., 299, 384
Messinger, M, 333, 351, 377, 420
Meyers, R. H., 351
Military Handbook MIL-HDBK-217–217F, 79 (2.7), 427, 506
Military Standard MIL-F-9409, 26
Miller, E., 186
Miller, G. A., 213, 339
Mood, A. M., 261
Moore, G. E., 5–8, 26, 147
Moranda, P., 258, 259
Motorola (ON Semiconductor), 492, 495, 496, 499, 501
Murray, K., 297, 308, 318
Musa, J., 232, 238, 251, 252, 253, 259, 260, 261

Neuman, J. von, 147
Newell, A. 146
Ng, Y. W., 516
Nishio, T., 24
Norwall, B. D., 26

# NAME INDEX

Orbach, S., 505
Osaki, S., 24

Papoulis, 104, 385
Paterson, D., 24, 120, 136
Pecht, M. G., 117
Pfister, G., 137
Pfleeger, S. L., 26, 207
Pham, H., 24
Pierce, J. R., 31
Pierce, W. H., 24, 194
Pogue, D., 26
Pollack, A., 26, 203
Pooley, R., 207
Pradhan, D. K., 24, 263, 265
Pressman, R. H., 207

Ralston, A., 505, 508
Randall, B., 5, 26, 147
Rao, T. R. N., 25, 34, 77
Rice, W. F., 366
Rogers, E. M, 26
Rosen, K. H., 66, 294, 312, 314, 361
Roth, C. H., 475, 488, 490
Rubinstein, R. Y., 505

Safie, F. M., 510
Sahner, R., 509
Sammet, J. E., 5, 26, 506
Satyanarayana, A. A., 300
Schach, S. R., 207, 218
Schmidt, H, 505
Schneider, 115
Schneidewind, 259, 269
Scott, K., 207
Shannon, C. 31, 147
Sherman, L., 132
Shier, D. R., 323
Shiva, S. G., 161, 475, 488, 490, 501
Shooman, A. M., 300, 305, 306, 308, 309, 318
Shooman, M. L., 4, 5, 25, 26, 90, 110, 111, 112, 135, 166, 172, 180, 185, 194, 199 (4.29), 202, 207, 214, 215, 225, 229, 230, 234, 238, 251, 252, 255, 256, 257, 258, 259, 260, 261, 262, 264, 265, 268, 284, 287, 288, 292, 295, 299, 337, 349, 351, 369, 384, 395, 409, 411, 420, 427, 428, 432, 435, 437, 439, 441, 446, 450, 460, 472, 505, 506, 507, 510

Shvartsman, A. A., 24
Siewiorek, D. P., 1, 25, 26, 27, 126, 129, 131, 135, 147, 158, 165, 169, 194, 207, 263, 270, 275, 427, 428
Smith, B. T., 25
Smith, D., 510
Sobel, D., 195
Stark, G. E., 258, 510
Stone, C. J., 384
Stork, D. G., 280 (5.6)
Stepler, R., 27
Stevens, P., 207
Swarz, R. S., 1, 25, 26, 27, 126, 129, 131, 135, 147, 158, 165, 169, 194, 207, 263, 270, 275, 427, 428

Taylor, L., 33
Tenenbaum, A. S., 320
Tillman, F. A., 351
Toy, W. N., 165
Trivedi, K. S., 23, 25, 509
Turing, A. M., 4, 27, 29 (1.25)

Van Slyke, R., 284, 301
Verall, J. L., 258
von Alven, W. H., 342
von Neuman, J., 4
Vouk, M. A., 189, 191, 192, 194, 268

*USA Today*, 27

Wadsworth, G. P., 384, 394
Wakefield, 510
Wakerly, J. F., 161, 475, 488, 490, 501
Wald, M. L., 27
Welker, E. L., 434
Wing, J. A., 210,
Wing, J. M., 283, 284
Wirth, N., 27
Wise, T. R., 366
Wood, A. P., 127, 130, 131
Workshop on Defect and Fault-Tolerance in VLSI Systems, 25
www.emc.com, 27
www.intel.com, 27
www.microsoft.com, 27

Yourdon, E., 270

Zuckerman, L., 27

# SUBJECT INDEX

Numbers in parentheses that follow page references indicate "Problems."

Aibo, 9
Air traffic control (ATC), 29 (1.20), 209, 215, 216, 217, 340, 341
Aircraft reliability, 15–18
Apportionment (*see* Reliability optimization, apportionment)
Architecture, Boolean algebra, 479–482
  decoder, 494–497
  DeMorgan's theorems, 409 (A1)
  flip-flops, 487–499
  gates, 37, 38, 478–480
  number systems, 475–477
    arithmetic, 477, 478
  parity-bit generators, 494
  set theory, historical development, 384–386
    union, 388–390
    Venn diagram, 386, 409 (A1)
  storage registers, 500, 501
  switching functions, 483, 484
    combinatorial circuits, product of sums (POS), 489, 490
      sum of products (SOP), 489–491
    integrated circuits (IC chips), 491–493
    maxterms, 484
    minterms, 483
    simplification, 484, 485
      don't-cares, 489
      Karnaugh map (K map), 485–488
      Quine–McClusky (QM) method, 488
    truth tables, 479–482
ARPA network (*see* network reliability)
Availability, concepts, 14, 15, 286–288
  coupling and decoupling, 183–186 (*see also* Markov models, uncoupled)
  definition, 14, 134, 135, 179
  Markov models, 117–119, 180–186, 454–461
  steady-state, 458–461
  typical computer systems, 16, 17
    Bell Labs' ESS, 134, 183
    Stratus, 134, 135, 183
    Tandem, 126, 127, 183
    Univac, 146, 147

Billy Bass, 9
Burst, 62
  code, decoder, 65
  decoder failure, 73–75
  encoder, 65
  errors, 32, 62
  properties, 64, 65
  Reed–Solomon, 72–75, 126

CAID, 119 (*see also* RAID)
Chinese Remainder Theorem, 66–71

**524** SUBJECT INDEX

Cluster, of computers, 135, 136
Coding methods, burst codes (*see* Burst)
  check bits, 35
  cryptanalysis, 29 (1.25), 30
  error-correcting, 2, 31
  error-detecting, 2, 31
  errors, 32, 33
  Hamming codes, 31, 44–47, 54
  Hamming distance, 34, 45, 46
  other codes, 45, 75, 76
  parity-bit codes, 35, 37
    coder (encoder, generator), 37, 38, 40, 42
    coder–decoder failures, 43, 53–59
    decoder (checker), 37, 38, 40, 42
    probability of undetected errors, 32, 39–42, 45, 52–53, 59–62
  RAID levels 2–6, 121–126
  Reed–Solomon codes (*see* Burst)
  reliability models (*see also* Probability of undetected errors)
  retransmission codes, 59–62
  single error-detecting and double error-detecting (SECDED), 47, 51–52
  single error-correcting and single error-detecting (SECSED), 47–51
  soft fails, 33
Cold redundancy, cold standby (*see* Standby systems)
Computer, CDC 6600, 11
  ENIAC, 4
  history, 4, 5
  Mark I, 4
Conditional reliability, 390, 391
Coverage, 115, 117
Cryptography (*see* Coding methods, cryptanalysis)
Cut-set methods (*see also* Network reliability, definition; Reliability modeling)

Dependent failures (Common mode failures), (*see* Reliability theory, combinatorial reliability)
Downtime, 14, 134 (*see also* Availability)

EMC (*see* RAID)
ESS (*see* Availability, typical computer systems)

Fault-tolerant computing, calculations, 12, 13
  definition, 1
Furby, 9

Global Positioning System (GPS), 195, 280 (5.6)

Hamming codes (*see* Coding methods)
Hamming distance (*see* Coding methods)
Hazard (*see also* Reliability modeling, failure rate), derivation, 222–224
  function of system, 94, 95
Himalaya computers (*see* Tandem)
Hot redundancy, hot standby (*see* Parallel systems)
Human operator reliability, 202

Laplace transforms, as an aid in computing MTTF, 169, 170, 174, 175, 468, 469
  definition, 462–464
  of derivatives, 465, 466
  final value computation, 170, 182
  initial value approximation, 469–471
  initial value computation, 173, 174
  of Markov models, 93
  partial fractions, 466, 467
  table of theorems, 468
  table of transforms, 465
Library of Congress, 10, 27 (1.1), (1.3), (1.19)

Maintenance, 146
Markov models (*see also* Laplace transforms, Probability)
  algorithms for writing equations, 113
  collapsed graphs (*see* merger)
  complexity, 453, 454, 461
  decoupling (*see* uncoupled)
  formulation, 104–108, 112–117, 446–450
  graphs, 450
  Laplace transforms, 461–468
  merger of states, 166, 453, 454
  RAID, 125
  solution of Markov equations, 106, 108, 115–117, 118, 166–179
  theory, chain, 404
    Poisson process, 404–407
    process, 404
    properties, 403, 404
    transition matrix, 407, 408
  two-element model, 450–453
  uncoupled, 172, 349, 350 (*see also* Availability, coupling and decoupling)
Mean time between failure (MTBF) (*see* Mean time to failure)
Mean time to failure (MTTF), 95, 96, 114, 115, 117, 140 (3.16), 169, 170, 174

## SUBJECT INDEX 525

constant-failure rate (hazard), 224, 225
   definition, 234
   linearly increasing hazard, 225
   RAID, 120, 123, 125
   tables of, 115, 117
   TMR, 151–153
Mean time to repair (MTTR), 112–119, 126, 127
Memory, growth, 7, 8
Microelectronic, revolution, 1, 4, 5
Microsoft, 5
MIL-HDBK-217, 79 (2.7), 427, 506
Moore's Law, 5–8

NASA,
   *Apollo*, 194
   Space Shuttle, 188, 194, 266–269
Network reliability, concepts, 13, 14, 31, 283–285
   ARPA network, 312
   availability, 286–288
   computer solutions, 308, 309
   definition, 285, 288
      all-terminal, 286
         cut-set and tie-set methods, 303–305
         event-space, 302, 303
         graph transformations, 305–308
      $k$-terminal, 286, 308
      two-terminal, 286, 288–301
         cut-set and tie-set methods, 292–294
         graph transformations, 297–301
         node pair resilience, 301
         state-space, 288–292
         subset approximations, 296, 297
         truncation approximations, 294–296
   design approaches, 309–321
      adjacency matrices, 312, 313
      backbone network-spanning tree, 310–312
      enhancement phase, 318–321
      Hamiltonian tours, 317, 327 (6.14), 328 (6.15)–(6.17)
      incidence matrices, 312, 313
      Kruskal's and Prim's algorithms, 312, 314–318
      spanning trees, 314–318
   graph models, 284, 285
$N$-modular redundancy, 2, 145, 146, 153–161
   history, 146, 147
   repair, 165–183, 454–461
   triple modular redundancy (TMR), 147, 148, 149–153,
      comparison with parallel and standby systems 178, 179
      Markov models, 166–170
      MTTF, 151–153
      voter logic, 161–165
      adaptive voting, 194
      adjudicator algorithms, 189–195
      comparison of reliability, 193
      consensus voting, 190–192
      pairwise comparison, 191, 193
      test and switch, 191
      voters, 154–161
      voting with lockout, 186, 188, 189
NMR (*see N*-modular redundancy)
$N$-version (*see* Software reliability)

Parallel systems, 2, 83, 97–99, 104 (*see also* Reliability optimization)
   comparison with standby, 108–111, 178, 179
   MTTF, 96, 114, 115
Polynomial roots, 165, 166
Probability, complement, 388
   conditional, 390–391
   continuous random variables, 395–401
      density and distribution function, 395–397
      exponential distribution, 397–399, 403, 433, 434
      Normal (Gaussian) distribution, 398, 400, 401, 403
      Rayleigh distribution, 398, 399, 403, 434
      rectangular (uniform) distribution, 397, 398
      Weibull distribution, 398, 399, 403, 434–438
   discrete random variables, 391–395
      binomial (Bernoulli) distribution, 393, 394, 403
      density function, 391, 392
      distribution function, 392, 393
      Poisson distribution, 185, 395, 396, 403–407
   Markov models (*see* Markov models)
   moments, 401–403
      expected value, 401, 402
      mean 402, 403
      variance, 403
Probability of undetected error (*see* Coding methods)

RAID, Advisory Board, 120
   EMC Symmetrix, 10, 27 (1.1)
   levels, 121–126

# SUBJECT INDEX

mirrored disks, 122
reliability, 119–126
stripping, 125
RBD (*see* Reliability modeling)
Redundancy (*see also* Parallel systems, Reliability optimization),
  component, 86–92
  couplers, 91, 92
  system, 86–92
Reliability allocation (*see* Reliability optimization)
Reliability analysis programs
  example, 514
  fault-tolerant computing programs, ARIES, 509
    ASSIST, 509
    CARE, 509
    HARP, 509
    SHAPE, 509
    SHURE, 509
  mathematics packages, Macsyma, 256, 505, 508, 509
    Maple, 256, 505, 508, 509
    Mathcad, 256, 505, 508, 509
    Mathematica, 256, 505, 508, 509
    Matlab, 505, 508, 509
  partial list, 512, 513
  risk analysis, CAFTA, 510
    NUPRA, 510
    QRAS, 510
    REBECCA, 510
    RISKMAN, 510
    SAPHIRE, 510
  software reliability (*see* Software reliability, programs)
  testing programs, 510–512
Reliability modeling (*see also* Reliability theory),
  block diagram (RBD), 413, 444
  cut-set methods, 292–294, 419, 420
  density function, 218–221
  distribution function, 218–221
  event-space, 288–292
  failure rate, 222–224 (*see also* Hazard)
  graph (*see* block diagram)
  probability of success, 219–221
  reliability function, 218–221
  system, example, auto-brake system, 442–446
    parallel, 440, 441
    $r$-out-of-$n$ structure, 441, 442
    series 438–440
  theory, 218–221

tie-set methods, 292–294, 419, 420
Reliability optimization, algorithms, 359–365
  apportionment 85, 86, 342–349, 366, 367
    Albert's method, 345–349
    availability, 349–351
    equal weighting, 343
    relative difficulty, 344, 345
    relative failure rates, 345
  communication system, 383 (7.31)
  concepts, 11, 12, 85, 86, 332–334
  decomposition, 337–340 (*see also* Software development, graph model)
  definition, 2, 4, 334–336
  dynamic programming, 371–379
  greedy algorithm, 369–371
  interfaces, 340
  minimum bounds, 341, 342
  multiple constraints, 365, 366
  parallel redundancy, 336
  redundant components, 336
  subsystem, 340–342
    bounded enumeration, 353–359
      lower bounds (minimum system design), 354–357
      upper bounds (augmentation policy), 358, 359
    exhaustive enumeration, 351–353
  series system, 335
  standby redundancy, 336, 337
  standby system, 367–369
Reliability theory (*see also* Reliability modeling)
  combinatorial reliability, 412, 413
    exponential expansions, 92–94
    parallel configuration, 415, 416
    $r$-out-of-$n$ configuration, 416, 417
    series configuration, 413–415
  common mode effects, 99–101
  cut-set and tie-set methods, 419, 420
  failure mode and effect analysis (FMEA), 418, 419, 443
  failure-rate models, 421–429
    density function, 422–425, 429–431
    distribution function, 423–425
    failure data, 421–425
      bathtub curve, 425, 426
      handbook, 425–427
      integrated circuits, 427–429
    hazard function, 422–424, 432–438
    reliability function, 423–425
  fault-tree analysis (FTA), 418, 445
  history, 411, 412
  reliability block diagram (RBD), 413

## SUBJECT INDEX 527

reliability graph, 413
Repairable systems, 111–117
 availability function, 111, 117–119
 reliability function, 111
 single-element Markov models, 115
 two-element Markov models, 112, 115, 116
Redundancy (*see* parallel systems)
 couplers, 91, 92
 microcode-level, 186
Rollback and recovery (recovery block), 191, 203, 268–275
 checkpointing, 274, 275
 distributed storage, 275
 journaling, 272, 273
 rebooting, 270, 271
 recovery, 271, 272
 retry, 273, 274
$r$-out-of-$n$ system, 101–104

SABRE, 135
SECDED (*see* Coding methods)
Software development, 203, 205
 build, 218, 221
 coding, 208, 214, 215
 error, 225–227 (*see also* Software Reliability, error models)
 fault, 225, 226
 graph model, 211–214
 hierarchy diagram (H-diagram), 211–214
 life cycle, 207–218
  deployment, 208
  design, 208, 211–214
  incremental model, 221
  maintenance, 208
  needs document, 207, 208
  object-oriented programming (OOP), 207
  phases, 208
  pseudocode, 226
  rapid prototype model, 208, 210, 220
  redesign, 208, 218
  requirements document, 208, 209
  specifications document, 208–210
  structured procedural programming (SPP), 207, 215
  warranty, 218
  waterfall model, 219
 process diagrams, 218–221
 reused code (legacy code), 210
 source lines of code (SLOC), 210, 211, 214
 testing, 208, 215–218
Software engineering (*see* Software development)
Software Engineering Institute, 268
Software fail-safe, 203
Software redundancy, 262
Software reliability, 203, 204
 data, error, 203, 225–227
  generation, 227–229
  models, 225–236
  removal, 227–229
   constant-rate, 230–232
   exponentially decreasing rate, 234–236
   linearly decreasing rate, 232–234
   S-shaped, 235, 236
 hardware, operator, software, 202
  independence, 202
 macro models, 262
 mean time to failure (MTTF), 238–241, 245–246,
 models, 237–250
  Bayesian, 261
  comparison, 249–250
  constant error-removal-rate, 238–242
  exponentially decreasing error-removal rate, 246–248
  linearly decreasing error-removal rate, 242–246
  model-constant estimation, 250–258
   from development test data, 260
   handbook estimation, 250–252
   least-squares estimation, 256, 257
   maximum-likelihood estimation, 257–258
   moment estimation, 252–256
  other models, 258–262
 $N$-version programming, 263–268
 programs, CASRE, 258
  Markov models, 507, 508
  reliability block diagram, 507
  reliability fault tree models, 507
  reliable software, 203
  SMERFS, 258
  software development, 205
  SoRel, 258
Space Shuttle (*see* NASA)
Standby systems, 83, 104
 comparison with parallel, 108–111, 178, 179
 redundancy, 2
Storage errors, CD, 62
 CD-ROM, 62
STRATUS, 122, 131–135
 availability, 134, 135
 Continuum, 134
Stuck-at-one, 147
Stuck-at-zero, 147

Sun, 136, 137
Syndrome, 51–56, 66

Tandem, 122, 126–131, 136
    Guardian, 127
    Himalaya, 126, 129
Technology timeline, 4
Telephone switching systems, 15, 16
Three-state elements, 92
Tie-set methods (*see* Reliability modeling; Network reliability, definition)

Triple modular redundancy (TMR) (*see* $N$-modular redundancy)

Undetected errors, 32
Uptime, 14, 134 (*see also* Availability)

VAX, 136
Voting (*see* $N$-modular redundancy)

Year 2000 Problem (Y2K), 205–208